The Rainbow of Mathematics

THE RAINBOW OF MATHEMATICS

A History of the Mathematical Sciences

Ivor Grattan-Guinness

W. W. Norton & Company
New York London

First American edition published 1998 under the title
The Norton History of the Mathematical Sciences

First published as a Norton paperback 2000

Published by arrangement with HarperCollins Publishers Ltd

Originally published in England under the title *The Fontana History of the Mathematical Sciences: The Rainbow of Mathematics*

Manufacturing by the Haddon Craftsmen, Inc.

Library of Congress Cataloging-in-Publication Data
Grattan-Guinness, I.
[Fontana history of the mathematical sciences]
The Norton history of the mathematical sciences: the rainbow of mathematics / Ivor Grattan-Guinness. — 1st American ed.
p. cm. — (Norton history of science)
Originally published: The Fontana history of the mathematical sciences. London: Fontana Press, 1997.
Includes bibliographical references and index.
ISBN 0-393-04650-8
1. Mathematics—History. I. Title. II. Series.
QA21.G695 1998
510'.9—dc21 CIP

ISBN 0-393-32030-8 pbk.

W. W. Norton & Company, Inc., 500 Fifth Avenue, New York, NY 10110
www.wwnorton.com

W. W. Norton & Company Ltd, 10 Coptic Street, London WC1A 1PU

1 2 3 4 5 6 7 8 9 0

'Man, we may say, appears to be not so much
a rational animal as an ideological animal'

To the affectionate memory of
SIR KARL POPPER
(1902–94),
master of conjectures and refutations

CONTENTS

The Rainbow of Mathematics

1

Pre-viewing the rainbow

1.1 *An exchange*

In the issue of the British journal *Nature* for 27 July 1905, the statistician Karl Pearson posed the following problem:

> A man starts from a point O and walks l yards in a straight line; he then turns though any angle whatever and walks another l yards in a second straight line. He repeats this process n times. I require the probability that after these n stretches he is at a distance between r and $r + \delta r$ from his starting-point O.

The answer appeared the following week in the next issue, given by the third Lord Rayleigh. The problem, he explained, 'is the same as that of the composition of n iso-periodic vibrations of unit amplitude and of phases distributed at random'; citing his own previous work, he gave the probability for very large n as

$$\frac{2}{n} e^{-r^2/n} r\, dr. \qquad (1.1.1)$$

The exchange provides examples of several features of this book. While occurring in the 20th century, the event is historical – around 90 years ago now. Two significant notions, not usually linked, were highlighted; for Pearson introduced the phrase 'random walk' in his note, and Rayleigh's answer gave birth to the term 'Rayleigh distribution' in statistics. In addition (though not a kind of point that will be stressed in this book), the latter name was a misnomer. S.D. Poisson had used this distribution from 1837 in connection with the accuracy of target shooting in artillery: the assumption of a normal distribution of error both horizontally and vertically relative to the bull's-eye leads to the normal distribution in polar coordinates, which is given by Rayleigh's expression. Lastly, applications of mathematics entered the exchange, and different ones: Rayleigh made clear that his result had come from studying systems of particles free to vibrate in a random manner. Indeed, randomness also interested contemporaries such as Henri Poincaré's student Louis Bachelier, who had written a thesis in 1900 analysing mathematically Brownian motion and its interpretation in economics; there, and in his book *Calcul des probabilités* (1912), Bachelier showed that when n was infinite such expressions involved solutions of Joseph Fourier's equation for heat diffusion, which had been known for a century.

Thus two little notes rapidly bring us to surprising or forgotten networks of ideas and theories. This is the history of mathematics at its best.

1.2 *Summary of the story*

This book recounts the growth of mathematical knowledge from the remote past to the time around the Great War of 1914–18. Individual chapters are cited as §3, say, and its fifth section as §3.5. Other works are cited in the manner '{Pearson 1905}' and '{Rayleigh 1905}' for the exchange above; details are given in the bibliography.

Ancient mathematics (§2) has no known origins; but

when it got going, it was much concerned with counting and space, heavenly bodies and physical structures; so we see the origins of arithmetic, geometry, trigonometry and mechanics. Mathematics was intimately linked with astronomy for all cultures. By the early centuries of the Christian epoch, the most developed cultures were those of the Greeks (to whom the bulk of §2 is devoted), the Indians and the Chinese.

The period of §3 is the millennium from the early Middle Ages, around the year 500, into the mid-16th century. The later Chinese and Indians continued to find nice things, and a most impressive tradition of Arabic mathematics developed, especially from the 8th century. European mathematics began to emerge principally during the 12th century; the next three centuries saw considerable advances, especially in the Italian and German states, with the discovery of Arabic and Greek mathematics (often in Latin translation). Early forms of algebra grew out of arithmetic; geometry found further uses in perspective theory; and dynamics came to prominence from the analyses of motion of bodies, both heavenly and earthly.

The rest of the book deals almost entirely with mathematics in Europe. The move of Europe to centre stage during the 16th century is the subject of §4, which goes up to around 1660. Trigonometry flowered, as did algebra, which was seen as the mathematics of the future by men such as Franciscus Vieta and René Descartes. Among applications, astronomy was still the most important one, although optics was developing.

The next 90 years are treated in §5. The two major figures are Isaac Newton and G.W. Leibniz, the calculus emerging as the main new branch; in addition, some important scientific societies and journals were launched. The second half of the 18th century is the concern of §6, with the calculus applied to mechanics (including engineering) constituting the main concern of mathematics. Various important functions and series were introduced in connection with these studies. The Continent provided most of the significant work; the chief figures were Leonhard Euler and J.L. Lagrange, with different and indeed competing views.

In §7 a survey is given of the new professional status of mathematics fostered in the early 19th century by the creation of new universities or equivalent institutions and the reinvigoration of certain old ones, and also a massive growth in publishing mathematics in books and journals. The French Revolution was the major single cause of change, and indeed France was by far the principal mathematical country between 1780 and 1830.

The work of the period from 1800 to 1860 is divided between three chapters. The calculus was extended by A.L. Cauchy into mathematical analysis, whence came complex-variable analysis, special functions and elliptic functions; this is covered in §8, together with some related work in geometries. Algebra became algebras, for the theory of equations was joined by differential operators, quaternions, determinants and algebraic logic (§9). Mechanics extended into mathematical physics with the mathematicization of heat, light, electricity and magnetism; a programme of molecular theories led by P.S. Laplace was the first attempt, but the definitive versions came from others, such as Joseph Fourier, often following different principles (§10).

The rise of Germany to become the main mathematical power from the 1860s to around 1900 is summarized in §11. However, mathematical practice had become very international, and would remain so. A considerable expansion in publishing is evident in these years, as is the founding of societies for mathematics.

The work from 1860 to 1900 is divided among four chapters. Mathematical analysis, real and complex, was taken up and refined by Karl Weierstrass and his many followers, leading to set theory and mathematical logic; mathematical statistics also came more to the fore (§12). Further algebras included axiomatized versions such as group theory and invariants, matrices and determinants, while geometry became geometries with the popularization of non-Euclidean versions; in addition, a rather geometrical approach to complex analysis was inspired by Bernhard Riemann (§13). All areas of mechanics and of mathematical physics expanded enormously, with potential theory prominent in both branches; new mathematical methods included vector analysis (§14–§15).

The final chapter of the chronicle, §16, covers all aspects of mathematics from 1900 (when David Hilbert gave a famous list of major problems in the subject) to the period around the Great War. Among the most prominent features are the rise of set theory, and the emergence of algebraic topology and of relativity theory. The War is my normal stopping point; only a few later developments are recorded. Some more arise in the final chapter, §17, an overall survey which discusses the remarkable range of mathematics, including the greatly differing histories of its various branches. The role played by mathematics in general cultural affairs, including recreation, the arts and religions, is also noted.

1.3 *Strategy and purpose*

Many general histories of mathematics have been written in the past; the present one differs significantly from many of them in several respects.

1. Many histories treat the ancient world, the Middle Ages and the Renaissance at some length, but later periods much less so relative to the advances in mathematics that took place in them; in particular, *the 19th century* is often treated in a perfunctory manner. Here, the balance in length is reversed, with that century taking nine chapters. There is probably some overbalance the other way; but, without intending any disrespect to the earlier times, mathematics since 1800 usually makes more contact with the modern mathematician or student, to whom this book is primarily oriented. In addition, since many other general histories do treat the earlier periods, sometimes very well, I refer the interested reader to such works.

2. Most histories fail to convey *the importance of applications to the physical world* as a source of mathematical progress. By contrast, it is a central feature of this book; thus the range of topics described here is much wider than is normally found in general histories. This point is especially significant for the mid-19th century onwards, when the growth of the mathematical profession led to some degree of snobbery with regard to pure as opposed to applied mathematics, an unfortunate attitude which many general histories follow. In particular, various areas of engineering mathematics are noted, from early map-making through 18th-century hydraulics to the establishment of aerodynamics in the 1900s. Hence the title of this book includes the umbrella term 'the mathematical sciences', and the word 'mathematics' is normally used in this more general sense.

3. As an important example of *different kinds of application*, I distinguish between two ways in which a mathematical theory M can be developed with a physical interpretation P in mind. The first is where the features of P control the scope and, especially, the limitations of M; in the second, guidance by these features becomes lost, so that M is only *notional* relative to P (for example, an effect is calculated which cannot be observed with the required precision, or physical constants are set equal to each other with no physical sense). In addition, I distinguish *general* kinds of physical situation from the more *specific* ones often found in engineering and technology, and M is oriented towards them (carefully or notionally). Outside applications we find so-called 'pure' mathematics, which includes Ms where no P is discussed although some lie in the background, for instance as original motivation.

4. *Probability and statistics* are often neglected in general histories or even omitted entirely. The main reason is the late arrival of these subjects on the mathematical scene: for the most part during the 19th century, and several important applications after the Great War (and so not treated here). I have tried to give due place to the early history of these two branches of mathematics, while also seeking

reasons for the lateness of their arrival and for the remarkable delay and disconnectedness of their development. They extended still further the range of mathematical concerns, for some new areas of application were found, especially in the social sciences.

5. While the major mathematicians dominate the story, I avoid a Great Man approach by also noting *lesser figures* who made essential contributions. In addition, tables give details of the careers and contributions of certain groups of mathematicians; not everyone so mentioned is necessarily discussed in the text, but the lists should provide useful guides to a reader of other, more detailed, histories. The dates of many mathematicians are given in the index.

6. I take *the word 'history'* to relate to the question 'What happened in the past?'; by contrast, mathematicians (and scientists in general, and even a distressing number of historians) take history to mean 'How did we get here?' The difference between these two questions is worth pondering. Answers to the second one draw *only* on those parts of the past that have led to our present situation; while a perfectly respectable form of research, they can give quite mistaken impressions about the aims and purposes of historical figures, and the priorities they saw in their own work. For example, theories of ratios gradually changed into one of equations in the hands of some later Greek (§2.26) and medieval mathematicians (§3.23), but treatments by ancients such as Euclid do not presage it (§2.18). Again, it looks as if Lagrange found Fourier series in the 1750s (§6.3); but, as we see in §8.7, that was not his aim.

7. Histories often record only apparently well established facts and details. By contrast, some major *changes in historical interpretation* are recorded here, especially in studies made in recent decades which have exposed new sources, broken into previously unstudied developments, or made new suggestions about the motives and contexts of well-studied periods.

8. One badly neglected aspect is *national differences.* Mathematics may be international, but interest in branches

and topics have differed widely between countries; in the 15th century medieval German algebra is subtly different from Italian, and a British version barely existed; in the 19th century the French and Germans did lots of mathematical analysis (not always by the same methods), while the English and Irish busied themselves with algebras; and so on.

1.4 *The rainbow of the mathematical sciences*

All the above aims are encapsulated in the subtitle of this book, which I consider here. The word 'rainbow' is intended not to convey an impression of mathematics as a merely two-dimensional spectrum with fairly determinate ends, but to suggest two other, more profound analogies.

The first analogy, which constitutes the principal lesson of this book, is with *the stupendous variety and vastness* of the range of activities in which mathematics has played a significant role. It enjoys a unique ubiquity in the history of ideas, and also in modern life.

The second analogy is in marked contrast. The history of mathematics is largely absent from the 'culture' of the educated public, historians and mathematicians included. The extent to which it is dismissed, abhorred, even derided, has to be experienced to be believed {Grattan-Guinness 1990c}. Like the rainbow, mathematics may be admired, but – especially among intellectuals – it must be kept at a distance, away from from real life and polite conversation. However, unlike the real rainbow, mathematics stays still when approached, and readily admits the active inquirer into its world of many colours.

1.5 *Some special terms*

In this book some terms have more restricted or rather different meanings from their everyday usage.

1. The word 'culture' refers to an ethnic or racial group occupying a geographical region (taking in maybe more

than one country), over a considerable period of time. A 'school' of mathematicians is so named only when pretty strict criteria are met: clear leader(s), well defined programme, regular avenues of publication, and so on. For looser collections of mathematicians, 'group' or 'community' is preferred. Individual persons are referred to as 'figures', to be distinguished from the 'Figures' which adorn the text as illustrations.

2. I stress the plurals 'geometries' and 'algebras' to emphasize the facts that, especially during the 19th century, different kinds emerged in each branch of mathematics: non-Euclidean and projective geometry complemented the Euclidean, and new algebras such as differential operators joined common algebra. During the 20th century the word 'algebras' has been adopted to refer to kinds of vector space in which (a) vector multiplication is defined; but this sense is deployed only in §16.19, when the origins of the theory are noted.

3. The adjective 'applied' is rather too wide-ranging, for it does not distinguish applications within pure mathematics from those to physical situations (for example, applying the integral calculus to number theory as opposed to applying it to elasticity theory). Thus I have 'pure', 'applied' (in the latter sense), 'applicable' (where no applications are made, although they helped motivate the research), and 'notional' as explained in item 3 of §1.3.

4. It is clear that major changes have taken place in mathematics, and one may be tempted to speak of revolutions occurring in the subject {Gillies 1992}. Unfortunately the word 'revolution' has been overused by historians of science, leading to incoherent discussion. Here a revolution will be understood to be a relationship between one or more mathematicians and a particular mathematical theory (or technique, or proof-method, or whatever) over a certain time period. Further, to be truly revolutionary, some kind of *change* of understanding has to be involved, with one theory being replaced, or at least demoted, by another

one. A Gestalt effect can be involved in comparing the new and old theories; the new way 'looks' at the phenomenon or situation differently from its predecessor. By contrast, really new (that is, non-replacing) ideas are characterized as 'innovations'.

5. In between these two processes lie 'convolutions', where new ideas intertwine and coil around old ones in the course of their rise to prominence. The notion of 'normal science' {Kuhn 1970} is essential here, in that a new normal theory or technique eventually supersedes a previous one. Normality, revolution, innovation and convolution can co-exist. For example, during the 1800s Fourier innovated the mathematical analysis of heat diffusion (§10.11), but by using normal methods of differential modelling; he also revolutionized upwards the status of methods of solving differential equations by trigonometric series (§8.7).

6. I distinguish between (developing) mathematical 'theories' and intuitive 'thinking' in the same area. Two branches of mathematics stand out. In addition to the various geometries as such, geometrical thinking has been evident from the earliest times and is often present in other branches; thus geometry in one form or another occurs in every chapter. The other branch is probability, where theory came rather late but thinking had been around for a long time.

7. One way in which thinking may approach the status of theory, especially in applications, is that it is 'desimplified'. A theory is often developed by making simplifying assumptions about phenomena in order to render a problem practicable; after useful theory is produced, some of these assumptions are abandoned and features are incorporated into the theory already established; and so on. Typical examples of (partial) desimplification include allowing for the rotation of the Earth in a problem in mechanics (but still assuming the rotation to be constant), or no longer assuming that the environment of the phenomena under study is at constant temperature.

8. Often allied to desimplification is 'structure-similarity', which arises when the structure of a pure mathematical expression is reflected in the physical situation to which it is applied (and 'structure-dissimilarity' when it is not) {Grattan-Guinness 1992b, p. 128}. To continue with Fourier, he required his trigonometric series to converge so that its sum would correspond to a finite temperature, but he did not go on to interpret the sine and cosine terms as a description of heat moving in waves (§10.11).

9. I distinguish various 'styles' within which a mathematical theory may be developed by a particular figure or group. Styles guide the formation of principles and proof-methods, and ways of reasoning in general, and the way in which a theory is applied in physical problems. The principal ones are geometric, algebraic and analytical, but others such as the axiomatic style also occur on occasion. A style is to be distinguished from any cognate branch of mathematics carrying the same name; for example, geometry can be done in a very algebraic way, while mechanics and mathematical physics have often been developed in a geometrical style, especially in the creation of new theories.

10. A style and a branch of mathematics may unite in an 'approach' to a problem or (re)formulation of a theory. For example, Archimedes' approach to geometry drew on a geometrical style and proof by contradiction twice over (§2.20), while Lagrange's algebraic approach to mathematics not only imposed a style but also followed a philosophy of reducing assumptions in formulating theories (§6.5–§6.6).

1.6 *Some essential limitations*

The range of the subject and the size of this book have imposed a number of limitations.

1. The obvious and major one is, of course, compression. Mathematics is not only so ubiquitous but also of the greatest ancestry, so that severe choices and omissions of

material have had to be made. In particular, I tend to describe (fairly) general theories rather than particular 'local' problems which often set the mathematician his tasks; unfortunately space is not available to explain them and their solutions comprehensively.

2. The story is normally confined to developments that were circulated or published and exercised influence (positive or negative) around the time of their creation. Thus European work dominates when it finally entered the fray. I usually ignore fine achievements, made in any culture and at any time, which for some reason did not gain the attention they deserved when published.

3. Apart from a few later examples, the main story stops with the Great War. This was chosen partly as a major event which affected the practice of mathematics (though not necessarily the theories) during and afterwards, and partly because to advance further would have compounded the problem of compression: to continue with the kind of treatment given here would demand at least another 200 pages to reach even the 1950s, and the ever-widening range of topics it would be necessary to take in is beyond the competence of any historian (certainly this one). In addition, arrival at recent times would give the impression that the book was intended as an introduction to today's mathematics, which is a perfectly worthwhile but *quite different kind* of writing.

4. While neighbouring sciences, especially physics and astronomy, appear in the story when appropriate, few details are provided, and only those aspects bearing directly upon mathematics are treated in any detail. For more discussion the corresponding literature should be consulted, such as the companion volumes in the Fontana History of Science series.

5. I have not necessarily chosen Great Theorems, results which today are well known; for they were not always so significant in their own times (and often did not take today's forms), the people after whom many are named are not the best historical candidates for their origination, and

so on. Similarly, many results and techniques were anticipated (at least partially) before becoming firmly established, and the reasons for the intervening neglect or oversight pose fascinating historical questions; however, only a few cases are examined, and then briefly.

6. Sometimes other disciplines drew on 'covert mathematics', in that little description or even direct evidence of its content or use is available. The lack of records arises partly from the absence or loss of sources, partly because the mathematical level sometimes did not advance beyond thinking to theory, and partly from historians' own lack of awareness of information in unfamiliar sources. Architecture, art and decoration embody a massive amount of such mathematics, from ancient times and seemingly all cultures: perspective and orientation, symmetries, implicit group theory in patterns, combinatorics, and so on. Other contexts, especially in ancient times, include games, methods of large-scale counting (including book-keeping), surveying in road design and battle plans, and sacred arithmetic and geometry. These limitations arise especially in §2 and §3, on ancient and medieval mathematics respectively.

7. While I comment on the modes of employment of mathematicians at various times and the institutions in which they did (or did not) work, and point to interactions between education and research (of which the history of mathematics provides many rich examples), I have not attempted any sophisticated analyses of the issues raised. Further, the accounts are usually confined to universities or their equivalent; school-level mathematics education is hardly discussed here, and in any case is not well studied historically. I also have not provided detailed accounts of biography and career, although short statements on some principal figures are given at their first appearance.

8. While no particular historical knowledge is assumed in the reader, familiarity with some modern version of basic mathematical theories has to be assumed. The highest level corresponds roughly to that of final-year university-level courses in mathematics, including basic calculus (including

the calculus of variations); basic properties in Euclidean geometry and the possibility of non-Euclidean alternatives; ideas in mechanics such as force, momentum and energy, and principles like Newton's laws; matrices and determinants; and fundamental notions of probability and of parameters in statistics such as median and standard deviation.

1.7 *The use of notations*

{Cajori 1928, 1929}

Choices of notation raise many questions of historical method. The use of a modern notation suggests a modern theory, which in the context of older mathematics is anachronistic. Specific points about notation are explained in the text where appropriate, but important and general cases need mentioning here.

Regarding numbers and numerals, I follow a normal convention of writing 'zero', 'one', ... 'ten', '11', ... in ordinary contexts. But when the development of number systems is itself the topic of discussion, more careful uses are required, and are explained. With matrices and determinants, for example, where the concept of an array of entries is an essential component, fundamental misinterpretations may arise when they are used in an earlier historical context. In such situations I have used these notations only to save space, with a warning that no modernizing is implied.

Among notations for the calculus, I write the economical symbol 'z_x' for the partial derivative of z with respect to x. This is the modern way of utilizing the symbol, largely following Cauchy (§8.6); I also use it for the ratio of differentials $dz \div dx$ in Leibniz's earlier approach (§5.5), and for the differential coefficient in Euler's revised version (§6.2).

Finally, I use the modern notations '$A := B$' and '$B =: A$' to denote the joining of a defined term A to its defining clause B. Sometimes they are accompanied by the warning that maybe the historical figure did *not* have a clear theory of definitions to accompany his formulation!

1.8 *References and bibliography*

In the text I refer to many major original books and papers by name and date. They are listed in the Bibliography, where the information given should be adequate to find them. The Bibliography starts with a short statement on general bibliographies and reference works. Works cited in the text are usually historical items, and I have given some preference to wide-ranging ones which contain excellent bibliographies of original and other historical writings. Russian texts are cited only in translation into a European language. English is the most common one, but German is also very important. The following source is one main reason.

The *Encyklopädie der mathematischen Wissenschaften* was a huge German project published between 1898 and 1935, which covered in six Parts almost all branches and topics of mathematics (§16.3). Many of its articles are technical reports rather than histories in the broader sense; but the mass of information provided is amazing, and the work as a whole will never be superseded. Articles from it are cited (mainly in §12–§16) in the manner '{EMW IIIC2}', meaning the third Part, Section C, second article. This is the easiest way to get around the very complicated system used there.

Another special source is the *Companion Encyclopedia of the History and Philosophy of the Mathematical Sciences*, a work of 1,800 pages which was prepared under my editorship and published by the London house of Routledge early in 1994. It attempts the same coverage as this book, but does so in much more expansive (but still compressed) style, with the help of 132 other authors. It would have been impossible for me to attempt the present book without this predecessor; I cite its articles by the numbering system used there, in the manner '{CE 10.9}', meaning Article 9 in Part 10.

These two sources increase very considerably the usefulness of the present book, and the concise forms of citation should not disturb the flow of the text. The references are given at the head of the bibliography.

1.9 *Organization and proper names*

Formulae, diagrams, tables, definitions and theorems are numbered afresh from the start of each section, so that we have formula (1.1.1) above, FIG. 5.10.1 below, and so on. Cross-references are made to these numberings. Quotations are translated into English where necessary (either from a reliable available version, or freshly made). Times before that of the alleged birth of Christ are rendered arithmetically ('–4th century', '–327'), in an interesting ordinal numbering system that lacks a zero.

Names of historical figures present some problems, especially when in ancient or medieval times even the identification is not always clear-cut. For a figure from any period who published in several languages and whose name was rendered differently in each one, I adopt a clearly established preference.

Names of figures from countries in the Middle and Far East are transliterated without recourse to the underdots and overbars used in specialist studies; so we have, for example, 'al-Khwarizmi'. Greek names are transliterated as, for example 'Pappos' rather than 'Pappus', which is far more common but is, however, the Latinized version. Russian names are given in their standard transliteration into English.

1.10 *Acknowledgements*

Many friends and acquaintances have helped me over the years, especially those who read draft sections or chapters of this book. Thanks are especially due to Roger Cooke, Jim Cross, A.E.L. Davis, John Fauvel and David Fowler. I am also indebted to Kirsti Andersen, Bruce Brackenridge, Tony Crilly, Andrew Dale, Joseph Dauben, J.V. Field, Günther Frei, Jeremy Gray, Niccolò Guicciardini, Peter Harman, Jan Hogendijk, Jens Høyrup, Khalil Jaouiche, Clive Kilmister, Detlef Laugwitz, Walter Ledermann, Albert Lewis (who also prepared the index), George Molland, John North,

Maria Panteki, Helmut Pulte, Karin Reich, Christoph Scriba and Imre Toth. My wife Enid saved me from some of my own typing, for which we are all grateful. Copy-editor John Woodruff exercised over the manuscript his high degree of intolerance for error and ambiguity, to the great benefit of the book now laid before you.

March 1996

2

Invisible origins and ancient traditions

2.1 *Without a start: eight ancient roots of mathematics*

The development of mathematics resembles a series of which there is no known first term. For there can be no means of finding out how mathematical thinking developed even in primitive humans, never mind among the animals; but presumably it began as a central part of thought itself, and before any manner of writing or inscription had been conceived. Indeed, if one considers the kinds of problems and needs of thinking beings from the start of their lives and cultures – for personal survival, communal existence, recreation and exercise, and religious contemplation – then at least the following eight types of intuitive thinking of a mathematical kind must have arisen:

1. counting and sorting, which led to arithmetic;
2. spacing and distancing, which led to geometry;
3. positioning and locating, which led to topology;
4. surveying and angulating, which led to trigonometry;
5. balancing and weighing, which led to statics;
6. moving and hitting, which led to dynamics;
7. guessing and judging, which led to probability theory;
8. collecting and ordering, which led to part–whole theory.

A principal theme of this book is the ways in which a culture turned (or failed to turn) intuitive thinking into theories – that is, bodies of mathematical knowledge – and with very different histories. There is no preferred order in the list above, except that arithmetic probably developed most quickly. For this reason it is usually given a principal place in mathematics; but this view seriously underrates the other seven types and their attendant branches of mathematics, and thus distorts the understanding of mathematics itself (compare §16.6). Several of them will also have started off quite quickly; but set theory and topology were extraordinarily late arrivals on the mathematical scene (§12.4, §13.17). Note further that algebra, the calculus and mathematical statistics are *not* among the ancient roots.

The word 'source' is used to refer to any kind of origin of information; 'text' covers artefacts, inscriptions on buildings, and incisions on objects and structures, as well as symbols written on some material such as paper or papyrus. A central difficulty for the historian is the inevitable reliance upon texts, the verbal side never being preserved. This obvious truth is of great significance for the understanding of the origins of (proto-)languages, including mathematical ones. It seems likely that spoken language emerged from signs and grunts, firstly in the form of one-word sentences such as 'Kill!' or 'Tree'; we still often talk this way, supplying clarity by means of tone, context and gesture. Gradually the unavoidable ambiguities were reduced by the addition, not only of further words, but also of grammars of various kinds ('no tree' or 'tree no', or whatever); and maybe it was only in written language that grammar and semantics began to take the sophisticated forms with which we are familiar.[1] Hence the early stages cannot be traced; mathematics without a detectable start.

Two deep and general points about ancient cultures are often underrated: that *people then saw themselves as part of nature*, and *mathematics was central to life*. These views stand in contrast to modern ones, in which nature is usually regarded as an external arena for problem-solving, and mathematicians are often treated as mysterious outcasts, removed from polite intellectual life. Hence the word 'research' does not really capture the attitude of ancient investigators.

Again, aspects of mathematics such as numerology and symbolic geometry are now derided as mere mysticism; but

[1] I am indebted to Sir Karl Popper for valuable discussions of this question. He pointed to Greek authors – writers, indeed – such as Plato and Isocrates stating that writing is not enough, and to Saint Augustine (+4th century) expressing surprise at seeing Saint Ambrose reading a book silently instead of out loud.

precisely for that reason they were far more highly esteemed in distant times, and played important roles in the development of various mathematical theories. Good evidence, to be found in many cultures, is provided by the giant representation of humans and animals, and by complicated designs such as spirals, coils and labyrinths to be found on temple walls; for they reveal profound intuitive understanding of curves and the similarity of shape to the real and smaller object, sometimes even of (a)symmetries and topological properties.

Some early stages in mathematical thinking must have employed distinctions which we do not use or appreciate. For example, we are accustomed to use integers (whatever we think they are) quite generally: six teeth, eggs, trees, virtues, or whatever. But some tribal cultures still use different words for the same number in different contexts, and ancient peoples very likely did the same; so "going general" must have occurred in a remote past of which we know nothing. Again, while próportionality or equality between mathematical objects will have been of major importance, *in*equalities must have played a great role: that there are *more* men in the enemy tribe than in ours, that the horn-animal is *nearer* to my mate than I am, and so on.

Finally, note the various senses of size. Take the question, 'Is this island large or small?' If you own it or want to tax its owner, then area is the prime consideration. But if you are sailing around it, then it is the perimeter that matters most. There is indeed evidence of this distinction in records involving the Mediterranean, the part of the world around which so much mathematics was born.

2.2 *The order of cultures*

The next section looks at very early evidence from sub-Saharan Africa and then from the Middle East (for convenience, I use these modern geographical designations). After that I consider, in roughly chronological order of emergence, the Babylonian, Egyptian and Greek cultures before moving to the Far East to note Chinese and Indian traditions. By far the main emphasis falls upon Greek mathematics, partly because of its intrinsic quality but mainly due to its dominating influence upon European mathematics, which is the main concern of the rest of this book. In terms of achievement *alone*, therefore, the treatment is unbalanced. The final section, §2.34, treats in general the question of influence of one culture on another. Among other cultures, the Japanese, Korean, Arabic/ Islamic, Hebrew and pre-Mayan are postponed until §3, for they achieved greater prominence after +500, which is the normal end-point of this chapter.

Histories have long been written, especially by Greeks and Arabs, on their own cultures and even other ones; for example, the Greek Plutarch on Egypt (§2.6). The practice was continued by Europeans, especially from the 17th century onwards, when Indian work also began to be studied. But historical research on ancient cultures *before* the Greeks (and also on the Far East) was much more patchy, and advanced only from the early 19th century with the discovery – and especially the decipherment – of ancient texts. A crucial advance was made by the French expedition to Egypt of 1798–1801, when major texts such as the Rosetta Stone were found and the first scientific studies were made. An important figure was a young civilian member of the campaign, Joseph Fourier. He led one of the two expeditions of discovery to Egypt, and upon returning to France and being appointed (against his will) Prefect of the French *département* of Isère based at Grenoble, he excited a passion for Egyptian studies in a young boy in the town. Jean Champollion-Figeac (1790–1832) was later to decipher the Rosetta Stone and obtain in the 1820s the first professorship in 'Egyptology' (as he called it), at the Collège de France in Paris. By then, Fourier had become a major figure in French science, principally through his work on heat diffusion and the solution of differential equations (§10.11, §8.7).

Champollion's achievements greatly helped to advance the understanding of other ancient cultures of that region, not least at the Collège, already a centre for the historical study of cultures in the Middle and Far East. Since then, new finds of artefacts and texts have stimulated successive generations to new interpretations; and recently the need for historians of mathematics to draw on allied disciplines such as archaeology and papyrology has come to be better appreciated. The resulting layers of historical (re-)interpretation are too complicated to receive due notice here, but I draw on some interesting recent ones, which have not received much attention in general histories of mathematics.

Among the most recent general books, especially recommended for ancient times are {Kline 1972; Katz 1993}; nice introductions to the period are given in {Neugebauer 1957} and {Giacardi and Roero 1979}. Among other works, several cultures are examined in special articles in the 15th volume (1978) of the *Dictionary of Scientific Biography*. Symbols and notations are widely covered in {Cajori 1928}; there is only space here for a smattering from an enormous range. In addition, the light treatment here of astronomy is augmented by the companion volume in the Fontana History of Science series {North 1994, Chaps. 1–5}, and by a magisterial catalogue of information provided in {Neugebauer 1975}. Despite all the historical work, words like 'seems' and 'maybe' occur below frequently – maybe not frequently enough . . .

2.3 *African traces: notches and tokens*

Let us begin, then, with the earliest sources that have been found and studied. Most of them are from Africa, which is generally regarded as the site of the origin of mankind. Mathematics has developed there among all tribes, in various ways and forms, with number-systems and geometry prominent {Zaslavksy 1979; CE 1.8}. Unfortunately, serious study of the history of ancient African mathematics south of the Saharan desert has begun only recently {Gerdes 1994}, and too little is known for a concise summary here; but one example can be given.

In a study of sources in Europe and the African Continent, {Marshack 1973} suggests that the heavenly bodies excited among its observers some of the earliest examples of written record, during the Upper Palaeolithic period and dated around 37,000 years ago. These records took the form of notches engraved on bones, which he has shown by microscopic analysis to fall into groups of similar kind. His interpretation is that the marks register the daily passing of the Moon through its various phases of full, half, crescent and new (FIG. 2.3.1). His arguments for ancient interest in the Moon are plausible; I would add the possibility that ancient woman noticed the correlation between lunar phases and her periods.

Three specifically mathematical features of the notches are worth stressing, for they will run through the whole

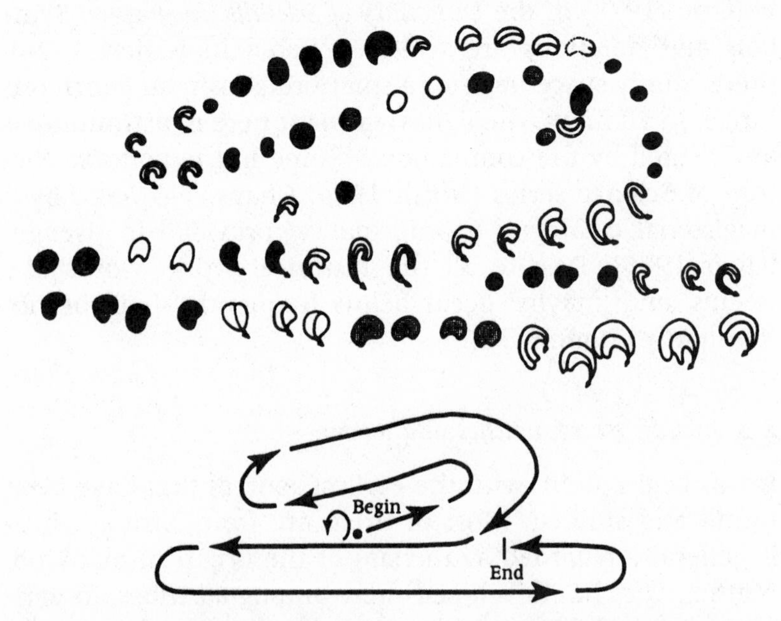

FIG. 2.3.1 *69 marks on a bone, with 24 changes of mark; and proposed order of reading.*
{Marshack 1973, p. 48}

chapter. Firstly, we note the early place of adding, and subtracting, in prehistoric arithmetic; multiplication and division emerged much later, and not always in general forms. Secondly, the importance of *the sign as such,* and of differences between signs, provide an early début for semiotics – that is, thinking and eventually theory about signs. Lastly, there is an element of mechanics here, not only in the intimate link between mathematics and astronomy but also in the means devised by the carver to mark the bone in the first place.

The next method of recording seems to have been by tokens, small fired clay objects which are found in quantity at archaelogical sites in the Middle East. Dating back to the –8th millennium, many were made in simple shapes such as spheres, cylinders, cones and tetrahedra; but those from the –4th millennium onwards include more sophisticated shapes such as bent coils and rhomboids, and even human or animal forms. Some have been found as collections inside hollow vessels (though unfortunately not always left there after finding), and {Schmandt-Besserat 1992} proposes that this location was their purpose; that is, they served as records of data, such as numbers of measures of grain or of members of a work-gang, with different shapes used for different contexts. Originally the manner of counting was by one-one correspondence, such as three cones to three consignments of grain, or four cubes for four workers in a gang; but then the sign language was enhanced by using, say, a larger cone for a larger measure of grain. Some tokens are perforated discs, which could be tied together in appropriate numbers; a few seem to be labels carrying a sign representing, say, wine. Other tokens carry notches or incisions, occasionally paintings.

The different shapes of tokens may have reflected, or caused, different words for the same integer (§2.1). But a greater generality is evident in the next stage, when tokens were replaced by tablets, and the physical cone was transformed into the cuneiform cone-shaped pictograph. In this way a community's arithmetic helped in the birth of its writing.

2.4 *The tokens and tablets of Babylon*

{Høyrup 1994, Chaps. 1–3; CE 1.1}

Historians usually trace the detectable origins of mathematics to two cultures. Seemingly the most ancient culture of all lived in a region stretching from the Mediterranean coast of Syria across modern Iraq to the Persian Gulf, especially the belt between the Tigris and Euphrates rivers which was called 'Mesopotamia' (Greek for 'middle river') during the −2nd millennium. (The other culture is the Egyptian, to which we turn in §2.6.) This culture seems to have named the northern and southern parts of the region as 'Sumer' and 'Akkad' respectively; later they were called 'Babylonia' (after the Greek name for the city of Babel, south of modern Baghdad) and 'Assyria'. To elide over considerable complications, I shall follow common historical practice and call the entire culture 'Babylonian', although Babylonia itself is not known to have been a major cultural or scientific centre.

The main phases in their mathematical development are 'the third dynasty of Ur' (or 'Ur III') and 'Old Babylonian' from the −20th to the −17th centuries, followed by five centuries of the Cassite period; and the 'Seleucid', from the −4th to the −1st centuries, for which new sources have come to light in recent decades. Of the mathematics in the intervening centuries, a period of subjugation and decline, little is known; but it is thought that the later periods were aware of the earlier ones.

Babylonia is rich in tokens, but the principal sources survive as collections of cuneiform symbols etched onto clay tablets which were then dried or fired. Most of them show the characteristics of teaching material. I shall call their authors 'scribes', although they may have used assistants to execute the texts. The scribes are not usually named, but by virtue of this literacy they may have been important persons.

Arithmetic was always to the fore, both in records and in exercises based upon data. The semiotic aspect is evident in various ways: for example, in that a small circle indicated (say) 6 units of grain while a large one marked 60 units. The Babylonians constructed a system of numerals based on 60; but one cannot confidently ascribe the choice of 60 to its many arithmetical factors, although 12 relates nicely to counting with the thumb on the joints of one hand; maybe it was related to their astronomy, with its 360-day year. By the Seleucid period and probably earlier, another system of numerals based on 10 was also being used; perhaps 10 was chosen *from* 60, because of the property $60 = 6 \times 10$.

Using a place-value system to base 10, the Babylonians wrote numerals from left to right, with signs resembling 'Γ' for the unit and '≺' for ten; thus 34 was represented by ≺ ≺ ≺ΓΓΓΓ. The absence of a zero left the notation as floating-point; the context would make it clear whether '34' was the equivalent of 34 or 340. Further, confusion between the equivalents of our 103 and 1003 arose, so at some stage they ceased to rely on leaving a blank space and introduced a symbol resembling '≼'. This, however, served only as a place-marker, and not as a genuine number that could be used in combination with other numbers.

As well as integers, the Babylonians used fractions based on 60ths, again in a floating-point way. The notation used by historians to represent such numbers was devised by Otto Neugebauer; for example,

'2, 6; 1, 34, 15' denotes the number

$$2 \times 60 + 6 + \frac{1}{60} + \frac{34}{(60)^2} + \frac{15}{(60)^3}. \qquad (2.4.1)$$

But one should avoid reading the Babylonian symbols as representing arithmetical operations.

Astronomy was a central part of Babylonian science, with the Earth placed at the centre of the Universe. The Babylonians seem to have invented the zodiac, the plane in which the heavenly bodies moved, split into 12 divisions of 30° each. Keeping records of the motions of the Moon was a major concern, in an elaboration of the tradition of marking bones (§2.3). Trigonometry is evident, but in a limited form; angles were measured only in the plane, and below or above the horizontal, not relative to any basic direction. But their analyses advanced to astonishing levels – in particular, they discerned and calculated variations in planetary and (apparent) solar motions. The modern graphical presentations of these arithmetical results are called 'zigzag', to convey the periodic effects that they had found; maybe they displayed them in a similar way in sand or on tablets.

2.5 *A characteristic Babylonian problem: the six brothers* {Caveing 1985}

Babylonian astronomy drew on geometry, which is present in many other contexts, for example in calculations of the flow of water in a canal. Many problems deployed geometry and arithmetic together; a nice case, which may date from the Cassite period, is shown in FIG. 2.5.1. A field $A_1B_1B_7A_7$, in the shape of a trapezium with the lengths of its sides indicated below the Figure, has to be partitioned among three pairs of brothers such that the members of each pair receive the same area, but each pair steadily less; the thick lines A_3B_3 and A_5B_5 show these latter differences. The solution falls out in nice integers; $CA_7 = 117$ and $B_7D = 45$, and the other required measurements are as given in the display. The last row gives the areas in proportion; the actual answers are irrational, since the breadth A_1C of the field is $18\sqrt{14}$. This interesting detail is not discussed in the original text; the Babylonians seem to have been unaware of irrational numbers, or unconcerned by them. Nor did the

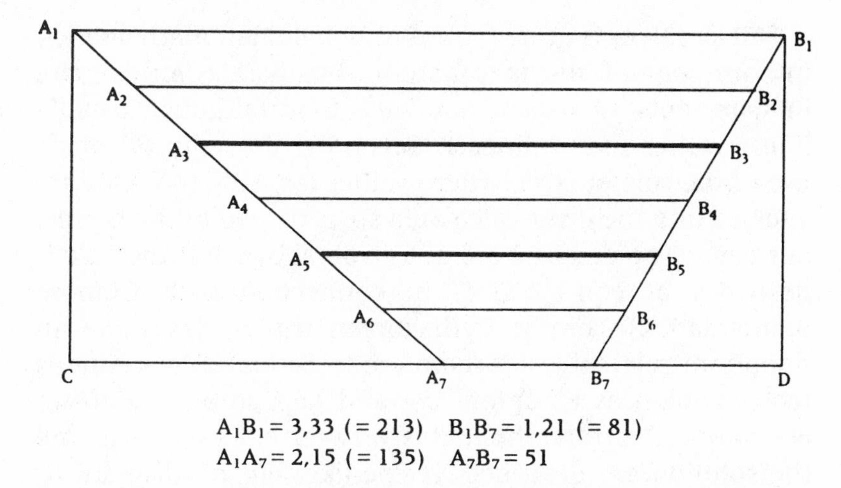

$A_1B_1 = 3,33\ (= 213)\quad B_1B_7 = 1,21\ (= 81)$
$A_1A_7 = 2,15\ (= 135)\quad A_7B_7 = 51$

FIG. 2.5.1 *The Babylonian 'six brothers' problem* (drawn to scale).

i	1	2	3	4	5	6	7
Length A_iB_i	213	183	147	123	93	75	51
Length A_iA_{i+1}	25	30	20	25	15	20	
Length B_iB_{i+1}	15	18	12	15	9	12	
Area $A_iB_iB_{i+1}A_{i+1}$	55	55	30	30	14	14	

scribe mention that the problem has two degrees of freedom in locating the 'boundary' lines A_3B_3 and A_5B_5, but a step in the instructions shows that he imposed three conditions for equal area *across* these boundaries:

$$2(A_iB_i)^2 = (A_{i-1}B_{i-1})^2 + (A_{i+1}B_{i+1})^2, \quad i = 2, 4 \text{ and } 6. \tag{2.5.1}$$

In addition, for $2 \leq i \leq 6$ each length A_iB_i and the average and the semi-difference of the neighbouring lengths form a triple of integers satisfying Pythagoras' theorem (as we call it: see eqn (2.12.1) below), such as (147, 153, 30) for $i = 3$; maybe the scribe devised the problem from these values in the first place.

This problem is quite typical of Babylonian mathematics, in both content and presentation. Firstly, it is an exercise in share-outs, of which they were fond (although usually linear rather than bilinear). Secondly, the solution made use of the rule of false, where values for A_3B_3 and A_5B_5 are guessed and the error calculated so as to lead to the correct answers. (The details were not given above, but the rule is described at eqn (2.31.4) in connection with Chinese mathematics.) Thirdly, Pythagorean triples play quite an important role; tables of them are to be found in a famous tablet known as Plimpton 322 and kept in a museum in New York. Fourthly, the text is very terse: a procedure and the solution are described, recipe-like, but no diagram or discussion is appended, not even of aspects such as assumed conditions. Doubtless the problem was created from the answers: it works out cleverly in integers, but is the exercise really worth doing? Maybe this is an early example in the history of mathematics (and maybe also of mathematical education) of knowledge for its own sake, notionally (§1.3) tied to real-life questions but in fact intended as a tactic to isolate (or train novice) scribes *away* from that real world and provide them with arcane knowledge denied to others.

My summary above is in modern notations; how was the problem handled in the Cassite time? From the 1930s onwards, Neugebauer, who published important editions of many texts, also popularized the view that common algebra underlay much Babylonian mathematics, including arithmetical solutions of geometrical problems such as this one, the lengths of the sides and transversals, products of such quantities, solving quadratic equations, and so on {Neugebauer 1957, Chap. 2}. His position was motivated by a search for the origins for the algebra that historians had argued for decades underpinned much Greek mathematics (§2.19).

My only concession to this view has been to use suffices, and then only to save space; for in recent years this interpretation has been challenged, especially by {Høyrup 1990}. In addition to the general danger of identifying number-words with number-symbols, he notes that different words were used to distinguish types of operation which common algebra merges under 'addition': for example, *adding* together the numbers of workers in two gangs was not treated in the same way as *adjoining* one geometrical region to another. Two companion forms of subtraction were employed (in each case with the lesser quantity taken from the greater). Multiplication occurred in four different ways: increasing the height of a cuboid by multiples, spanning a rectangle by its base and altitude, repeated doubling of an area, and repeated addition of numbers. Further, the word 'side' referred to *an area with a unit width,* like an idealized brush-stroke. Artists still understand this use of the word, as one sense of the phrase 'drawing the line'.

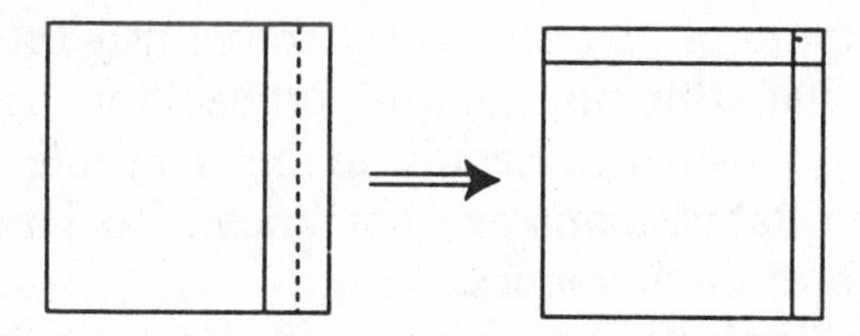

FIG. 2.5.2 *Completing the square with the 'side', Babylon-style.*

FIG. 2.5.2 shows an important example of the procedures that a scribe seems to have followed (some tablets carry diagrams). He adjoined a side to a square, slit the side down the middle, and moved half of it round to the top; then he 'completed the square' (as we still say) by introducing the small square. An L-shaped region was thus created; the Greeks were to call it a 'gnomon' and grant it much significance, for both mathematical and astronomical reasons (§2.22).

This example shows that common algebra does not help our historical understanding of Babylonian mathematics. If an algebra was used, it *must* have observed the eight different operations recorded above. And even these are not quite enough; for in a magical moment one scribe constructed a rectangle with sides equal to the *areas* of squares already obtained. The distinction between dimensions was thereby broken, although not permanently.

2.6 *The tablets and papyri of Egypt*

{Gillings 1972; CE 1.2}

Contemporary with the Babylonians, the Egyptians were developing a remarkable culture over to the south-west. Coming to pictography out of a language of tokens, they left relatively rich sources of information on their mathematics, especially carvings on buildings and writings on papyrus scrolls. While the two cultures have many features are common, there are significant differences also. For example, the counter game preserved in the reliquaries from the tomb of Pharaoh Tutenkhamun (−14th century) suggests that thinking about probability had started, although no evidence seems to be available about the details; the Babylonians are not known to have initiated this branch of mathematics.

Champollion's decipherment of the Egyptian language in the 1820s (§2.2) was a major breakthrough in understanding all ancient cultures in the Middle East, and Egyptology has always held a special place. Perhaps the richest single text is the 'Rhind mathematical papyrus', named after its English purchaser of the 1850s and published several times since; a particularly nice edition is {Rhind Papyrus 1987}. Three different kinds of writing have been detected, all written from right to left (the opposite direction from Babylonian texts).

Arithmetic is the most evident branch of mathematics, with a system of integers based upon 10 which advances to a million. As with the Babylonians, the Egyptians recorded data, and posed questions and even puzzles, maybe as educational exercises; however, their system of numbers was not based upon place values. They gave some numbers special status; in particular, according to the commentary 'On

Isis and Osiris' on their culture by the Greek historian Plutarch (+1st century), they extolled 18 for the equality of area and perimeter of the rectangle of 6 units by 3:

$$18 = 6 \times 3 = 6 + 3 + 6 + 3. \qquad (2.6.1)$$

An example from their architecture is a vestibule in the Temple of Hathor at Dendera, which contains 6 × 3 columns in two squares of 9.

Another important number was 42; its various properties included the number of the lost books of learning, of gods who assess the dead, and of the sins that one might commit. At least one text on a temple wall has 42 columns for its text.

The Egyptians also adopted unit fractions $\frac{1}{n}$ for various small values of n, and also $\frac{2}{3}$, as basic numbers. They probably did not understand these fractions in the way that we construe rational numbers; they might have been ratios, such as gradients of slope (that is, vertical drop to horizontal displacement) like the 1:6 and 1:7 of the faces of their temples {Robins and Shute 1985}. To avoid anachronism, historians customarily write, say, their $\frac{1}{5}$ as '$\overline{5}$', and $\frac{2}{3}$ as '$\overline{\overline{3}}$'. The Egyptians attached great significance to unit fractions; and a hieroglyph known as 'Horus-eye', which is thought to be fairly late and to record fractions of a particular measure of grain, shows them in inverse powers of 2 from $\overline{2}$ up to $\overline{64}$ (FIG. 2.6.1).

FIG. 2.6.1 *Egyptian 'Horus-eye' fractions, with analysis.* {Gillings 1972, p. 211}

Egyptian texts include tables of values of $\overline{\overline{n}}$ and $\overline{n}$ as sums of $\overline{m}$ for appropriate collections of values of m and n. The theory is less cumbersome than is often thought, although complications arise since many numbers have more than one summation. It is not clear that other kinds of number aroused their interest. For example, various claims are made about the presence of π and the golden section (§17.20) in the design and proportions of the pyramids, but they may have arisen from numerical coincidences, or implicitly from rolling a barrel along the ground exactly once.

The Egyptians also married arithmetic to geometry in connection with land measurement, although the calculations were sometimes approximate. For example, their formula for the area A of a rectilinear region with sides of lengths a, b, c and d, in that order, was:

$$A = \bar{2}(a + c) \times \bar{2}(b + d), \tag{2.6.2}$$

which is obviously too large; the correct expression requires each term to be multiplied by the sine of the angle between the corresponding two sides. Since the Egyptians had some feeling for trigonometry through their astronomical work, their failure to draw on it here is surprising, especially as the formula seems to be late. Or maybe the tax-gatherers were rapacious.

2.7 *Egyptian pyramids from various points of view*

The famous pyramids and other buildings provide evidence of Egyptian mechanics and astronomy. The mechanics of structures was well grasped; the temples at Luxor and elsewhere are extremely stable constructions, with huge safety factors evident in the size and close spacing of their massive columns. Such knowledge must have been developed early on; the three great pyramids at Gizeh date back to the –3rd millennium, and the Sphinx may be much older still. Astronomy is also involved there, for those pyramids are located a little out of line in exactly the same configuration as the three stars that make up the 'belt' of the constellation of Orion. Further, the large stepped pyramid at Saqqara is located at the place corresponding to their view of 'Sothis', the bright star for which we use the Greek name 'Sirius' and whose rise just before dawn in the 12th month determined their lunar year. The Egyptians were building heaven on their Earth – and they must have developed good trigonometry to do so.

On the more theoretical side, one of the Egyptians' most impressive formulae, present in the Moscow Papyrus and elsewhere, gave the volume of the frustum of a pyramid having square section and faces of equal slope. The manner of its discovery is a mystery, but the following line of thought, suggested by a Chinese analysis (§2.31), seems plausibly Egyptian. The algebra below saves space, and is not essential.

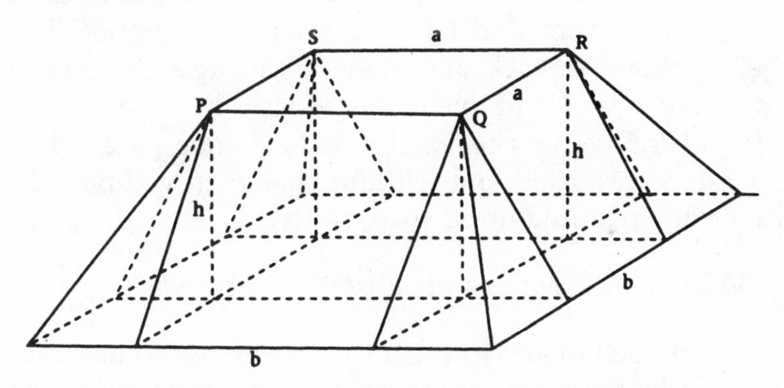

FIG. 2.7.1 *A dissection of the frustum of a pyramid.*

Divide the frustum, as shown in FIG. 2.7.1, into a cuboid, four triangular slices and four corner pyramids. Lift up the slices at back and front and stack them upside down against those to left and right with edges PS and QR respectively; this creates a larger cuboid, of volume *abh*. Slide the corners together somewhere to make a pyramid, of which the volume was already known to be $\overline{3}h(b-a)^2$. Fine so far, and the h is taken care of; but pity about the $\overline{3}$. So lay out the rectangular base of the larger cuboid *three times*, and also the base of the pyramid, and move one region, as shown in the plane section of FIG. 2.7.2, to obtain

$$\text{Volume of frustum} = \overline{3}h[a^2 + ab + b^2]. \qquad (2.7.1)$$

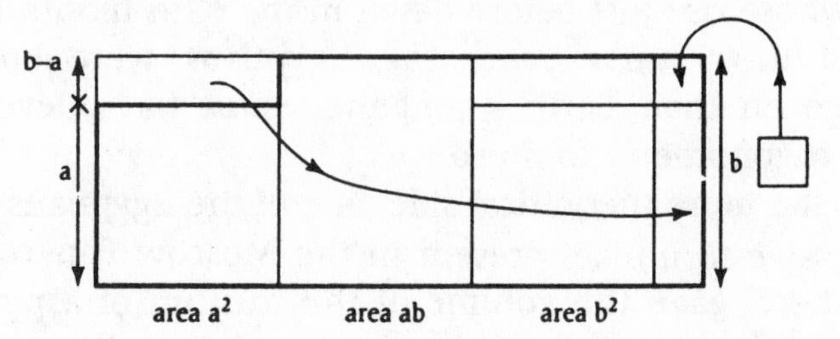

FIG. 2.7.2 *A possible derivation of the Egyptian frustum formula.*

The Babylonians also had a formula, easier to determine than (2.7.1) but more complicated to use. This time, lift up the slices at front and left and stack them upside-down against those at back and right with edges SR and QR respectively, and 'complete the square' by stacking at R three of the corners to make a cuboid (one pyramid will have to be reshaped, but without change in volume). This leaves one pyramid on its own, so that:

$$\text{Volume of frustum} = h[\tfrac{1}{2}(a+b)]^2 + \tfrac{1}{3}h[\tfrac{1}{2}(a-b)]^2. \qquad (2.7.2)$$

This construction and formula can be extended more easily than that of the Egyptians to frusta with a rectangular base, or ones in which the slope at left and right is different from that at back and front.

2.8 *Influence across cultures: the case of 3, 4, 5*

My conjecture for eqn (2.7.2) uses the method of completing the square, attributed to the Babylonians (FIG 2.5.1), to obtain b^2 in (2.7.1); but I do not have to assume that the Egyptians *learnt* it from them. This point raises the perplexing question of influence (or lack of it) across cultures, which is discussed in general in §2.34. Here are three examples concerning the right-angled triangle with sides in the ratio 3:4:5.

1. Pythagoras's theorem, known also to the Babylonians, has not been found in any surviving Egyptian texts; yet Plutarch explicitly attributed to them the fertility interpretation that the erect father Isis 3 and the horizontal mother Osiris 4 produced the son Horus 5. Now, Plutarch may have been led to incorrect history by a Pythagorean tradition (§2.11) in which these integers are respectively associated with masculinity, femininity and marriage; or the Egyptians may really have known this theorem, but the evidence has disappeared since Plutarch's time.

2. Some recent historians (for example, B.L. van der Waerden) have argued that the stones at Neolithic sites such as Woodhenge in England were laid out in elliptical or ring forms using this triangle, and even some higher Pythagorean triples (for example, 12, 35, 37) as reference arrangements. However, {Knorr 1985} has wondered whether such precision was attempted in the first place, and has given alternative design plans based upon only straight lines and arcs of circles.

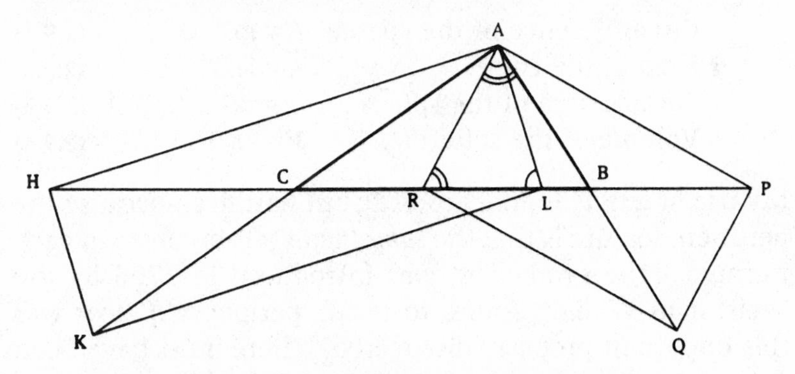

FIG. 2.8.1 *The 3:4:5 triangle and its rectangles.*

3. FIG. 2.8.1 shows a beautiful and simple property of this triangle. The sides AB and BC, in the proportion 3:5, generate the rectangle APQR which is itself in the proportion 2:1; similarly, the sides AC and BC, in the proportion 4:5, produce the rectangle AHKL in the proportion 3:1.

(Further, the square on CA and AB is of course 1:1.) The result is so simple that it could have been known in ancient times, and it does seem to have been around in the Middle Ages among the cathedral builders (§3.21); but it is completely unknown today, and (more to the point) there seems to be no trace of it in any ancient mathematics. A more general theorem holds: for any integer n there is just one right-angled triangle with integral sides which can generate the rectangle with sides in ratio $n:1$, but 3:4:5 is the only triangle where *both* rectangles have integral ratios. As far as I know, this result is due to your humble reporter, in February 1994!

2.9 *Circumferences, areas and volumes: which π?*

{Smeur 1970; Seidenberg 1972}

The above examples are problems about rectilinear regions; here is a major question concerning curvilinear ones. Consider the following properties of the circle and the sphere of diameter D:

Circumference of the circle:	$C = \pi D$;	(2.9.1)
Area of the circle:	$A = \frac{1}{4}\pi D^2$;	(2.9.2)
Surface area of the sphere:	$S = \pi D^2$;	(2.9.3)
Volume of the sphere:	$V = \frac{1}{6}\pi D^3$.	(2.9.4)

My use of letters is anachronistic but harmless, because the pertinent feature is that the *same* factor π is involved in each formula. (The symbol 'π' was introduced in 1706 by the Welshman William Jones, to recall 'periphery'.) How was this important property discovered? There must have been many failures and partial achievements; a Chinese instance is mentioned in §2.32. Although the question is of greater mathematical importance than many of the numerous relationships involving π that are discussed by historians, surprisingly little is known about its history prior to Archimedes, who knew of all four properties. Maybe the following four insights helped.

1. Areas increase with the square of length, and volumes with the cube of length. This would not be hard to spot from playing around with squares and cubes, and guessing it to be true of curved regions could follow.

2. When π is eliminated, formulae arise between objects of *different* dimensions. For example, a Babylonian text contains a result for the area of the semi-circle which, in the above notation reads:

$$\tfrac{1}{2}A = \tfrac{1}{4}(\tfrac{1}{2}C \times D), \tag{2.9.5}$$

and other results of this kind would have built up the quartet of formulae above. How did they arise? One possibility is shown in FIG. 2.9.1, where the circle has unit diameter.

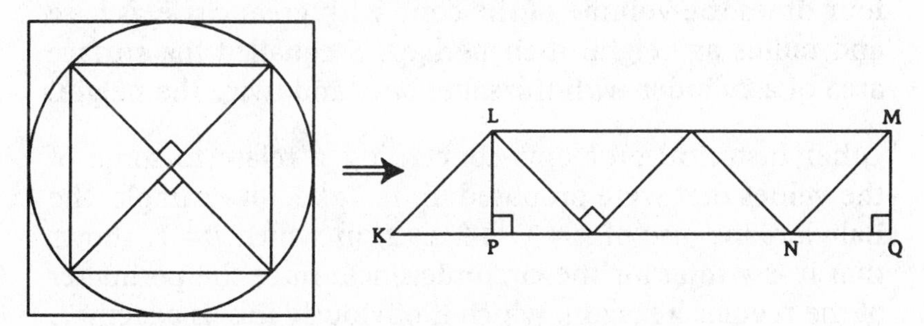

FIG. 2.9.1 *Areas of the circle and of the rectangle.*

The two diagonals of the inscribed square (ignore the escribed square for the moment) create four right-angled triangles; rearrange them to form the parallelogram KLMN, and switch triangle KLP to position NMQ to make the rectangle PLMQ. Its area is then that of the inscribed square, which is not far off that of the circle; and if the procedure is repeated with more sectors (a regular inscribed 8-gon, 16-gon . . .), then the difference becomes ever smaller. But then a formula like (2.9.5) for the circle is emerging; for the length of the base PQ approaches $\tfrac{1}{2}C$, and the height BP comes closer to $\tfrac{1}{2}D$ with more sectors as it becomes steadily more vertical.

3. The formula for the volume of a pyramid, mentioned before eqn (2.7.1), may have been used to obtain similar

formulae relating surfaces and volumes. For example, some forms of this intuitive 'mechanical' argument due to Archimedes (compare §2.21) may have been known earlier. Divide the sphere into an infinity of thin pyramids, each with apex at its centre and of height $\frac{1}{2}D$, and sum their base areas to produce its surface and thereby derive the relationship

$$V = \tfrac{1}{3}S \times \tfrac{1}{2}D, \tag{2.9.6}$$

then use item 1 to assume eqn (2.9.3), and (2.9.4) follows at once.

4. Archimedes was to report the following line of thought concerning eqn (2.9.3) for S. Knowing that V was four times the volume of the cone with great circle as base and radius as height, then perhaps S equalled the surface area of a cylinder with the same base and twice the height.

Other historical problems concerning π relate to some of the values that were proposed for it. Take, for example, the Babylonians' use of $\pi = 3$. FIG. 2.9.2, in which $D = 1$, shows that it is wrong for the circumference, since the perimeter of the regular hexagon, which is obviously less, is exactly 3. (Of course, the value might have been used as an *intentional* simplification.) However, the first diagram in FIG. 2.9.1 (ignore the diagonals this time), again with $D = 1$, presents a possible case for $\pi = 3$ *for its area*, since the inscribed and escribed squares have respective areas $\frac{1}{2}$ and 1, and it is not

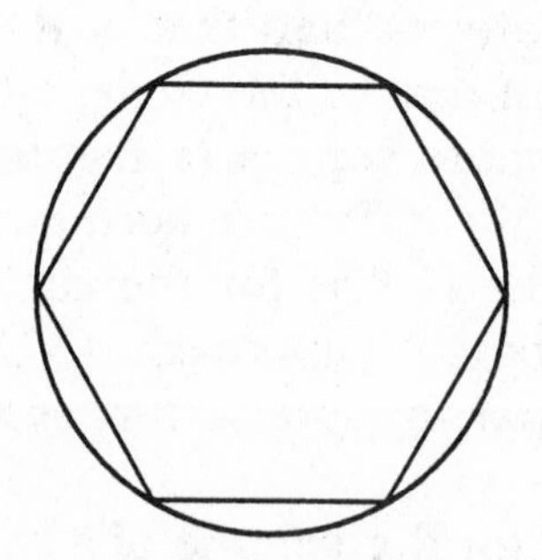

FIG. 2.9.2 *Circumference*: $\pi = 3$?

unreasonable – at least, without very careful drawing – to guess that the circle has the mean value of $\frac{3}{4}$. Reasoning like this has been attributed to Bryson, a rather obscure Greek figure of the –4th century.

FIG. 2.9.3, also for the areal π, comes from Problem 48 of the Rhind Papyrus (§2.6). The octagon, whose corners are the points that divide the sides of the escribed square into thirds, has area 63/81; add on a bit and the circle has area 64/81. As a number this is equivalent to $(1-\overline{9})^2$, and so uses a unit fraction; as a value it corresponds to $\pi = \frac{256}{81}$, only 0.6% in excess. Alternatively, the guess may have come from winding a rope into a circular spiral and noting its length.

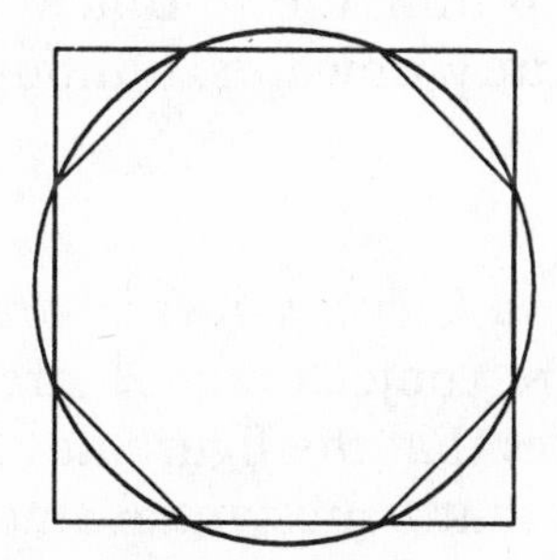

FIG. 2.9.3 *Possible origin of the Egyptian value $\frac{256}{81}$ for π.*

By such means relationships between lengths, areas and volumes may have emerged, and the roles for π clarified. The issues were still alive in Greek times; and to that culture we now turn.

2.10 *Ionic sections: personnel in Greek mathematics*

{CE 1.3–1.5}

Of all ancient cultures, the Greeks had by far the largest influence on the development of mathematics in Europe, which is our main concern in this book. So they dominate this chapter, with this as the first of 19 sections; nevertheless, only some features are described.

Table 2.10.1 gives unadorned details of the principal squadron of figures. They are listed for their contributions or influence, though not necessarily both; for example, Theon is there mainly for his edition of Euclid's *Elements*. Their work outside mathematics or astronomy (for example, in physics) is not noted. The main cities of activity are mentioned only to distinguish a figure from someone else with the same name but from another location. During the –4th century, an exceptionally active period, many figures were based in Athens, and afterwards they were often in Alexandria (indeed, in all periods Greeks spent time or at least travelled in Egypt); but, for example, Archimedes was born and worked in Syracuse in Sicily. Quality of work decreased during the so-called 'Byzantine' period, from the +4th century onwards, although astronomy was still actively pursued.

TABLE 2.10.1 *Principal known figures in Greek mathematics.* Almost all dates are conjectural; '?' stresses the point. '±' after a date indicates that the figure flourished around that time. 'W' indicates that a substantial amount of writing has survived, in apparently original form and/or from a Greek transcriber or Arabic translator; for the others we have only fragments or nothing. No distinction is made between original contributions by a figure and his commentary upon work of another one; and no reference is made to work in other areas (for example, physics).

The main topics are abbreviated as follows:

ari	arithmetic, number theory	mot	motion
ast	astronomy, cosmology	mus	music
car	cartography	nos	numbers (integral, rational, etc)
con	conics		
geo	geometry (not conics)	opt	optics
his	history, editions, commentary	prf	proof-methods, logic
		prp	proportion theory
inc	incommensurability studies	tcp	three classical problems
mec	mechanics	trg	trigonometry

–620?––540?	Thales	ast geo trg
–560?––480?	Pythagoras	ari mus nos
–490?––420?	Zeno of Elea	mot prf
–480 ±	Democritos	ast geo inc
–460?––400?	Theodoros	nos inc
–425 ±	Hippocrates of Chios W	ast geo tcp
–429––347	Plato W	ari geo prf
–417––370	Theaetetos	geo inc nos prp
–408––355?	Eudoxos	ast geo inc prp tcp
–400––347?	Eudemos	his prf
–384––322	Aristotle W	mec mot prf
–375 ±	Archytas	geo inc mus nos tcp
–350 ±	Dinostratos	geo tcp
–350 ±	Menaechmos	con tcp
–330 ±	Aristaeos	con
–310––230?	Aristarchos W	ast trg
–300 ±	Euclid W	con geo inc mus nos opt prf prp
–300 ±	Autolycos W	ast geo
–287?––212	Archimedes W	geo inc mec mot tcp
–280––200?	Eratosthenes	ari ast car tcp
–250––175?	Apollonios W	ast con opt tcp
–220 ±	Nicomedes	geo tcp
–190––120	Hipparchos	ast trg
–110 ±	Diocles	opt tcp
+50 ±	Hero W	geo mec opt tcp
+80 ±	Nicomachos W	ari mus nos
+100 ±	Menelaos	ast geo trg
+100–+178	Ptolemy W	ast car mus trg
+285–+330	Iamblichos W	ari his
+250 ±	Diophantos W	ari nos
+320 ±	Pappos W	ast inc geo his mec prf tcp
+330–+405	Theon of Alexandria W	ast his opt
+370–+415	Hypatia	ast geo his
+410–+485	Proclos W	ast his tcp
+480 ±	Eutocios W	geo his mec tcp

Several figures seem to have been partly motivated by educational needs; the word 'sophist', applied to Greeks in various disciplines, means 'teacher of young men', and indeed 'mathematics' itself comes from *mathemata*, meaning 'matters to be taught'. Some centres acted like Institutes of Advanced Study, especially Plato's 'Academy' (after *Academos*, the man or demigod in whose honour was named the garden where they met) in north-east Athens.

Apart from groups or schools such as this, and relatives such as the brothers Menaechmos and Dinostratos, or Hypatia and her father Theon, contacts even between contemporaries were limited. But influence spanned generations, for some figures wrote extensive commentaries on predecessors: especially Pappos, and also Iamblichos, on Nicomachos; and Hypatia on Theon. Eudemos seems to have been primarily a historian; his work was used by Proclos.

Several other figures are in the wings, maybe for the lack of surviving texts rather than lack of achievements. Even the identity of some of them, and the attribution of certain texts, is uncertain. Questions of the authenticity of the diagrams in these texts are especially tricky.

Historical work has focused much on these issues; other areas of scholarship have included textual and linguistic matters, revised dates for authors and their works, and new editions of texts and translations into Latin and (mostly) European languages. New texts are found from time to time; often they are medieval Arabic translations or commentaries on Greek originals which are still missing. When Europe began to wake up from the 12th century onwards, many Latin editions were prepared which served as important sources of influence and research (§3.12). Greek mathematics was treated primarily as historical material only from the mid-17th century; since then there have been several flurries of activity. In particular, after the French Revolution there appeared the first volume of the second edition of J.E. Montucla's *Histoire des mathématiques* (1799) and François Peyrard's translations of Euclid and

Archimedes. Late-19th-century work in German was reported in the first volume of Moritz Cantor's vast *Vorlesungen* on the history of mathematics (1880, third edition 1907}, and research on sources and editions by the Danes H.G. Zeuthen and J.L. Heiberg (especially his edition {Euclid 1886–1916} of Euclid's works); they heavily influenced English editions and studies by T.L. Heath and French ones by Paul Tannery, among others. Neugebauer led further work from the 1930s; but alternative interpretations have emerged in recent times (§2.19).

Euclid's *Elements*, written about −300, has always been a popular source text, even if the positions of other figures are somewhat distorted thereby; I cite it by Book and (for example) Proposition in the form '{5.prop.8}'. In addition, I write the ratio of two quantities a and b as '$a:b$', and a proportion proposition asserting the sameness of this ratio with that between quantities c and d as '$a:b::c:d$'. The significance of these names and symbols is explained in §2.18.

2.11 *Early Greeks: Thales and the Pythagoreans*

{Burkert 1972}

Of the ancient roots of mathematics, the fastest developments in the Grecian Empire came in arithmetic (especially with Pythagoras), geometry (especially with Thales) and mechanics including astronomy (presumably for all Greeks). Of Thales virtually nothing can be asserted with confidence, but he is credited with basic theorems of the triangle (so making him also a father figure for trigonometry), and with some method of predicting eclipses.

More information is available about the Pythagoreans, both in and outside mathematics. However, even the earliest sources are from several centuries after the alleged events occurred, and it is not always clear whether a particular idea is attributed to Pythagoras or to a follower. This said, it seems relatively certain that they knew this theorem, named after the leader:

THEOREM 2.11.1 *The square on the side of the hypotenuse of a right-angled triangle equals the squares on the other two sides.*

Since they are said to have extolled number, and are credited with theories of even and odd integers much as is presented (maybe extended) in Book 9 of the *Elements*, they must also have noticed that, with any common factors removed, triangles satisfying the theorem must have sides of both odd and even integer lengths. The distinction between even and odd integers seems to have been important, for they gave them male and female status respectively; for example, according to one tradition they dedicated the angles of an equilateral triangle to gods and those of a square to goddesses. They also regarded as important the 'triangular' numbers, integers which can be written in the form $1+2+3+\cdots$ and represented geometrically by triangles of stones laid on the ground; the sequence starts 3, 6, 10, . . .

Pythagoras's 'philosophy' (apparently his word) laid great emphasis on vibrations of strings, especially the integral proportions of lengths to produce basic musical intervals, starting from the ratio 1:2 for the octave and 2:3 for the perfect fifth. The co-presence of odd and even integers means that the two can never meet; however, the ratio of 12 fifths and 7 octaves gives the very near miss $2^{19}:3^{12}$, or 524,288:531,441, only about 1.35% under, which is known as the Pythagorean 'comma' (literally, 'cut'). A wide range of discussions of music theory ensued throughout Greek culture, with ratios as a central feature. The Greeks devised several methods of dividing the octave, and developed various views on concord and discord; in addition, they discussed in detail rhythm, musical modes and melody {Barker 1984, 1989}. Their work strongly influenced the study of musical temperament in the European Renaissance (§4.16).

In this context the Pythagoreans launched the notions, and maybe even the names, of the 'arithmetic', 'geometric' and 'harmonic' means of two magnitudes; their follower

Archytas so reported in the –4th century. Moreover, his/their definitions take a form using proportions which is far more elegant than our normal way with algebraic expressions; for them the magnitude x was the mean of a and $b < a$ according to whether

$$\begin{aligned} a-x : x-b &:: a:a \text{ (arithmetic)} \\ \text{or} \qquad &:: a:x \text{ (geometric)} \qquad (2.11.1) \\ \text{or} \qquad &:: a:b \text{ (harmonic),} \end{aligned}$$

where '::' means 'the same as' (§2.18).

Among magnitudes, the quartet 6, 8, 9 and 12 gained a prominent place in music and elsewhere. For 8 and 9 are respectively the harmonic and arithmetic means of 6 and 12; and we have also the octave (6:12), the fifth (6:9), the fourth (9:12 and also 6:8) and the whole tone (8:9). Plato discussed these and many other ratios, especially in his *Republic* and *Timaeos*; and some of his own numerological forays may have been stimulated by properties of intervals, since he drew upon musical notions and used ratios of integers of the form $2^m 3^n 5^p$ {McClain 1978}. Later, Nicomachos rehearsed some of these ideas in a *Manual on Harmonics* in which, among other things, the names of the means in (2.11.1) were publicized.

2.12 *The discovery of incommensurable magnitudes*

Another famous finding attributed to the Pythagoreans is usually formulated thus:

THEOREM 2.12.1 *The number* $\sqrt{2}$ *is irrational;*

but this formulation is anachronistic in various ways. Firstly, '(ir)rational' have become normal adjectives in European languages, due to Latin translations; but they give a wrong impression, and the Greek words '(a)logos' are better rendered as 'word(less)', and '(a)rhetos' as '(in)expressible'. Secondly, the theorem concerns numbers, whereas when the Greeks referred to it (which was not often) they used geometrical phrases such as 'the incommensurability of the side and the diagonal' of a square. (Aristotle so described it, but without reference even to a square.) Finally, the discovery is alleged to have provoked a crisis in the foundations of mathematics of the time; but commentators such as Aristotle himself do not mention one, and the idea may be an ahistorical interpolation by some later Greeks, or even a misunderstanding. For example, the story that the discoverer of the theorem was drowned at sea for his pains seems to a misinterpretation of his *amazement* at it; an annotation to some manuscripts of Euclid's *Elements* imagined that the discoverer would be 'carried under into the sea of creation, and is being overwhelmed by the unstable surges of it' {Euclid 1886–1916, Vol. 5, p. 417}. Thus, far from experiencing a crisis in foundations, the early Greeks may have enjoyed a time of great mathematical voyages.

The classical proof of this theorem can also be misunderstood. It proceeds by contradiction: assume that side s and diagonal d are indeed commensurable and without common factors, and show by squaring each side and the usual substitutions that s and d are both even. However, the version of the proof to which Aristotle alludes, which was added much later at the end of some editions of the *Elements*, does not show that s and d are both even and odd. Instead, it refers to *the different numbers of prime factors* in $2s^2 = d^2$: presumably that $2 \times s^2$ has an odd number of factors while d^2 possesses an even number, so that no ratio $d{:}s$ can be found for $\sqrt{2}$.

Both proofs assume that an integer factorizes *uniquely* into prime numbers. Maybe this property was regarded as self-evident; C.F. Gauss was the first to see a need for a proof (§9.2).

2.13 *The importance of incommensurability*

{Knorr 1975; Fowler 1987; CE 6.3}

Whatever the way the Pythagoreans interpreted Theorem 2.12.1, it refuted the belief (*if* they ever held it) that all features of the Universe could be expressed in terms of ratios of integers. The next ratio theory, seemingly due to Theaetetos and Eudoxos, was based on a procedure now usually called the 'Euclidean algorithm' (§2.18); its founders called it 'anthyphairesis' or 'antanairesis', meaning 'reciprocal subtraction'. Of two magnitudes of the same kind (lines, for example) the lesser one L is subtracted from the greater one G an integral number of times until a remainder R is found; the non-zero R, by definition less than L, is now subtracted from D until a new remainder R' is found; and so on. For example, 8 goes 4 times into 37 with remainder 5; 5 goes 1 time into 8 with remainder 3; 3 goes 1 time into 5 . . .; that is, in the notation adopted by historians,

$$37:8 = [4, 1, 1, \ldots]. \qquad (2.13.1)$$

But what does '. . .' mean? The process may stop in a finite number of steps (if in one step, then L is a factor of G); but if it continued for ever, then G and L were said to be 'incommensurable'. This latter case led to some very remarkable mathematics, especially when the sequence of remainders was periodic. Book 10 of the *Elements* contains a massive survey of 13 kinds of incommensurable ratio of magnitudes constructed from sums and differences $(\sqrt{m} \pm \sqrt{n})$ for various kinds of magnitudes m and n. A beautiful example was

$$\sqrt{2}:1 = [1, 2, 2, 2, 2, 2, \ldots], \quad \text{or} \quad \sqrt{2}:1 = [1, \overline{2}] \qquad (2.13.2)$$

in the modern notation adopted to indicate a periodic sequence.

Despite this publicity given to it by Euclid, anthyphairesis is not well studied even today; the nearest analogue is the (arithmetical) theory of continued fractions. Operations are awkward to handle; there is no easy way of determining the anthyphairesis of the sum or the difference of two ratios. This probably contributed to its later eclipse by fractions, which can be handled in a far more straightforward fashion (see §2.26 on Diophantos).

Another important shape with a Pythagorean background is the regular pentagon (FIG. 2.13.1). The context was

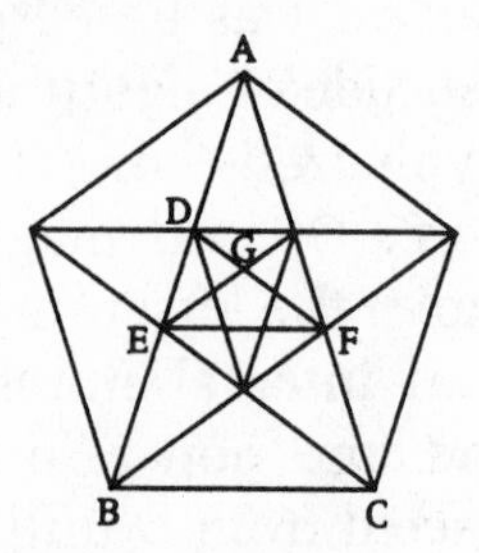

FIG. 2.13.1 *'Division in extreme and mean ratio' in the pentagon.*
{Knorr 1975, p. 30}

'division in extreme and mean ratio' (hereafter 'DEMR'), the special case where anthyphairesis on the magnitudes (usually lines) a and its lesser companion b yields a fresh pair of magnitudes in the same ratio as their parent lines {6.def.4}: that is,

$$a:b::b:(a-b) \text{ (and also } a:b::(a+b):a). \;\therefore\; a:b=[\overline{1}], \tag{2.13.3}$$

another beautiful result (the corresponding numerical ratio is 1.608 . . . : 1). In the pentagon, DEMR plays a role corresponding to $\sqrt{2}$ for the square: it is the ratio of AB and BD, where the 'mean' D so divides the 'extremes' A and B {Herz-Fischler 1987; CE 12.4}. It is quite ubiquitous in this Figure, for the remainder AD also equals the diagonal of the inscribed pentagon DEF, and so on, again to infinity if one is so minded. As with anthyphairesis, this ratio slumbered long; but it eventually it gained some attention under the name of the 'golden section' (§17.20).

2.14 *The five regular solids*

Complementing these studies of the plane were analyses of solids in space. Theaetetos was an important pioneer, to be followed especially by Archytas, Eudoxos, Archimedes and (much later) Menelaos; but curiously – and to the regret of Plato in his *Republic* – *general* interest in solid geometry was limited. One case study was duplicating the cube, discussion of which we postpone to §2.28, where it is examined together with two other 'classical problems'.

The greatest achievement lay in the study of the regular solids, perhaps motivated by the study of crystals. The mathematical theory divides into two aspects {Waterhouse 1972}. One is the study of particular solids, which may have started with Pythagoras. The other aspect is the general theory, including a satisfactory definition of regular solid and a proof that the list of them is complete; much of it may be due to Theaetetos in the –4th century, while Plato's name is attached to the solids for noting (in the *Timaeos*) that the definition needs to state not only that the faces be regular polygons congruent to each other which meet at the same dihedral angle, but also to require each vertex to touch the circumscribing sphere so that the solid be convex.

The 'answer' is the five solids known as regular polyhedra. Their construction, and various related features

Solid	cube	tetra-hedron	octa-hedron	dodeca-hedron	icosa-hedron
Faces	6	4	8	12	20
Vertices	8	4	6	20	12
Edges	12	6	12	30	30

(including DEMR), formed the subject of the final Book 13 of Euclid's *Elements*; he ended by showing that there were no more regular solids by means of an argument based upon properties of planar and solid angles {addendum to 13.prop.18}.

If the definition is relaxed to allow each face to be *some* regular polygon, then 13 more solids can be added to the list, a remarkable achievement due to Archimedes. If the solid need not be convex, then four star-shaped ones can be added: two were found in the 16th century by Kepler, who also showed that Archimedes' list was complete; the others came from Louis Poinsot early in the 19th century.

2.15 *Philosophical influences*

The Greeks drew upon reasoning, proof and definition in their mathematics; and, to a degree unique among ancient cultures, they even pondered upon possible mathematical worlds, expressing themselves in the subjunctive mood. They also used indirect proof, especially proof by contradiction in Theorem 2.12.1. Its *début* in Euclid's *Elements* occurs at {8.prop.1}, in a theorem about integers in geometric progression; Theorem 2.21.2 below is a double example from Archimedes. The method now often carries the medieval Latin name *reductio ad absurdum*.

An important innovator of this method was the philosopher Parmenides (–5th century), who stressed the need for indirect proof. His pupil Zeno followed the same philosophy with his famous paradoxes, which lead, by such arguments, to apparent absurdities. His own purpose cannot be established, since no texts from him have come down to us. But all his paradoxes were seen as brilliant counter-examples to naive ideas about gradual change, and they helped to stimulate concern in Greek mathematics and philosophy with measurement and continuity in/of space and time, and their bearing upon the status of infinitesimals {Toth 1991}. An example from Archimedes is mentioned in §2.21.

There has been massive discussion over the centuries of the paradox according to which Achilles cannot catch up with the tortoise when starting from behind because he has to pass through all the intermediate locations the tortoise has traversed. However, almost all of this scholarship is irrelevant, for in the formulations that have come down to us (from Aristotle, for example) *it is a valid argument*; 'Achilles is still running' {Grattan-Guinness 1974}, for it is *not* stated that either contestant is moving with uniform velocity, so that each could be slowing down all the time.

One student of philosophical matters, and also a source on Zeno, is Plato. Apart from specific topics such as musical intervals, triangular numbers and the regular solids, his influence upon mathematics was rather oblique, though fairly substantial: he advocated geometry as the study of forms, and stressed the objective status of integers and mathematical objects in general (which has given the name 'Platonism' to such kinds of philosophy of mathematics). In his Academy in Athens mathematics was greatly extolled, and the contributions of Eudoxos, Theaetetos and others were presented and analysed {Fowler 1987}. It is said that the motto 'let no man unversed in geometry enter here' was written over the main entrance; however, the earliest surviving evidence for this dates only from the +6th century.

Another member of the Academy, and commentator on Zeno, was Aristotle, who later formed his own 'Lyceum' of pupils. His philosophical influence upon mathematics was mixed. His treatment of collections of things, handled by dividing a whole into parts (item 8 in §2.1), was used by mathematician colleagues. But his 'syllogistic' ('thinking') logic played little part, even in such a severe text as Euclid's *Elements*, or among other mathematicians who thought carefully about necessary and/or sufficient conditions for the truth of theorems. Alternatives such as Stoic logic do not seem to have been used in mathematics, either. Indeed, in the centuries to follow possible uses of syllogisms in Euclid were *very rarely* explored (§9.3) (and modern analysts such as {Mueller 1981} deploy different logics). Mathematics and logic started and stayed at arm's length (§17.9).

However, Aristotle's views on mechanics guided much work on equilibrium and motion. He held, for example, that the Universe is finite and spherical, that the application of 'force' (*dunamis*) displaces a body by a distance proportional to time (akin to a notion of velocity) and that objects in motion seek their natural place in equilibrium. An important example was his theory, proposed in *On the*

Heavens, that these 'heavens' were the site for unending uniform circular motions of bodies around the stationary Earth. These motions took place within *real* 'spheres', actually 'shells' of some thickness to allow eccentric rotations; and the transport was effected by the 'aether' (his word, derived from *aei thein*, 'runs always'). This metaphysics was long to influence astronomy {Aiton 1981}, especially after its adoption by Ptolemy (§2.25).

Aristotle's general preference for empirical notions was also philosophically influential, as was his predilection for secure knowledge (*episteme*) over speculation (*doxa*). However, it was perhaps prohibitive on Aristarchos who, according to Archimedes, proposed in the −3rd century a Sun-centred system for the Universe but without being able to furnish 'proper' foundations.

Although all these issues were important, one should not imagine that Greek mathematics was besotted with logic and (not the same thing) axioms. As with all ancient cultures, much more emphasis was laid upon constructions, algorithms, and the posing and solving of problems. The latter included 'porisms', rather obscurely described by Pappos as an intermediate stage between a problem and a theorem proper. In some cases a porism seems to be a halfway house in the sense of an indeterminate infinitude of solutions en route to a final unique solution, or a definite collection of them. Unfortunately, works on porisms have been lost, including the one by our next author.

2.16 *The layout of Euclid's* Elements

The time is ripe for a look at the most famous work of Greek mathematics, which was written by Euclid around −300. The word 'element' referred to things laid out in a row; an earlier use, maybe intended by earlier writers of 'Elements' (which are all lost), denoted any proposition deployed in the proof of another one. Euclid's work consisted of 13 Books; those known as 14 and 15 were by later authors. Each was composed of numbered propositions (theorems, often mixed in with problems) and also, where necessary, definitions, postulates and axioms, which are also numbered in more modern editions. He treated numbers and, especially, geometric magnitudes, together with their ratios and proportions relating pairs of ratios; I shall use the word 'quantities' to cover all these. The order of the Books is rather curious: for example, although numbers were deployed from the start, they were formally presented only in Book 7.

Although Euclid did not discuss logical features, he was quite sophisticated on questions of definition and proof, and careful about stating necessary or sufficient conditions for theorems. The rigorous character comes over at once, for Book 1 opens with 23 (purported) definitions of geometrical objects, such as point and acute and obtuse angle. 'A line is breadthless length' {1.def.2} was perhaps so formulated to reject the Babylonian understanding of a line as an area of unit width (§2.5).

There followed a quintet of 'postulates'. The first three allowed these basic constructions to be effected: drawing a straight line through any two points; extending the line continuously; and making a circle of any centre and 'distance' (or radius). The fourth postulate admitted 'That all right angles are equal to one another.' The last one bore upon parallelism: Euclid defined 'parallel straight lines' as those which being produced indefinitely in both directions, do not meet one another in either direction', and then formulated a postulate in terms of *non*-parallelism:

> That, if a straight line falling on two straight lines make the interior angles on the same side less than two right angles, the two straight lines, if produced indefinitely, meet on that side on which are the angles less than two right angles.

This 'parallel postulate', as he did *not* call it, was to perplex many successors for centuries (§3.7, §6.13); he showed no qualms. Indeed, it is striking that shortly before him Aristotle had occasionally considered geometries in which the sum of the angles of a triangle took values other than two right angles {Toth 1967}. Maybe Euclid had *decided* to develop a geometry on the basis of the parallel postulate, which most clearly exhibited *episteme*.

This opening was completed by another quintet: of 'common notions', to cover all types of quantity. They included 'Things which are equal to the same thing are also equal to one another', and 'The whole is greater than the part'.

The main topics (and examples or features) in subsequent Books may be briefly summarized thus:

1. geometry of triangles and circles, transformation of regions (Pythagoras's theorem);
2. geometry of the rectangle and square;
3. circles: their tangents and diameters;
4. regular n-gons in and around circles (n= 3–6, 15);
5. basic properties of proportions about geometric magnitudes: basic theory (alternation theorem (2.18.2));
6. similarity of rectilinear figures in the plane (DEMR);
7. basic definitions and properties of integers (operations, primes, divisors);

8,9. properties of powers and products of integers (infinity of primes {9.prop.20});

10. (in)commensurable and (ir)rational lines and regions ('binomial' and 'apotomes', etc.: description of anthyphairesis);
11. solid geometry: solids, spheres, angles, parallelepipeds;
12. regions and solids: polygons, cones, cylinders;
13. the five Platonic solids (§2.14).

The character of this mathematics has been much debated. My reading of its quantities {Grattan-Guinness 1996b}, given in the next three sections, draws upon modern reinterpretations which are explained in §2.19.

2.17 *Euclid's theory of numbers and magnitudes*

Euclid probably conceived of numbers in a more 'concrete' way than is normal today. In the *Elements* they were the positive integers from 2 onwards, each being a 'multitude' of the unit 1, which was not itself a number {7.defs.1–2}. He had no zero, although some theorems about tangents skirted around the concept of a zero angle. 'Multitude' was an informal term referring to the number of objects in a collection (for example in {7.def.3}, to make up a number from units); 'equal in multitude' {5.prop.1} denoted a one–one correspondence between collections. The operations of addition, subtraction ('the lesser from the greater' to yield a positive value) and multiplication were defined, together with comparisons of equality and inequality (greater and lesser). Also defined were properties of integers such as even and odd, prime, coprime and perfect; ensuing theory included properties of the least common multiple and greatest common divisor, but not the unique factorization of integers.

The only other numbers used were the unit fractions half, third, and so on; presumably they came from Egyptian stock (§2.6), or via the important musical intervals 1:2, 1:3 and 1:4. Euclid did not formally define them, and used no other rational numbers; 3 was 'part' of 12 by being a factor of 12 and 'parts' of 7 when not {7.defs.3–4}, but the *numbers* 3/12 or 3/7 were not involved. Hence he had no irrational numbers, either; instead, incommensurable *magnitudes*, specified by unending anthyphairetic sequences (2.13.1).

Rather unhappily, Euclid named as 'plane' and 'solid' the products of two and of three integers, and 'square' and 'cube' those powers of an integer {7.defs.16–19}, and gave the corresponding geometrical interpretation. The drawback is that geometry is really the province of another theory, that of his geometrical magnitudes. There were ten kinds of these, in five rectilinear/curvilinear pairs (curvilinear meaning circular or spherical, and including mixed cases such as the semicircle): lines, straight or curved; planar regions, with straight or curved perimeters; surfaces in space, flat or curved; solids, with straight or curved surfaces; and planar or solid angles. With the exception of a theory of exhaustion noted in §2.21, recti- and curvilinear magnitudes were carefully separated. Further, the differences in dimension between lines, regions and solids was fundamental.

Euclid never spoke of the 'radius' of a circle or a sphere, but of its dia-meter (*diametros,* or 'through-measure'). Moreover, a parallelogram also had a diameter; namely, either of its diagonals.

Euclid called a geometrical object of zero dimension 'sign' (*semeion*), a practice which most successors followed, replacing the former word 'point' (*stigme*). But the Roman scholars Martianus Capella and then Boethius were to return to 'point' (*punctum*) in their commentaries on Greek mathematics (§3.11), and since then it has become standard. But 'sign' is preferable; perhaps Euclid liked it because it emphasized the fact that signs are not subject to the operations of other geometrical magnitudes, which we now consider.

As with numbers, magnitudes *of the same kind* could be added or subtracted ('the lesser from the greater' again); and they could be equal to, greater than or less than each other. Among the methods of construction, special importance was placed on 'application of areas', in which a planar region such as a rectangle was constructed equal in area to some given rectilinear region and with a given line as base.

The principal difference in operation between numbers and magnitudes concerned multiplication. Firstly, magnitudes could be multiplied by numbers (but not vice versa); among fractions $\frac{1}{2}$ was used the most, especially in bisecting an angle or a line {1.props.9–10}. Secondly, and this is usually overlooked, *nowhere in the* Elements *did Euclid multiply magnitudes together,* either of the same or another kind. Many products cannot be defined anyway (for example, planar region by planar region, or line by angle), and in any case the result is always of a different kind from that/those of its components; so they were *all* avoided. Thus, when Euclid 'described' the square on the line AB {1.prop.46}, he proved that the construction gives a square (previously defined); but he did not invoke the property that its area is AB × AB. In other words, *the square on the side is not the side squared*: it is a *region*, and Euclid was not concerned with its *area*. Hence the very next two propositions, Pythagoras's and its converse, were presented as a property of square regions, and proved in the famous 'windmill' diagram by establishing congruences of regions; they did not involve squares of lines on the sides of a right-angled triangle. An analogous theorem in arithmetic would be, say,

$$25 + 144 = 169, \textit{ not } \quad 5^2 + 12^2 = 13^2. \qquad (2.17.1)$$

2.18 *Euclid's theory of ratios and proportions*

Instead of multiplying magnitudes, Euclid handled the ratio of two of them of the same kind, or of two numbers; the main purpose of the theory was to *compare* ratios of numbers and of magnitudes, which could be of *different* kinds. I shall write the ratio between quantities a and b as 'a:b'. Ratio was a primitive notion, perhaps understood as a generalization of musical intervals and analysed by means of anthyphairesis; for example, the ratio 1:3 differs in type from the unit fraction $\frac{1}{3}$. Ratios were compared in proportion propositions; the relations were 'the same as', 'greater than' or 'less than' {the somewhat unclear 5.defs.3–7}. His avoidance of 'equals' means that we cannot use the standard symbols for (in)equality, so I follow the practice started by William Oughtred in the early 17th century of symbolizing sameness as '::'.

Writing a, b, c and d for magnitudes (lines, say – and *not* their lengths!) we never find in Euclid an equation such as

$$a \times d = 2b \times c, \text{ but instead the proportion } a{:}2b{::}c{:}d. \qquad (2.18.1)$$

A principal merit of ratios is that they avoid difficulties with what we call 'units and dimensions'; if a and b are quantities of the same type or kind, then the ratio is *dimensionless* and so can be compared with other ratios in proportions.

By contrast, in the equation (2.18.1), one has to know the type of quantity to which ad and bc are equal – and if it involves vague notions like force, then the question is best avoided. For this reason, ratios were often used in the early stages of theories, especially in applied mathematics, up to modern times (for example, electrostatics in the 19th century). The main drawback with ratios concerns calculation: the factor K by which a and b stand in ratio is not stated, and may even be wrongly stated relative to the units of measurement. Since Euclid was not concerned with calculation in the *Elements*, he did not have to worry about any K.

An important general theorem is that of alternation (not Euclid's name) {5.prop.16}:

$$\text{if } a:b::c:d, \text{ then } \quad a:c::b:d. \qquad (2.18.2)$$

Surprisingly, he required only that the pairs a and b, and c and d, be each of the same kind, not that the whole quartet to be so; hence (2.18.2) could assert the sameness of two 'mixed ratios' (my name), where a pair of ratios of the same kinds of different magnitudes (angle:volume, say) are related. This feature of the theorem is often regarded as a slip, but this was not so: Euclid used mixed ratios in Book 12, where he presented his version of Eudoxos's theory of exhaustion (§2.13) of curvilinear regions by rectilinear ones. For example, he proved that 'circles are to one another as the squares on the diameters' {12.prop.2} by constructing regular polygons inside each circle and using (2.18.2) to show, by double proof by contradiction, the sameness of the ratios of each polygon to its parent circle.

Euclid defined ratios only between two quantities. A sequence of quantities were in 'continued proportion' if each consecutive pair took the same ratio; in addition, 'mean proportionals' were sometimes sought between two quantities such that the whole sequence was in continued proportion (§2.28).

Another operation in the *Elements* was that of 'compounding' the ratio $a:b$ with $b:c$ to produce the ratio $a:c$

{5.def.9}, the 'duplicate ratio' of the original pair; the process may be repeated on $c{:}d$ to produce the 'triplicate ratio' $a{:}d$ after three stages, and so on. This operation again has obvious affinities with adjoining musical intervals. It is not to be confused with multiplication, although there is an obvious structural similarity. Denote compounding by '·'; then, say,

$$(8{:}9)\cdot(9{:}43)::8{:}43 \text{ differs in type from } (8/9)(9/43)=8/43, \tag{2.18.3}$$

for the outcome of the first proposition is a ratio, while the second one yields a number – and, moreover, a general rational number, which does not occur in the *Elements*. Compounding underlay the theorem that the ratio of two equiangular parallelograms is the compound of the ratios of the respective pairs of sides {6.prop.23}; it substituted for the forbidden multiplication of sides to determine an area.

Another example of the difference between numbers and magnitudes is that Euclid defined proportionality between four of them in distinct ways. For integers (A, B, C and D, say), the theory of multiples, part and parts was invoked {7.def.20}, but for any integers M and N (in the case of part, $N = 1$),

$$A{:}B::C{:}D{:}= \text{ if } MA=NC, \text{ then } MB=ND; \tag{2.18.4}$$

or $[A, C]=[B, D]$ in the notation (2.13.1) of anthyphairesis. (I recall from §1.7 that ':=' means 'equal by definition'.) But for magnitudes (a, b, c and d, say) the following theory of (in)equalities, which historians attribute to Eudoxos, appeared {5.def.5}: for any integers m and n,

$$a{:}b::c{:}d{:}= \textit{ if } ma > \text{or} = \text{or} < nc, \text{ then } mb > \text{or} = \text{or} < nd. \tag{2.18.5}$$

In such ways did Euclid make very clear that numbers and magnitudes were different types of quantity – not only in the matter of such details but because *arithmetic dealt with the discrete and geometry with the continuous*. I suspect that this contrast led to Book 10, where he presented a most elaborate theory of (in)commensurable and (ir)rational lines and regions; he may have taken over some such theory of numbers and reformulated it, since only geometry could guarantee the continuum of magnitudes upon which it rested.

2.19 *Was Euclid an algebraist?*

The reader accustomed to standard accounts of the *Elements* may well be surprised by most of the summary just given, for it clashes with the standard interpretation. Starting with some Arabic and then European algebraists in the Middle Ages, and finally expanded about a century ago by the historians Zeuthen and Paul Tannery (§2.10), the *Elements*, especially the Books concerning magnitudes, was interpreted as a disguised version of common algebra with variables, roughly after the manner of René Descartes in the 17th century (§4.18). In this view the more simple Books handle identities, usually of quadratic or bilinear forms; many of the later constructions, especially the application of areas, are versions of the extraction of roots from quadratic or quartic equations.

Zeuthen and Tannery gave this interpretation the name 'geometric algebra' – that is, they saw the mathematics as *really* an algebra dressed in geometry rather than the other way round. Some version of this interpretation was long adopted in general and even specialist accounts of the *Elements*, and of much other Greek mathematics; an influential figure is Heath, whose English editions and translations are still widely used. In the 1930s it was continued by Neugebauer and van der Waerden, who looked for the antecedent algebra that must have lain behind Babylonian arithmetic (§2.4).

However, especially since the 1960s, the view has grown that this reading of the *Elements* is historically unbelievable. In particular, in a polemical but important paper {Unguru

1975} has shown conclusively that, while there are obvious similarities between Euclid's procedures and geometric algebra, the view is seriously anachronistic; some other historians {e.g. Mueller 1981} have recently come to similar standpoints. Among the criticims put forward, the most significant for me is non-multiplication of magnitudes, since it bans writing x^2 for squares (or xy for rectangles, and so on) within the context of common algebra. Other criticisms include the absence of any historical evidence or discussion of algebra from that time (it begins to emerge in Greek mathematics only in the +3rd century, especially with Diophantos (§2.26), and then to a limited extent); and loss of essential information, such as not distinguishing a rectangle from a parallelogram or not discriminating the adjoining of, say, two rectangles at top, bottom, left or right. If one wishes to make algebras at all, it clear that three *different* ones are required, each restricted to positive values:

1. an orthodox one for integers, with division limited to unit fractions: related by (in)equality;
2. addition and subtraction of magnitudes, usually of the same kind, together with multiplication by numbers: also related by (in)equality;
3. compounding as the only operation on ratios: related by sameness and inequality in proportions.

2.20 *Archimedes and his axiom*

In the century following Euclid, the major figure was Archimedes, one of the greatest stars in the Greek constellation. He produced profound work on the geometry of the circle, spiral, sphere, cylinder, conoid and spheroid; on stability in statics and hydrostatics; and on military engineering.

As with Euclid, many texts have come down to us, but several different works are involved and their chronology is difficult to determine. A major revision has been proposed by {Knorr 1978b}, based in part on the question of influence from predecessors. He notices that Archimedes drew much upon a theory of proportions attributed to Eudoxos but different from (2.18.5) for magnitudes used by Euclid. Based upon the iterated bisection of a magnitude (halving, quartering, . . .), it involved relationships between a magnitude of a given kind and those of smaller size. But *measurably* smaller or not? This theory was perhaps the first after the Pythagoreans' wondrous surprise over $\sqrt{2}$ to try to cope with incommensurability. Its main theorem, which Archimedes gave in the *Spherics*, may be stated as follows:

THEOREM 2.20.1 *Given two magnitudes a and b, $a>b$, and another one c, there exists a fourth one, d, between a and b in size and commensurable with c.*

The proof worked by bisecting c until a magnitude e was obtained which was less than $a-b$. If e was commensurable with b, then so was $b+e$ and all was done. If not, then multiples of b were taken until $(n-1)e<b<ne$ for some n; then ne was the magnitude sought.

The process of making smaller magnitudes may have led Archimedes to emphasize a property which is usually named after him. He stated it as a 'lemma' (which means an *assumption* in ancient Greek, not 'subsidiary theorem' as we now use it) in his earlyish *Quadrature of the Parabola*: 'Of unequal areas, the excess by which the greater exceeds the lesser is capable upon being added to itself of exceeding the preassigned bounded area.' He did not confine himself to bisection (or rather its converse, doubling, quadrupling, . . .); any multiple n might do. The modern formulation states that, of two magnitudes l and g,

$$\text{if } l<g, \text{ then for some integer } n, Nl>g \text{ if } N\geq n; \qquad (2.20.1)$$

but this version carries with it an arithmetical connotation possibly foreign to geometer Archimedes. He saw his principle as applicable to geometrical magnitudes of all kinds; the reference to areas in his formulation arose in connection with quadrature, which we now consider.

2.21 *'Geometrical' and 'mechanical' methods in Archimedes*

One of Archimedes's main theorems in *Quadrature of the Parabola* is

> THEOREM 2.21.1 *The area of any segment of a parabola cut off by a chord equals 4/3 of the area of the triangle with the same height and this chord as base.*

He gave two quite different proofs, one mechanical and the other geometrical; the diagrams in FIG. 2.21.1 show each

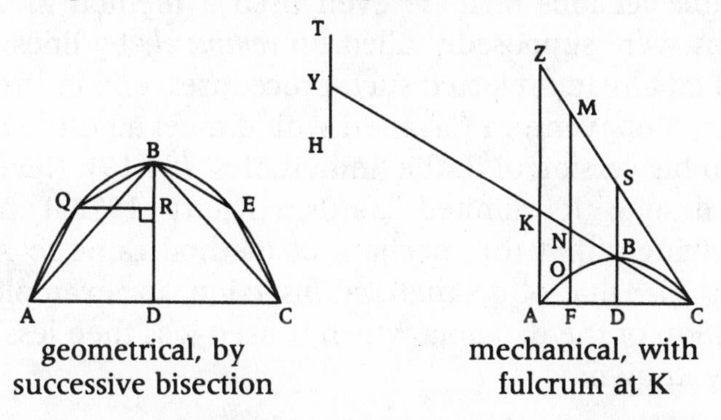

FIG. 2.21.1 *Archimedes' constructions on the parabola.* {after *Knorr* 1978*b*, *pp.* 225, 226}

one for the segment ABCD with axis BD. Without going through all the details, the main contrasts are as follows.

1. The defining properties for the parabola were different:

$$\text{geometrically, } BD:BR::AD^2:QR^2; \qquad (2.21.1)$$
$$\text{mechanically, } MF:OF::YK:NK$$

for any line OF of the segment parallel to BD.

2. The geometrical method worked by successively bisecting the parabola with points B, then Q and E, . . ., thereby exhausting the region with triangles; the diagram shows the first two stages, and a more elaborate example will be presented in a moment. The mechanical method proceeded by analogues from statics, in which lines or regions were transported by congruence and balanced up as if they were weights. TH was defined as the line parallel to BD such that K was the midpoint of YC, and OF was

transferred so as to lie along TH and balance MF about K as fulcrum.

3. In this text, and especially in a reflective piece called *The Method* which came to light only in the 1900s (§16.3), Archimedes made quite clear that the mechanical construction was only a heuristic device to aid the presentation and discovery of results; the presence of notions outside geometry prevented it from standing as a respectable proof. In some versions of it he even used a method in which regions were supposedly filled up *completely* by lines; Aristotle had already rejected such procedures, and in the 17th century Bonaventura Cavalieri would meet a hostile reception to his version of it, the 'indivisibles' (§4.19). Nevertheless, despite its limited status, {Knorr 1978a} argues convincingly that the mechanical method came to Archimedes *after* that of geometrical bisection; for example, the definition of the parabola which it used was then less commonly known.

Archimedes' most famous use of successive bisection comes in his early *Dimension of the Circle.*[2] First of all, he proved eqn (2.9.5), in this form:

THEOREM 2.21.2 *The circle C equals in area the right-angled triangle K with base equal to its circumference and altitude equal to its radius.*

For proof he applied Eudoxos's method of bisection by constructing sequences of inscribed and escribed regular polygons (FIG. 2.21.2), where at each stage the angle subtended at the origin was exactly half that of the previous segment. But with this sandwich effect Archimedes may have extended Eudoxos's conception; it certainly goes beyond the use made of exhaustion in Euclid's Book 12. Thus, while some of Archimedes' methods were pre-Euclidean, his concerns went *beyond* the Elementary bounds of

[2] The warning given in §2.1 about the trustworthiness of texts is well exemplified by Archimedes' *Dimension of the Circle*, for the surviving version is an edited version of the lost original {Knorr 1986a}. However, it seems likely that he is represented fairly here.

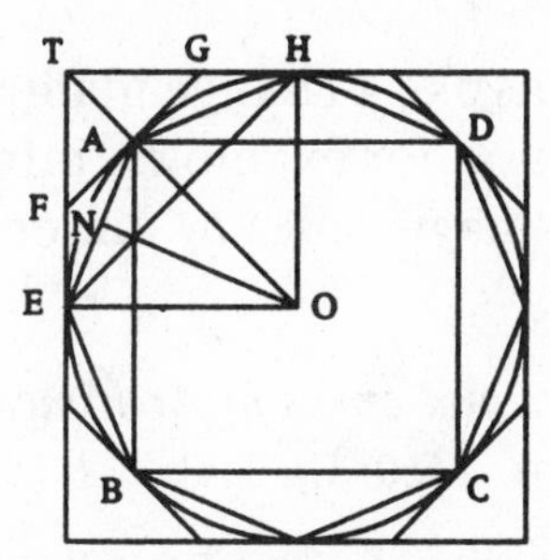

FIG. 2.21.2 *Archimedes' sandwiching of the circle.*

Euclid; he also allowed a curved line to equal a straight one, and multiplied geometric magnitudes. In addition, he proved theorems on the (approximate) lengths of lines and areas of certain regions; the most celebrated, also in *Dimension of the Circle,* stated that the circumference of the circle 'exceeds by less than the seventh part, but more than ten seventy-first' parts the triple of the diameter. This result is usually described as his upper and lower bounds $3\frac{1}{7}$ and $3\frac{10}{71}$ for π, but to him it was a result about *ratios,* not fractions; in this respect he is at one with Euclid. He found these values by analysing respectively the circumferences of the escribed regular hexagon and the inscribed regular octagon; he is also credited with closer approximations to π. In *On Sphere and Cylinder* he proved theorems of the form of Theorem 2.21.2 involving the three other roles for π described in §2.9.

In his proofs, Archimedes quite frequently followed the Greek tradition of double proof by contradiction (§2.15). In the Theorem above, assume first that $K < C$, and then that $K > C$, and in each case derive an impossibility from an inequality obtained from the exhaustion process. This procedure may surprise us; for since A.L. Cauchy and Bernhard Riemann in the 19th century we expect the argument to be finished off by treating the circle as the common limiting value of the sequences (FIG. 12.6.1). But this theory was not on the Greek menu, probably because the limit process itself involved a *change of kind of geometrical object,* from rectilinear regions to the curvilinear circle. In other cases limits would invoke a change in dimension (from ever shorter lines to a point, for example), infringing the distinction of geometrical magnitudes *by* dimension.

2.22 *Some mechanics and optics to Archimedes' time*

{Pedersen 1993, Chaps. 2–10; CE 1.4}

Some of Archimedes' important contributions to mechanics were general. For example, in *Plane Equilibria* he used bisection procedures to prove basic theorems on the balance of the lever, while in *On Floating Bodies* he expounded some properties of hydrostatic equilibrium – the story of his cry 'Eureka!' (literally, 'I have found!') and post-ablutionary dash through Syracuse is the most famous example, though probably an apocryphal one. He understood that parts of a fluid may affect contiguous parts (compare Leonhard Euler on pressure in §6.10). But in addition he designed mechanical gadgets. A celebrated example is his 'screw', a device for raising water by turning a helix about its axis. Maybe he was inspired here by geometry, or he may have taken over the device from the Egyptians. However, the military need for offensive and defensive machines of various kinds yielded the most public manifestation of his genius. The two Punic Wars between Rome and Carthage were fought during his lifetime, and indeed caused its end. The Syracusans repelled a Roman invasion in −213, apparently with much help from his mechanical contrivances; but the city fell the following year, and he is reported to have died by a Roman sword while meditating upon a geometrical sand-picture.

Long before Archimedes' time there must have been many fine engineers, drawing at least upon notions in statics, geometry and trigonometry. The 'seven wonders of the world', such as the Colossos statue at Rhodes, must have been remarkable structures. Six of them are gone, but the mathematical knowledge that went into the great *surviving* artefacts is covert in the sense of item 6 of §1.6, so we can only guess at its extent. For example, the construction of the tunnel on the island of Samos, attributed to Eupalinos in the –5th century, suggests great skill in both surveying and the statics of materials {Rihill and Tucker 1995}, and the famous Parthenon in Athens shows much subtlety of perspective in the details of its design {Dinsmoor 1923}.

Among other branches of applied mathematics, optics inspired results for the geometry of the plane concerning caustic curves, the envelopes produced by families of neighbouring rays of light. Optics also benefited solid geometry, especially with properties of spheres and paraboloids. In 'catoptrics', the geometry of reflected light (the word means something like 'broken vision'), a prominent topic was the theory of 'burning mirrors', where rays of sunlight striking the paraboloid could cause conflagrations at the focus – indeed, that use of the word 'focus' (Latin for 'fireplace') was to be proposed by Kepler precisely because of this physical property (§4.12). Diocles wrote an important work on these mirrors {Diocles 1976}, drawing on the property that any point on a parabola is the same distance from the focus as from the directrix; indeed, he is the oldest known Greek source for it. Archimedes was credited with remarkable experiments using large mirrors, although the earliest known testimony dates from several centuries later.

Astronomy included observing motions of the heavenly bodies and speculating on their shapes, relative positions and origins. Astrological concerns were routinely in evidence, with the ecliptic split Babylon-style into its 12 zodiacal divisions. Usually the Earth itself was located centre stage; but, as was noted in §2.15, Aristarchos placed the Sun there, with the Earth in a circular orbit about it as centre.

The motions of the 'planets' (a Greek word, meaning 'wanderers') were regarded as real, carried around within transparent solid spheres. This was the usual physical explanation offered; otherwise a kinematic interpretation was adopted. The normal assumptions were that the heavenly bodies moved in circular orbits at uniform speed;

spheres were to the fore, or cylinders for the flat-Earthers. However, it was well known that the observed motions were not uniform, so this assumption had to be desimplified. For example, Eudoxos introduced 'hippopedes', a class of figure-of-eight curves named after the shape of the device used to tether a horse by its feet. He imagined a sphere centred on the Earth and rotating about an axis, and also a second sphere rotating with the *same* uniform angular velocity in the *opposite* direction about a *different* axis; a hippopede was the result of combining these two motions. While limited success was gained in the astronomical aim of modelling the stationary points and retrograde motions, the mathematical skill was considerable; for example, it constituted an early example of studying curves in space. Among other uses of mathematics, anthyphairesis may have been applied to the computation of calendars, combining seven-day weeks with unequal months.

In planetary studies, Eratosthenes devised a simple way of calculating the diameter of the Earth (FIG. 2.22.1). At the

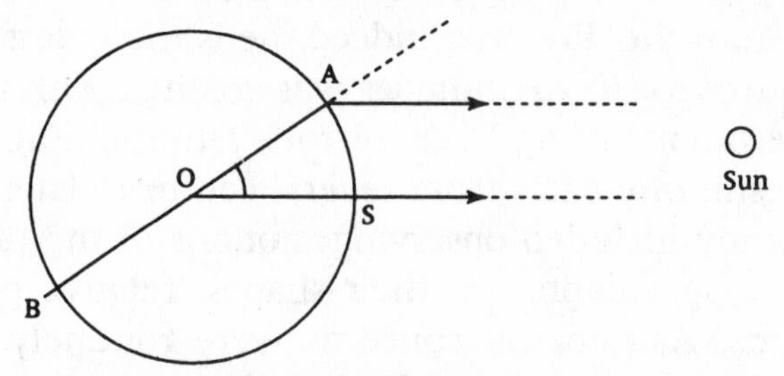

FIG. 2.22.1 *Eratosthenes' calculation of the diameter of the Earth.*

summer solstice the Sun was directly overhead at S (Syene, the modern Aswan) on the Tropic of Cancer, and off by around $7\frac{1}{5}$ degrees due north at A (Alexandria, where he was head librarian). Taking the distance AS to be 5,000 'stades' (about 1580 kilometres), he found the equivalent of 12,520 kilometres for the diameter and 39,370 kilometres for the circumference of the great circle through A and S. At about 2% below modern values, the estimations are very good; but compensating errors may partly be responsible, for in fact the cities are not on the same longitude and the proposed distance between them looks like a convenient guess. Indeed, if he performed the calculation at all, he may have intended only to produce an approximate value {Goldstein 1984}. These considerations suggest that topography and geography were still at a rather early stage of development, although the basic trigonometric conception was sound.

A contribution attributed to Menaechmos in the –4th century used the L-shaped 'gnomon', a Greek word meaning 'that enabling to be known'. One reference of the word was to the upright sundial, where a triangle casts its shadow upon a horizontal surface, suitably marked to record the corresponding time of day.[3] FIG. 2.22.2 shows the effect in the northern hemisphere. The gnomon GNM casts the line GR at sunrise, GT at some arbitrary morning time, GD at

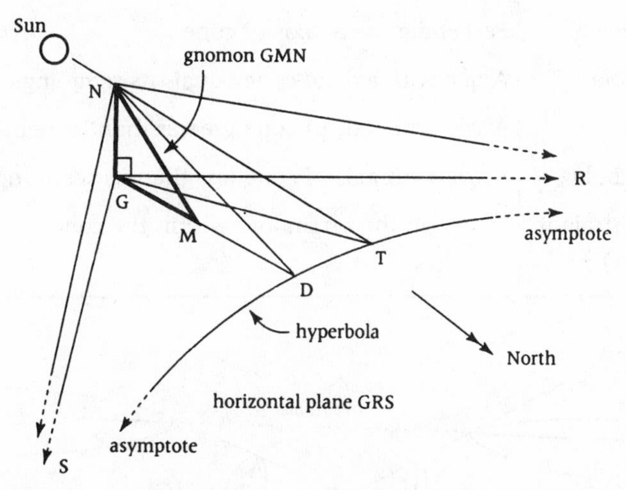

FIG. 2.22.2 *Perspective view of the gnomon and the hyperbola.*

[3] Another application of the word 'gnomon' was to L-shaped regions constituting or adjoined to diagrams in theorems in geometry; they appear as early as the Babylonian completion of the square (FIG. 2.5.2), and are quite often found in Greek geometry.

midday, and GS at sunset. The curve RTDS is a hyperbola, with GR and GS as asymptotes and D as a pedal point; for it is formed as a plane section through the cone created by the Sun's motion – which to Menaechmos was real.

2.23 *Apollonios on the 'symptoms' of the 'conic sections'*

Menaechmos's construction may have come from the insight that the hyperbola could be *defined* as a certain type of section of a cone. The definitive presentation of this approach was given in the late –3rd century by Apollonios of Perge in Asia Minor (modern Turkey). The theories of the circle, parabola, ellipse and hyperbola were unified as 'conic sections' (FIG. 2.23.1, where a right double cone is used, and the axis of each curve is also shown). His

Curve	Modern specification of planar section
Circle	Perpendicular to axis of cone
Parabola	Angle with axis of cone equals its semi-angle
Ellipse	Angle with axis of cone greater than its semi-angle
Hyperbola	Angle with axis of cone less than its semi-angle
(Two straight lines)	(Through the origin and within the cone)

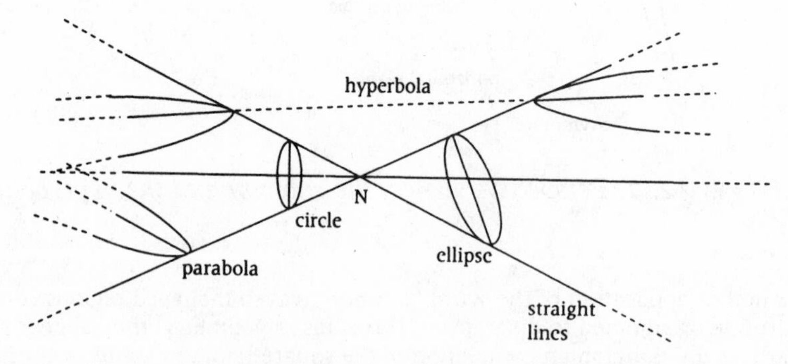

FIG. 2.23.1 *Apollonios's construction of the 'conic sections'.*

procedure was somewhat more complicated than is shown here, for it involved a fixed plane through the cone containing a fixed line F.

F also played a role in the way in which Apollonios used the method of application of areas (§2.17) to define each curve also by its respective 'symptom', or relationship between a certain rectangle R and a particular square S for any point on it. Each symptom reflects the name that its curve acquired. FIG. 2.23.2 shows the 'parabola'; the rectangle contained by F = ZL and ZM equals the square on LK (which is constant in direction). The trio of definitions and names relate thus, together with a modern equation in coordinates x and y with constants a and p, all specifiable on the cone:

Curve	parabola	ellipse	hyperbola
Meaning	exact 'application': $S = R$	'falling short': $S < R$	'excess': $S > R$
Equation	$y^2 = px$	$y^2 = px(1 - x/a)$	$y^2 = px(1 + x/a)$

Note that solid geometry provided the basis of Apollonios's definitions – maybe in reaction to Plato's lament about the neglect of this part of geometry (§2.14). They largely supplanted earlier definitions, of which the most

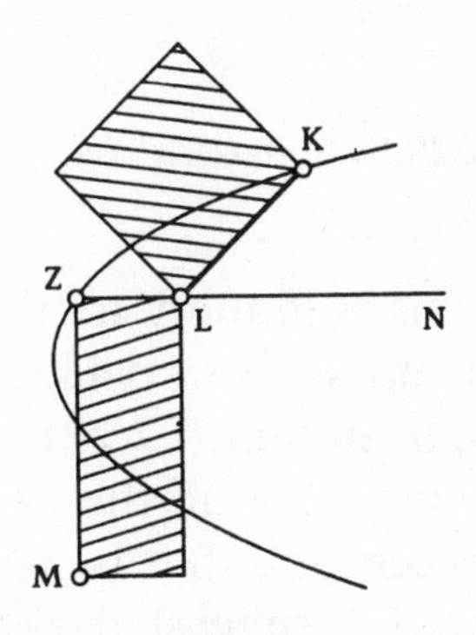

FIG. 2.23.2 *Apollonios's 'symptom' for the parabola.*

popular were geometrical properties corresponding to the above equations. One advantage over other definitions was their utility in the discovery and proof of theorems, in that a result for the ellipse (say), proved with the help of properties of the cone, could suggest something similar for the hyperbola.

In Apollonian terms, which Menaechmos may also have understood, the hyperbola in FIG. 2.23.1 is a plane section though a cone with N as vertex. From a modern mathematical standpoint projective geometry is also in the air here, but apparently not for Apollonios, although his later commentator Pappos was to bring it out a little more (§2.27). This subject was not to develop until the 17th century, with figures such as Gerard Desargues (§4.20). Apollonios also tackled more abstract questions such as the number of curves which can touch a given one: more general thoughts of this kind were to inspire Gaspard Monge early in the 19th century (§8.3) and Hermann Schubert near its end (§16.18).

In addition, Apollonios looked at metrical properties of the conic sections. An interesting case was to find the point P on a parabola for which the distance to a fixed point A on its axis is a minimum. The answer, that AP is the normal at P, makes him a grandfather of the calculus methods which Descartes and Pierre Fermat were to launch in the 17th century (§4.19).

2.24 *Hipparchos and the geometrical trigonometry of astronomy* {CE 1.4}

Apollonios turned his attention to the geometry of the heavens, and came up with two ideas to desimplify the assumption of circular motion (§2.22). Firstly, he proposed that the planets still went round the Earth but that the circular orbits were 'eccentric': that is, with their centres *not* located there. Then he suggested that the orbits were 'epicyclic': that is, circular but on a 'deferent' (meaning 'carrying') circle, the centre of which was itself travelling on a circle around the Sun. In both cases all motions were taken to be uniform, although not at the same speed. He may not have tested out his ideas against observation, but they were to have a great influence on his successors.

The next major contributor was Hipparchos, in the −2nd century. His main insight was to relate part of an orbit (epicyclic or not) to an angle that it subtended at some point, such as the Earth or the Sun. He studied properties of the 'chord' chd α of $\angle$BOC = α in FIG. 2.24.1, defined for

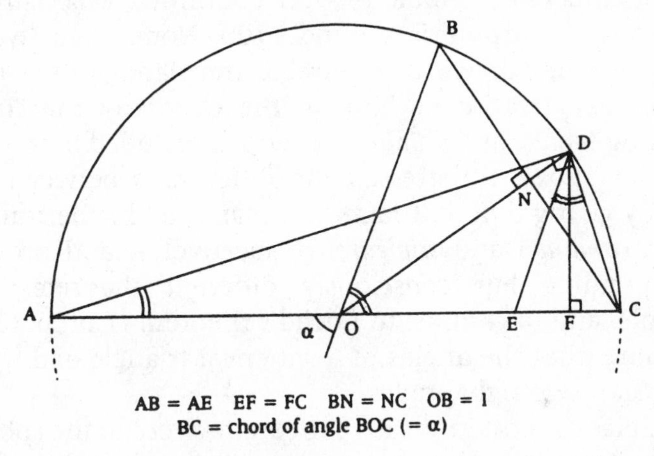

FIG. 2.24.1 *Possible construction for Hipparchos's bisection of ∠BOC.*

a circle of unit radius as the length BC of the line joining the ends of the arc subtended by the angle. One of his most important theorems related an angle to its half-angle, from which he was able to compile tables of chords. He may have found it by bisecting $\angle$BOC with radius OD. Since triangles ACD and CDF are similar, then

$$\text{CD:AC::CF:CD; that is, } \text{chd}^2\tfrac{1}{2}\alpha = 2[2 - \tfrac{1}{2}\text{chd}(\pi - \alpha)] \qquad (2.24.1)$$

in modern terms. From his proportion he was able to compile one of the first tables of values of chords for a sequence of bisected angles. With results such as this, geometric trigonometric *theory* began to emerge; for, since the modern definition of chd α is $2 \sin \frac{1}{2}\alpha$ (compare eqns (4.8.1)), the theorem corresponding to (2.24.1) is the formula for $\cos \alpha$ in terms of $\cos \frac{1}{2}\alpha$. Hipparchos used this angle–chord trigonometry extensively in his astronomy, which covered not only the motions and distances of the heavenly bodies but also planetary sciences such as geography, in which the planet (usually the Earth) is not treated as a point. These problems led him to develop spherical as well as planar trigonometry.

Both parts of trigonometry were continued, especially by Menelaos of Alexandria around +100. None of his works has survived, but we have Arabic translations of a Hipparchos-like treatise by him on the chords of the circle, including tables; and a *Sphaerica*, which included both parts {Björnbo 1902}. He stressed the differences between the parts by giving different names to planar and spherical triangles (*trigunon* and *tripleuron* respectively), and proving corresponding but consciously different theorems; for example, and in contrast to Euclid's theorem {1.prop.32} in the plane, that the angles of a spherical triangle add up to more than two right angles.

Menelaos's most innovative theorems were in the spherical part, concerning properties of arcs on great circles of the sphere, another notion which he highlighted. Two examples are shown in FIG. 2.24.2; I conjecture that he

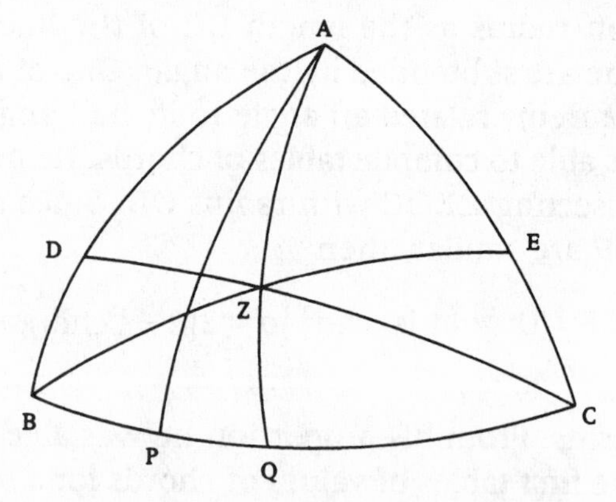

FIG. 2.24.2 *Menelaos on spherical triangles and cross-ratio.*

thought them out in terms of '·' for compounding ratios (2.18.3) of chords corresponding to the arcs.[4] In these terms the first one reads:

THEOREM 2.24.1 *Let AB, AC, BE and CD be four arcs of great circles on the sphere, each one shorter than a semicircle. Then*

$$chd\,CE : chd\,EA :: (chd\,CZ : chd\,ZD) \cdot (chd\,DB : chd\,BA). \quad (2.24.2)$$

Secondly, elsewhere he drew on the property that for (now) *any* quartet of arcs of great circles drawn through A and cut by any circular arc BPQC,

$$(\text{chd BC} : \text{chd CQ}) \cdot (\text{chd PQ} : \text{chd BP}) \text{ is constant.} \quad (2.24.3)$$

To us this proposition states that the cross-ratio of curves on the sphere was constant. He did not claim it as his own finding; he may have been inspired by a corresponding result in Apollonios for straight lines and for lines in the plane. Once again the potential for projective geometry was to sleep until the 17th and 19th centuries (§4.20, §8.4).

2.25 *Ptolemy's 'compilation' on astronomy*

Meanwhile, during the +2nd century Ptolemy profited from these and other theorems of his predecessors in his astronomical studies. His *Almagest* was actually called 'the great compilation'; the familiar short title is a corruption by medieval Latin translators of the Arabic word *almegiste*, meaning 'the greatest'. It consisted of 13 Books, the same number (by coincidence?) as in Euclid's *Elements*. Intended as a comprehensive statement of astronomy and its associated trigonometries {Berggren 1987}, covering both orbits and

[4] The question of the original form of Menelaos's theorem is another fine example of the lack of original texts mentioned in §2.2. Editions or commentaries on him from the Arabs and the West expressed it in terms of fractions, products and equations, but I suspect that my (2.24.2)–(2.24.3) is more faithful to him.

distances of heavenly bodies, and also including tables, it turned out to mark the peak of Greek astronomy. He built upon predecessors for various mechanical conceptions, upon Aristotle for his notion of a finite spherical Universe with the Earth at the centre, and upon Hipparchos and Menelaos for mathematical methods (indeed, he is our authority for their results (2.24.1) and (2.24.2)).

In order to preserve the Greek tradition of uniform circular motions and the reality of the spheres, Ptolemy had to resort to further desimplifications to cope with observed variations in planetary motion. In his theory of motion (FIG. 2.25.1), a planet P rotated on an epicycle e, a circle whose

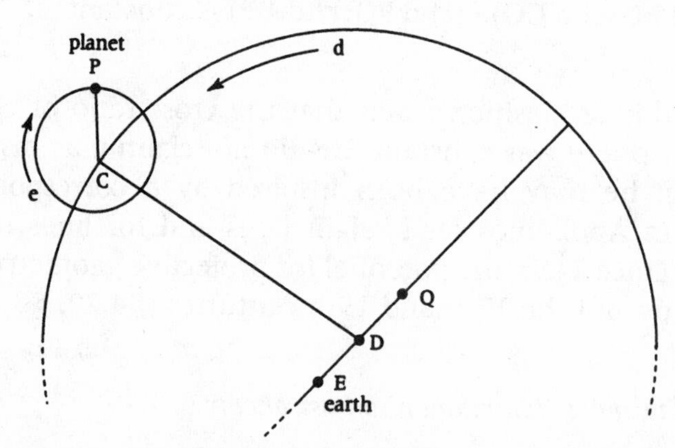

FIG. 2.25.1 *Ptolemy's construal of planetary motion around the Earth* (not to scale).

centre C was carried upon a deferent circle d with centre D on the diameter of the circle joining its apogee and perigee (respectively its furthest and nearest locations from E). In contrast to the real spheres, these curves were imaginary objects, brought in solely to develop the theory. But even with them, his account of the motion of the Moon, which was more complicated still, predicted far greater variations in its apparent size than was observed. This defect was to lead Copernicus to doubt the Ptolemaic theory and develop his own different astronomy, as we shall see at FIG. 4.10.1.

For the motions of the outer planets, Ptolemy thought

up an idea which was to be important for Kepler (§4.11) and other successors. Abandoning the assumption of uniform rotation of a planet on its epicycle, he assumed that there was a point Q he called the 'equant' on ED, such that ED = DQ, and relative to which the planet P rotated uniformly (that is, $\angle PQE$ increased constantly with time).

Among other works of Ptolemy to have survived, the *Harmonics* was a substantial study, of considerable influence in Europe from the late Middle Ages. In it he laid out a most elaborate theory of intervals, and their bearing upon concords and discords, in the tradition established by the Pythagoreans (§2.11) though different on various details. He seems to have intended to compile a comprehensive survey of the applied mathematics of his time (mechanics, geography, optics, and so on), and some surviving texts seem to have been written for this purpose.

Ptolemy maintained an Alexandrian tradition in applied mathematics which had been greatly enhanced in the previous century by Hero, the most eminent successor to Archimedes in the construction of mechanical devices. His textbook on mechanics for engineers and contructors included the durable idea of five simple 'powers', or machines: winch, lever, pulley, wedge and screw. Another interest was surveying, to which he contibuted this influential theorem:

THEOREM 2.25.1 *Let Per be half the perimeter of the triangle ABC in* FIG. 2.25.2; *then*

$$(\Delta ABC)^2 = Per(Per - AB)(Per - BC)(Per - AC). \quad (2.25.1)$$

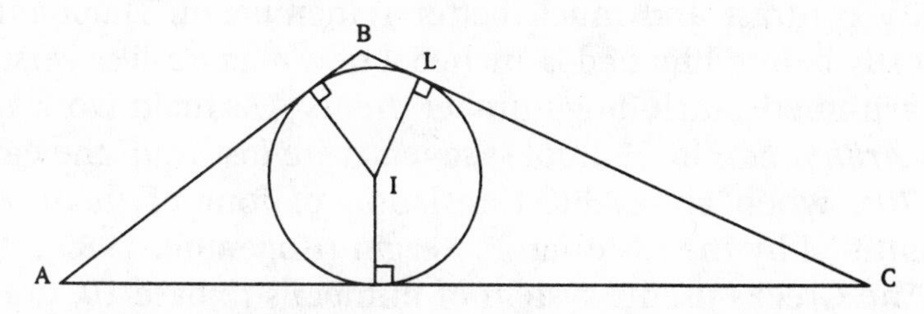

FIG. 2.25.2 *The inscribed circle for Hero's formula.*

The theorem is named after him, although it is now known that priority belongs to Archimedes. The beautiful proof {Heath 1926, Vol. 2, p. 87}, based on the insight that the area equals the product of Per with the radius IL of the inscribed circle, showed that there was still plenty of life in planar geometry, even to the extent of multiplying together *four* lines, a procedure lacking geometric justification.

2.26 *Primarily Diophantos*

Two roughly contemporary figures of the +3rd to +4th centuries show differences emerging in the interests in number theory. Plato's follower Iamblichos wrote a 'theology of arithmetic' which brings out clearly the status of integers as *invariants*: for example (and an important one in Greek culture), the 'heptad' 7 manifests itself as the notion common to the interval of the fourth 3:4, the geometric sequence 1, 2, 4, the two shortest sides of the right-angled triangle 3:4:5, the possible directions of movement (the seventh one is circular), the ages of man according to the doctor Hippocrates of Cos, and so on (and on). While he does not enjoy a good reputation as a commentator on predecessors, he exemplifies well the strong tradition of numerology in Greek mathematical culture {Roscher 1904}. He also commented with enthusiasm on an *Introduction to Arithmetic* composed two centuries earlier by Nicomachos, where the Euclidean representation by lines and squares (§2.17) had been replaced by an abstract approach, with 'mystical' features such as triangular numbers well to the fore, and applications made to musical intervals.

By contrast and much better remembered, Diophantos shortly before him had launched a new algebra-like version of arithmetic, including number theory. His main work was an *Arithmetica*, in 13 Books; seven were lost until the early 1970s, when an Arabic translation of four of them was identified by the historian F. Sezgin {Hogendijk 1985}.

The Greeks used a system of numerals to base 10, drawing upon the letters of their alphabet (which included three letters that have since been discarded). For clarity they sometimes attached an overbar or prime-stroke: thus 'α′' for 1, ..., 'θ′' (9), 'ι′' (10), ..., 'κ′' (20), 'ι′κ′' (21), and so on. Diophantos increased this use of symbols, although it is not clear which new ones are due to him or to commentators, or whether some were abbreviations made by scribes. At all events, in his work the letter 'ς' denoted the unknown value (this was an innovation in itself), 'Δ^{Υ}' its second power (*dunamis*), and 'K^{Υ}' its *kubos*. For powers beyond the third (another innovation), up to the sixth, he used concatenations of these names and symbols; for example, '*dynamo-kubos*', symbolized 'ΔK^{Υ}', for the fifth, with 'square-cube' for the fifth power of a number. In addition, 'χ' marked the reciprocal of a power, and '$\mathring{M}$' a constant term. Addition was shown simply by concatenating the numbers and/or terms; subtraction he denoted by '⩚' (not a negative number or term as such). He worked with rules of multiplication which we now write as

$$- \times - = +, \quad - \times + = -, \quad + \times - = -; \qquad (2.26.1)$$

he symbolized equality by 'ισ'. Thus, for example, the equation $13x^4 + 2x^3 - 5x - 9 = 4x^2$ could be written

$$\text{'}\Delta^{\Upsilon}\Delta\iota\gamma\ K^{\Upsilon}\beta ⩚ \varsigma\varepsilon\ \mathring{M}\theta\ \iota\sigma\ \Delta^{\Upsilon}\delta\text{'}. \qquad (2.26.2)$$

In his number theory Diophantos largely pioneered the study in Greek mathematics of 'indeterminate analysis', as his European successors were to call it: equations, usually linear or quadratic, with unknowns, normally one or two, which took integral or other rational values in their solution(s). His methods included transformations of unknowns, and gathering together terms of the same power. Some of the simpler problems and solutions carried a Babylonian aroma (§2.4), probably by conscious import; but in many cases he was entering new territory for number theory.

In some superficial respects, such as the influence of geometry and the absence of zero and of negative numbers,

Diophantos's arithmetic resembled Euclid's. However, the differences from Euclid were fundamental; in particular, he used fractions rather than anthyphairesis, and indeed may have helped in the decline of the latter. His preferred proof-methods also differed. He ushered in a more critical phase in being aware that a problem may not have a solution: a remarkable case stated that the cube of an integer could not be expressed as the sum of two other such cubes, and in the 17th century it was to inspire Fermat to a famous generalizing conjecture (4.22.2). Often he assumed the solution to be known before seeking conditions; sometimes he knew the answer to be wrong and used proof by contradiction to correct it. In the +4th century this kind of proof would influence Pappos, to whom we now turn.

2.27 *Treasures in Pappos*

The name *Collection* is given to a surviving octet of books by Pappos commenting upon Euclid, Ptolemy's *Almagest* and, especially, Apollonios. In places the texts seem to be more thrown together than collected, perhaps not by him; but they are valuable for information on works which otherwise are little known or even lost.

In Book 7 Pappos listed a 'Treasury of analysis' from his forebears ({Pappos 1986} is an excellent recent edition). Here the word 'analysis' relates to his discussion of methods of proof in mathematics, which is one of his own most influential contributions. The most conspicuous logical feature of the various proof-methods is the (in)valid inference from one statement to the next; and two opposite directions of derivation had long been evident (to Archimedes, for example). Pappos adopted this distinction, and extended it to discriminate between two methods of heuristics. In the 'analytic' method one assumed the desired proposition P and inquired 'backwards' from it until finding proposition(s) A already proved or assumed; the 'synthetic' method went 'forwards' from A and arranged in proper order the consequences of the assumptions until P was reached {Maula 1981}. This distinction is basically quite clear; the Babylonian concoction of the six brothers problem in FIG. 2.5.1 and solution by recipe is a previous such case. However, the later normal interpretation of his discussion as concerning *proofs* rather than heuristics, and the associations of analysis with algebra and synthesis with geometry, are far less happy, as we shall note in connection with Franciscus Vieta (§4.7) and J.L. Lagrange (§6.11).

The most important predecessor was Apollonios, from whose work Pappos was inspired to develop further the notion of the cross-ratio between four points in a manner which Desargues would call 'involution' in the 17th century (§4.20). A famous theorem named after Pappos states that if three points A, B and C were taken on a straight line, and three points X, Y, and Z on a coplanar line, then the trio of points where AY met BX, AZ met CX, and BZ met CY, were also on a straight line.

One of Pappos's most remarkable problems is this: given any even number ($2n$, say) of lines, draw any line from a given point P to cut each line at the same angle at points C_r such that the intercepted lengths PC_r satisfy the following relationship involving compounded ratios:

$$(PC_1 : PC_2) \cdot (PC_3 : PC_4) \cdot \cdots \cdot (PC_{2n-1} : PC_{2n}) \text{ is constant.} \qquad (2.27.1)$$

(For an odd number of lines, a constant line replaces PC_{2n}.) If this construction is possible, then P lies on a curve; but which ones, as n varies? The answers extended the class of curves studied in Greek geometry; they were to be a major feature in Descartes's advertising of his algebraic methods in his *Géométrie* of 1637 (§4.18).

In addition, the compounding in (2.27.1) extended the range of techniques in Greek geometry by showing that the second term in a ratio did *not* have to be repeated as the first term in the next ratio, as in Euclid's (2.18.3); thus the operation began to look like multiplication. Pappos gave this reason for the extension. Apollonius had defined conic sections and other curves by such definitions using four lines, specifying that the ratio of the product of one pair and the other pair be constant and treating each product as a parallelogram. Six lines could be handled similarly, with the products taken as parallelepipeds, but there the method had to stop. However, with compounding read as in (2.27.1), *any* number of lines could be taken {Pappos 1986, pp. 118–25}. A similar transition in the use of ratios would develop in medieval European mechanics (§3.26).

Despite these valuable contributions, Pappos is still best remembered as a commentator. So are the later figures: Theon on Euclid, Hypatia on her father Theon, Proclos on Ptolemy trying to prove Euclid's parallel postulate, and various other matters mathematical and philosophical. The light of originality was dimming in the Grecian Empire.

2.28 *Hippocrates on neusis and the 'three classical problems'* {Knorr 1986b}

This review of Greek mathematics ends with a return to the −5th century to consider the principal contributions of Hippocrates of Chios, which were important sources of inspiration to his successors. One of his concerns was the curves that became known as 'lunes' after the shape of the crescent Moon. FIG. 2.28.1 shows the lune EKBHZ in the case where two chords inside the inner arc EZH fit three

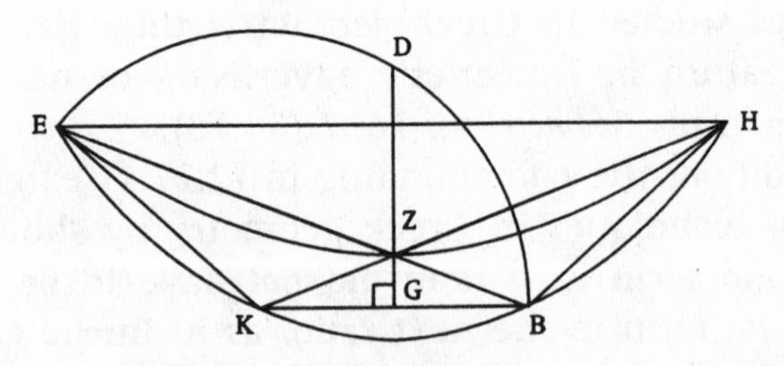

FIG. 2.28.1 *Hippocrates' construction of a lune: the birth of 'neusis'.*

inside the outer arc EKBH. Each chord subtends the same angle at the centre of its circle; hence the lune equals the rectilinear region EKBHZ.

This statement of the property is synthetic – what about the analytic means of constructing the curve? One possible procedure starts with drawing the given line KB and its perpendicular bisector at G, and then an arc with centre K and radius KB to meet the bisector at D. Then E is determined by choosing the line through B for which EZ equals the given length of the other chord.

This last move was a major innovation in its own right. Hippocrates is reported to have described EB as 'inclining towards B' and of the given length; the construction became known as 'neusis' (Greek for 'inclination' or 'verging'), and was most fruitful for his successors.

A popular use of neusis was found in connection with other contributions made by Hippocrates to the 'three classical problems', as they have become known: squaring the circle, trisecting the angle and doubling the cube. While he did not invent them, he appreciated early that the required constructions could not be executed by means of straightedge and compass alone to produce lines and arcs of circles. The consequences led to many major innovations by him and several of his successors; I briefly note some landmarks.

1. *Squaring the circle* seeks a square equal in area to a given circle; for some students, the problem may have been seen as circling the square. It relates closely to Hippocrates' construction of the lunes, and neusis played a role later, for example in a construction due to Archimedes using his equiangular spiral (which he had shown to be an important curve in geometry in other contexts). Among earlier solutions, it seems to have led Dinostratos in the –4th century to extend the range of known curves by defining the quadratrix (FIG. 2.28.2) as the points of intersection F of the radius AE, rotating with uniform angular velocity from vertical AB to horizontal AD, with the horizontal line HK falling with uniform speed from BC to AD.

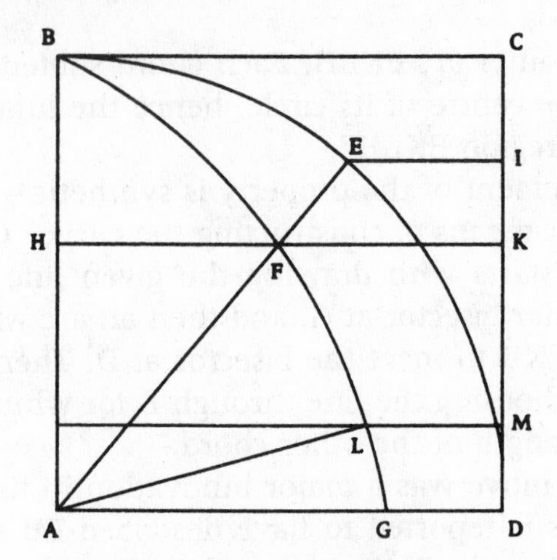

FIG. 2.28.2 *The quadratrix, possibly due to Dinostratos.*

2. *Trisecting the angle* asks for a method of dividing an angle into three equal parts. Some solutions also drew upon the quadratrix, for the uniform rotation of the radius allowed ∠FAG to be trisected by ∠LAG, where L was specified by the property 3DM = DI. Later solutions include Nicomedes, late in the –3rd century, introducing the conchoid (FIG. 2.28.3), defined via the neusis that the intercept TH cut by the fixed 'canon' line AB off the line TE drawn from any point T to the fixed 'pole' E was constant in length (the 'interval'); depending upon this value, the curve can

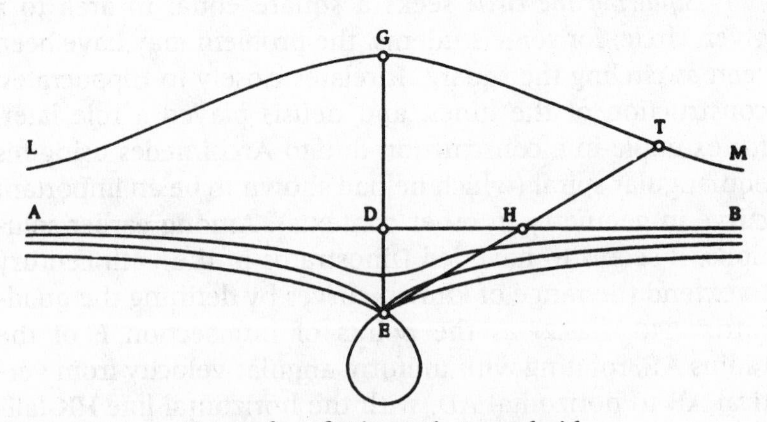

FIG. 2.28.3 *Examples of Nicomedes' conchoids.*

take a cusp at E or a loop there. It is an early example of a curve with an 'asymptote' (a Greek word, meaning 'not touching'). This construction of the conchoid easily allowed a line of given length to be intercepted by two non-parallel lines, so that the trisection could be effected. Later, Pappos tried various solutions, one using the quadratrix and some neusis constructions, another involving the intersection of a circle with a hyperbola.

3. *Duplicating the cube* seeks to construct a cube twice the volume of a given one {Knorr 1989, Part 1}. It might have been posed as an exercise in the anthyphairesis of $\sqrt[3]{2}:1$ in order to determine the side, but it was normally handled as a search for magnitudes a and b in mean proportion between 1 and 2. However, no simple sequences emerged. Remarkable but complicated solutions involving clever constructions in three dimensions were found in the –4th century by Archytas (semi-cylinders) and Menaechmos (conics; compare FIG. 2.23.1). Then Nicomedes showed that his notion of intercept, described above, simplified the determination of the mean proportionals. Some later solutions by mean proportionals drew upon neusis: by Apollonios and then Hero with a certain construction within a triangle; and in between by Diocles using a type of closed curve, similar to the ivy leaves so often illustrated in Greek art. From the 17th century this curve was commonly called the 'cissoid' (Greek for 'ivy-shaped'), and some of Diocles' own successors may have noticed the resemblance {Knorr 1986b, pp. 233–63}.[5]

Problems such as this trio, and various others, had led Greek mathematicians gradually to extend the class of curves, surfaces and solids from rectilinear examples such as triangles and cuboids, through curvilinear ones like conic

[5] Today the cissoid is often defined as the diamond-like shape obtained from neusis constructions on the parabola. Another such curve is the 'cardioid' (from the Greek word for 'heart'), defined in the same way as the conchoid (FIG. 2.28.3) but with a circle though the pole acting as canon instead of the straight line AB.

sections, to spirals and more general cases such as those produced by Pappos's condition (2.27.1). The study of geometrical properties, starting out naturally enough with angles, lengths and areas, had come to include tangents and normals, and tangent planes. These three problems were classical indeed, especially in the rich variety of solutions offered for them and the contributions they made to the extension of curves. Work on them was to be continued by the Greeks' Arabic and European successors, with algebra and eventually analysis introduced to back up the geometry.

2.29 *Roman numerals and technology*

{Stahl 1962}

The rest of this chapter is devoted to other ancient cultures; their lesser influence upon later European mathematics (§2.2) shortens the accounts. The natural first case is the Romans, especially with their empire in place from Caesar Augustus in −27 to the late +4th century. Disliked for their militarism and cruelties, in consequence they are *enormously underrated* in the history of science in general. Their mathematics must have been wide-ranging to have made possible, for example, map and chart projection on prodigious scales (Strabo of the ±1st centuries is important, following Eratosthenes), quantity surveying and bookkeeping of similar range, hydraulics for flooding amphitheatres in preparation for games between crocodiles and prisoners, and some naive probability theory when betting on the *victores* in gladiator bouts. Four aspects will have to suffice here.

1. The Roman system of numerals is still with us, a familiar use being the numbering of preliminary pages in books. Perhaps following their predecessors the Etruscans, the Romans reacted against the Greek practice of using the letters of their alphabet in sequence for the integers (§2.26); like the Greeks, they had no zero. Their system was not place- but order-valued, with the letters usually ordered in

the forward direction of their written language; contrary to later European practice, they rarely placed a lower-value numeral before a higher one to represent subtraction, and so spared themselves zigzags such as 'MCCDXLIV'.

The origins of this system are not well understood, but in 1906 Giuseppe Nicasi (1858–1915), a historian of folklore,

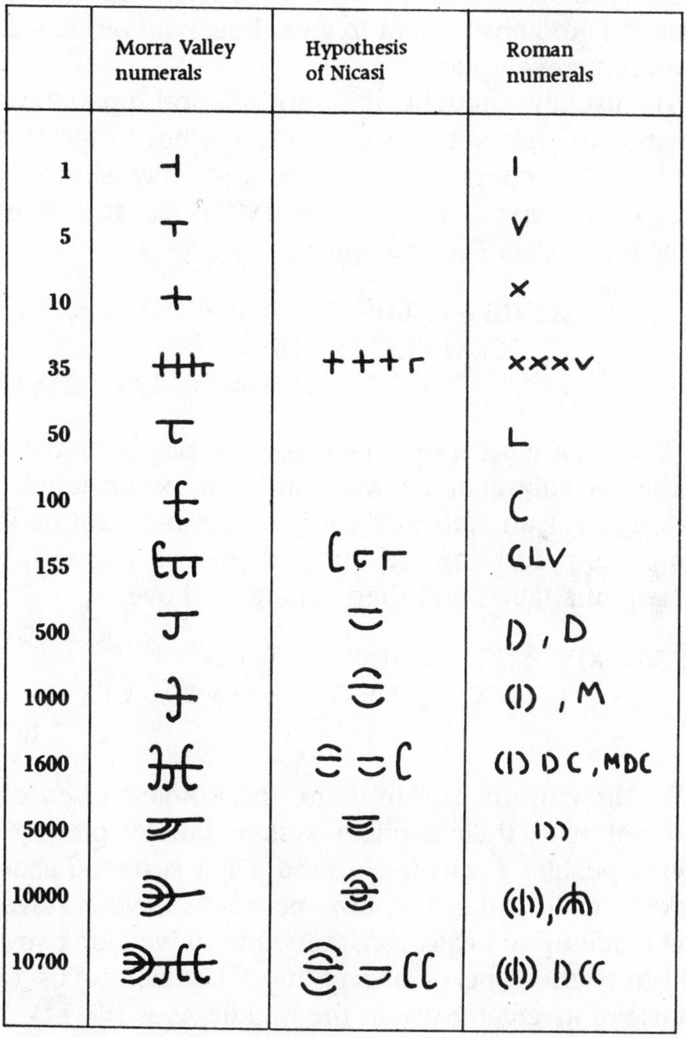

	Morra Valley numerals	Hypothesis of Nicasi	Roman numerals
1	⊣		I
5	T		V
10	+		X
35	[illegible]	+++Γ	XXXV
50	[illegible]		L
100	[illegible]		C
155	[illegible]	[illegible]	CLV
500	[illegible]	[illegible]	IↃ, D
1000	[illegible]	[illegible]	CIↃ, M
1600	[illegible]	[illegible]	CIↃ IↃC, MDC
5000	[illegible]	[illegible]	IↃↃ
10000	[illegible]	[illegible]	CCIↃↃ, [illegible]
10700	[illegible]	[illegible]	CCIↃↃ IↃCC

FIG. 2.29.1 *Nicasi's hypothesis on the origin of Roman numerals.*
{Giacardi and Roero 1987, 33}

proposed a nice hypothesis. In his time, in the Morra valley near Spoleto, north of Rome, the citizenry were still using an ancient system from which Roman numerals might have been derived. Combinations of signs were used, including half-signs for half-values. For example, 2 is ⊣|, while 20 is ⧺; 10 is +, and 5 is ⊤. Nicasi's idea was that these compound symbols were split off into simpler, separate ones, which the Romans rotated to give their symbols; FIG. 2.29.1 shows some examples.

It is usually thought that arithmetical operations in Roman numerals are very complicated, but in fact concatenating the letters can produce relatively simple procedures {Detlefsen and others 1976}, as the Romans probably realized. For example,

$$\begin{aligned} &\mathrm{CXLVIII} + \mathrm{CXLIV} + \mathrm{LXXII} \Rightarrow \\ &\quad \mathrm{CCLXLXLXXIVVIIIII} \Rightarrow \\ &\qquad \mathrm{CCLLLVVIIII} \Rightarrow \mathrm{CCCLXIV} \end{aligned} \tag{2.29.1}$$

(the two +'s are my interpolations; they may not have written the two subtractions). Multiplication can be handled by breaking the numerals into components and treating them term by term, forming each product (maybe originally with the help of a table) and then adding as above:

$$\begin{aligned} &\mathrm{XLVI} \times \mathrm{XIV} \Rightarrow (\mathrm{XL} + \mathrm{V} + \mathrm{I}) \times (\mathrm{X} + \mathrm{IV}) \Rightarrow \\ &\qquad (\mathrm{XL} \times \mathrm{X}) + (\mathrm{XL} \times \mathrm{IV}) + \cdots \Rightarrow \mathrm{CD} + \mathrm{CLX} + \cdots . \end{aligned} \tag{2.29.2}$$

Like the cultures before them, the Romans often calculated not with their number system but by placing and moving pebbles ('calculi') around a flat surface ('abacus') marked out in squares. A few specimens have survived of abaci made up of beads constrained to move along straight line slots – the type of abacus most familiar to us today because of its greater use in the Middle Ages (§3.14).

2. As part of their administrative exactitude, the Romans developed a refined calendar. 'Calends', originally meaning the distributions of time into periods, also named the first

day of the month; a middle day (the 13th or 15th) was the 'ides', and nine days earlier occurred the 'nones'. In puzzling contrast to their avoidance of subtraction in arithmetic, any day was named by counting *back* from the next epochal day (for example, the 11th day before the calends of the next month). Consistently with their lack of zero, the counting started *from* the day of counting; for them Monday was three days before Wednesday, not two days.

3. A field in which the Romans were determinedly strong was architecture, together with its associated technology. The best remembered figure is Vitruvius Pollio (−1st century), whose *De architectura* covered not only construction and decoration of buildings but also town planning, hydraulics, mechanics of machines, sundials and waterclocks. He even treated some mathematical topics such as duplicating the cube and Pythagoras's theorem, and popularized anecdotes such as Archimedes' 'Eureka' (§2.22). His work became the principal Roman mathematical text in the Middle Ages and later, but he was conservative on many architectural issues, and his influence on compatriots may have been negative as well as positive.

4. One topic treated at some length by Vitruvius was land measurement, which not surprisingly was a major Roman concern {M. Cantor 1875}; Hero's Theorem 2.25.1 was a staple result. Some applications dealt with spatial problems, though the methods used drew more on stereometry than on spherical trigonometry.

2.30 *Mathematics in the Far East, and its languages*

Now begins a survey of mathematics in India and China. Although short, it is unique in continuing in the next chapter, for these cultures maintained strong mathematical traditions from at least the −3rd century until after the first influx of European mathematics in the late +16th century. Before commencing the survey, a special difficulty for non-readers of the languages deserves attention: preferred manners of translation.

The task of translating from one language to another one is not just a problem of conveying the content and intention of the original text: where the two languages have markedly different syntactic and grammatical structures, much of the sense is inevitably lost. Various attempts have been made to tackle this problem, especially in educational works. In particular, in classics there has long been a tradition of 'interlinear translation' from a classical language (C) into a modern one (M), where the word order of C was rearranged to suit M but the translation of each word (-group) imitated the semantic structure of C. To take a Roman example, phrases from Virgil's *Aeneid* might be rendered in English as follows:

mecumque	fovēbit	Romānus,
and-with-me	shall-cherish	the-Romans.

When C is a non-European language, the problem of sense is still more intractable. To tackle it for Chinese mathematics, {Hoe 1978} devised a method of translation, somewhat similar to the above, in which each character is translated, by one or maybe two word-roots in English, in an order that follows the original one as closely as possible. The result gives those of us ignorant of Chinese a new means of grasping its syntax. For example (one of Hoe's), take a right-angled triangle with altitude, base and hypotenuse of lengths which I letter a, b and h. Then the algebraic relationship that we symbolize as

$$b^2 + [h + (a - b)] = bh, \qquad (2.30.1)$$

and might begin to write in words as 'the square on the base plus the sum of the hypotenuse and . . .', comes out in his translation thus:

BASE-squared plus HYP-DIFF-SUM (2.30.2)
equals BASE times HYP,

strikingly similar to a modern computer programming language. Among other advantages, Hoe's rendering catches the ambiguities of the original which (2.30.1) or any verbal analogue lacks completely; his proposals deserve to be applied to other texts in non-European languages. His particular context was a remarkable Chinese text of the + 13th century (§3.2); here we consider some earlier work of that culture.

2.31 *The rods and scrolls of ancient China*

{Martzloff 1987 (in French), Li and Du 1986 (in English); CE 1.9}

The origins of Chinese mathematics go back at least to the –3rd millennium, with some number systems and geometry. Among early instances of more developed theory, the 'Book of Changes' (*Yi jing*) of around the –7th century draws upon combinations in calculating the number of different forms of trigram and hexagram; that is, stacks of respectively three and six broken or unbroken lines (for example, ☵). Reliable records date from around the –3rd century, especially after the unification of the country in –221. However, eight years later Emperor Qin Shi Huang (he of the terracotta soldiers) decreed that all books be burnt; even if he was not fully obeyed, much knowledge was lost.

The point of division between this section and §3.2 is the end of the South Dynasty in +581. Mathematics seems to have been developed for its physical applications, and disseminated for training and examining elites such as the civil service and engineering corps. Their technology provided the Chinese with an advantage over most other cultures (for both themselves and their historians), for in addition to bones and the surfaces of vases available elsewhere, they had writing material in the form of turtle shells, bronzes, bamboo strips and, later, sheets of silk and paper.

The ancient Chinese surpassed other contemporary cultures in methods of calculation {Lam and Ang 1992}. They used collections of small rods, often made of bamboo but sometimes of materials such as animal bones, wood or

Position	1	2	3	4	5	6	7	8	9
Units Hundreds Ten thousands	𝍠	𝍡	𝍢	𝍣	𝍤	𝍥	𝍦	𝍧	𝍨
Tens Thousands	𝍩	𝍪	𝍫	𝍬	𝍭	𝍮	𝍯	𝍰	𝍱

FIG. 2.31.1 *The Chinese system of numerals using rods.*

paper. Two groups of signs, each sign employing up to five rods, represented the integers from 1 to 9 (FIG. 2.31.1) and were used in tandem to represent any integer. Negative integers were distinguished from positive ones by using rods of two colours, or by placing a rod obliquely across the first non-zero digit. Number signs were read like word signs, from right to left; thus, for example,

−2,753 was rendered as 𝍪 𝍦 𝍭 𝍢̸ (2.31.1)

Fractions were marked by placing the signs for the numerator above those for the denominator. However, since they left only a blank space for zero, like the Babylonians (§2.4), they had to be careful in distinguishing 704 from 70,004, say, and still more 7 from 700.

The Chinese developed a complete theory for the arithmetical operations on both integers and fractions; but they went further in various ways. For example, they could solve the general quadratic equation, and also some special forms of equations of higher degree which arose from problems such as calculating the amounts of different qualities of rice from given mixtures. Their finest innovation was a method of solving linear equations with positive coefficients. Here is an example about prices of good (x) and poor (y) land, presented in modern and ancient Chinese forms. I follow their layout of the equations from right to left and downwards, but for clarity I represent each zero by a square, a symbol that they used sometimes, presumably when tired of watching those spaces:

$$x + y = 100 \qquad 300 \qquad 1 \qquad |||\square\square \qquad | \qquad (2.31.2)$$

$$300x + (500/7)y = 10{,}000 \Rightarrow 500/7 \qquad 1 \Rightarrow |||||\square\square \qquad | \qquad (2.31.3)$$

$$\overline{||}$$

$$10{,}000 \quad 100 \qquad |\square\square\square\square \quad |\square\square$$

The solutions were found by applying arithmetical operations to numbers in a column so that some entries in the array became zeros; the corresponding actions were to move rods around in order to create blank entries. Thus the Chinese developed parts of matrix theory, which Europe was not to develop until the 19th century (§13.5–§13.6).

Some of the solutions used the method known also to the Babylonians and the Greeks, and named 'the rule of false' when Europeans deployed it in the late Middle Ages (§3.18): guess a wrong answer and correct by using ratios. In this case 'double false' is used, as there are two unknowns. The solution $x = 20$, $y = 80$ is correct for (2.31.2) but gives the surplus $S = 1{,}714\frac{2}{7}$ for the expression in (2.31.3); similarly, $x = 10$, $y = 90$ yields in (2.31.2) the deficit $D = -571\frac{3}{7}$. So the right answers are obtained by calculating arithmetically the equivalent of our algebraic expressions

$$x = \frac{10S + 20D}{S + D}, \quad y = \frac{90S + 80D}{S + D}. \qquad (2.31.4)$$

This is a simple case; typically, the Chinese applied these rules to much more complicated systems of equations.

2.32 *'The Nine Chapters' and its commentators*

By the +1st century enough mathematics had been achieved in China for compilations to be made. The most important one, *Jiuzhang suanshu*, is most commonly known in English as 'The Nine Chapters of the Mathematical Art'

{Lam 1994}.[6] It contained 246 problems in arithmetic, geometry (including the formula (2.9.5) for the area of the circle) and the proto-linear algebra just discussed, together with an account of the methods and algorithms needed to solve them; however, there were no general proofs of pertaining theorems. An emphasis on physical applications is apparent in the titles of the constituent Books, among them 'Field measurement' (a major stimulus for mathematics in China), 'Cereals' and 'Fair taxes'. As an inspiration for further study for its culture, the 'Nine Chapters' at least equalled Euclid's *Elements*; the last Book title quoted above suggests that it is worthy of modern attention.

A major commentator was Liu Hui (+3rd century). FIG. 2.32.1 gives a possible configuration for his proof of

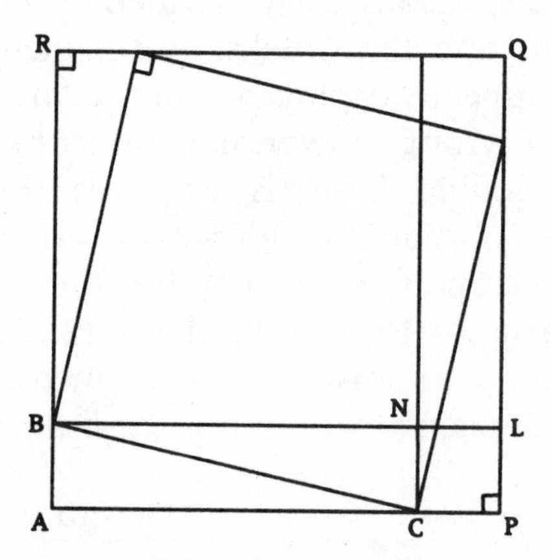

FIG. 2.32.1 *A possible diagram for Liu Hui's proof of Pythagoras' theorem.*

Pythagoras' theorem, which seems to have been based upon the fact that the squares on BN, CN and CB have

[6] I use the modern Pinyin method of transcription of Chinese words. The author of {CE 1.9}, J.C. Martzloff, prefers to translate *Jiuzhang suanshu* as 'Computational Prescriptions in Nine Chapters'. The most important parts of this work will be included in an annotated and selected edition of 'The Ten Books of Mathematical Classics' (§3.2), in preparation by J.W. Dauben (New York) and a team of Chinese colleagues.

some regions in common, and the rest of them have to be moved around and/or split up in congruence style. Or maybe his proof rested on this simple fact: the four triangles outside the square on CB can be swung together to fill the rectangles BNCA and TQLN, which are the regions outside the squares on CN and BN. It seems odd that this proof is not in Euclid's *Elements*.

Liu Hui also found the Babylonian formula (2.7.1) for the volume of the frustum of a pyramid, baldly stated in the 'Nine Chapters', by an analysis based on the dissection shown in FIG. 2.7.1; it inspired my conjecture in §2.7 that the Babylonians and Egyptians may have known similar methods. Liu found an approximation to π by inscribing regular polygons in a circle; with one of 3,072 sides he found the value 3,927/1,250, which is in excess by only 0.01%. He proceeded as did Archimedes with FIG. 2.21.2, but independently of him; he did not sandwich the circle with exterior polygons, and he also obtained the circle by a limiting process rather than by Greek-style double proof by contradiction. In addition, he tried to find the volume, as given by eqn (2.9.4), of the sphere by dissecting it into an *infinity* of parts {Fu 1991}; however, he could not find a satisfactory means, although like Archimedes he came to a form of the 'indivisibles' of Cavalieri (§4.19). This partial success with π is a good example of its different roles stressed in §2.9.

In Chinese trigonometry an important method of calculating height or altitude was called 'double differences' (FIG. 2.32.2). Two poles of the same height were set up vertically at positions P_1 and P_2 on horizontal ground, and their tops Q_1 and Q_2 were aligned with the observed point Z to determine points C_1 and C_2 at ground level. The lengths P_1P_2, C_1C_2 and P_1Q_1 (or P_2Q_2) were measurable, so that both the height AZ and the horizontal distance AC_1 could be calculated. The method was used to good effect in preparing maps, and a variant of it using a vertical reference line was developed to calculate depths.

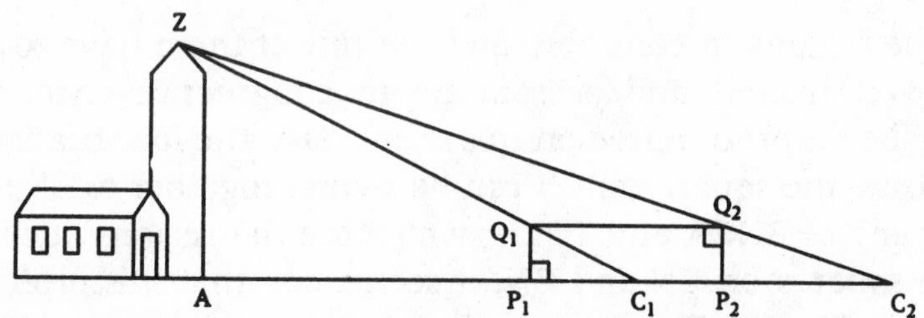

FIG. 2.32.2 *Chinese trigonometry: the method of 'double differences'.*

2.33 *Indian mathematics and astronomy*
{CE 1.12}

The cultural beginnings in India seem to date from around the middle of the –3rd millennium; our termination at around +500 stops us short of the Muhammadan period and delays to §3.1 major figures such as Brahmagupta. Buddhism seems to have encouraged the transmission of scientific knowledge in general.

I confine this survey to two main traditions. One was maintained by the Jain(a)s, non-conformists to Buddhism, distinguished by a strong spirit of religious enquiry and speculation and by its use of their own dialect, Prakrit. Their concern with arithmetic advanced as far as speculations about the actual infinite(s); they also found some elementary properties of permutations and combinations.

The other main tradition was the Vedic, working within Buddhism and writing in Sanskrit. Practical concerns meant that computations and algorithms usually took precedence over general theory. But religious motivations are notable in, for example, geometry, where a principal question concerned the shapes of sacrificial altars. As one consequence, Pythagoras's theorem appears in Indian mathematics as a property of a rectangle and its diagonal rather than of a triangle and its hypotenuse.

In arithmetic, an important notable aspect of Indian systems of numerals is their pioneering use of a symbol for zero. Initially (known written evidence dates from the +3rd century) it was used just as a place-marker, like the

Babylonian '≼' (§2.4); but by the +5th century it was functioning as a genuine number, combinable with other numbers. The first symbol used was a circular blob, and then a small circle, from which '0' was to evolve in many cultures. The motivation for the latter symbol is also not known, but as so often with ancient origins it may have been a fertility sign – a representation of the vagina as the nothing out of which things are born. Other possible sources include the Moon as it appears again in its monthly new phase, and 'o' as the first letter of the Greek word *ouden* for nothing.

The Vedic poems, *Vedas* ('lores'), contain many arithmetical properties in their numerous verses, possibly inspired by musical intervals and tones {McClain 1976}. Calculations were normally carried out on boards covered in chalk or dust. A remarkable range of arithmetical manipulations was developed, including even some anticipations of common European algebra {Shankarachaya 1965}. For example, they multiplied and divided numbers in a manner natural to the place-value representation: units by units, then tens by units and units by tens, and so on. In our notation this means that $29{,}357 \times 8{,}062$ is effected as follows:

$2 \times 7 = 14$, write 4 carry 1;	29,357	
$(2 \times 5) + (6 \times 7) + 1 = 53$,	8,062	
write 3 carry 5;	———	(2.33.1)
$(2 \times 3) + (6 \times 5) + (0 \times 7) + 5 = 41$,		
write 1 carry 4; . . .	236,676,134	

Division of integers, and the special case of finding the square root of an integer, may be executed by the converse algorithm. Fourier rediscovered these methods for himself in the 1820s, but unfortunately they have never become established in Europe, where the usual methods are far more cumbersome.

The *Vedas* also contain much astronomical and astrological lore. Indian astronomy seems to have depended quite heavily upon imports, especially from the Babylonians around the –5th century and the Greeks in the +2nd and +4th centuries {Pingree 1978}. Moreover, these transfers were rather tardy; in particular, Ptolemy's theories do not appear to have arrived before +500. (The historian benefits, however, for the Indian versions are important testaments on the Greek theories themselves.) As with their mathematics in general, they handled computations and algorithms more than general properties: positions of planets and the Moon (including the Babylonian zigzag method (§2.4) of recording variations), eclipses, control of waterclocks for measuring periods of daylight, use of the gnomon, and so on. In trigonometry, which they practised extensively in both its planar and spherical parts, they pioneered the use of the sine of an angle (conceived of as a line) instead of the Greek chord (2.24.1), and compiled tables of its values.

2.34 *Similarities and differences: transmission across cultures?*

The previous section recorded examples of conscious influence from one ancient culture to another. We have noted many other instances of similar theories in different cultures, such as in §2.8 concerning the right-angled triangle with sides in the ratio 3:4:5. Another example is the rule of double false (2.31.4), of which forms occur in most cultures apart from the early Greeks. So the question arises of whether we interpret such similarities as coincidence or influence.

Influence entails contact between cultures, either in person or by transmission of theories in some written form. This possibility can be entertained to some extent for countries adjoining the Mediterranean sea, such as Babylon → (or ↔?) Egypt → Greece → Rome. Surviving examples of balance weights suggest that some uniformity developed in metrology among Mediterranean countries {Petruso 1985}.

Apart from known imports such as astronomy to India from Greece, the manner of contact between the Middle and the Far East is harder to determine. When cultures such as the Japanese and the Korean are included in the picture, it becomes more complicated still, especially as the content of their early mathematics is not well understood (and so has not been appraised in this chapter). However, the trading of silk from China and spices through India to the Middle East doubtless brought mathematical knowledge in its baggage. The history of this traffic is not well known, but Aristotle had heard of silkworms, so that some Chinese mathematics could have been in Greece by the –4th century. Later, Alexander's invasion of India in –327, just before Euclid's time, could have brought in some mathematical commodities. But *clear* evidence of such introductions from abroad is very scanty.

When influence is not assumed or claimed, common origins can be considered. The role of ritual has been advocated by {Seidenberg 1962a, 1962b}. Giving some emphasis to India, he notes several instances where similar attitudes or pursuits are evident, from general inter-cultural features such as a preference for geometry over arithmetic to more specific aspects such as an interest in very large numbers. He also stresses religious imperatives as a cause of common views. Examples are the Greek concern with doubling the cube, which has a resonance in the Indian concern with the double cube as the preferred shape for altars, and squaring the circle in the construction of cubical or hemispherical burial mounds (cubing the sphere probably turned out to be too hard a problem to solve). Common formulae such as those for the volume of the frustum of a pyramid (eqns (2.7.1) and (2.7.2) leads one to wonder why several cultures built pyramids in the first place. One common reason is the construction of funeral pyres, and in India also lavishly decorated temple towers (*gopuram*) on a rectangular base; in addition, Babylonian ziggurats or Egyptian stepped pyramids look like stacks of frustums, and may have been constructed that way. In all such cases the motivations could have included not only the practical one of calculating the amount of earth to be moved, but also the importance of knowing a basic property of a sacred object.

Turning to branches of mathematics, the case of arithmetic is quite tricky. It is probable that both names and signs for integers, and theories of their operations, developed quite quickly in each culture, and therefore largely independently {Cajori 1928: CE 1.16}. They were needed for commercial purposes such as weighing and paying, not just everyday counting {Menninger 1969}. Hence we find a plethora of notations and name-systems, some of which are still in place, such as the curious handling of names for 70, 80 and 90 in French. The standard European system of numerals seems to be a mixture of different ones in various cultures: how did this happen? Again, why did so many cultures adopt a 10-based system for arithmetic rather than, say, 12? It seems that counting parts of the body was *not* a major motivation, for in any particular language the words for parts of the body do not always closely align with those for integers.

Various differences in the conception, handling and notation of fractions are apparent, as are differences in the treatment of arithmetical factors such as 3 being part of 12 {Benoit and others 1992; CE 1.17}. We have seen already a great difference between Egyptian and Greek unit fractions and Greek ratios, and noted in eqn (2.33.1) the Vedic way of dividing; other examples include Liu Hui making a strong analogy between dealing with fractions, and the Chinese method (eqns (2.31.2) and (2.32.3)) of solving two linear equations.

To complicate the situation still further, differences between cultures are no easier to interpret than similarities; for they do *not* entail independence of origin. Firstly, as is well known to anthropologists under the name of 'stimulus diffusion', the product of one culture or country may take a *variant* form when imported by another one. Further, *influence can be negative*, as a reaction against something – but then the differences might well be deliberate. Euclid's *Elements* provides examples: its truly geometrical character evident even in its arithmetic, its careful distinction of geometrical dimensions and insistence that lines have no breadth, and its avoidance of calculating any length, area or volume. All these features characterize a Greek tradition which broke with Egyptian arithmetical thinking, where numbers were not subordinate to geometry (although normally related to the physical world), and where in eqn (2.6.1) an area of 18 was equated with a perimeter of 18.

To sum up, probably more contacts were made between cultures during those many centuries than we can ever now know or discover; on the other hand, one has to be cautious in attributing influence in particular cases. No definite answers are available, nor can any be expected. But in the differences and negative reactions as well as in the positive influences we find a remarkably wide range of achievements, in most of our eight ancient roots of mathematics (§2.1), from an unknown start with grunts, shouts, marks on bones, and tokens. It furnished a rich and varied legacy for medieval mathematicians to adopt and adapt in their own various cultures.

3

A quiet millennium: from the early Middle Ages into the European Renaissance

The first part of this chapter continues with the Far East, taking in two further cultures; then the Mayan tradition in Central America is considered. The second part (§3.5–§3.8) is a concise summary of the fine Arabic tradition, which ran from around the 8th to the 15th century. The transition to Europe is made in §3.9–§3.11 by noting Hebrew mathematics and mathematics in the Bible, and recording the rather miserable performance in Europe from the end of the Roman Empire up to its revival during the 12th century. The rest of the chapter treats European mathematics up to the mid-15th century. What is meant here by the word 'Europe' is explained in §3.12; the phrase 'the West' is used occasionally to refer to modern contexts, in which other areas, especially the USA, are to be adjoined. {Lindberg 1992} gives a fine up-to-date survey of our (lack of) understanding of the growth of science in Europe throughout the Middle Ages, and of its various backgrounds.

One theme common to most cultures during this period is the development of arithmetic, especially in connection with the emergence of common algebra. Much information is given in {Tropfke 1980}, while {Cajori 1928} deals remarkably well with the many notations.

The degree of contact between cultures and countries was still often limited; so many questions of priority and influence are extremely difficult to address. Our understanding is again blocked by much covert mathematics, where little textual evidence is known: book-keeping, architecture, fortification, maps and charts, ship construction and operation, and so on. With these caveats, let us pick up again in India.

3.1 *Later Indian mathematics and astronomy*
{Pingree 1978; CE 1.12}

The Subcontinent suffered invasions from the Middle East soon after the rise of Islam in the 7th century, and also in the 10th, 13th and 14th centuries. They were partly repulsed, and the effect on mathematics seems to have been modest. Imports of Greek pre-Ptolemaic astronomical theories (§2.33) from the 5th century made much more impact, continuing the close connections between mathematics and astronomy. The Indians laid great emphasis on meticulous computations of parameters for planets: their mean motions, latitudes, and mean and true longitudes, distances and sizes. The work required much geometry and trigonometry; sines were to the fore, including tables of their values. Major figures across the centuries frequently intermingled the two disciplines in their treatises, which were often written in Sanskrit verse: Aryabhata I (5th–6th century), Aryabhata II (12th), Bhaskara I (7th), Bhaskara II (11th) and Brahmagupta (early 7th). This work spread to Tibet, along with the rise there of Buddhism {CE 1.13}.

Greek influence is evident in some other branches of Indian mathematics. For example, Brahmagupta took up Diophantine equations of the second degree (§2.26), solving, for example, in positive integers

$$Cu^2 \pm D = v^2, \ u \text{ and } v \text{ unknown, } C \text{ and } D \text{ given.} \quad (3.1.1)$$

Independently and many centuries later, Fermat would reopen enquiries in Europe with eqn (4.22.1).

The Indians also formed linear equations in several unknowns (their language suggests that coloured beads were used to indicate different unknowns), and polynomials in one unknown up to the square–square power, like Diophantos, which they solved, at least for special cases. Of especial note is Brahmagupta's efforts to find the smallest integer N that yields given remainders A and B after division by given integers R and S respectively; in our notations,

$$N = Rx + A = Sy + B, \text{ with } A > B; \quad \text{hence } Rx + (A - B) = Sy. \tag{3.1.2}$$

This gave him a method of solution strongly reminiscent of anthyphairesis: in the notation of (2.13.1) and using his numerical example, $A = 10$, $R = 137$, $B = 0$ and $S = 60$, so that

$$137{:}60 = [2,3,1,1,1]. \tag{3.1.3}$$

The value of N was found in effect by using these successive remainders to solve the second of eqns (3.1.2) for x and y and finding N as a sum of multiples of them. This study may have been partly motivated by the construction of calendars: the integer unit was the year, R and S marked the lengths of astronomical cycles, and the solutions determined coincidences (that is, the years when the cycles would commence together).

3.2 *Later Chinese computations*

{CE 1.9}

In European mathematics, (3.1.2) was to be called 'the Chinese remainder theorem'. The name is justified, for a special case had been known in China as early as the 3rd century, and study more general than the Indians' was undertaken later, especially in the 13th century by Qin Jiushao. When there were more than two relations, the Indians had taken them in pairs and effected successive division by the first of eqns (3.1.2); but Qin massed them together and solved them by working from the least common multiple of $R, S, \ldots$. His procedure, far more sophisticated than those of his Indian predecessors, suggests an independent development. However, the *problem* might have been introduced from India, maybe as part of a limited adoption in China of Buddhism.

In general and by intention, China kept itself aloof from foreign connections, despite such religious contacts and the continued trade in commodities such as silk and spices. The teaching of mathematics was increased under the Sui and Tang dynasties (6th–9th centuries); a compilation known as 'The Ten Books of Mathematical Classics', which included 'The Nine Chapters' (§2.32), was made, and approved as a textbook by the Imperial Academy (a governing body for education). This encouragement of teaching reached a peak in the 13th century during the Song dynasty; Qin Jiushao's achievement just noted is only one example.

Among other examples, the Chinese system of numerals was enriched by a sign for zero (an Indian-style small circle, in fact, but not *known* to have been imported), so that they possessed a full place-value system of notation. The proto-matrix theory of eqns (2.31.2)–(2.31.3), effected by the use of rods on the counting-board, was extended to linear equations in four unknowns and polynomials to the 14th degree by Zhu Shijie in a book known in English as 'The Jade Mirror of the Four Unknowns' (the West has never used titles like that, sadly!); this was the work for which Hoe proposed his character-by-character translation (§2.30). Study of combinations in connection with extracting roots of polynomial equations produced this triangle of integers which is named 'Pascal's' after its later finding in France (§5.19): in their and our notations, respectively,

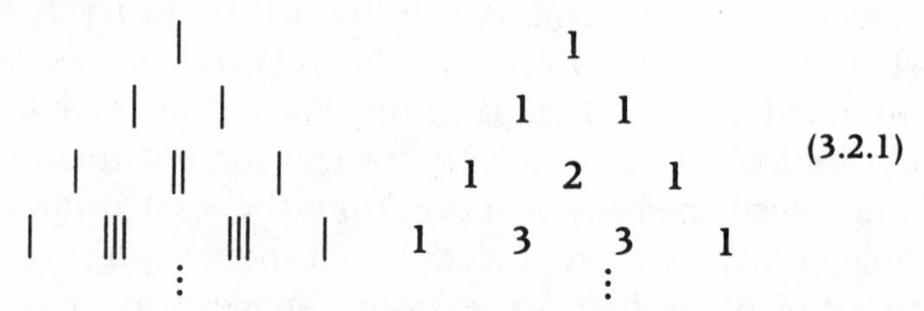

Elsewhere in mathematics, spherical trigonometry was well developed, including formulae for approximating the values of half-chords of arcs and for inverse sines. Much of the motivation came from astronomy and surveying, as usual.

The Ming dynasty, which lasted from the mid-14th to the mid-17th century, is much admired for its porcelain and glazing, which shows evidence of a fine sense for design and geometrical shape. Some encyclopedias were compiled to preserve and diffuse mathematical knowledge. But mathematics in general was declining; however, the earlier achievements had been appreciated and adopted elsewhere in the Far East, as we shall now see.

3.3 *Computational mathematics in Japan and Korea* {CE 1.10–1.11}

The Chinese seem to have set both style and content of much mathematics elsewhere in the Far East to an extent which ancient Greek mathematics was to enjoy in the Middle East and then in Europe; Korea and Japan were especially heavily influenced. As in China, mathematics was taught principally for the training of bureaucrats, civil servants and officers; in Korea, institutions included an Agency for Calendars. The staple branches were arithmetic and the theory of equations, with geometry and some trigonometry provided to aid astronomy, geography and surveying. Large-scale compilations were prepared from time to time, in broad imitation of the Chinese 'The Nine Chapters' (§2.31), which itself was a standard text in Japan until at least the 9th century. Computation was preferred over general theories, and the abacus and counting-board were much used.

Both the Chinese and the Indians had studied magic squares – arrays of numbers with the property that each row, column and diagonal added up to the same value; FIG. 3.3.1(a) shows the third-order case, which in fact is the only solution apart from common multiples of the integers. Probably inspired by Chinese work, the Korean mathematician Ch'oe Sok-chong (1646–1715) found the 'square' in FIG. 3.3.1(b), in which the integers from 1 to 30 are arranged in five hexagons such that each sextet adds up to 63.

(a)

4	9	2
3	5	7
8	1	6

(b)

(a) The unique case of order three (b) A Korean extension

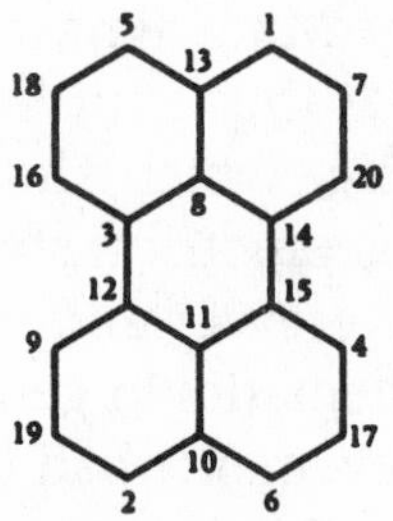

FIG. 3.3.1 *Magic squares.*

Chinese mathematics came to Japan on occasion through Korea; in the 1590s after Japanese invasions of Korea (which China defended). But in the 15th century the Japanese had already started to free themselves from Chinese domination and develop their own indigenous tradition. They called it *Wa-san,* which might be translated as 'Japan calculation'. Takakazu Seki (1640?–1708), in extending methods of solving equations, anticipated parts of the theory of determinants, and he and others summed various infinite series. Some of these arose from their study of π; they usually approximated to it in its role (2.9.1) in the circumference of a circle by operating on the perimeter of an inscribed n-gon for various values of n – even larger values than had been used by Liu Hui, up to 2^{44}. These kinds of research continued until the later 19th century, when European influence arrived for good.

3.4 *The glyphs and codices of the Mayas*

{Lounsbury 1978; CE 1.15}

All this achievement in the Far East was gradually to be replaced by European mathematics from the early 17th century onwards, when commerce and Christianity began to make ever deeper inroads (although the Jesuits' translation of Euclid's *Elements* was ignored). But in Meso-America, a region comprising the Central America of today together with Peru, Columbia and Ecuador, the indigenous cultures were ruthlessly eliminated, and few Spanish invaders tried to understand or record the thoughts of a heathen population. However, in recent decades anthropologists and archaeologists have made valiant efforts to reconstruct the marvellous traditions of many centuries that were all but obliterated. Here we consider the Mayas, who occupied Southern Mexico and some regions to its south. They left records as 'glyphs' (carvings and pictures on temples and other buildings, like the Egyptians), drawings on vessels of various kinds, and a few 'codices' (books with pages folding zigzag fashion, like screens); a little folklore was remembered by their descendants.

As elsewhere in the ancient and medieval world, Mayan astronomy, calendars, arithmetic and geometry advanced to the level of sophisticated means of calculation and construction. For example, the huge square-based pyramid at Chichen Itza in Mexico is oriented so that its sides face the Sun at the spring and autumn equinoxes. Shades (as it were) of the Egyptian pyramids at Gizeh (§2.7); the issue of similarity across cultures (§2.34) arises again.

But several features of Mayan mathematics are unique, especially their system of numerals. Based upon 20, it used a blob symbol '•' for 1. Blobs were laid out in a row to count up to ••••, after which a streak '—' was used for 5, stacked up to ≡ (or vice versa in layout, blobs up and vertical streaks along); zero was represented by an ellipse-like symbol possibly intended to represent the Moon. Larger integers were given other symbols, even lovely portraits of heads or entire persons.

Astronomy was a major motivation for Mayan arithmetic. They developed an elaborate scheme of days in units of 20, 360, 360×20 and $360 \times 20 \times 20$, related to an almanac of 13 cycles each of 20 days. FIG. 3.4.1 shows a tier of data from a codex which is thought to date from the 12th century, although the system in which it is recorded had probably been in use for a long time. The transcription alongside interprets it as part of the introductory explanation of a record of the passage of Venus, divided into

60V	55V	50V	45V
4	4	4	3
17	9	1	13
6	4	2	0
0	0	0	0
6 Ahau	11 Ahau	3 Ahau	8 Ahau

FIG. 3.4.1 *Mayan records of the passage of Venus, with interpretation.*
{Lounsbury 1978, p. 785}

periods of five synodic Venusian years ('V') each of 584 days. Each number is read downwards, and the tiers are read from right to left; thus the second number is $(4 \times 7{,}200) + (1 \times 360) + (2 \times 20) + (0 \times 1)$, starting on the last day ('Ahau') of the third cycle. This codex is read tier by tier from the bottom upwards.

In a most attractive hypothesis, {Diaz-Bolio 1987, 1988} argues that many Mayan doctrines were inspired by a rattlesnake which is found widely in that region and named by them 'Ahau Can' (the latter word means both 'serpent' and 'four'); Western natural history knows the reptile as *Crotalus durissus*. Its importance is attested by many decorations showing it speaking to priests. The criss-cross patterns on its skin may have inspired Mayan geometry and then architecture. For example, both the plan and the elevation of the pyramid at Chichen Itza imitates this design, while the faces are stepped in the manner of both the design and the motion of the animal itself; so does the layout in rectangles of their calendars. Contemplation of this pattern may even have helped Mayan geometry in the first place – and perhaps also some sacred arithmetic, with 13 chosen as the most frequent number of rattles found in the snake's tail.

Europeans also invaded the north of the continent in some quantity during the early 16th century, and very soon began to apply commercial arithmetic to the purchase of citizens in North Africa from local chiefs and kings, and the later sale of those still alive to entrepreneurs and landowners across the Americas. They too made little effort to conserve the culture of either the slaves or of the indigenous tribes. Nevertheless, the latter have managed to maintain a repertoire of mathematical theories, not only in arithmetic, geometry and astronomy (there is a remarkable 'clock' in the Big Horn Mountains in Wyoming, for example) but especially in connection with skills such as archery and in games of chance involving the throwing down of rods and sticks decorated in various ways {Culin 1907}.

3.5 *The rise of Arabic mathematics*
{Yushkevich 1976b; CE 1.6}

The Spanish and Portuguese began to voyage to the New World in the late 14th century, after they had freed themselves from centuries of control by the Arabs, whose mathematics is now our subject. I shall refer to it as 'Arabic' simply in reference to the main common language; and I use the modern phrase 'Middle East' to denote most of the region, as there were too many changes of frontier and dominion over the centuries to be described here. From the 7th century, the armies of Islam spread across North Africa, even south of the Sahara Desert; they also occupied most of the Iberian peninsula, and the island of Sicily. The dominating nationalities were usually Persians and Turks; the differences between their mathematics are not treated here, and in fact need be more closely studied for science in general.

Scientific activity developed especially from the 8th century, and was greatly encouraged by the establishment of a new capital at Baghdad in the middle of the 8th century. A 'House of Wisdom' was instituted there early in the 9th century to translate and circulate Greek texts, and then to extend the knowledge found in them {Høyrup 1994, Chap. 4}. Islamic religion was often a strong inspiration; all the scientific works produced there began 'In the name of God . . .'.

Table 3.5.1 lists the principal mathematicians, in the short versions of their names by which they are best known (I recall from §1.9 that a simple transliteration of Arabic into English has been used). The modern names of their countries of birth (or of chief activity if elsewhere) are mentioned, and also their main areas of mathematical interest. At least another score of names could be added, especially of figures who concentrated mainly on astronomy and/or philosophy. Even the list of topics in the Table hints at a great range of achievements, perhaps wider than that of any culture since the Greeks, from whom they learned much and to whom they added much, showing perhaps more general interest in practical mathematics. They were most influenced by Euclid, Archimedes, Apollonios, Ptolemy, Menelaos and Diophantos; Thabit ibn Qurra was the most significant translator, but before him al-Hajjaj rendered first Euclid and then Ptolemy into Arabic around the early 9th century. They drew also on mathematics from India, but apparently little from the Far East. After seven remarkable centuries they declined, seemingly as a result of wars and religious hostility to original thinking within

TABLE 3.5.1 *Principal Arabic mathematicians from the 9th to the 15th centuries.*

Few dates of birth or death are definitive: '?' emphasizes the point, while ' − ' and ' + ' respectively indicate over- and underestimates. ' ± ' indicates a period of time during which the figure flourished.

alg	algebra	mec	mechanics
ari	arithmetic, number theory	mus	music
		opt	optics
ast	astronomy, astrology, geography	phi	philosophy, proof-methods, logic
cal	calendars	ser	summation of series
eqs	equations (roots and/or solving linear)	trg	trigonometry
		trs	translations, editions
geo	geometry		

820s ±	Al-Khwarizmi; Uzbekistan	alg ari ast eqs
810s ± ?–870s	Banu Musa (brothers Muhammed, Ahmed and Al-Hasan); Iraq	geo mec mus trg trs
836–901	Thabit ibn Qurra; Iraq	ari ast geo mec phi ser trs
850?–930	Abu Kamil; Egypt	eqs geo
858?–929	Al-Battani (= Albategnius); Iraq	ast geo trg
960s ±	Al-Khazin; Iran?	alg ari ast geo
940–997?	Abu'l-Wafa; Iran	ari ast geo trg
973–1050	Al-Biruni; Afghanistan	ari ast cal geo trg
?–1036?	Abu Nasr Mansur ibn Iraq; Afghanistan	ast geo trg
980–1037	Ibn Sina (= Avicenna); Iran	ast geo mec phi
1000 ±	Al-Quhi (*or* al-Kuhi); Iran	alg ast geo
1000 ±	Al-Baghdadi, Abu Mansur; Iraq	ari phi
1010s ±	Al-Karaji; Iran	alg ari ser
965–1041	Ibn al-Haytham (= Alhazen); Egypt	ast geo opt ser
1045?–1123	Al-Khayyam (= Omar Khayyam); Iran	ari cal eqs mus phi
1070 ±	Al-Mu'taman; Spain	ari geo phi
1120s ±	Al-Khazini; Iran	ast cal geo mec
?–1174	Al-Samawal; Iran	alg ari eqs ser
1130s ±	Jabir Ibn Aflah (= Geber); Spain	ast geo trg
1180 ± –1214	Sharaf al-Din al-Tusi; Iran	alg ast eqs
1201–1274	Nasir al-Din al-Tusi; Iraq	ari ast geo trg
?–1320	Kamal al-Din al-Farisi; Iran	ari ast opt
1430s ±	Al-Kashi; Samarkand	ari ast eqs geo trg

the region and pressure from without such as the reconquest of the Iberian Peninsula. However, al-Kashi produced a fine comprehensive encyclopedia of mathematics in the first half of the 15th century. He was a member of a group of astronomers in Samarkand in Central Asia, led by a sultan known as Ulugh Beg ('Great Prince'); they also compiled trigonometric tables.

Some Arabic work was transmitted into Europe from the 12th century onwards (§3.12). Not all of the best or the latest material arrived; indeed, some of it has been studied in detail only in the last 30 years.[1] Therefore, in keeping with the restriction of this book to influential and European work, only a short summary of Arabic mathematical achievements is given here, in the next three sections. There is a preference for contributions which were picked up in medieval Europe; transmission *within* the culture, from period to period or region to region, is not discussed, and in many cases it is not well understood.

3.6 *On Arabic arithmetic and algebra*

Texts in these branches were prepared to teach administrators, tax collectors, and similar functionaries.

1. In *arithmetic*, various systems of numerals were used for positive integers. They took the 'astronomers'' numbers mainly from Ptolemy; letters of their own alphabet were used, and fractions were expressed Babylonian-style (§2.4) in powers of 1/60. Also used were versions of the Indian numerals; one of these versions we use today, calling it the 'Hindu-Arabic' system. The Arabic role lay mainly in popularization, which was initially effected early in the 9th

[1] In view of the mass of recent work on Arabic mathematics, a new general history is sorely needed for the benefit of non-specialists. In the meantime, good service is provided by many articles in the *Dictionary of scientific biography*, and papers appear regularly in *Historia mathematica*, *Arabic sciences and philosophy*, and elsewhere. Enormous quantities of manuscript material remain to be studied; they are found not only in the Middle East but also in Pakistan, India and Europe.

century by al-Khwarizmi, one of the earliest members of the House of Wisdom. Their symbols included 'ʘ', called 'sifr'; after Latinization this word was to bequeath to English both 'cipher' and 'zero'. Eventually a decimal expansion scheme was adjoined.

2. A common *method of reckoning* with integers used 'dust numbers', in which numerals were written in arrays in sand or dust and rubbed out when transferred to other positions. For an example see FIG. 3.14.1.

3. Some Arabs became ready to work with *rational and irrational numbers,* though not with complex ones and seemingly not with negatives either. Ratios of integers were considered in connection with the problem of determining musical intervals; al-Khayyam was one such author in the 11th century, drawing on Greek sources (§2.11) as well as on Arabic predecessors.

4. Some figures explored *number theory*. Possibly building on Iamblichos, Thabit ibn Qurra found a clever rule for determining 'amicable numbers', that is, pairs of integers in which each is the sum of the proper divisors of the other; his successor of the 14th century, al-Farisi, found the pair 17,296 and 18,416. Euclid's theorem {*Elements,* 9.pr.36} that the sum of a geometric progression of integers was perfect (that is, self-amicable) was extended by Ibn al-Haytham. Among other topics, attention was paid to magic squares of order larger than the simple 3×3 case shown in FIG. 3.3.1(a); spiritual considerations seem to have inspired the interest.

5. *Solving polynomials* became a speciality after its introduction by al-Khwarizmi in a book on equations written early in the 9th century. Avoiding negative numbers, he (re-)wrote a quadratic equation such that all its coefficients were positive; for example (one of his, but using our symbols with indices),

$$4x^2 = 40x - x^2 \quad \text{was 'restored' to } 5x^2 = 40x. \qquad (3.6.1)$$

His name for this operation was *al-jabr,* a word used in the title of his book, which medieval Europeans were to convert into 'algebra'. When necessary (not in this example) he found the roots by completing the square using algebraic analogues of the geometrical methods of the Babylonians and Greeks (FIG. 2.5.2). Later figures such as al-Quhi, al-Khayyam and Sharaf al-Tusi made some progress in solving cubic and higher-degree equations, or at least approximating to roots. When Diophantos's work became known in the late 9th century, indeterminate analysis (as in eqn (3.1.1)) was taken up.

6. Ignorant of Diophantos and thus of his symbols (2.26.2), al-Khwarizmi introduced a *word-system* to indicate powers of the unknown; it was normally adopted by his successors, and was to be taken up in Europe (§3.15). He called the first power 'thing' (*shay*) or 'root' (*jizr*), followed by 'power' (actually *mal*, a certain amount of money) for the square, and then 'cube' (*ka'b*). Higher powers were denoted by compound words such as 'power–cube'; as with Diophantos, the three dimensions of space prevented the introduction of further new words.

7. Al-Khwarizmi's extensive survey of methods in this and other contexts gave his work a strongly *computational* and practical flavour – or, to use the term introduced by medieval European mathematics in his honour, 'algorismic' (§3.14) and later 'algorithmic'. This aspect was extended by many successors, especially in finding a wide variety of approximations to the roots of an equation, and related results such as values close to π.

8. This activity extended to aspects of *algebra as such* (hence the distinction between 'eqs' and 'alg' in TABLE 3.5.1). Consideration of $(a+b)^n$ (in our notation) led al-Karaji, al-Samawal and others around the 11th century to the triangle (3.2.1) of coefficients, and to an analogue of changing the unknown in an algebraic expression. Certain finite series, such as $\sum_r r$ and $\sum_r r^2$, were summed by methods including a rudimentary form of mathematical induction; a few of them arose in connection with evaluating areas. Some of the Greek inheritance was algebraized; in particular, Euclid's theory of incommensurable

magnitudes (§2.13) became one of irrational roots in al-Karaji's hands, while al-Kashi (a major figure in this subject) rendered the classical problem of trisecting the angle (§2.28) in terms of approximating to a root of a cubic equation. Sharaf al-Tusi was to study that topic in general, after the manner of al-Khwarizmi on quadratics. The reading of Euclid as a hidden algebraist (§2.19) also began to be proposed.

3.7 *On Arabic geometry and trigonometry*

These branches of mathematics developed and interacted powerfully.

1. In *theoretical geometry,* Euclid's parallel postulate (§2.16) came under scrutiny, and various attempts were made to prove it or replace it by more self-evident assumptions {Jaouiche 1986}. For example, in the 11th century al-Khayyam tried to obtain a contradiction from the assumptions that the sum of the angles of a planar quadrilateral was less, or greater, than four right angles. Two centuries later, Nasir al-Tusi also tried; though unsuccessful in his aim, his contribution helped give birth to non-Euclidean geometry in Europe, especially after it was translated into Latin (§5.26). Archimedes showed his influence in various respects, including the sandwiching of circles and other curves between inscribed and escribed polygons and proving theorems by double proof by contradiction (§2.21). A variety of geometrical solutions to the three classical problems were attempted {Knorr 1989, Part 2}.

2. Geometry played a central role in *optics,* which was still concerned mostly with properties of rays and the eye rather than with the nature of light. Building largely on Ptolemy, Ibn al-Haytham made important contributions in the early 11th century to the analysis of reflection and refraction, and to burning mirrors; this work was taken up relatively quickly in Europe after Latin translation two centuries later (§3.24).

3. Both geometry and optics were intimately liked with *trigonometry*; both parts became very popular, with comprehensive treatises coming from al-Biruni around 1000 and from Nasir al-Tusi in the 13th century. Chords were imported from Greece (especially Ptolemy), and sines from India. The planar theory was enriched by the other five functions, including cosine and secant, and the versed sine in the geometrical form (radius-cosine). In the spherical branch, results like Menelaos's THEOREM 2.24.1 on arcs of great circles of a sphere were taken up and extended; a major result was:

THEOREM 3.7.1 *In the spherical triangle HLM of* FIG. 3.7.1,

$$\sin h : \sin H :: \sin l : \sin L :: \sin m : \sin M \qquad (3.7.1)$$

('sin h' refers to the sine, construed as a line, of the angle subtended by arc h at the centre of the sphere).

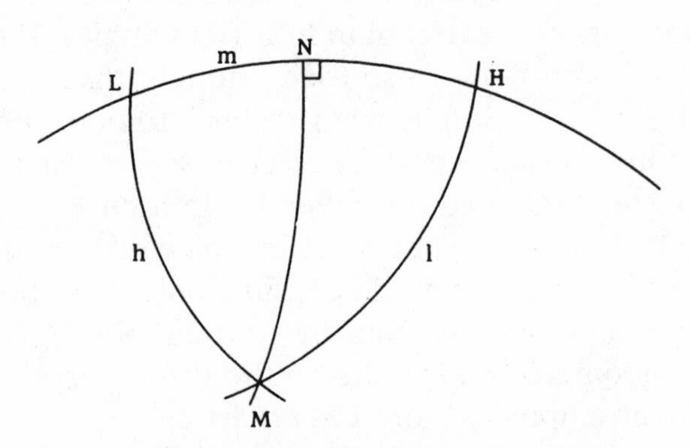

FIG. 3.7.1 *A sine theorem in spherical trigonometry.*

The proof was based upon drawing the perpendicular arc MN from M onto HL, which split HLM into two right-angled spherical triangles. One of its creators, Abu'l Wafa, applied trigonometry to astronomy, which we now consider.

3.8 *On Arabic astronomy and mechanics*

These branches of mathematics had social and religious roots.

1. Among the fine contributions by al-Biruni to trigonometry, outstanding was his solution of *the 'qibla' problem* {CE 1.7}. The task is to determine the direction in which the Muslim must face in order to face Mecca when praying at the decreed times; with M as Mecca, L as the location of the worshipper and N as another main point such as the North Pole, the *qibla* is ∠NLM. THEOREM 3.7.1 was essential to al-Biruni's construction; he devised a general method of solution using a succession of right-angled spherical triangles. Tables were prepared giving this angle for the latitude and longitude of many main locations; other tables provided accurate timings for prayer based upon values of solar parameters such as the altitude and azimuth angles, all measurable by astrolabe or quadrant {King 1973}.

2. This task was one of many cases where geometry and trigonometry interacted with *astronomy*, a science to which the Arabs contributed much theory, computation and instrumentation {North 1994, Chaps. 7–9}. An early start was made by al-Khwarizmi in the 9th century, with a *zij*, or handbook, largely following Ptolemy's theories and containing tables calculated usually by Indian methods. This mixture set the style for many of his successors. On the theoretical side, Arab astronomers adopted the notion of real shells within which the planets moved (§2.25). Thabit ibn Qurra made a significant modification when he decided that Ptolemy had been wrong to think that the equinoxes precessed at a constant rate, and that the ecliptic plane was fixed relative to the stars; adapting an idea due to Theon of Alexandria, he let the outer sphere oscillate and thereby allow both parameters to vary. His theory was called *trepidatio* ('agitation') when it was adopted in Europe. Among later work, in the 13th century Nasir al-Tusi achieved a good desimplification of epicycle theory by allowing a planet to move on a circle which was rotating inside a

larger one; thus he avoided Ptolemy's violation of unform motion. One later benefit was a lunar theory which improved upon Ptolemy's excessive variation in the apparent size of the Moon (§2.25).

3. *Astronomical tables* were constructed for various purposes; not only the calculation of the *qibla* but also, in particular, the prediction of planetary positions necessary for astrology. Many were compiled by the group at Samarkand under Ulugh Beg. Some methods of non-linear interpolation were used, drawing on basic formulae in spherical trigonometry; like a corollary of (3.7.1), they often involved the product of two sines.

4. Astronomy was a major part of *mechanics in general,* with which the Arabs were perhaps less concerned. Aristotle provided some of the main views on rest and motion, while Archimedes was an influence on the technical side, such as hydrostatic balances and the science and balance of weights. Two principal figures in this area were Thabit ibn Qurra in the 9th century and Ibn Sina in the 11th.

5. A rich area of application was *architecture and building.* An interesting example is the execution of *muqarnasat,* the intricate networks of niches on vaults and columns, which required careful use of geometry and trigonometry {Dold-Samplonius 1993}. Other examples were the calculation of the volume and surface area of a *qubba,* the shelled dome atop a mosque. When bounded by two hemispheres, the analysis needed only the appropriate formulae and a good value for π, with acknowledgement to Archimedes; but in the 15th century al-Kashi carried out a much more sophisticated analysis of the case when the *qubba* was bounded by surfaces or revolution, slicing it up by a sequence of horizontal planes and treating each one as a truncated cone.

6. *Design and decoration* also drew upon geometry and some trigonometry; but much covert mathematics must have been involved. The designs on the walls of mosques and palaces, and the designs on carpets, exhibit superimposed patterns based upon pentagrams, hexagrams, and so

on {Albarn and others 1974} – to the modern eye, implicit but complicated group theories of rotations and of (a)symmetries. Numerology is also evident; some mosques that I have seen exhibit beautiful plays of 7's against 8's (for example, the Dome of the Rock in Jersualem, built in the 7th century). The special artistic Arabic script written on the walls of mosques displays various geometrical properties. In the 10th century the scholar and calligrapher Ibn Muqla initiated a design of Arabic lettering that followed rules of geometric proportion.

3.9 *The Hebrew tradition*

{CE 1.14}

Again I use the word 'Hebrew' to identify a culture by its language. Its members were apparently always Jewish, originally living in or near the region of modern Israel but dispersing over the centuries across the Persian, Turkish and Christian Empires, and intermingling there, sometimes uneasily. Influenced by Arabic mathematics, and relatively rapidly, its mathematics was developing during the 8th century. The first encyclopedia was produced by Abraham bar Hiyya ha-Nasi (?–1136), who worked in Spain; he covered arithmetic, geometry, optics and music. He included the following delightful way of determining the area of a circle, in a version like (2.9.5): fill it with smaller concentric ones (FIG. 3.9.1 shows seven, and the centre), cut them along a radius, and roll them out straight to form an isosceles

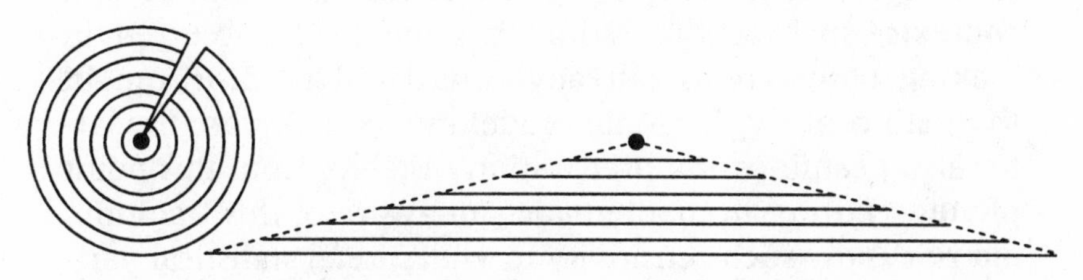

FIG. 3.9.1 *Abraham bar Hiyya ha-Nasi's determination of the area of a circle* (drawn to scale).

triangle; alternatively, hold them at the radius and roll out to give a right-angled triangle.

Students of Greek mathematics included Rabbi Moses Maimonides (1138?–1204). While more philosopher than scientist, he drew upon both Apollonios and ibn al-Haytham in his study of conics. The French-born mathematician Levi ben Gerson (1288–1344) took his interest in arithmetic and algebra as far as proving theorems on combinations and permutations by mathematical induction, which he understood in a remarkably general way (compare item 5 of §3.6).

Influence could flow *from* Hebrew work, as when al-Khwarizmi wrote a book on the Hebrew calendar. Ben Gerson is best remembered for his 'Jacob staff', a bar with cross-piece for measuring the angular separation of stars, which was widely adopted. Latin translations were made of writings of all the figures named above. In particular, bar Hiyya's encyclopedia was soon rendered into Latin, maybe in the 1140s, by his compatriot Plato of Tivoli, with a consequence noted in §3.15; bar Hiyya became known in Europe under the Latin form 'Savasorda' of his name.

In two areas, Hebrew mathematics is quite distinctive. Firstly, to a degree unique among ancient and medieval cultures, it was sensitive to the widespread occurrence of *variations in values*. From at least the −1st century, Hebrew texts give many examples, such as preparing building-blocks in construction, determining the boundaries of territory, and measuring them by means of ropes. This interest led them to pioneer aspects of probabilistic and statistical thinking {Rabinovitch 1973}. Astronomy provided several contexts, such as the errors that inevitably occurred in making observations (already considered by al-Biruni in the 11th century) for Maimonides and ben Gerson; the latter also examined the preparation of tables from this point of view. European mathematics, unaware of this tradition, did not show such sensitivity to widespread statistical variation until early in the 20th century (§16.22).

Secondly, more intensely than in any other culture,

Hebrew literature laid emphasis on the alphabet of its language, and drew upon mathematical features such as *numerology*. The Talmud (4th–6th centuries) taught that the 22 letters of the Hebrew alphabet, alternately designated male and female, represented the separate spiritual essences emanating from God. The first two letters, *aleph bet*, meant 'learn wisdom'; the 21 letters after *aleph* fell into three septets denoting the mysteries of grace, mercy and strict justice. Concerning *gematria*, the doctrine of assigning integers to letters and determining the integer for a word as the sum of those corresponding to its letters, theirs is the most ancient system known, dating from at least the –8th century; a variety of forms was developed until the Middle Ages {Judaica 1971}. And both numerology and gematria are very evident in a Hebrew story whose chronicling, in Chaldean, Hebrew, Greek and then Latin, came to have unsurpassed influence on the thought of European Christians.

3.10 *The threes and twos of Christianity*

When the Holy Bible was established in the late 4th and early 5th centuries in a series of synods in Rome, Hippo Regius and Carthage in North Africa, and Constantinople, the New Testament recounted the 'Orthodox' account of this story. Based upon 3's, it told of Jesus living for 3×10 years in obscurity before prosecuting a 3-year ministry and advocating the Trinity of God, Father and Holy Ghost prior to dying in a 3-some on a cross but coming back to life 3 days later and showing himself 3 times to his 3×4 disciples. The New Testament contained 27 books, doubtless deliberately: 3^3, 3 to the cube, the faith of the Trinity proclaiming the measure of God across the three dimensions of space, 'as high as heaven' and 'deeper than hell', 'longer than the earth, and broader than the sea' {Job 11: 8–9}.

Numbers and arithmetic are evident in many places in both Testaments. A notorious case, undoubtedly gematriac, is 666 as the Number of the Beast {Revelations 13: 18}.

Many interpretations have been given; a very convincing recent one takes it to be a criticism of the closed nature of certain religious orders, arithmetically as the sum of the numbers of letters associated with the stages of initiation: 60 (*Samekh*, the grade of full initiation as a priest) + 200 (*Resh*, his entrance into the sanctuary) + 400 (*Taw*, final graduation) + 6 (*Waw*, a letter appended to the other three to make a word) {Thiering 1993, p. 514}. The number of Psalms is not 150 just to have a nice large collection – and moreover, the German practice of counting the incipit (the opening dedication) as the first verse makes much more numerological sense of the resulting numbers of verses than does the Anglo-Saxon custom of leaving it unnumbered. 40 is prominent, for the days that Moses spent alone in the desert, for the days between Christmas Day and Candlemas (and also for the period of Ramadan in Islam); the arithmetical connections probably lie not only in its factors but also in these relationships:

$$40 = 37 + 3 \text{ and } 37 \times 3 = 111, \text{ the Trinity number.} \quad (3.10.1)$$

A very striking case from the New Testament is when Jesus appears before the Disciples after the Resurrection, and Peter catches 153 fish {John 21: 11}. The catching of fish is highly symbolic on its own; and 153 is the triangular number of 17, which was the number of Jewish sects, orders and societies at that time. This arithmetical property was noted by a North African, Saint Augustine of Hippo, one of the Founding Fathers of the Church in the 4th century. His advocacy of numerical relationships became influential; for example, 12 ($= 3 \times 4$) as the second 7 ($= 3 + 4$).

Some other branches of mathematics are evident in the Bible, not always happily. For example, Solomon's house seems to have been partly equipped on the assumption that $\pi = 3$ for the circumference of the circle {I Kings 7: 23}.

With the establishment of the Bible, the rival 'apocryphal' version (as it became known) of the life of Jesus went into apparent oblivion. It asserted that Mary Magdalene was the wife of rabbi Jesus, who survived the crucifixion; so 2-ness is prominent, together with 11 derived from the digit string 1–1. This difference in integers arose from time to time; in particular, some of the schisms in the Middle Ages between the Russian Orthodox and Roman Catholic Churches involved the doctrinal issue of giving the blessing with three fingers (Kiev) or with two (Rome).

Another type of difference between the two Christianities was semiotic. The orthodox dagger-like cross '†' shows the dominance of the vertical male over the horizontal female (compare §2.8 on this matter); but the Apocryphists' equal-armed cross '+' represents sexual equality, and the lower horizontal bar of the Cross of Lorraine, '‡', symbolizes their claim that Jesus did not die on the cross.

Some texts outside the Bible remained known and even influential. A mathematical example lies in the phrase 'in measure, number and weight', used routinely by Christian authors writing on mathematics, which comes not from the Bible but from a major pre-Christian Hebrew text known as the 'Book of Wisdom' {11: 21}.

From the 3rd century onwards, orthodox Christianity, based on a Hebrew story and worshipping the Jew Jesus, also led many campaigns of anti-Semitism. The 'Wandering Jews' (a Christian phrase of the Middle Ages) were not to feature significantly in European mathematics until the beginning of the 19th century (§7.4). Let us turn now to its performance up to the 12th century.

3.11 *European mathematics in the early Middle Ages?*
{Butzer and Lohrmann 1993}

During these centuries of activity in the Near and Far East, deep intellectual slumber seems to have fallen over Europe. Christianity was the dominant religion, and Latin the common cultural language, the Bible having been rendered into Latin from the Greek. Europe was divided into various kingdoms, many of which were interested in acquiring one another. Much of western Europe belonged to the Later

Roman or Byzantine Empire, whose capital had been moved during the 4th century from Rome to Constantinople, the city built on the shore of the Bosphorus as an extension of the ancient city of Byzantium. After its fall in 1453 to the Osmanli Turks, most of the Eastern part of Europe belonged to the Ottoman Empire.

The Greek classics were disseminated in Europe and commented upon in Latin by various figures, especially the pagan Martianus Capella based in Carthage across the 4th and 5th centuries, and the Roman Christians Anicius Boethius and Flavius Cassiodorus during the century following. They all advocated the seven liberal arts in the training of monks, comprising the trivium of *grammatica, rhetorica* and *logica,* followed by the highly mathematical quadrivium of *arithmetica, geometria, harmonia* and *astronomia.*

Boethius, who may have coined the word 'quadrivium', wrote influential texts on *Arithmetica* and *Musica,* both strongly following Nicomachos (§2.26). When translating Euclid and other Greek geometers, he rendered their word *semeion* for 'sign' (§2.17) in good Roman down-to-earth style as *punctum;* since then all languages have used their equivalent of 'point'. He followed Plato's enthusiasm for mathematics, and his interests included the Pythagorean interpretation of numbers as essences, triangular numbers, and even the theorem that the nth triangular number was the number of different ways of picking two objects out of a collection of n. He also advocated and applied Aristotle's syllogistic logic with fervour; yet, like the Greeks themselves, he did not link logic with mathematics in any intimate way.

Astronomy was granted quite a high status, especially when it bore upon Christian concerns such as the calendar. This topic came under *computus* – the art and science of calculating the time. The determination of Easter was of special significance; for example, it formed part of the 'victory' secured in 664 by the Roman over the Celtic Churches at the Synod of Whitby in Yorkshire, England. Soon afterwards the English philosopher the Venerable Bede (673–735) was to be a major figure in calendar theory; he became known as 'Beda Computata'. He proposed another method for determining Easter which was adopted on the Continent, and his widely read 'Ecclesiastical History of the English Nation' (to translate its title from his Latin) popularized the system of numbering years before and after the supposed birth of Christ that we still use. He also publicized the following system of counting integers on the fingers up to 10,000:

1–9	left hand, second to fourth fingers erect, bent at the middle joint, or bent flat down;
10–90	left hand, thumb and index fingers in various combinations of position;
100–900	right hand, second to fourth fingers in the above positions;
1,000–9,000	right hand, thumb and index fingers in the same combinations of position.

This system naturally worked to base 10, which it will have helped establish, at least in England. In the 9th century King Alfred, who commissioned and maybe effected a translation of Bede into Anglo-Saxon, divided his country into 'hundreds' (probably of land occupied by 100 families) and each hundred by tens into 'tithings'. A few place-names still retain these names.

During the late 8th century the Frankish Kingdom was formed, and under Charlemagne it became known also as 'Carolingian'; he himself aspired for it to become 'Roman'. It stretched over most of Western Europe, including Italy down to Rome (but excluding the Iberian Peninsula, mostly still under Arab control). Anxious to encourage learning in general, Charlemagne appointed, as a sort of Minister of Education, Bede's follower Alcuin of York (735?–804); he advocated teaching of the seven liberal arts, and is credited as the author of a collection of arithmetical and geometrical 'Propositions for Sharpening the Minds of Youth'. Astronomy was pursued, drawing a little on Arabic and Indian sources; the work included calculating the distances and dimensions of Sun, Earth and Moon.

After Charlemagne's death, his kingdom was divided among three sons in the 810s, whereafter its intellectual status declined. Late in the century Gerbert of Aurillac, the first Frenchman to be elected Pope (Sylvester II, from 999 until his death five years later), raised its educational aspirations by teaching from Greek, Roman (Boethius and Martianus Capella) and Arabic sources, and using the counting-board for arithmetical operations. He also corresponded extensively, often on mathematical and scientific topics, advocating the quadrivium.

As usual, there must have been much covert mathematics which has left little or no record. But clearly the level and quality of activity in Europe was very modest in comparison with that in the Middle or Far East, and the most influential figures were not primarily mathematicians or scientists at all. In Greece, part of the Byzantine Empire and the object of periodic invasion by Slav or Balkan tribes, mathematics seems to have been moribund, although during the first half of the 9th century Leo the Mathematician at Thessaloniki tried to regenerate interest in Greek classic texts. But they were to flower again in the intellectual soil of Europe from the early 12th century, when the level of learning there began at last to rise.

3.12 *The awakening of Europe from the 12th century onwards*

{Kaunzner 1987}

It is necessary now to be more specific over geography. 'Europe' will now refer to the British Isles and the Continent east from France and the Low Countries across to Poland, Austria and Hungary; those regions had various other adhesions in the Middle Ages. It will also include the Iberian Peninsula, which was gradually reconquered from the Arabs from the mid-12th century; Portugal began establishing itself as a separate country from the late 13th century. But Scandinavia and Russia will be excluded, for mathematics was not yet cultivated there to a significant degree. Indeed, in Russia the Orthodox Church maintained its hostility to science in general, and in any case the region fell vassal to the Mongols during the 13th century.

The rise of mathematics was inspired largely by acquaintance with the Greek and Arabic classics. After neglect for seven centuries, copies were found, read, understood at least in part, and discussed or even translated into Latin. The history of these translations, commentaries and paraphrases is extremely long and complicated, running well into the 15th century. It is not easy to determine when works by a particular author became known; the case of Pappos will be noted in §3.24. Indeed, at the time the authenticity of text was not easily established; for example, some early Latin 'Euclids' actually contained interpolations by commentators (§3.23), and he himself was sometimes misidentified with the philosopher Euclid of Megara of the –5th and –4th centuries.

The first translations of scientific works, mainly of Arabic astronomy, seem to have been prepared in Germany in the first half of the 11th century by Hermann the Lame, followed in the 13th century by Albertus Magnus and Albert of Saxony. The most prolific translator of Greek mathematics was William of Moerbeke, and of Arabic, Gerard of Cremona, who worked mostly in Spain in the 12th century, as did the Englishman Robert of Chester, who also translated the Koran. Hermann of Corinthia was another important translator of Arabic. Copies of these translations, or transcriptions of the original texts, became available; their number is uncertain, as a considerable proportion seem now to be lost. The introduction of printing in the mid-15th century gradually increased the number and authority of the texts; the arrival of copper-plate engraving in the 1470s gave a similar boost to the production of maps and charts (which usually set east rather than north at the top, a medieval practice from which comes the term 'orientation').

Increasing trade and other contacts with the Middle East, by land and across the Mediterranean Sea (§3.14), made Europe aware of the Hindu-Arabic number system. It also aided the diffusion, and especially the translation, of Arabic mathematical texts – at first more than of Greek ones, for Arabic culture was active while the Greek was dead. However, Greek work gradually gained in popularity, perhaps because of the translation of material in other disciplines such as philosophy, and also through events such as the flight of scholars, especially to Italy, after the fall of Constantinople in 1453. Another reason was that, outside occupied Spain and Sicily, Arabic was little known or studied in Europe – hence translations were essential. In addition, European sympathy for Arabic science may have been discouraged by the mutual slaughter during the three centuries of Christian Crusading invasions of the 'Holy Lands', ended by the Mameluks in the 13th century.

Several significant Arabic achievements seem not to have been transmitted, including decimal expansions of numbers, roots of equations of degree higher than the quadratic, and mathematical induction in a fairly general form. Others may have been taken over; at all events, some 'interesting similarities' between European and Arabic authors can be discerned in various subjects. Even then, European mathematicians took a long time to match their Arabic predecessors, in some areas not before the 17th century. Al-Khwarizmi gained most attention, principally through translations of his book on equations made in Spain by Gerard of Cremona and Robert of Chester in the mid-12th century, and the impact of his ideas upon Fibonacci (§3.15). Other favoured writers include al-Kindi, Thabit ibn Qurra, Abu Kamil, Ibn al-Haytham, Jabir Ibn Aflah, and a few others regarded now as more minor. Some figures working mainly in astronomy (including astrology and horoscopes, as usual), optics and/or philosophy were relatively popular; for example and in the Latin names by which they became known, Albategnius, Avicenna (Ibn Sina),

Avempace (Ibn Bajji) and Averroës (Ibn Rushd) of the 11th and 12th centuries.

Among Greek mathematicians, Archimedes emerged as the most inspiring source of mathematical knowledge and maybe even research {Clagett 1964–1984; CE 2.1}; Euclid served both as main guide to geometry and arithmetic and as paragon of rigour; and Ptolemy became the dominant astronomer. Other figures well received included Apollonios, Hero, Nicomachos, Menelaos and Diophantos. (Europeans became aware of, say, 'Hipparchus' rather than 'Hipparchos'; I shall continue to use Greek transliterations.) Arabic (and some Hebrew) versions were the first sources for certain Greek texts; indeed, they remain the *only* sources for parts of Apollonios and Diophantos, the original texts having been lost.

Mathematics was also influenced by philosophy, principally through Aristotle. Many of his works were translated by 1200, and he was sometimes referred to simply as 'the philosopher' {Laughlin 1995}. His views were widely studied, especially on cause and effect, continuity and change, place and motion, and general rules of demonstration and argument. The links were closest in mechanics, as we shall see in §3.30. But even he faced some Christian opposition, especially for having advocated the pantheistic view that God was to be found in Nature.

Up to the late 15th century familiarity with Plato was more limited; then it gradually increased, partly though advocacy and translation in Italy by the Jesuits. This double sway led philosophical positions allied to Plato and Aristotle to emerge through to the Renaissance, although their relationship to the original figures is not always clear {Copenhaver and Schmitt 1992}. One important position was especially sympathetic to the explanatory power of mathematics in theorizing; its name, 'Neoplatonism', was introduced later.

Among other cultures, some Indian and Hebrew works came into Latin, Greek into Hebrew, and so on; Boethius himself, and also Bede, were re-edited. By contrast, the mathematics of the Far East seems to have remained largely unknown, although Chinese silk and other goods still came in from 'Cathay'.

Monasteries and their schools continued to provide an important base in Europe for scholarship, but the search for learning led to the creation of universities or colleges for men out of existing schools, closely aligned to the Christian Church (*universitas* was Latin for 'guild'). Mainland Italy led the way with a university at Bologna during the 11th century, and within 200 years nearly a dozen more had been established. By then some other countries had followed suit: the German-speaking lands with Vienna, Heidelberg and Cologne; Poland at Cracow; Portugal with Lisbon, and Spain with Valladolid and Salamanca; France at Paris and four provincial cities; and Britain at Oxford and then Cambridge. The seven liberal arts played an important role in policy, although the mathematical quadrivium proved harder to impart than the more humanistic trivium.

3.13 *Principal figures in Europe*

Table 3.13.1 lists the most important European mathematicians for the period covered by the rest of this chapter. Their work in other disciplines is not noted. The name of each figure is given in the form which is now most commonly used in English; at that time many carried a Latin name, and some are still known only by a Christian name and a town or city with which they were associated. Jordanus de Nemore is especially fugitive, for his (or her?) nationality is not known; in addition, the location of 'Nemore' has not been determined, and it may not be coincidence that *nemo* in Latin means 'nobody'.

The figures are grouped by the country of principal activity; this was not necessarily the country of birth. A few spent substantial periods in different countries, partly to see copies or translations of classic texts, and partly to teach. Higher education was then understood primarily as the mastery by the student of standard *texts* rather than the learning of *subjects*, and this uniformity led to a 'right of teaching anywhere' in all disciplines. Further, certain *established* translations were often used; so the contributions of some figures listed lay mainly in the preparation of such texts.

While there are a few teacher–pupil relationships, the only group as such is the 'Oxford Calculators', as they became known, active in the early 14th century and centred on Merton College. The principal mathematical figures were Bradwardine, Heytesbury and Swineshead (maybe up to three men of this name), with Richard of Wallingford not far away at St Albans Abbey. After their era, mathematics almost disappeared in Britain for nearly two centuries (§4.3–§4.4). Another member of this group was the logician and philosopher John of Dumbelton. Figures of this type are omitted from the Table, although their interests related to mathematics in some ways, usually via Aristotle and especially concerning geometrical magnitudes and mechanics. Predecessors in the 13th century include fellow Britons William of Ockham and John Duns Scotus, and the Catalan Ramón Lull.

A further complication arises from the frequent presence of 'Anonymous', whether as author or translator. The warning given in §2.1 about not interpreting ancient mathematics as research in the modern sense applies to the Middle Ages also, for the Europeans and probably for the other cultures already discussed. Often they saw themselves as contributing to the store of wisdom in God's House, and valued learning above originality – indeed, figures such as John of Gmunden (early 15th century) were chary of advancing beyond classical knowledge at all.

The four-century period surveyed here straddles the boundary between the Middle Ages and the Renaissance – a boundary whose location has been much discussed. Both the phrase 'Middle Ages' and the word 'Renaissance' were Italian inventions of the early 15th century, the former put forward in a derogatory sense to contrast with both the period of ancient glory and the current 're-birth'. But modern historical scholarship – including on mathematics – has steadily undermined the claim of a discontinuity of thought and activity across the supposed divide.[2] I use 'Renaissance' in referring only to some increase in performance and a greater sense of individual contribution and originality, especially in connection with Italy, the leading country for mathematics in the 15th and early 16th centuries.

From now on this book almost always tells a story about Europe. {E. Grant 1974} is a fine source book on its medieval science.

TABLE 3.13.1. *Principal European mathematicians from the 12th to the early 15th centuries.*

The grouping by country is that of main activity, not necessarily birth. The order within each country is chronological, as far as this can be determined. the signs '?', '+', '−' and '±' are used as in Table 3.9.1. 'E' indicates that a reliable modern edition or translation of many of the works of a figure has been published.

ari	arithmetic, number theory	opt	optics
ast	astronomy, cosmology	phi	philosophy, proof-methods, logic
cal	calendars		
eqs	equations, proto-algebra	ser	summation of series
geo	geometry, perspective	trg	trigonometry
mec	mechanics	trs	translations, editions
mus	music, harmony		

Britain

Adelard of Bath (1116±–1142±) ast trs
Grosseteste, R. (1168?–1253) ast cal opt phi
Bacon, R. (1219?–1292) geo mec opt phi E
Bradwardine, T. (1290?–1349) ari geo ser trg
Richard of Wallingford (1292?–1336) ast trg E
Heytesbury, W. (1330±) mec phi
Swin(e)shead, R. (1350±) ari geo mec

[2] On the supposed division between the Middle Ages and the Renaissance I owe a great deal to discussions with the late Charles Schmitt. Menso Folkerts (Munich) is preparing an index of European manuscripts in medieval mathematics {see his 1981}.

France

Hugh of Saint-Victor (1096?–1141) ari ast geo
Sacrobosco (= John of Holywood) (1190±–1256) ari ast
John of Murs (1290?–1360?) ari geo mus
Buridan. J. (1300?–1358?) ast mec phi
Oresme, N. (1323?–1382) ari ast geo ser trs E
Chuquet, N. (1445?–1488) ari eqs ser E
Fine, O. (1494–1555) ari ast geo trs

Germany and Austria

Albertus Magnus (1193?–1280) mec opt phi E
Albert of Saxony (1316?–1390) mec phi
John of Gmunden (1380?–1442) ari ast cal trg
Nicholas of Cusa (1401?–1464) geo mec phi E
Peurbach, G. (1423–1461) ari ast cal
Dürer, A. (1471–1528) geo E
Ries, A. (1492?–1554) ari eqs E

Poland

Witelo (1230?–1275+) geo opt

Italy

Leonardo of Pisa (= Fibonacci) (1170?–1250?) ari geo ser E
Campanus de Novara (1210–1296) ast trs
William of Moerbeke (1220?–1286–) geo trs
Alberti, L.B. (1404–1472) geo mec mus opt
Piero della Francesca (1420?–1492?) geo trg
Regiomontanus (= Müller, J.) (1436–1476) ast cal trg E
Pacioli, L. (1445?–1517) ari geo trs
Leonardo da Vinci (1452–1519) geo mec opt E
Maurolico, F. (1494–1575) ast geo opt trs

Spain

Gerard of Cremona (1114–1187) ast trs

Nationality unknown

Jordanus de Nemore (1220±) ari eqs geo mec E

3.14 *Practical medieval calculations: 'algorists' versus 'abbacists'* {D.E. Smith 1908; Barnard 1917; CE 2.4}

Arithmetic gradually grew away from traditional geometrical connotations. Boethius's essay on it, which was widely used in Church schools throughout the Middle Ages, played a role, for it was grounded upon Nicomachos's rather abstract approach (§2.26). But the main motive was the increase in trade, which encouraged the development or revival of practical and efficient methods of calculating with numbers, not necessarily integers. Sufficient work was available to furnish a living for these arithmeticians, as assistants to merchants and landowners and also as teachers. The availability of printing during the late 15th century gave some of them greater chances to make money as authors; their fondness for commercial examples may have carried several meanings!

Several methods of calculation were embodied in a game called 'rithmomachia'. A chess-like checkerboard was used, with each player holding 'counters' (an interesting word) with various integral values. Each player attempted to capture his opponent's pieces by a most complicated collection of rules involving the arithmetic, geometric and harmonic means of the values of his own counters. Of uncertain origin (Boethius is sometimes credited), the game became popular at least in Britain, possibly as a teaching aid {Evans 1976}.

Among methods of finding unknown values, the 'rule of three' became popular, and its less well remembered companion rule of five. In later algebraic terms, these rules correspond respectively to solving one proportion for an unknown x, and to two for x in y, as in

$$7 : 9 :: 5 : x, \text{ and in } x : 5 :: 124 : 4 \text{ and } x : y :: 6 : 4. \qquad (3.14.1)$$

Until the 14th century, various types of method of arithmetical calculation were in use, and indeed in competition. The Arabic method of 'dust numbers' (item 2 of §3.6) was quite popular; normally it drew upon the Hindu-Arabic system of numerals (hereafter 'HAN'). But two other methods did not draw upon numerals, at least in the methods of manipulation: systems of finger-counting, like Bede's described in §3.11 though often less elaborate; and forms of a counting-board (FIG. 3.18.1 below) known under the general name 'abacus' but not necessarily the rows of beads constrained to move along straight wires (§2.29) to which we now usually restrict the word. Practitioners of all these methods were called 'algorists', in honour of al-Khwarizmi.

Then a gradual change began to take place, especially in Italy, to methods whose practitioners came to be called 'abacists'. The choice of name (§3.15) is very unfortunate, for it did not use any kind of abacus; I shall follow a modern practice by spelling the word 'abbacists'. Working normally in HAN, their main innovation was *to preserve all the workings in calculation*. Their numeral arithmetic challenged the algorists' position arithmetic.

	3 7 2		3 7 2
	4 6		4 6
	———		———
×300:	1 3 8|7 2		3 8 0 0
	4 6		
×70:	1 7 0 2|2		3 2 2 0
	4 6	×2:	9 2
	———		———
×2:	1 7 1 1 2		1 7 1 1 2

FIG. 3.14.1 *Algorist and abbacist methods of multiplication.*

FIG. 3.14.1 compares an algorist's procedure using dust numbers with the abbacist's method. The example is 372 × 46 – in that order, with the first number multiplied into the second. The dust method cannot be properly represented because numbers were erased or moved in location at each stage: I imitate the processes by moving '46' along, and my vertical bar marks off the digits to its left, which belong to the final answer, from those to the right which await multiplication into 46. Apart from these features, the Figure is quite faithful to history, including the placing of digits from right to left in order. The calculation could go up to large integers – even to a *millione*, the Italian for 'great thousand'.

The abbacists developed a range of attractive methods of multiplication; FIG. 3.14.2 shows a trio of them (left as exercises to the reader over a few glasses of wine). They came

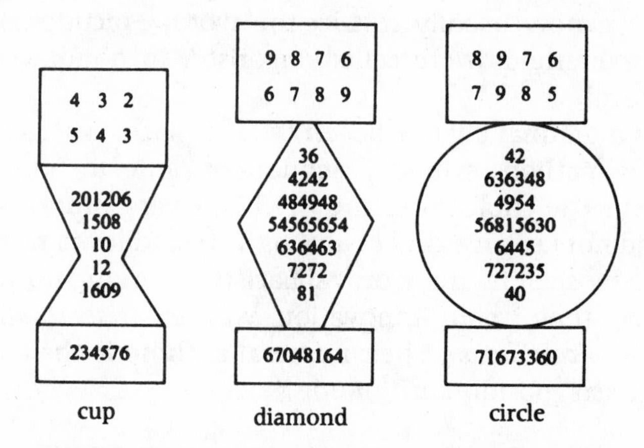

FIG. 3.14.2 *Three methods of abbacist multiplication* {CE, p. 205}.

later to reverse the order, multiplying the second number into the first one, as we struggle to do today when the battery in our calculator has run flat. The prize for fluency in calculation, however, surely still belongs to the Indian Vedic method (2.33.1) of centuries earlier.

3.15 *Fibonacci on arithmetic and equations*

The first important abbacist was Leonardo of Pisa, better known by his nickname 'Fibonacci' ('son of Bonacci'). Taken by his businessman father across the Mediterranean Sea to North Africa and the Middle East, he learnt mathematical as well as business methods and developed the former in a manuscript called *Liber abbaci* (thus giving rise to the name 'abbacist' noted above). Available for circulation in 1202, when he may have been in his early thirties, it was reworked by 1228. It is best remembered today for its introduction of Fibonacci numbers; but this detail was passed over for three centuries, and the topic is postponed until §17.20.

Many other parts of Fibonacci's book, however, gained a much readier public. Written in Latin much under the inspiration of Latin editions of al-Khwarizmi and of bar Hiyya, and other Arabic authors such as Ibn al-Haytham, it contained a detailed explanation of HAN and the methods of calculation, including multiplication. But its foundations often still lay in geometry: completing the square, for example, was justified by the appropriate diagram. However, Fibonacci also presented a cubic equation whose roots could not be determined geometrically by straightedge and compass alone, and thereby exposed an important limitation of Greek procedures.

Fibonacci also included more theoretical topics, such as rules, of both Arabic and Hebrew origin, for determining whether an integer is divisible by a smaller one (7, say, or 11). He presented the rule of three (3.14.1) for solving proportions, and many problems on solving systems of linear equations, based on Diophantos and al-Khwarizmi. The motivating problems were often notional, such as the calculating the length of time to eat a sheep needed by a lion, a leopard and a bear dining together in civilized disregard of each other's rates of munching.

Several of Fibonacci's problems related to trade and exchange, again sometimes notionally. For example, seven men wished to buy a horse by the curious procedure that seven subgroups of three begged the other four to contribute certain given proportions of their own funds; the task is to determine the sizes of all seven funds. So the problem may be rendered as seven equations, one of which is

$$x_4 + x_5 + x_6 + \tfrac{1}{5}(x_7 + x_1 + x_2 + x_3) = y, \qquad (3.15.1)$$

the price demanded. However, this formulation is anachronistic precisely because algebra *was born out of* such exercises; Fibonacci wrote words and numbers and saw himself as working within abbacist arithmetic. Note that the coefficients were not all different, thus rendering solution easier; further, y was not given, so that one degree of freedom was available (many other problems used only specific integers). All unknowns occurred in each of the equations here; but some problems were not so formulated, itself a small innovation. Fibonacci admitted negative integers in the solutions (as debts, for example), also somewhat novel {Sesiano 1985}.

3.16 *Stages towards the algebraization of arithmetic*

Fibonacci continued in Latin the Arabic practice of expressing an unknown quantity and its powers in words (item 6 of §3.6). He used *res* ('thing') for the unknown, *census* ('register') for its square, and *cubus*; he also concatenated these words to name higher powers (*census census*, and so on, though not far). In a later work called *Liber quadratorum* (1225) he tackled number theory, solving some Diophantine equations (3.1.1) of the second degree, and he also summed a few finite series.

Some algorists wrote at this level. For example, in manuscripts *De numeris datis* ('On Given Numbers') and an 'Arithmetic' of the early 13th century, Jordanus de Nemore treated properties such as factors and divisibility of integers as well as the rule of three. He also called the positive integers 'natural numbers', not from any manifestations in nature but as the results of successive repetitions from 1. But gradually the abbacists' approach to arithmetic prevailed; a notable example was John of Murs in 14th-century France, who developed a system of numerals to base 60. Manuscripts on arithmetic and its operations by abbacist methods began to be circulate after Fibonacci, especially in Italy and Germany; then they became printed books, the first appearing in 1478 at Treviso in the Venetian Republic.

Soon afterwards, Luca Pacioli published his large *Summa de arithmetica* (1494), including general remarks on numbers, especially integers, before turning to double-entry book-keeping and other topics. Little in the book was original, several sections coming from Fibonacci, and some of those arriving via an abbacist text by Pacioli's master, Piero della Francesca; but his was a *printed* work, and so gained considerable influence from this novel mode of diffusion.

Pacioli's and some other works of the time also included forays in the direction of algebra, a new branch of mathematics which is our next subject. As we saw with eqn (3.6.1), the word 'algebra' was created in honour of the restoration technique of al-Khwarizmi; the subject convoluted out of arithmetic and into an independent branch of mathematics in three stages:

1. using words to designate both unknown quantities and operations such as 'square root of', but still relying much on geometry for proofs and demonstrations of laws and properties;
2. replacing or supplementing words by abbreviations and single letters or symbols such as '*x*' and '√', and reducing the dependence upon geometry; and
3. allowing letters to denote *variables* as well as unknowns, and also accepting negative numbers (maybe even complex ones); and systematically using an index notation for powers, thereby re-admitting arithmetic in the rules of their combination ($x^m/x^n = x^{m-n}$, and so on).

The last two stages came largely in the early and late 16th century respectively, as we shall see in §4.6–§4.7 and §4.18. The first stage is treated now, in two rather distinct traditions in Italy and Germany – another example of national differences in the development of mathematics, especially with texts produced in their own languages.

3.17 *Italian origins of algebra: towards the* cosa

{Franci and Toti Rigatelli 1985; CE 2.3–2.4}

The Italian line is well exemplified by Pacioli: think in general about arithmetic, and try to solve equations. Many of the problems took the form of one or two linear equations in unknowns, often with a commercial character (nominally, anyway): to calculate the unit cost of a given quantity of figs, say, or the weight of the box which held them. For this latter type of unknown special words were used, especially *fusti* ('stem of a plant') and *tara* (from the Arabic word *tarha*, 'what is thrown away'); sometimes the quantity constituted an unknown in a problem, otherwise its value was given. Pacioli also used, but probably did not invent, the symbols '$\tilde{p}$' (abbreviating *piu*) for addition, '$\tilde{m}$' (*meno*) for subtraction and '℞' for taking the positive square root; higher roots were symbolized by appropriate strings such as '℞℞.*cuba*' for $\sqrt[7]{}$.

This approach to algebra, now in print thanks to Pacioli, had been under way for some time, since Fibonacci. It continued in manuscripts written in Italian which seemed to have had some circulation in Italy at the time but which only now are being published in quantity.[3] A typical (and typically anonymous) example from the late 14th century, recently edited as {Simi 1994}, rehearses the two parts of '*alcibra*'. Firstly, state the 'rules of sign' of the unknown 'thing' (*cosa*) and its powers – *ciensi, chubo, ciensi di ciensi* . . ., in names translated from Fibonacci's Latin. Secondly, present the formulae for the roots of some simple equations. The only symbol was 'R' for 'radice' ('root'), introduced by Fibonacci; otherwise the text was composed entirely of words or HAN. I use modern symbols for brevity, with the usual warning about anachronism.

Taking as known the four arithmetical operations, this author stated laws for handling square roots, from simple ones such as $\sqrt{m}\sqrt{n} = \sqrt{(mn)}$ to expansions such as

$$(n + \sqrt{m})(n - \sqrt{m}) = n^2 - m \qquad (3.17.1)$$

[3] Laura Toti Rigatelli and Raffaella Franci (Siena) direct an extensive project to publish manuscripts of texts on early Italian algebra (including {Simi 1994}), and also on the practical geometry described in §3.22.

and even rational expressions like

$$\frac{n+\sqrt{m}}{p+\sqrt{q}}=\frac{(n+\sqrt{m})(p-\sqrt{q})}{p^2-q}. \qquad (3.17.2)$$

The formulae for roots covered the linear and quadratic equations, but the equations were classified after al-Khwarizmi as at (3.6.1), so that all coefficients, quantities under the square root and the roots themselves were positive; for example,

$$ax^2+bx=c \text{ had only } \sqrt{\left(\frac{b}{2a}\right)^2+\frac{c}{a}}-\frac{b}{2a} \qquad (3.17.3)$$

as root. Other equations included

$$ax^4=bx^2,\ ax^3=b \text{ and } ax^3=bx^2+c. \qquad (3.17.4)$$

The verbal form of the first formula in (3.17.3) reads as follows: 'Quando i ciensi e lle cose sono iguali al numero' ('when the squares and the things are equal to the number') – that is, without explicit mention of the coefficients.

Unfortunately, this author thought that this formula for the quadratic solved a cubic also! By Pacioli's time such mistakes had usually been sorted out: it was recognized that solving the general cubic equation was a Big Problem. But the leap to the second stage of algebra, and the answer to this Problem, would await compatriots such as Girolamo Cardano and Raffaello Bombelli (§4.6).

3.18 *German origins of algebra: the* Coss *tradition*

{Glaisher 1921; CE 2.3}

Starting early in the 15th century, the Germans took over the Italian word *cosa* for the unknown, in the form *Coss*; its practitioners became known as *Cossisten*. Their aim was basically the same: to explain the laws of arithmetic in some generality, solve equations, use methods such as the rule of three when appropriate, and so on. They were also called 'calculating-masters', who produced 'calculating books' –

Rechenbücher – hence the English word 'reckoning'. They built on both Greek and Arabic traditions read in Latin editions, although one of their principal figures, Adam Ries, thought that the Arabs had lived first!

The Cossists expressed equations largely in verbal form; for example, in 1481 the anonymous author of a manuscript entitled *Deutsche Algebra* expressed the first formula in (3.17.3) as 'Eyn dingk vnd zall gleich eynem czensz' ('a thing and number equals a square'), again without explicitly mentioning the coefficients. Some authors specified only the relationships between degrees of the *Coss*; for Christoph Rudolff in 1525, 'Wann zwo quantitetn natürlicher ordnung einander gleych werden' ('when two quantities in natural order are equal to each other') covered, say, $13x^5 = 4x^4$ as well as $2x = 6$. They also preferred the abbacist way with arithmetic; a popular textbook of 1504 showed their triumph over the algorists (FIG. 3.18.1).

But the Cossists differed from the Italian abbacists in some ways. In their explanations of Roman numerals and counting-boards, they were associated to some extent with the algorists; indeed, some books, such as popular ones by Sacrobosco and later by Peurbach, confusingly carried the word *algorismus* in their titles. Often they were professionally linked to writing, making some of their living also by teaching calligraphy. Perhaps for this reason they were rather more ready than the abbacists to use symbols for powers: 'Φ' for a number, a special sign resembling 'r' (from *res*) for the *Coss*, '3' for its *czensi*, 'T' for *chubi*, '33' for the fourth power, and so on, up to maybe the 13th power. Sometimes they put powers into commercial problems: Rudolff had a shopper buy $\sqrt{6}$ lengths of cloth for $\sqrt[4]{15}$ guilders, which suggests a rather notional transaction.

For arithmetical operations, the Cossists often used words such as 'plus, 'minus', 'mer', and 'minner', but the symbols '+' and '−' appeared in some manuscripts. Their origins are not at all clear, but '+' might have been a scribal abbreviation for *et* in Latin, while '−' may well have carried the connotation of negativism in general: for example,

FIG. 3.18.1 *Happy abbacist, sad algorist (as in Greek drama?): propaganda in Georg Reisch's* Margarita philosophica, 1504.

The algorist has these numbers on his board: on our left, 2 + 30 + 50 = 82; on our right, 1 + 40 + 200 + 1,000 = 1,241.

some medieval logicians used it to denote 'not'.

These two signs were first published in 1489, in a Cossist work entitled *Behēde vnd hubsche Rechnung auff allen kauffmanschaft* ('Smart and Pretty Reckoning for All the Merchantry') by the Leipzig professor Johannes Widman.

Strikingly, he applied them to different kinds of quantity: for example, '6 eggs − 2 pence for 4 pence and 1 egg' ('6 Eyer − 2∂ pro 4∂ + 1 ey') in a problem seeking the unit price of eggs. The signs were given other functions; in the rule of false (2.31.4), '400 + 22' denoted not 422 but 'guess 400, excess 22' {Glaisher 1921, pp. 19–22}.

The use of symbols was to increase in the 1540s among successors such as Michael Stifel (§4.5). Again, Adam Ries had calculated the value of an unknown by the rule (3.14.1) of three; but his Cossist son Abraham reworked the problem in terms of an algebraic equation.

3.19 *Chuquet on roots and exponents*
{Flegg and others 1984}

Another noteworthy figure worked in France (which, in contrast with its later renown, was at this time a rather minor country for mathematics): Nicolas Chuquet, author of a manuscript entitled *Triparty, ou la science des nombres* and completed in 1484, when he may have been in his late thirties. His follower Etienne de la Roche copied substantial parts of it into his book *Larismethique nouvellement composée* (1520), and it seemed to gain some of its deserved influence this way.

Chuquet's three parts dealt in turn with arithmetic, 'roots of all kinds' and some basic algebra which he called 'rules of primes'; the latter word was his equivalent to *cosa*. In the first part, he took a remarkable standpoint for the time: integers, negatives, rationals, irrationals were all on a par as numbers; any of them could be taken as an 'exponent' of an integer; and the sequence of positive integers as powers corresponds in this interesting way to the integers themselves:

$$\left.\begin{matrix} 0 & 1 & 2 & 3 & 4 & \dots & 14 & 15 & \dots \\ 1 & 2 & 4 & 8 & 16 & \dots & 16{,}384 & 32{,}768 & \dots \end{matrix}\right\} \quad (3.19.1)$$

He seems not to have picked up the whiff of logarithms,

which were to develop in the early 17th century (§4.15). But he did form a notational system for non-positive exponents: '7°' means $7x^0$ to us, and '$\bar{m}7^{1\bar{m}}$' is our $-7/x$.

Chuquet made an original contribution to approximating to the roots of an equation by simply placing between two rational-number guesses the number given by the sum of its numerators divided by the sum of its denominators, and repeating the process as often as desired: having guessed, say, $\frac{3}{7}$ and $\frac{9}{14}$, try $\frac{12}{21}$, then try $\frac{15}{28}$, $\frac{21}{35}$, ... Following Fibonacci, he also proposed a method for calculating the cube root of a number.

In the second part Chuquet used, and maybe invented, the symbols '$\bar{p}$' for addition and '$\bar{m}$' for subtraction; he also extended Fibonacci's symbol 'R' for roots to write, for example, 'R^4' for the fourth root. He wrote compound expressions as follows:

$$\text{'}R^4\underline{21\bar{m}R^2 147}\text{' for } \sqrt[4]{21-\sqrt{147}}. \tag{3.19.2}$$

Chuquet's work, like that of his Italian and German contemporaries, was to further the cause of algebra, which we take up next in §4.5. But one quite significant omission limited the theoretical side for all participants: only positive roots for polynomial equations were considered, as with the quadratic (3.17.3). As a result, the relationship between the degree of a polynomial equation and the number of its roots (of any kind, including complex ones) was not addressed – the 'fundamental theorem of algebra', by which name it would later be honoured (§4.18).

3.20 *Sacred arithmetic*

{Hopper 1938; Dornseiff 1922; CE 12.5}

Numerology, especially when inspired by Christianity, continued to be an important motivation for the study of integers; virtually all the early European figures mentioned in §3.10–§3.11 advocated doctrines of some kind. An important newcomer was Dante Alighieri (1265–1321), who developed a system based upon his own male 'completed' 10 (the Pythagorean triangular number 1 + 2 + 3 + 4), as against the 9 of his beloved Beatrice. The Neoplatonist philosopher Nicholas of Cusa gave the same interpretation to 10, and sought to elevate oneness and otherness into a cosmological principle.

From the 7th century until figures such as Hugh of Saint-Victor in the 12th, numerology featured in the (highly complicated) issue of the grades and orders in the hierarchy of the Christian Church. The preferred number was seven, but it varied with frequent changes in organization, so that the scholars became confused. Hence the phrase 'at sixes and sevens' was born {Reynolds 1979}.

The link with music was strengthened, for composers of the 14th century such as Philippe de Vitry and Guillaume de Machaut, and of the 15th like Guillaume Dufay and Jakob Obrecht, wrote choral music of astounding complexity precisely because it was based upon numerologically significant numbers of notes and phrases, and rhythmic patterns. Sometimes the texts included the name 'Boethius', in honour of his dictum that 'music is number made audible'.

Gematria (§3.9), which controlled certain texts set by composers, seems also to have played a role in cathedral building. An important case is Chartres Cathedral, which was rebuilt in its present form around 1300. It was given the motto *assumptio virginis beate mariae* ('assumption of the Blessed Virgin Mary'), which takes the integer 306 in the so-called 'lesser canon' of Latin gematria. This integer is also twice 153, whose place in ancient Jewish history was noted in §3.10. It is also an important length in the building itself when measured in Roman feet, a unit commonly used in medieval architecture. To this activity we now turn.

3.21 *Sacred geometry: the art of the masons*

{Lund 1921; CE 12.5}

One of the greatest achievements of medieval science and technology was the 'art' of designing and building medieval cathedrals, carried out by teams of masons, each team working under a master. We stand in awe of the marriage of artistic beauty and technological immensity; but often we know little about their basic design, and the reason lies in secrecy – covert mathematics *par excellence*. Even a source like the drawing-book of Villard de Honnecourt (1190s ±) does not reveal the lore beneath the images. The building of God's House was the highest form of craft, and embodied religious world-views and spiritual traditions.

Many proportions seem to have been determined by the Greek division (2.13.3) into mean and extreme ratio, with inscribed pentagons and pentagrams frequently used. Another major feature of Masonic practice was the Pythagorean 3 : 4 : 5 right-angled triangle, which rather oddly became known as the 'Masons' square'. The properties shown in FIG. 2.8.1, that the two rectangles constructed on the sides as diagonals have sides in the ratios 2 : 1 and 3 : 1, may have been the source of the importance that the cathedral-builders placed on these ratios, which were known respectively as *ad quadratum* and *ad triangulum*. The basis normally taken for 2 : 1 was the isosceles triangle with base equal to altitude, and for 3 : 1 the equilateral triangle. Classical origins of the importance of these ratios include Solomon's Temple, built in the dimensions 60 : 20 : 30 of cubits ({I Kings 6: 2}, but without explanation of the rationale); and Book 5 of of Vitruvius's *De architectura* (§2.29), in which he mentioned them as recommended proportions in public buildings, and even related them to musical intervals (again without explanation). The successors to this secret knowledge from the 17th century were to be the Freemasons, who maintained some of the rituals and tools of the real Masons (§17.20).

3.22 *Cutting stone: the craft of the masons*

{Sakarovitch 1997; CE 2.2, 8.7}

Somewhat more is known of the scientific and mathematical methods used in the construction of buildings. One reason why major buildings from this period have proved so durable is that the stresses and crushing loads borne by their fabric were several times below critical levels – oversafe to a modern viewpoint which follows (or defends itself on) the tag that 'any fool can make a building which stays up: it takes an engineer to make one which *just* stays up'. It seems likely that loadings were determined by thinking rather than theory – for example, from simple proportions based on intuitive ideas about moment of forces and couples. However, important innovations were also made; in particular, the flying buttresses at Chartres Cathedral in the early 13th century (see {James 1979, 1981} for an anatomy of this building).

Two methods of cutting stone to make components of arches became popular (FIG. 3.22.1). The first, which became known as 'squaring', was based on the double projection of a solid object onto two perpendicular planes. Imagine a

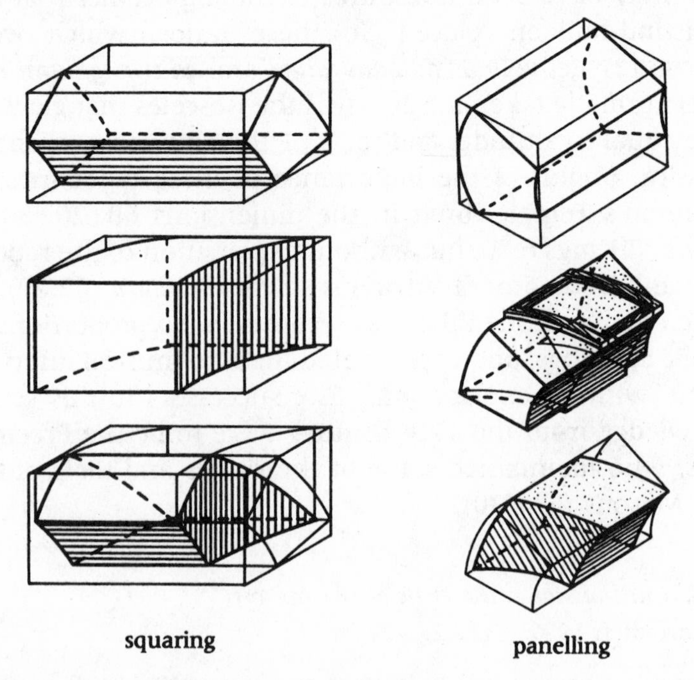

FIG. 3.22.1 *Two ways of cutting stone.* {Sakarovitch 1997}

cuboid of stone set in a horizontal plane. The component was determined as the surface of intersection of appropriately shaped horizontal and vertical cylindrical surfaces. This method could be wasteful of material and work-time, so there emerged an alternative that was more economical but required more preparation, which became known as 'panelling'. The component, imagined to be within the smallest possible parallelepiped P, was determined by working out in advance the shape of the horizontal section (or 'panel') in P and also locating precisely the vertical cylindrical surface. The procedure was not only geometrical: the static balance with neighbouring components also had to be considered.

These activities formed parts of a subject expressed by the title of an unoriginal but well-received compilation: the *Practica geometriae* (1230?) of Hugh of Saint-Victor. Manuscripts and eventually books of that title appeared in quantity, intended to help the constructors of buildings of various kinds. The authors presented the geometry of the circle and triangle, but with values assigned to the various components, or to be calculated for them: explicitly the analogue of the practical arithmetic described in §2.14. The applications included not only land measurement and surveying but also the construction of buildings. The part concerning the measurement of solids, including the volume of a cask or cistern, was known as 'stereometry'.

3.23 *'Speculative' geometry*

Books on practical geometry often began with a few flourishes towards Euclid: points having no parts, lines no breadth, and so on. Fibonacci produced a *Practica geometriae* around 1220, but he went much further in Euclidean considerations of geometrical figures and magnitudes, and other 'speculative' questions. General acquaintance with Euclid in Europe was encouraged especially by a version of the *Elements* made in the 1290s by Campanus of Novara; it was also the first to be printed, in 1482. However,

Campanus had elaborated upon the translation made by Adelard of Bath in the 12th century, so that the text was more garbled than its earliest readers realized.

Interest in 'speculative' questions, whether from Euclid or elsewhere, was relatively limited; but solutions to the three classical problems gained some attention, from both Greek and Arabic sources. Archimedes' *Dimension of the Circle* led to a variety of attempts to square the circle; the theorem (2.9.5) relating area, circumference and radius suggested that perhaps one could construct a rectilinear figure of equal area with the radius as one side {Knorr 1989, Part 3}. Thomas Bradwardine and Albert of Saxony pursued this possibility.

In solid geometry, fresh efforts were made to duplicate the cube; but perhaps the most spectacular work concerned the five regular solids and the extensions by Archimedes (§2.14). A substantial work was completed around 1488 by Piero della Francesca, whose great fame as a painter has partially eclipsed our knowledge of his considerable mathematical gifts. Here he united both talents, basing much of his extensive analyses on division by mean and extreme ratio of the edges of the regular solids (for a few earlier manifestations of this ratio, see FIG. 2.13.1). His work was made popular – indeed, largely plagiarized – by his follower Pacioli, in the book *De divine proportione*, written in Italian and published in 1509, the year in which Pacioli's Italian translation of the *Elements* also came out. It inspired wonderful drawings of the regular solids by Pacioli's friend Leonardo da Vinci; FIG. 3.23.1 shows two of the Platonic ones.

Underlying some of these speculations on geometry was the question of the rigour of proof in Euclid, especially relative to Aristotelian logic and theory of demonstration. One of the very few figures to address it explicitly was Alessandro Piccolomini (1508–78), a member of a distinguished noble Italian family. In 1548 he argued that mathematics may not reach the heights hoped for in physical sciences, since its objects have no *essential* properties: bodies have to

FIG. 3.23.1 *Two Platonic solids, illustrated by Leonardo da Vinci for Pacioli.*

be heavy, for example, but triangles need not be equilateral {Mancosu 1992}.

3.24 *The geometry of light rays*
{Lindberg 1976; CE 2.9}

Optics was still concerned largely with sight rather than light, so its links with both geometry and trigonometry remained intimate. European interest developed during the early 13th century, especially at Oxford with Robert Grosseteste and his follower Roger Bacon. Another leading authority was the Pole Witelo, author of *Perspectiva*, a comprehensive manuscript on the subject; he seems to have been influenced by Ibn al-Haytham, although he gave no reference. He presented enough geometry for some universities to use his book in the quadrivium in place of Euclid's *Elements*; and, on the grounds of the marked similarity of content and even wording of certain theorems, {Unguru 1974} argues that he must have been drawing upon Pappos, either directly or through some intermediary such as his friend the translator William of Moerbeke. If so, then this was a contact that occurred two centuries before the publication of the Latin translations of Pappos which are usually regarded as the vehicles of his introduction to the West. This is a good example of the uncertainty of dating the transmission of classic Greek texts.

Witelo was also interested in pinhole cameras, which had been used since antiquity as a means of projecting images onto a screen. A puzzling property was that they cast a circular image of the Sun even if the aperture in the camera was, say, triangular. To argue against this apparent refutation of the assumption that light travels in straight lines, writers from Bacon through Leonardo da Vinci to Francesco Maurolico in the mid-16th century analysed the ways in which *cones* of rays seemed to behave at the edges of such apertures; the mathematical interest lay in the solid angles of the cones, and also in relationships between the intensity of the image and the distance of the screen from the aperture and their orientation to each other.[4] The phenomenon of rectilinear propagation of light was saved; insights into diffraction were not to come until the early 18th century.

3.25 *Painting and perspective*

{Andersen 1987; CE 12.6}

In entitling his work *Perspectiva,* Witelo followed the tradition set by Boethius in using that word as a Latin translation of the Greek *optike*. But gradually 'perspective' came to carry the connotation which it still retains. The subject, an intimate alliance of geometry and optics, became quite a speciality in art and architecture – indeed, it helps in

[4] The case of Leonardo da Vinci is particularly interesting. Not only did he emphasize properties of projected pinhole images {Thro 1994}, but also, according to the very convincing argument put forward in {Picknett and Prince 1994}, around 1492 he made the Turin Shroud by casting on cloth the projected image of the body of a cadaver, and also of his own head.

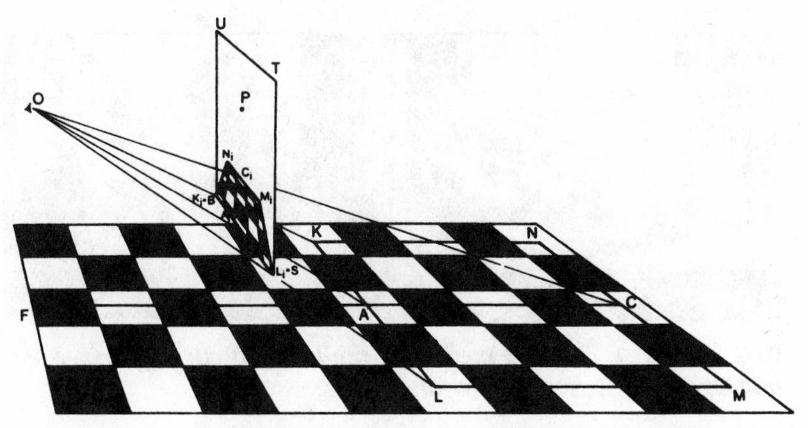

FIG. 3.25.1 *A version of Alberti's theory of perspective* {Andersen 1987, p. 120}.

distinguishing the Renaissance from the Middle Ages, especially for Italy.

An important pioneer was the mathematician and architect Leone Battista Alberti, who produced manuscripts on *La pittura* by the mid-1430s, when he was in his early thirties. Curiously, he included no diagrams, but FIG. 3.25.1 is a faithful presentation of his ideas. The painter, his eye (point) at O, wished to render a credible perspectival version of the segment KLMN of the horizontal chequered pavement on the bottom part of his vertical canvas BSTU situated above it. He considered the visual pyramid arriving at O – the rectilinear equivalent of the cones of light studied by opticians – and construed the segment in the painting as the section of the pyramid cut by the plane of the canvas.

Alberti described two procedures for constructing this intersection. One of them was taken up and beautifully illustrated by Albrecht Dürer in his *Underweysung der Messung mit dem Circkel und Richtscheyt* (1525, 1538). These 'Instructions on Measuring with the Circle and Straightedge' used a vertical reference grid between subject and artist to help prepare accurate perspectival images on a canvas similarly marked out and laid on a horizontal table (FIG. 3.25.2). Dürer was using a sort of arithmetical coordinate

FIG. 3.25.2 *Dürer's method of rendering accurate perspectival images with the help of girl and grid.*

geometry, without the algebra that various European contemporaries were struggling to form.

Alberti's other procedure was based upon perspective; several followers either elaborated upon it or developed variants. An important text of the 1480s was the *De prospettiva pingendi* ('On Perspective for Painting') by Piero della Francesca, in which he provided rules for producing the kind of perspectival images that his own art exhibited so wonderfully; but apparently his readers found him pretty hard to follow! To throw difficult objects into perspective, he used a method of plan and elevation which was to be elaborated in Gaspard Monge's descriptive geoemetry (§8.3).

Piero and other artists also used simple procedures such as locating main features of a painting on a basic curve, like the circumference of a circle or the perimeter of an isosceles or equilateral triangle. The cover of this book carries part of a wonderful example of all these applications of the geometry of the beautiful. Entitled *Baptism of Christ,* it is thought to be early – but already masterful {Lavin 1981}. FIG. 3.25.3 shows it in full in half-tone, with some more features marked.

The painting was the central part of a triptych, mounted centrally behind the altar of a church in the town of Sansepolcro (shown between Christ and the tree); thus Christ is placed centrally also. It is in the proportion 2 : 3, the unit

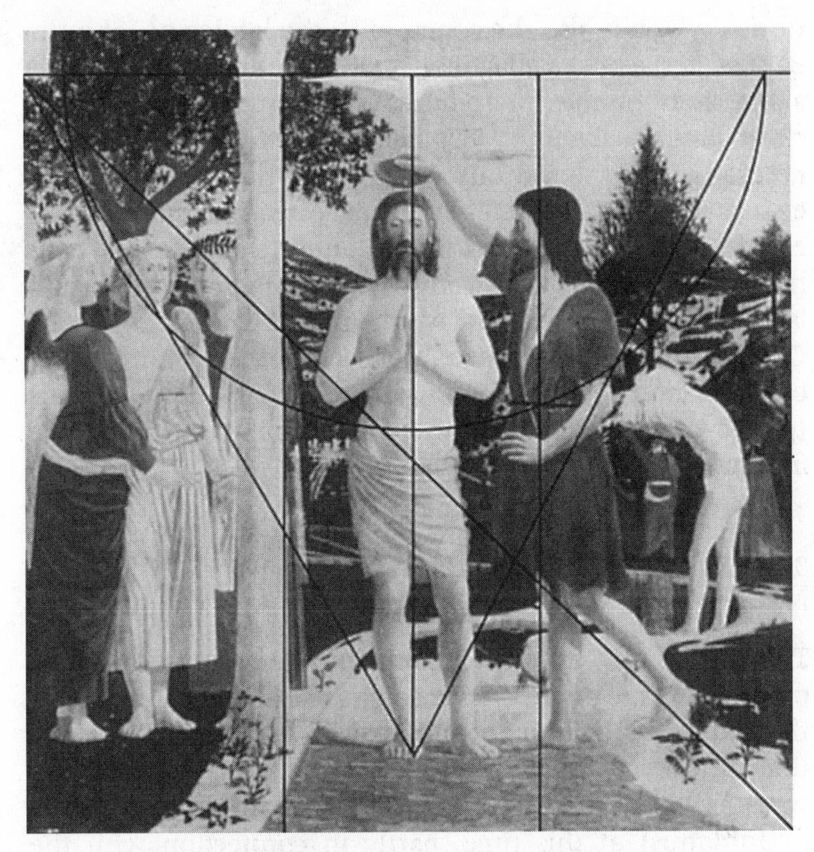

FIG. 3.25.3 *Geometry and proportion in Piero della Francesca's* Baptism of Christ.

being the standard *braccia* (arm) of that time (its full size is 167 × 116 cm/65.7 × 45.7 inches). Christ stands at $\frac{3}{2}$ *braccia,* half the full height. The painting is also divided into three parts vertically, by the tree and John the Baptist; and into two parts horizontally by the line of the eyes of the three angels.

The form F of the painting comprises a nearly square rectangle R and the upper half of a circle C above it (and thus symbolizing the heavens?). The centre D of C lies on the line of the wings of the dove, which mark the circle's horizontal diameter, as is emphasized by a neighbouring cloud. The full circle has its lowest point at the stomach S of

Christ, which is also the centre of R; the left leg of John the Baptist lies along a diagonal of R. The arms of Christ lie along lines of the five-pointed star inscribed within the circle having diameter SD, half that of C. This property naturally leads to various cases of division by mean and extreme ratio (FIG. 2.13.1); other cases include Christ's elbows so dividing R horizontally, and the chalice above his head so dividing F vertically. The most prominent equilateral triangle is formed by the upper corners of R and Christ's right big toe; there are several other triangles, less obvious. As usual, the artist wrote no EasyGuide manual (please-replace-after-use); the viewer looked, thought and interpreted (or not . . .).

3.26 *The emergence of an arithmetical theory of ratios*

{Sylla 1984}

The rest of this chapter is concerned largely with trigonometry and mechanics, with some special attention paid to the Oxford Calculators of the early 14th century. This section notes a gradual change in ratio theory which their work brought out; it was occurring with other figures (such as Jordanus) at this time, partly in connection with the developments in arithmetic and algebra described earlier.

The change had begun in classical times. Recall, for example, that Pappos had defined a curve by a condition (2.27.1) involving compounding ratios *in general*, not under the restriction that Euclid had imposed at (2.18.3). Letting, for brevity, letters denote quantities, we have

$$\text{Euclid:} \quad (a:b)\cdot(b:c)\cdot(c:d)\cdot\ldots \qquad (3.26.1)$$

but

$$\text{Pappos:} \quad (a:b)\cdot(c:d)\cdot(e:f)\cdot\ldots.$$

I characterized Euclid's ratios as generalized musical intervals, different in kind from numbers and geometrical magnitudes, and gave his form of compounding as evidence. In contrast, the new kind of ratio brings compounding much closer to arithmetical multiplication, and ratios to rational numbers; lines were replaced by lengths, as it were, and they could be multiplied and divided. It appeared even in some corrupt medieval versions of Euclid's *Elements*, perhaps because of the Arabic acceptance of rational numbers mentioned in §3.6.

The Oxford group used it when discussing properties of lines and arcs of circle in spherical trigonometry. Richard of Wallingford, for example, executed procedures corresponding to

$$\text{if } b = c : a, \text{ then } c : b = a \text{ and } a \times b = c. \qquad (3.26.2)$$

Menelaos's Theorem 2.24.1 on compounding chords, and several related results, were widely used, normally in the new way; so were inequalities between ratios. Other cases include this deduction concerning magnitudes:

$$\text{if } g : h = k : l, \text{ then } g \times l = h \times k, \text{ and vice versa.} \qquad (3.26.3)$$

Here the '=' sign accurately reflects the language; indeed, the sign itself was sometimes used.

This change brought a convolution in the relationship of arithmetic to both algebra and geometry as rational numbers gained more prominence. Nicole Oresme is a noteworthy figure. In his essay *De proportionibus proportionum* of the mid-14th century he converted Euclid's (in)commensurable geometrical magnitudes into (ir)rational ratios, clearly thinking of them as numbers. (His title exemplifies a rather unfortunate change in use of the word 'proportion' to refer to a ratio rather than a proposition relating two of them.) He applied this interpretation of ratios to many areas of mechanics, such as the motions of planets in circular orbits, with time and distance along orbits measured from the occasion of a conjunction of a pair of planets.

Oresme went on to propose the novel idea that *any* number can be represented by a geometrical line – or, maybe, length. His 'Questions on the Geometry of Euclid' actually concentrated upon arithmetic, considering even infinite series of numbers. Indeed, a well remembered contribution is the surprising property of Euclidean unit fractions that the harmonic series is divergent, which he proved by means of the inequality

$$\begin{aligned} &1 + \tfrac{1}{2} + \tfrac{1}{3} + \tfrac{1}{4} + \tfrac{1}{5} + \tfrac{1}{6} + \tfrac{1}{7} + \tfrac{1}{8} + \cdots \\ &> 1 + \tfrac{1}{2} + (\tfrac{1}{4} + \tfrac{1}{4}) + (\tfrac{1}{8} + \tfrac{1}{8} + \tfrac{1}{8} + \tfrac{1}{8}) + \cdots. \end{aligned} \tag{3.26.4}$$

However, Euclid's theory of ratios continued to be popular, as part of his growing general influence. Thus the two theories of ratios were running alongside each other, but the differences between them were not always clearly stated. 'Difference' (*differentia*) itself between two ratios is one case: between 9 : 8 and 10 : 9, it was 10 : 8 in Euclid's style but 1 : 72 in the new arithmetical one; in the new style, 8 : 9 from 9 : 10 could not be handled at all since that kind of difference was negative. The use of similar or even the same terms was rather confusing: an important instance was duplicating (that is, the square of the ratio, *à la* Euclid) as opposed to doubling (that is, twice the numerator). Confusion between the two theories was to last for a long time, even to the first edition (1687) of Isaac Newton's *Principia* (item 8 of §5.10).

3.27 *The steady growth of the trigonometries*

{von Braunmühl 1900; CE 4.2}

Use of the new theory of ratios in Menelaos's important theorem on chords exemplifies the steady development of trigonometry, in both its parts. Under Arabic influence the sine tended to replace the chord as the main notion (compare §2.24). All functions of angles were still usually understood geometrically, as lengths or lines, not as numbers or ratios.

A simple but popular use of trigonometry was as a method of measuring heights by using reflection off water or a mirror. In FIG. 3.27.1 the observer with eye at Q determined the height AZ from the similar triangles PQR and AZR joined at the reflector R. The underlying theorem was probably learned from Euclid's *Optica*. The method was

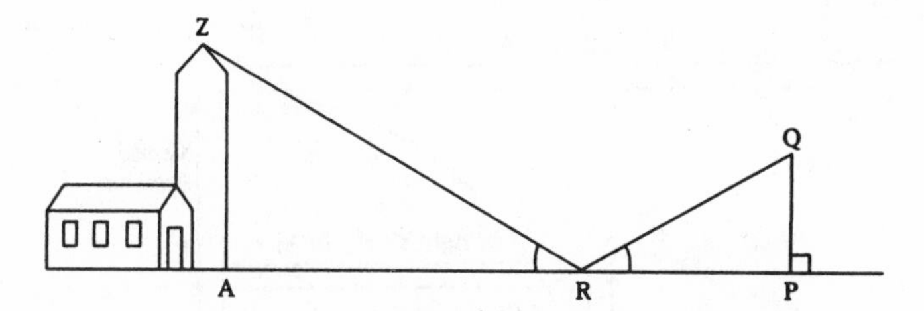

FIG. 3.27.1 *European trigonometry: the method of triangles in proportion.*

widely taught in Europe over the centuries; it contrasts interestingly with the Chinese method of double differences (FIG. 2.32.2).

The tangent and cotangent functions were introduced by Arabs such as Abu'l Wafa in the 10th century, and rather slowly made their way into European trigonometry via translators such as Adelard of Bath. They arose largely in connection with lengths of shapes made by shadows, such as is cast by the gnomon in FIG. 2.27.2; the respective Latin names *umbra recta* and *umbra extensa* were used. Sometimes they were specified for an angle *A* by proportions relating them to the sine and cosine functions:

$$\tan A : 1 :: \sin A : \cos A :: 1 : \cot A. \qquad (3.27.1)$$

FIG. 3.27.2 shows the relationships between the four functions, and their common names of the time in English; it probably does not capture any conception held during the Middle Ages, since the tangent and cotangent were often treated separately from the sine and cosine.

The spherical part became especially significant, with Menelaos's theorem playing an important role. One main use lay in the design of maps and charts, which from the 1470s onwards were printed by copper-plate engraving {Tooley and Bricker 1976}. This was, after all, the period of Columbus sailing the ocean blue in 1492 (and thinking that the West Indies was Japan). Ptolemy's *Geographica* received many translations and commentaries, in print first in 1475. Indeed, it was an important German editor of Ptolemy,

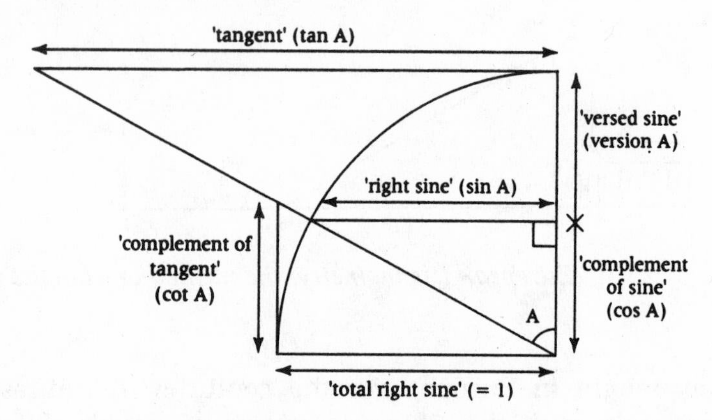

FIG. 3.27.2 *The principal main functions in medieval planar geometrical trigonometry.*
The arc is of radius 1.

Martin Waldseemüller (1470–1518), who produced a book and world map entitled *Cosmographica* in 1507 and engraved the name 'America' on his representation of the new continent, in honour of the explorer Amerigo Vespucci.

Richard of Wallingford in England was a leading trigonometrist of the early 14th century; but perhaps the most zealous figure was Regiomontanus, especially during his last years 1461–76 in Italy. In a remarkable concordance of aim and intention with a treatise on 'The Figure of the Sector' written two centuries earlier by Nasir al-Tusi, Regiomontanus's manuscript *De triangularis omnimodis* not only contained all known theorems and techniques in both parts 'for triangles of all kinds' but also presented some new ones. His most durable innovation was to establish the sine theorems (as we now call them) for both parts: For Δabc in the plane and ΔABC as three great circles on the sphere,

$$\text{planar: } ab : bc :: \sin c : \sin a; \tag{3.27.2}$$

$$\text{spherical: versin } BAC : [\text{versin } A - \text{versin } (B - C)] :: \sin 90^\circ : (\sin b \times \sin c). \tag{3.27.3}$$

(Compare the latter with the Arabic equation (3.7.1).) My

expression 'sin 90°' was then the 'total right sine' (Regiomontanus's phrase) line on the right angle (FIG. 3.27.2); he gave it various large values, seemingly so that the calculated quantities could have integral values, avoiding the bother of representing fractions. One of these values, 10^7, was often used by successors (see, for example, §4.15).

Although Regiomontanus wrote soon after the start of printing, his text was not published until 1533. But as the first such book, it greatly helped trigonometry become a major branch of mathematics, as we shall observe in the next chapter. He and others continued to compile tables of values, and of other functions including the versed sine. All quantities were still geometrical – lengths, not numbers or ratios.

Another attractive feature of trigonometry was a type of theorem to which the Greek name *prostaphairesis* became attached. In planar trigonometry they state that

$$\overset{\cos}{\underset{\sin}{}} a \overset{\cos}{\underset{\sin}{}} b = \tfrac{1}{2}[\cos(a+b) \pm \cos(a-b)]; \qquad (3.27.4)$$

various versions obtain in the spherical part. Publicized by Regiomontanus's student Johann(es) Werner (1468–1527?) and others, they provided means of adding and subtracting quantities (hence the name, which means 'plus or minus'), instead of multiplying and dividing. This gave another nudge towards the notion of logarithms (§4.15).

3.28 *Astronomy after Ptolemy*

{Pedersen 1993, Chap. 18; CE 2.7}

Ptolemy's theories were dominant: the eight real spheres remained in place to contain the motions of the planets and stars on great circles. But various modifications were proposed. Cosmologists added three spheres for the world-soul, the mind and God Himself to give a total of eleven (which, surely not by coincidence, is the number of circles in the great labyrinth in the nave of Chartres Cathedral).

A profound modification was proposed by Jean Buridan and Nicole Oresme, when they speculated that the Earth rotated about its polar axis. Fewer spheres and slower rotations were among the benefits; but they did not convince themselves, and the idea slumbered for a century until revived by Copernicus (§4.10). Perhaps the combined authorities of the Bible, Aristotle and Ptolemy were too great to allow the stationary Earth to move.

Two elementary textbooks, written by popular algorists (§3.18), became widely used. In the first, *De sphaera* by Sacrobosco, the real spheres were given prime place. Completed in the 1220s, it was first printed in 1472. Curiously, that was also the year of publication of the other work, the *Theoricae novae planetarum* by Peurbach, a rather more extended account following the same lines but including 'New theories' such as Thabit ibn Qurra's theory of *trepidatio* (item 2 of §3.8). Peurbach, court astrologer to Holy Roman Emperor Frederick III, had produced it in 1454, when he was in his early forties, in connection with lectures given at a college in Vienna. The audience had included the young Regiomontanus, who supervised its posthumous publication. Many further editions and commentaries of both works followed over the decades, in both manuscript and print, and not only in Latin – those produced in Paris in the 1510s by Oronce Fine helped the rise of mathematics in France. A recent English translation is provided in {Aiton 1987}.

In his last years Peurbach had been preparing an abridged version of Ptolemy's *Almagest*. Regiomontanus finished it for him, but it came out only in 1496, twenty years after his own death. Then it was to have an important negative influence on Copernicus, as we shall see in §4.10.

3.29 *Jordanus on equilibrium on the inclined plane*
{Pedersen 1993, Chaps. 15–17; CE 2.6}

This and the next two sections are devoted to mechanics, in which the Europeans showed perhaps most originality. The subject was often construed, like optics, as lying between mathematics and physics. The philosophical influence of Aristotle was strong, partly through Arabic authors such as Averroës and Avempace, who also provided fruitful ideas of their own. Christianity occasionally motivated questions: for example, in the 15th century Nicholas of Cusa, who spent his last years as a close friend of Pope Pius II, pondered on infinite velocity in connection with the possibility of angels being in two places at once.

In statics, Jordanus de Nemore made a profound innovation in studying bodies at rest under gravity on the inclined plane. His Greek and Arabic predecessors had tried but failed to find a relationship between the weight and planar inclination of *one* body; so he considered a *pair* of bodies, one set on each side of a double inclined plane and linked together by a string across a pulley at the top T (FIG. 3.29.1). He drew upon two general assumptions, each one very fruitful in its own right (the symbols are mine):

1. *slight* displacements could be made from the equilibrate state; and
2. the weight of a body lay in inverse ratio to its vertical displacement, in that raising W by height h was equivalent to raising, say, $2W$ by $h/2$.

Now link a third body of weight Z to the first two by strings, and by assumption 1 move it down a little from C_1 to C_2 and thereby raise weight W_A from A_1 to A_2, where

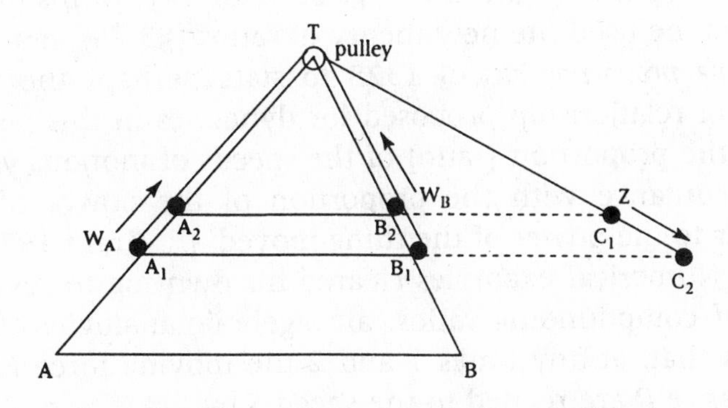

FIG. 3.29.1 *Jordanus de Nemore on the equilibrium of bodies on the inclined plane.*

A_1C_2 and A_2C_1 are horizontal lines. Repeat this displacement of Z to raise W_B from B_1 to B_2. Then, by assumption 2,

$$W_A : Z :: A_1A_2 : C_1C_2 \text{ and } W_B : Z :: B_1B_2 :: C_1C_2. \quad (3.29.1)$$

Hence, as required,

$$W_A : W_B :: A_1A_2 : B_1B_2 :: AT : BT. \quad (3.29.2)$$

Jordanus's idea of bodies sliding on planes helped lead successors to add the inclined plane to the quintet of 'simple machines' that Hero had proposed long before (§2.25).

Among other authors, the Oxford Calculators used their new theory of ratios (§3.26) in various situations in statics; in particular, they applied the interpretation (3.14.1) of the rule of three to the balance of forces in static equilibrium, and to the science of weights. These studies formed an important step in the use of ratios between quantities other than numbers and geometrical magnitudes. But their main contributions came in the study of motion.

3.30 *Proportion and configuration in mechanics*

Bradwardine distinguished between dynamics and kinematics (to use modern names) in Aristotelian terms of distinguishing the causes of motion from its effects. In an effort to 'put to flight' the 'fogs of ignorance' of his predecessors, he used the new theory of ratios (§3.26) in a *Tractatus de proportionibus* of 1328 to state perhaps the most general relationship proposed for dynamics in this period; that 'the proportion [ratio] of the speeds of motions varies in accordance with the proportion of the power of the mover to the power of the thing moved' {E. Grant 1974, p. 302}. Numerical examples cleared his own fog to reveal a law of compounding ratios; an algebraic analogue of the law is that, at any times 1 and 2 the moving force F and resistance R are related to the speed S by

$$F_2 : R_2 = (F_1 : R_1)^{(S_2 : S_1)} \quad (3.30.1)$$

It would be easy for us, from our modern viewpoint, to state a general law using a multiple of S in the exponent; but such formulations did not come until the 17th century, partly from backgrounds such as this one (compare the form of Newton's equation (5.11.1) for the path of a particle under a central force).

The Oxford Calculators also applied their theory of infinite series to problems of motion. Their follower Oresme also deployed it during the 1330s to criticize the assumption made by Aristotle that a body had to travel an infinite distance in an infinite time; he took the case of steady decrease in motion corresponding to the convergent series

$$1 + \tfrac{1}{2} + \tfrac{1}{4} + \tfrac{1}{8} + \cdots = 2. \qquad (3.30.2)$$

However, nobody seems to have remarked on the fact that Zeno's supposed paradox of Achilles and the tortoise (§2.15) was satisfied by this situation.

Oresme's own version of eqn (3.30.2) was geometric, for he introduced the pleasant idea of the 'configuration of qualities and motions', in which velocity was graphed as a function of time, so that the region underneath it represented distance traversed. This was a novel relationship between dimensions, used, it seems, to avoid giving velocity the status of a fourth dimension (hence his word 'qualities'). He pictured the three different categories of motion which Oxford Calculators such as William Heytesbury had recently distinguished. He divided the last category of motion according to whether one or several curves were required in the configuration:

Oresme's 'quality'	Modern equivalent(s)	Oresme's 'configuration'
'uniform'	constant velocity	rectangle
'uniformly difform'	constant de- *or* acceleration	right-angle triangle
'difformly difform'	non-constant velocity	region under a continuous curve

Oresme also re-proved geometrically various theorems on the mean value of motions which had been stated at Oxford in words. One, perhaps due to Heytesbury and which they all recognized as important, asserted (in my algebraic abbreviation) that if a body began moving with speed B and increased with uniform difformity to take speed E after a time t, then it travelled a distance s given by

$$s = [B + \tfrac{1}{2}(E - B)]t. \tag{3.30.3}$$

Such results helped in the measurement, or 'quantification', of the qualities; but they also had potential value for future foundations of dynamics. In particular, if $B = 0$ and speed is a linear function of t, then there follows the law of free fall: $s = \frac{1}{2}gt^2$, with g as the acceleration towards the Earth (as Galileo was to show: §4.14). However, he did not pursue this approach but instead adopted and adapted an Aristotelian notion due to his teacher, Buridan: the 'impetus' of a body, understood as its tendency to continue in motion, either in a straight line or a circle, and in which more velocity required more impetus.

This emphasis on straight and circular motions was maintained even to the presentation of unrealistic diagrams (straight line, then arc) of the paths of projectiles; European understanding of air resistance did not seem to match the quality of their Arabic predecessors. This weakness is surprising, for projectile theory became a leading topic. Military needs led the way here – not merely slings and arrows, but much larger gadgets such as the trebuchet for hurling large missiles long distances.

3.31 *Machines – with mathematics?*
{CE 2.8}

We come by this last example to machine theory, a major part of European mechanics. Indeed, the Latin word *mechanica* was read both as a noun to refer to the study of bodies in rest and motion, and as an adjective to allude to the art of constructing and using machines. Much thinking in mechanics must have been involved: balance and equilibrium, components of forces, exchange and transmission of force(s), and the limitation and even dangers of impact, From elsewhere, geometry and trigonometry must have permeated the preparation of designs. Yet evidence is slight and the mathematics covert – sometimes for reasons of secrecy imposed by the interests of nation or state, no doubt. Machine theory was often represented only by the striking but uninformative image of a weight rising or falling at the end of a rope attached to a machine. Again, many devices for handling water were devised, yet no substantive science of fluid mechanics emerged.

The clearest manifestation of professional engineering is found in Tuscany during the early 15th century, where Alberti collaborated with Mariano di Jacomo Taccola (1381–1453?) and the architect Filippo Brunelleschi (1377–1446). But the finest talent was a compatriot successor, who profited greatly from the tradition they had launched: Leonardo da Vinci {Truesdell 1968, Chap. 1}. For a half century from the 1470s he regularly filled his notebooks with comments on and diagrams of transmission systems such as gears and cranks, and machines like bicycles and aerostats. In much related work he pondered over beams and arches in structures, laws of falling bodies, and the properties of eddies and vortices in flowing water. Quite often he thought in terms of ratios: the depth of penetration of a projectile as against the speed of its launch, say. Yet the work often resembles dreams of his mind; questions appear in his texts more than answers, and sometimes observations rather than experiments are depicted. He reads like an author of riddles. Furthermore, there were very few readers of this most secretive of men at the time, and then only of some of his forays.

3.32 *Comparisons: imports into Europe*

This elusive scenario makes Leonardo a suitable concluding figure for this chapter. Nevertheless, he seems to contradict its title: was it a *quiet* millennium, with him around? Surely he was also typical of his time as a Renaissance *ecce homo*, arising not in medieval shadows but as an individual, and with a pretty clear biography from known date of birth to known date of death.

Yes, change had come, but at the end of a quiet *millennium*. During that very long period the Chinese and the Indians had continued rather sedately, the Japanese and Koreans rose somewhat, and the Arabs flowered in the greatest single body of mathematical work (though even there somewhat fitfully down the centuries). Moreover, unfortunately only parts of all these achievements came over to the Europeans, whose collective intellect had lain extraordinarily dormant for centuries. Progress in the early 13th century is remarkable: the inauguration of several universities (eight in the 1220s alone), the building of cathedrals such as Chartres, and so on. But advance was not relentless: there seems to have been a lull between 1360 and 1420, after the Oxford Calculators and Oresme and before the main activities of algorists and abbacists. Epidemics across Europe such as the Black Death must have been a major cause.

Many plaudits should go to the translators for providing the soil for the growth that did occur in Europe. We can barely imagine the travails and devotion of figures such as Hermann the Lame, William of Moerbeke and Gerard of Cremona, toiling over one battered Greek or Arabic text after another to think and write out a translation on parchment or scroll when light or lighting permitted and ink was available.

Of the various ancient roots of mathematics (§2.1), arithmetic, trigonometry and geometry flourished most vigorously, with arithmetic beginning to yield the mathematics of letters that was gradually growing into algebra. Mechanics has also stirred, largely on the coat-tails of astronomy, which continued to hold a prime place among the sciences. Religions were prominent inspirations for all cultures: the sacredness of numbers and shapes which man must understand thanks to the beneficence of 'God the creator who has bestowed upon man the power to discover the significance of numbers', as al-Khwarizmi put it in the opening flourish of his book on equations {E. Grant 1974, p. 107}.

On the institutional side, the establishment of universities in Europe, with mathematical subjects dominating the curricula for the quadrivium, gave new opportunities for thought and employment. But the most profound change was the invention of printing (centuries after the Chinese) in the mid-15th century, and then engraving, which provided means of communication unimaginable hitherto. Financial concerns severely restricted the opportunities for publishing books in mathematics above the elementary level of the arithmeticians; but gradually they began to appear and gain readers, some of whom became authors themselves – at first of books alone, and later of articles in the new scientific journals. For the rest of this volume, which concentrates almost entirely on developments in Europe, the printed word will play a steadily greater role in the prosecution and profession of mathematics, in its ancient and newer branches.

4

The age of trigonometry: Europe, 1540–1660

This period has been chosen in order to start with the innovations of men such as Girolamo Cardano in algebra and Nicholas Copernicus in astronomy; it ends with aspects of the work of René Descartes and Pierre de Fermat. Up to §4.16, the development of topics is followed usually to around 1630. (Cardano initiated a few aspects of probability theory, in connection with games of chance; but the subject did not advance until the mid-17th century, and so it is postponed until §5.19.) The rest of the chapter treats the next 30 years.

4.1 *Trigonometry and algebra*

The editor of Copernicus's *De revolutionibus orbium coelestium* (published in 1543), G.J. Rheticus (§4.10), first issued the chapters dealing with planar and spherical trigonometry in the previous year, as a book with a title beginning *De lateribus et angulis triangularum*. This little-known point exemplifies a central feature of the period covered in this chapter, stated in its title: the prominence of trigonometry, which during this period surpassed even the status it had gained in medieval times and grew to be a major branch of mathematics ({von Braunmühl 1900–1903}, still the only detailed history of the subject). Spherical trigonometry developed more than the planar part, in response to the requirements of navigation, cartography and surveying. Tables remained a major tool: Regiomontanus's tables of sines and cosines (§3.27), very recently printed in 1533, were reproduced alongside Copernicus's chapters down to the printing errors.

Another important text, based in part on Rheticus's own contributions, was the treatise *Trigonometria* (editions in 1595, 1600 and 1612) by Bartholomeo Pitiscus. He introduced the word in this title, and covered both the spherical and the planar parts. He gave prominence to the trigonometric functions (for example, he furnished the formulae for $\sin(A \pm B)$) and included tables of values for the six main functions. He also helped to publicize the sign '°' for degrees, which had been in circulation for some time. It might have been intended as a zero marker; that is, in the more modern notation, '7°31′47″' reads 7 degrees plus 31 minutes (as degrees^{-1}) plus 47 seconds (as degrees^{-2}).

Every figure of note in this period used trigonometry, and some of them contributed to its progress. Algebra, though it was applied to trigonometry and also geometry, did not gain comparable status until the early 17th century. Its only major 'research' area was in the theory of equations, and then principally as an 'art' for calculating their roots {J. Klein 1968}. Otherwise it was still seen as an extension of arithmetic, that essential but workaday branch of mathematics, often limited to discrete quantities rather than the continuous realm treated in geometry and trigonometry.

However, from the 1620s the status of algebra within mathematics rose quite considerably. Its *methods,* and not merely its results, came to be studied: applications of a new kind to geometry were effected in pre-calculus; and number theory was extended, with integers and ratios treated as mathematical *objects.* Such differences mark out many of the last five sections of this chapter from the preceding ones.

4.2 *Personnel*

An example of the growing status of trigonometry and algebra is provided by Spain and Portugal. Their gradual release from Arab occupation in the 12th and 13th centuries was summarized in §3.12. The reintroduction of Catholicism gave new power to the Society of Jesus, which was founded in Rome in the 1530s; some Jesuits were involved in the spread of mathematics teaching, and during the 1580s a Mathematical Academy was founded in Madrid. Popular topics included solving equations, map- and chart-making, and perspective theory (for the painter Diego Velázquez, for example); indeed, there must have been a considerable body of mathematics, much of it Arabic, deployed in design and architecture. However, neither country produced any of the major mathematicians whom we now consider.

Table 4.2.1 lists the main details of career and mathematical activity of the nine men who enjoyed the widest range and deepest influence during this period. Some similarities are worth noting among the obvious differences. Above all, the very low professional status of mathematics is evident from the training these figures received and employments they followed. This was partly caused by political difficulties and strife: for example, Queen Elizabeth's 'police state' in England. (The union with Scotland in 1605 did not lead to corresponding fusions of intellectual traditions, either then or later.) The Thirty Years' War (1618–48) deeply affected the Continent, especially the German states. Only in Italy were there sufficiently strong universities for talent to be employed; the careers of Cardano and Galileo are quite similar. The contrast with France is especially clear, where another resemblance is discernible in the extra-mathematical careers of Vieta and Fermat. But the professional needs even in Italy can surprise us – for example, one important reason why Professor Galileo taught mathematics and (Ptolemaic) astronomy at Padua University was to enable the students at its important Medical Faculty to achieve competence in the preparation of the astrological charts that were used to prescribe correct treatments of illness or ailment.

Other figures important for specific branches of mathematics will be introduced in due course; but three background figures are noted here. Firstly, the Danish astronomer Tycho Brahe (1546–1601) brilliantly improved the quality of observational data, which were to be particularly important for 'Keplerus mathematicus' (as Kepler signed himself). Secondly, Thomas Harriot (1560?–1621) showed ability and range comparable with that of any contemporary, with notable innovations in numerical mathematics, the application of algebra to geometry, navigation (§4.8), ship-building, ballistics and optics {Shirley 1974}. However, in England even Cambridge was scientifically mediocre (and his home town of Oxford even worse), and Harriot's influence was confined to contacts with some continental contemporaries (such as Kepler on astronomy and optics) and with British ship-builders, and with a group of second-raters who published some of his algebraic ideas after his death (§4.17). Finally, Marin Mersenne (1588–1648) not only led a circle of *savants* in Paris but also served as an international correspondent around the scattered scientific community, diffusing others' ideas and even publishing some of them in his two-part *Harmonie universelle* (1636–7). He also contributed to some topics, most notably acoustics and harmony (§4.16).

Until well into the 17th century the growth of algebra

TABLE 4.2.1 *Principal figures; careers and main activities.* Only main dates of published works are given; those in brackets are of creation when it was significantly different. '+' indicates that work was done on the topic marked, but not in any one major text. ? indicates lack of reliable information.

Name	Copernicus, N.	Cardano, G.	Vieta, F.	Stevin, S.
Dates	1473–1543	1501–1576	1540–1603	1548–1620?
Nationality	Polish	French	French	Dutch
Training	Italy	Italy	Poitiers	?
Subjects studied	Law, Astronomy	Mathematics Medicine	Law	?
Residence(s)	Poland	Pavia, Bologna	Paris	Holland
Employer(s)	Administration	Professor	Lawyer, Parliament	Prince, Army
Algebra		1539, 1545	1591–1594	1585
Probability		(1570s)		
Trigonometry	1542	+	1571	
Geometry			1593	
Pre-calculus			1586	
Hydraulics		1550		1586
Mechanics		1550, 1557		1586
Astronomy	1543	(1600)	+	1608
Optics				
Other scientific interests		Technology, Medicine	Calendar	Technology, Music, Navigation
Main edition	1970–	1663	1646	1955–1968

presented a challenge to printers, who often found it difficult to lay out formulae and equations in a clear way. Often they were run on with the text instead of being displayed, so that features of theory (patterns common to several series expansions, for example) were obscured. When typesetting grew in sophistication, the communication of mathematics benefited from improved presentation.

For these and other reasons, prospects for the sale of mathematical publications were poor – hence the importance of Mersenne's letter-writing. The most popular type of mathematical book dealt with practical topics: arithmetic and algebra, geometry and trigonometry. But production increased from the 1590s; Table 4.2.2 lists the main books

Galileo, G.	Kepler, J.	Descartes, P.	Fermat, P.	Pascal, B.
1564–1642	1571–1630	1596–1650	1601–1665	1623–1662
Italian	German	French	French	French
Italy	Tübingen	Travels	Toulouse	Father
Mathematics, Physics	Astronomy	Science	Law	Mathematics
Pisa, Padua	Germany	Netherlands, others	Toulouse	Self-supporting
Duke of Tuscany	Emperor Rudolph II	Various	Lawyer, Parliament	
	1627	1637	+	1665
	+		(1654)	1665
	1596, 1611	1637	(1630s)	1640
	1611	1637	(1630s)	1659
1612, 1632				
1586, 1638		1644		
1610, 1613, 1632, 1638	1596, 1604, 1609, 18–21	1637		+
1610	1604, 1611	1637	+	
Navigation, Logic	Astrology, Music	Philosophy	Number theory	Calculator, Philosophy
1890–1909	1937–	1897–1913	1894–1912	1954

(from the mathematical point of view) published from that time to 1650. The most prolific authors were Galileo and Kepler, and the list suggests that they were in competition; however, while their mutual concern with astronomy was

TABLE 4.2.2 *Principal books in(volving) mathematics, 1590–1650*

These books are chosen for their mathematical significance, not for science in general. The dates are of (nominal) publication. When titles have very well known English versions, these have been used. The reference indicates the section where a book, or at least its context, is discussed.

YEAR; AGE	AUTHOR	SUBJECT(S); ACCOUNT
1591; 51	Vieta	algebra; §4.7
1596; 25	Kepler	astronomy; §4.12
1609; 38	Kepler	astronomy; §4.11
1610; 46	Galileo	astronomy, optics; §4.14
1611; 40	Kepler	optics; §4.12
1614; 64	Napier	logarithms; §4.15
1615; posth.	Vieta	trigonometry; §4.7
1618–21; 47	Kepler	astronomy; §4.11
1619; 48	Kepler	astronomy; §4.12
1620; 68	Bürgi	logarithms; §4.15
1621; 50	Kepler	astronomy; §4.12
1623; 59	Galileo	astronomy, §4.12
1624; 63	Briggs	logarithms; §4.15
1627; 56	Kepler	astronomy, logarithms; §4.15
1631; posth.	Harriot	algebra; §4.17
1632; 68	Galileo	astronomy, mechanics; §4.14
1635; 37?	Cavalieri	pre-calculus; §4.19
1637; 41	Descartes	algebra, geometry; §4.18
1637; 41	Descartes	optics; §4.21
1636–37; 49	Mersenne	music, etc.; §4.16
1638; 74	Galileo	astronomy, mechanics; §4.14
1639; 48	Desargues	geometry; §4.20
1644; 48	Descartes	mechanics; §5.8
1647; 49?	Cavalieri	pre-calculus; §4.19

important, Galileo disliked Kepler's mysticism, and seems to have paid limited attention to Brahe's work.

Latin remained the chief language, both for original work and for translations from Greek and Arabic; but others began to compete with it (for example, Galileo in Italian, and later Descartes in French), and translations began to appear into and even between European languages. Some figures published in both Latin and in a European language, and for that reason (as well as the normal casual attitude of the time) their names took different spellings; I use the most common form. Let us start the survey with a country

(SHORT) TITLE COMMENTS
In artem analyticem isagoge
Mysterium cosmographicum, 1st ed
Astronomia nova; 1st two laws
Siderius nuncius
Dioptrice
Mirificium logarithmorum; English translation 1616
Ad angularium sectionum theoremata
Epitome astronomiae Copernicanae
Harmonice mundi; 3rd (harmonic) law
Arithmetische . . . Tabuln; written in the 1590s
Mysterium cosmographicum, 2nd ed.
Il saggiatore; comets
Arithmetica logarithmica
Rudolphine tables
Artis analyticae praxis; written in the 1600s?
Dialogue . . . world systems
Geometria indivisibilibus
Géométrie
Optique
Harmonie universelle
Two new sciences; inclined plane, pendulum
'Brouillon' project
Principia philosophicae
Exercitationes geometriae sex

speaking a minor scientific language in the mid-16th century, where, however, efforts at improvement were being made.

4.3 *Dee on the scope of mathematics*
{Fauvel 1987}

A popular text for translation was Euclid; it even appeared in English in 1570 thanks to Henry Billingsley, in a beautiful edition including some pop-up diagrams. A very ambitious 'Præface' was furnished by John Dee (1527–1608), a talented figure then in his 44th year {Dee 1570}. He welcomed the translation as an aid to school education, dismissing fears that university instruction in classical languages would be harmed.

Not confining himself to geometry, Dee also surveyed the mathematics of his time, concluding with 'Groundplat' (thus, literally, a survey!) in the form of a synoptic table. His scheme started off with arithmetic and geometry, but did not explicitly feature trigonometry or algebra (partly reflecting the retarded state of mathematics in Britain). However, these branches were embodied in some of his other categories, which included architecture (prominently discussed), 'Perspective', 'Musike', 'Statike', 'Cosmographie' (a mixture of spherical trigonometry and geography characteristic of the time), 'Astrologie' (still a brother to astronomy, and a strong interest for Dee), and 'Navigation'. Some of his other category names were not adopted (for example, 'Zographie', which attended to 'the Intersection of all Visuall Pyramids, made by any plaine assigned'). This wide range of topics well suited the creator of the phrase 'British Empire'. His considerable posthumous influence is noted in §5.4.

4.4 *Grinding out numbers*

Dee's confidence must have been boosted by some decades of progress in Britain, where the Middle Ages had ended at last. One active area was practical arithmetic. As we saw in §3.14, calculations on paper had largely replaced use of the abacus and dust-number methods. Careful reckonings were often needed, especially where money was involved. Did the duke's surveyors accurately measure my land before computing my tax liability? How many kilns'-worth of bricks should my staff order for the building of my new stables, when each kiln costs so much money?

An important figure in improving instruction in practical mathematics in Britain was Robert Recorde (1510?–58), a pioneer of commercial arithmetic and also of distance learning {Yeldham 1936}. His books were written in English; the most influential were *The Grounde of Artes* (1544) and *The Whetstone of Witte* (1557). His first title used 'art' in the normal connotation of that time to mean the useful production of results. The second title, itself witty, shows a greater ambition: the abbacists' word *cosa* (§3.17) became *cos*, the Latin word for 'whetstone', and art was to be raised to the level of wit, that is (or to wit), to knowledge.

In these and other books, Recorde ran through many of the standard methods such as long division, the rule of three and the use of proportions. As with many authors with similar aims, he included many worked examples, in contrast to the severe style of Euclid. However, he also made durable innovations, especially the sign '=' of equality, chosen 'bicause noe 2. thynges, can be moare equalle' than such parallel lines.

During his career Recorde practised as a doctor; he also ran the Bristol Mint, which produced Irish coins. This activity led him to analyse a particular problem which he called 'alligation', the solution of which had broader consequences for mathematics {Williams 1995}. Here is one example: if a coin-maker has gold available in six different carat mixtures of fineness, what proportions of them should he choose to obtain a mixture with a given intermediate measure *M*? Answer: pair off each smaller measure with a larger one, and assign to each member of the pair the magnitude of deficiency (D) or excess (E) *of its mate* relative to M: for example,

Given measures and desired measure M:	15	16	18	20	22	23	24	
D (−) or E (+):	−5	−4	−2		+2	+3	+4	(4.4.1)
Chosen pairing (by letter):	a	b	c		a	b	c	
Required proportions:	2	3	4		5	4	2	

The possibility of other pairings admits different solutions, showing that the problem belonged to 'indeterminate analysis', to use the phrase of the next century when algebraic formulations were adopted (§5.14). In these terms the underlying identity is

$$(M-D)E+(M+E)D=M(E+D). \qquad (4.4.2)$$

The generality of this method is limited: the number of given data must be even, and M has to lie in the middle. However, authors attested at the time that the computational needs (modest in this simple example) helped the Hindu-Arabic numeral system to replace Roman numerals. This point deserves some emphasis, for the change had been opposed on ethical grounds. Those responsible for accounts wished to preserve the Roman system because, say, 'v' added to 'iii' gave the sign 'viii', checkable for honesty or accuracy, whereas 5 plus 3 gave '8', which as a sign bore no similarity to '3' or to '5'.

However, the battle was lost, and the new numerals became widely used in Britain during this period. The use of a symbol for zero had profound cultural consequences {Rotman 1987}. 'William Shake-speare' wonderfully exploited nothing as something very important in *King Lear* (1600s), and more comically in *Much Ado About Nothing* (1590s) and elsewhere; and he will not have missed the semiotic link between '0' and the shape of a theatre, a 'Wooden Oh'.

4.5 *Stifel and Stevin: Continental arithmetic and algebra*

Meanwhile, on the Continent, arithmeticians continued the advance towards the second stage of algebra described in §3.16: using symbols for quantities and for operations, and increasing the autonomy from geometry. An important follower of the German *Coss* tradition (§3.18) was the pastor Michael Stifel.[1] His principal work was his *Arithmetica integra* (1544), published when he was about 60 years old, followed the next year by a less ambitious *Deutsche arithmetica*, in which he explicitly avoided Latinized words in his German. Here he popularized the use of '+' and '–', and of letters (often Cossist symbols) to represent powers (for example, 'FFFF'). He also introduced the square root sign '√' and even extended it to, say, '11√38' for $\sqrt[11]{38}$. This type of consideration brought him to 'exponents' (his word), which he realized could take negative values: for example, $\ldots, 4, 2, 1, \frac{1}{2}, \frac{1}{4}, \ldots$ took exponents $\ldots, 2, 1, 0, -1, -2, \ldots$ (compare Chuquet's (3.19.2)). The general notion of the logarithm was within Stifel's grasp; but it was to be seized upon by others, as we shall see in §4.15.

The representation of numbers by numerals led to an exotic collection of notations for powers, ratios and roots {Cajori 1928, pp. 271–328}. One reason was the common perception that integers, rational and irrational numbers were not normally regarded as the same kind of object; but such nervousness was swept aside by Stevin in 1585 (his 38th year) with the booklet *De Thiende* ('The Tenth') and especially the book *La Pratique d'arithmétique*. Bunching all

[1] Stifel vented his hatred of Catholicism in gematric form: he tried (but failed) to make 666 the number of 'leo decimus' and thereby associate the Beast (§3.10) with Pope Leo X. He was close friend of Martin Luther – who, however, disparaged gematria. Another numerologist of this tendency was Johannes Faulhaber (1580–1635), perhaps the last major Cossist; in the 1620s he interacted with the young Descartes.

types of number together, he sought a common notation, principally on the grounds that magnitude would then be easier to determine. This was a good point – try telling whether, for example, $\frac{53}{28}$ is less or more than $\frac{89}{47}$. He found a solution in a form of decimal notation in which, for example,

2913.7986 was rendered as '2913⓪7①9②8③6④'. (4.5.1)

(At that time '.' was often used only as a separator of symbols.) Stevin was not the first to suggest a general notation, but he made especially clear why it was needed. Further, as with Stifel though from a different point of view, he also prepared the ground for logarithms (§4.15).

4.6 *Cardano and Bombelli: complex numbers, more or less.* {Freguglia 1988, CE 6.1}[2]

All figures mentioned so far considered equations; Recorde, for example, saw them as a means by which knowledge surpassed art. A principal research topic, especially among the Tuscans, was determinating the roots of cubic equations and of higher-order equations. However, negative numbers were still widely distrusted, so any method for extracting roots had to be modified for the various cases of positive and negative coefficients in the equation (compare (3.6.1) for al-Khwarizmi). In addition, the language was largely symbolic but partly verbal; for clarity, I shall normally use the modern version, which is anachronistic by about a century (§4.18).

The cubic was solved by Scipione del Ferro (1465?–1515) around 1500 at Bologna, and then by Niccolò Tartaglia in

[2] Among a rich historical literature on these developments, the *Bollettino di bibliografia e di storia delle scienze matematiche* (1866–87) contains various valuable articles in its 20 volumes. Later, many papers were written by Ettore Bortolotti from the 1920s onwards, and by Gino Arrighi after the Second World War; but unfortunately almost all of them are buried in Italian periodicals which are difficult to find in other countries.

1535; but neither man gave out the details. However, Cardano heard about the success and, after eventually prising the secret out of Tartaglia, he published the method in his *Ars magna* (1545, his 45th year). A tremendous dispute then ensued.

The crucial insight for cubics was this: by a linear transformation of the unknown the term in x^2 could be removed, so that no generality was lost by taking only forms such as

$$ax^3 + bx + c = 0, \quad ax^3 + c = bx, \quad ax^3 = c + bx \qquad (4.6.1)$$

with $a, b, c \geq 0$. For the first of (4.6.1) with $a = 1$, which Cardano would have written something like

'1.cubus.p.b.pos.p.c.aeq.0', (4.6.2)

a root was given by

$$x = \sqrt[3]{-\frac{c}{2} + r} + \sqrt[3]{-\frac{c}{2} - r}, \text{ where } r := \sqrt{\frac{c^2}{4} + \frac{b^3}{27}}. \qquad (4.6.3)$$

A remaining mystery was the determination of the other two roots (the existence of three being taken for granted) by using the appropriate powers of a complex cube root ω of unity to multiply the two cube-root terms of (4.6.3) respectively by ω and ω^2 and by ω^2 and ω; this was tidied up only in 1771 by A.T. Vandermonde. A deeper issue, evident already with quadratics, was the strange situation that if a cubic with real coefficients possessed only real roots, then still one had to wander into complex numbers to find them ((4.6.5) below is an example). The links with geometry were somewhat weakened thereby, but they remained important; Cardano demonstrated the completion of the square and indeed of the cube by arguments drawing upon diagrams like FIG. 2.5.2.

In his book Cardano also included the first method for solving the quartic equation Q, due to his protégé Lodovico Ferrari (1522–69). (He assumed that Q lacked a term in x^3, but this is not necessary.) Introduce a new unknown y in such a way that the equation converts to a form F given by $(x^2 + \cdots)^2 = p(x, y)$ (p a quadratic in x and y); assume that $p(x, y)$ is also a square; find that y has to be a root of a cubic in x, and is therefore calculable; take the square roots of each side of F, and so reduce it to a quadratic in x. This method added to the contretemps after 1545.

In later work, especially a book called *De aliza* (1570), Cardano tried hard to resolve (or avoid) not only complex but even negative (to him 'alien') numbers, arising in isolation rather than just in subtraction {Tanner 1980}. His starting point was that '+' multiplied and divided into itself but that '−' changed signs. He tried to develop an algebra of '+' and '−' to cover root extraction as well as the four standard operations, with rules such as (in a symbolic notation)

$$- \times + = -; \ - \times - = - \ \text{(sic)};$$

$$\therefore - \div - = \pm \text{ and } + \div - = ? \qquad (4.6.4)$$

Cardano seems to have given his book its curious name as a result of mishearing the Arabic word *a'izzâ*, meaning 'risky' or 'doubtful'; and this allusion is typical of his (and others') respect for the Arabic origins of algebra. His conception of algebra was oriented around its applications to arithmetic and geometry, and the subject was still largely concerned with calculating or approximating to numerical values. But an important move towards autonomy for algebra was effected by the engineer Rafael Bombelli, in his treatise *Algebra*, published in 1572, his 46th and last year. A few years earlier, a colleague had discovered in the Vatican Library some hitherto unknown transcriptions of texts by Diophantos on indeterminate analysis. Bombelli's reading of these manuscripts, and his adhesion to the strong neo-humanist sentiment of the time of recovering the wisdom of antiquity, shaped the writing of the treatise. Thus his algebra became both more abstract in its focusing upon *properties* of numbers rather than numerical values, and more autonomous in its pointing towards a new branch of mathematics {Parshall 1988} – hence, no doubt, his striking choice of the one-word title.

As part of this change, Bombelli changed jargon from the normal *cosa* and *census* for x and x^2 respectively (§3.17) to the translations *tanto* and *potenza* of the Diophantine 'number' and 'power'. He worked though the various cases (with only positive coefficients) of cubics and quartics. He showed that the cube roots would reduce to the linear form $(a+bi)$ (where I write 'i' for $\sqrt{-1}$); moreover, he distinguished i from $-$i, as respectively 'p.[iu] di m.[eno]' and 'm.[eno] di m.[eno]' (a ring of Cardano here). To take one of his examples (not new with him),

$$x^3 = 15x + 4, \quad \text{root } x = \sqrt[3]{2 + \mathrm{i}11} + \sqrt[3]{2 - \mathrm{i}11} \quad (4.6.5)$$
$$= (2 + \mathrm{i}) + (2 - \mathrm{i}) = 4.$$

I have used modern symbols; he extended algebraic language to denote operations by single letters, so that

$$\sqrt[3]{2 + \mathrm{i}11} \text{ was written 'Rc}\underline{|2\text{pRq}|0\text{m}121|}\text{'}; \quad (4.6.6)$$

that is, the (or a) cube root of 2 plus (positive square root of [0 minus 121]); compare Chuquet at (3.19.2).

This expression exemplifies another matter which much exercised Bombelli: the role of complex numbers in algebra. It must have helped to convince him that algebra had to be made more independent of geometry; he did not complete the parts of his book containing algebraic solutions to geometrical problems, and may not have been sure how to proceed. But in a manuscript on the classical problem of trisecting the angle (§2.28), he formed a cubic equation equivalent to the formula for $\sin A$ in terms of $\sin (A/3)$, and showed that it possessed three real roots, each corresponding to a geometrical construction impossible to effect by straightedge and compass {Bortolotti 1923}. His analysis showed how the new algebra could lead to a better understanding of classical geometrical problems.

4.7 *Algebra as analysis: Vieta's expanding vocabulary*

Among Bombelli's followers, the most significant and enthusiastic was Franciscus Vieta. He contributed to Diophantine analysis and to algebraic treatments of the Greek classical problems, and introduced a (rather clumsy) algorithm to approximate to a root of a polynomial. (It is not known for certain that Vieta had *read* Bombelli; but their views are consonant.) A major work was the short *In artem analyticem isagoge* ('Introduction to the Analytic Art') of 1593, published in his 54th year. It is translated in {J. Klein 1968}.

Vieta started with the art of 'zetetics' ('seeking the truth'), followed by the 'poristic art' (an unhappy choice of term, to make a link with synthesis), and finished off with the 'rhetic art' (where applications were made to particular cases). Thus he began by taking an unknown constant as if its value were known, and obtained its value in terms of actually known constants. In this he was imitating the Greek proof-method by 'analysis', of assuming a theorem that was to be proved as already proved and using it to derive results already accepted as true or axiomatic. Now, as we saw in §2.27, Pappos did not associate either 'analysis' or its contrary 'synthesis' with any branch of mathematics, and this new link of analysis with algebra was to confuse the meaning and lead to obscurity as well as light (§6.10, §8.5).

Vieta played fanfares for the autonomy of algebra; in particular, he did not rely on geometrical diagrams or proofs to establish algebraic results {CE 6.9}. As part of his 'symbolic logistic' he advocated the use of upper-case vowels such as '*A*' for unknowns ('scalar quantities') and consonants such as '*Z*' for knowns ('comparative quantities'); he treated equations on a par with proportions, and stated them without any specific numbers at all. Yet traces of geometry remained. When naming powers, in separate collections of adjectives, he followed tradition in allowing for the three dimensions of space: respectively 'latus' (Latin for 'side') for 'radix' (A^1), 'quadratum' (A^2) and 'cubus' (A^3), but then 'quadrato-quadratum' (A^4), 'quadrato-cubus' (A^5), . . . , 'cubo-cubo-cubus' (A^9); similarly, 'longitudo' or 'latitudo' (Z^1), 'planum' (Z^2), 'solidum' (Z^3), 'plano-planum' (Z^4), 'plano-solidum' (Z^5), . . . , 'solido-solido-solidum' (Z^9). Multiplication and division respectively raised and lowered dimensions. Square roots were notated 'L' for 'latus': L36 = 6. Only quantities of the same dimension could be added or subtracted; for example, $(AAA - ZA)$ was legitimate only if Z were planum.

After Vieta, various words, not necessarily his, were used to name powers. In particular, 'potentia' ('power') itself referred *only* to the square, and even took a companion verb, meaning 'equal in power': thus

$$\text{'}a \text{ potest } b \text{ et } c\text{' stood for } a^2 = b^2 + c^2. \tag{4.7.1}$$

Elsewhere, Vieta himself treated certain aspects of geometry, such as the order of contact of a tangent to a curve and the idea of a circle as an infinitude of tiny contiguous lines and angles. He also showed that the construction of the lune given by one inner chord and four outer ones (FIG. 2.28.1 shows two and three respectively) could be reduced to solving a cubic equation in the cosine of a related angle. Again, in making an Archimedean exhaustion of the circle by polygons of 4, 8, . . . sides, he found one of the loveliest of all formulae for π:

$$\frac{2}{\pi} = \sqrt{\tfrac{1}{2}}\sqrt{\tfrac{1}{2} + \tfrac{1}{2}\sqrt{\tfrac{1}{2}}}\sqrt{\tfrac{1}{2} + \tfrac{1}{2}\sqrt{\tfrac{1}{2} + \tfrac{1}{2}\sqrt{\tfrac{1}{2}}}}\sqrt{} \ldots \tag{4.7.2}$$

Vieta's position between algebra and geometry was also evident in his innovations in *algebraic* trigonometry. His *Canon mathematicus* (1571–9) and later writings did much to highlight that part of trigonometry concerned only with angles, which became known in the 18th century as 'goniometry', after the Greek word 'gon' for 'angle'. He produced tables of values for both planar and spherical triangles, in which he drew on new formulae such as

(in our rendition)

$$\sin(60° + A) - \sin(60° - A) = \sin A \qquad (4.7.3)$$

$$\text{and } \operatorname{cosec} A \pm \cot A = {}^{\cot}_{\tan}\tfrac{1}{2}A.$$

4.8 *Trigonometry on land and sea*

Vieta's work on trigonometry related to his life-long interest in geography and cosmography. Economic and political movements from the late 15th century onwards greatly increased the need for maps and charts. The mathematical problem here is one of the greatest of all: how to represent shapes (countries, islands, oceans, and so on) on a curved surface (that of the Earth) as shapes on a plane (the map or chart) without losing or distorting too many of their properties. It had become evident that the question split in two, in that on land the needs that arose through ownership, taxation and travel necessitated the preservation (if possible) of shape and size, while at sea the dominant requirement was for simple and reliable means of navigation and orientation; the former needed lengths, while the latter sought angles. So the distinction had grown up between maps and charts to fulfil the desired aims – maps for marking out our country and those of the enemies, charts for tracking our and their ships.

Of the two aims, that of charting grew in especial importance, with the major European countries rivalling each other to cross the seas and invade ('discover' was the normal term) the 'New World'. In Spain and then Britain, for example, the significance of sea-travel was beautifully reflected in the 'ruff', the man's white collar whose undulations represented the ocean waves. In Elizabethan England, the ruff was also called the 'pride'; it was worn over the 'falling-band', the white lace collar worn over the shoulders and breast. This is one source of the phrase 'pride comes before the fall'.

Navigation had usually been based upon the 'planar' chart, for which the surface of the Earth was taken to be completely flat, with latitude and longitude lines equally spaced grid-fashion instead of converging to the poles; it was used in conjunction with the magnetic compass and depth sounding-line {CE 8.18}. Satisfactory as an approximation for shortish journeys such as around the Mediterranean Sea, it was too crude to meet the navigational requirements of the longer voyages now being undertaken; so astronomy and trigonometry were brought in. Latitude was determined by observing the Sun by day or the Pole Star by night. For longitude the west coast of Africa luckily provided a good guide for that part of the Earth's surface, but elsewhere the navigator was often literally on his own, with the erratic motions of the Moon and perhaps an unreliable timepiece as his only aids.

So new types of chart were produced. The most distinguished was the *Nova et aucta terrae descriptio* (1569, his 57th year) by Gerardus Mercator, cosmographer to the Duke of Cleves, one of the many independent German states of the time. He seems to have projected the spherical Earth onto a circumscribed cylinder (which was then unrolled onto a plane) not by the usual direct projection from the centre but in a special way in which latitudes and longitudes were mapped into a rectangular lattice of straight lines. The straight path of a ship, the 'rhumb line' or 'loxodrome' (from the Greek for 'inclination'), followed a constant direction on its surface; it was mapped onto a line which conformally cut other lines, and lines of the lattice, at *constant and correct* angles. Thus accurate estimates of direction could be made rapidly (though at first sailors found the method hard to understand).

Chart-makers needed an accurate method for locating and laying out the parallels of latitude, and a clarification of Mercator's methods; these were supplied by the English navigator Edward Wright in 1599 (with the encouragement of Dee, who had met Mercator and had obtained some of his globes). FIG. 4.8.1 shows a quadrant of the Earth, centre C and poles N and S, with a slice of the equator E_1E_2 and the corresponding slice T_1T_2 at latitude λ; T_1 is projected to

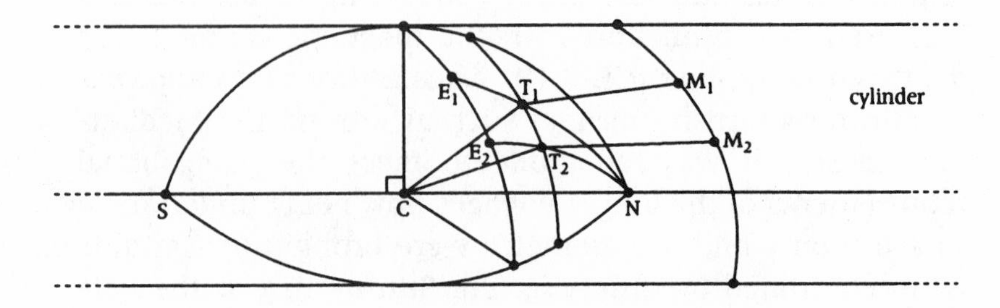

FIG. 4.8.1 *Mercator's cylindrical-type chart projection.* A quadrant of the Earth, centre C and axis NS, is shown; $\angle E_1CT_1 = \angle E_2CT_2 = \lambda$.

M_1 on the cylinder, and T_2 to M_2. Mercator showed that

$$E_1E_2 = T_1T_2 \sec\lambda \quad \text{and} \quad \delta|T_1M_1| = \sec\lambda\, \delta\lambda, \qquad (4.8.1)$$

where I use 'δ' to indicate a slight change of length or angle, and the modulus signs in '$|T_1M_1|$' to indicate distance *on the chart.* (The name 'secant', as well as 'tangent', was introduced by Thomas Finck in his *Geometria rotunda* (1583).) The first of the above equations gives the factor by which T_1T_2 should be expanded when laid on the chart. The second is presented for convenience like a differential equation; such equations did not exist then, so tables of the values of its 'integral' were compiled (by Harriot, among others) for the parallels of latitude to be located {Pepper 1968}.

On the Mercator chart, sizes, shapes and distances were greatly distorted, ever more so as latitude increased; but that was not a concern. However, to this day it is the normal so-called 'map' of the world, despite these malformations. FIG. 4.8.2 contrasts it with the recent map produced by Arno Peters, in which the shapes and relative sizes of areas are preserved much better.

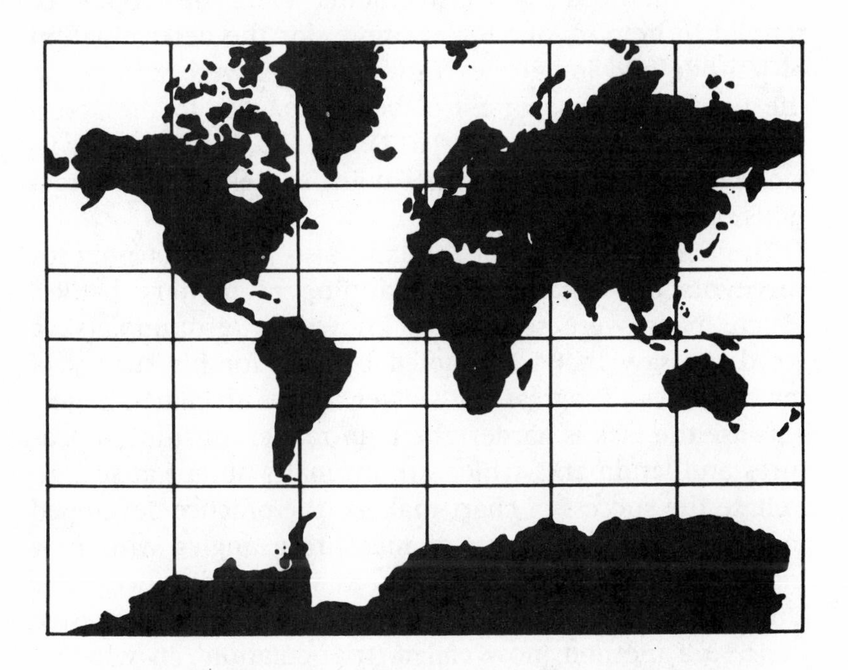

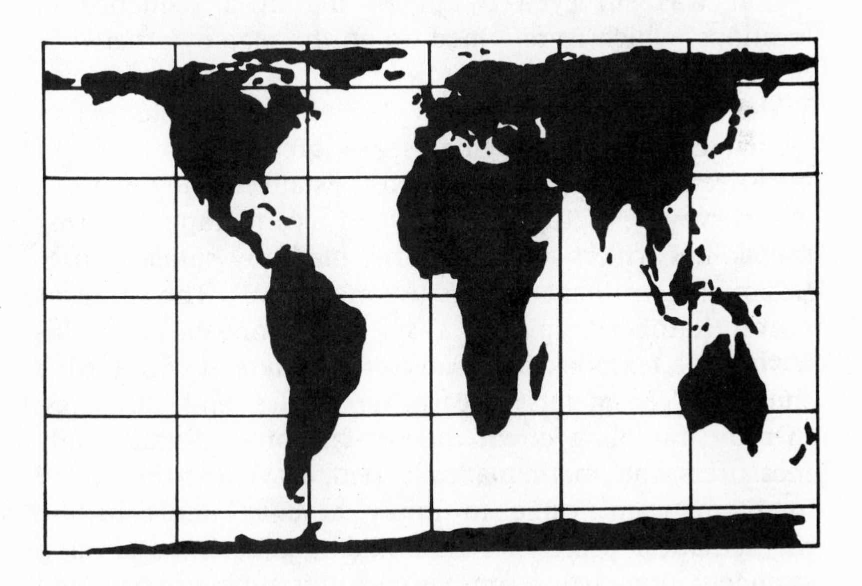

FIG. 4.8.2 *The Mercator and the modern Peters projections of maps of the world.*

Other methods and instruments were developed or refined to help sailors; for example, for the determination of latitude and longitude. Various essays were written on the theory and use of astrolabes. One, due to the Jesuit Christopher Clavius (1537–1612) in 1593, is notable for its pioneering use of the decimal point to help indicate fractions of a second.

The needs for maps of the land also grew, not only for surveyors but also for the planning of military battles. (Such maps were sometimes known as 'groundplats', a word we saw in §4.3 adopted by Dee for his survey of mathematics.) Progress was slower than with charts, largely because the task is harder: the map-maker must locate features and landmarks which are normally absent at sea. To imitate the success of chart-makers, the practice developed in the late 16th century of measuring angles with new instruments such as theodolites, as well as distances. The Dutch map-maker Gemma Frisius (1506–1555) suggested in 1533 a method, now called 'triangulation', in which a region was surveyed by measuring at a sequence of locations, which were joined up on the map by a network of triangles.

The new professionals in Britain called themselves 'geometers', in the original sense of the word, 'Earth-measurers' by angle as well as distance. They worked in 'surveying', 'topography' and 'cartography', using theodolites, rulers and quadrants made by similarly professional instrument-makers {Bennett 1991}. The surveyor Aaron Rathborne placed a splendid frontispiece in his influential textbook *The Surveyor in Foure Bookes* (1616) showing 'geo-meters' measuring angles and distances, thereby trampling down the old-fashioned distance-only measurers, the 'mathematicalls' (FIG. 4.8.3). Thus the word 'mathematician' came to imply a *lower* status in the mathematical profession than did 'geometer'; and in a strangely long-lasting consequence this snobbery persisted for more than two centuries.

FIG. 4.8.3 *The geometers triumph over the 'mathematicalls': the frontispiece of Aaron Rathborne's* The surveyor, *1616.*

4.9 *Differing formulations for mechanics*

We turn now to mechanics and astronomy, which passed though crucial phases during the period covered by this chapter and the next. First, some general comments are necessary.

At least the following collection of distinctions was involved, but often each of them was dimly perceived or missed altogether: space, place and distance; rest and inertia; motion and change of motion; uniform or non-uniform motion; speed, velocity and acceleration; motion in a straight line, or a nearly straight line, or a circle, or some more complicated curve(s); mass, weight, volume, density; force, endeavour, impetus, conatus and other terms out of the medieval bestiary (§3.30); and continuity and discontinuity, with consequences for collisions (themselves of various kinds). Underlying the discussions of such matters lay questions of generality (for example, over types of body), causation (at the time, objects usually would '*be* moved' rather than our 'move'; but by what?), and of absolute or relative motion.

In §4.10–§4.14 a small selection of important and/or typical contributions is presented, from the work of Copernicus, Galileo and Kepler; Descartes's contribution is taken up in §5.8. Many more minor but distinguished figures are regrettably omitted. But even within this sample, a remarkable *variety* of ideas as well as of influence is found. Good general histories include {Koyré 1957}. Introductory surveys are available in the companion history of astronomy in this Fontana series {North 1994, Chaps. 10–12}, and in the appropriate volumes of the series *The Cambridge History of Astronomy*. {Barbour 1989} surveys the distinctions between absolute and relative motion made, often unclearly, from the Greeks to Newton. He also proposes the nice word 'motionics' as the analogue to 'kinematics' for dynamics – that is, with no assumption of cause.

Progress in statics lay mainly in general theories of balance and equilibrium. Stevin publicized a beautiful thought

FIG. 4.9.1 *Stevin's wondrous illustration of equilibrium on the inclined plane, 1586.*

experiment by showing it on the title pages of his books from 1586 onwards (FIG. 4.9.1). The inclined plane and loop of weights are in equilibrium; they remain so when the dangling part is removed; hence we may deduce the law of components of reaction on the plane, and 'the wonder is no wonder' (as he wrote there). This use of weights recalls Jordanus de Nemore's analysis at FIG. 3.29.1; Stevin might have seen the published version of this analysis, which appeared in 1533. But most of the principal problems came from dynamics.

4.10 *Copernicus: astronomy in circles*

{Neugebauer 1968; CE 2.7}

A significant scientific innovation in this period was the 'Gestalt switch' proposed by Copernicus: the heliocentric hypothesis that the Sun rather than the Earth was the stationary member of our planetary system. Some of his doubts about Ptolemy's theory seem to have been inspired by reading in Peurbach's abridged account of the *Almagest* (§3.28) of the variation in apparent size of the Moon, which was far greater than as observed (§2.25). Other doubts may have come from reading Arabic astronomers. He may have started his own book in the mid-1510s, when he was in his early forties; but the refined and definitive theory 'on the revolution of the celestial spheres' came out only in 1543 – in his dying days, and thanks to the editorship of Rheticus (§4.1). The book began with an anonymous preface stressing the hypothetical character of the theory, written by Copernicus's associate Andreas Osiander; he has earned much opprobrium for it, but it is a fair appraisal.

To be more precise, Copernicus argued for both the heliocentric hypothesis and also for the rotation of the Earth about its axis. He was well aware of their complementary character, pertaining respectively to celestial and terrestrial mechanics, for he referred to the 'twofold' character of motion of objects on the Earth. He also admitted, as a third motion, a conical rotation of the Earth about its axis, in order to compensate for its rotation about the Sun and so keep the stars in the same (apparent) positions. The compensation was not exact; the difference was used to explain the precession of the equinoxes. Not the first person to suggest any of these motions, he seems to have come to both largely though his own cogitations. The first one was by far his main concern.

A principal consequence of the theory was that some motions of the planets across the sky, especially the periodic reversals of direction now called retrograde motion, could be interpreted as artefacts of relative motion with respect to the moving Earth, rather than their real paths (with respect to a stationary Earth), as Ptolemy seems to have assumed. The issue of absolute versus relative motion was thereby heightened.

Copernicus did not seem to follow Ptolemy's Aristotelean

belief in the existence of real spheres (or shells) within which the heavenly bodies moved; for him *all* spheres, circles and epicycles were geometrical constructions {Aiton 1981}. However, many details of his theory were similar or even identical to those of his predecessor; indeed, even the structure of his book resembles that of the *Almagest*. Here are a few examples. He too allowed each planet to have its own centre of rotation, and placed it on an epicycle. He used third-order circles for some motions, and he deployed similar geometrical constructions to analyse motions, often with a strong trigonometric element. He also avoided discussion of the cause of motion: for example, one of his arguments in favour of circular orbits was the aesthetic claim that they were appropriate for spherical bodies like planets.

Many of the structural features exhibited by Copernicus's theory also imitated those of his predecessor. For an outer planet P (that is, one beyond the Earth O) the Ptolemaic deferent and epicycle became respectively the orbits of P and of O, and for an inner planet the orbits of O and of P. On the left of FIG. 4.10.1 stands a Copernican representation of the motion of P: the centre C of its epicycle was travelling

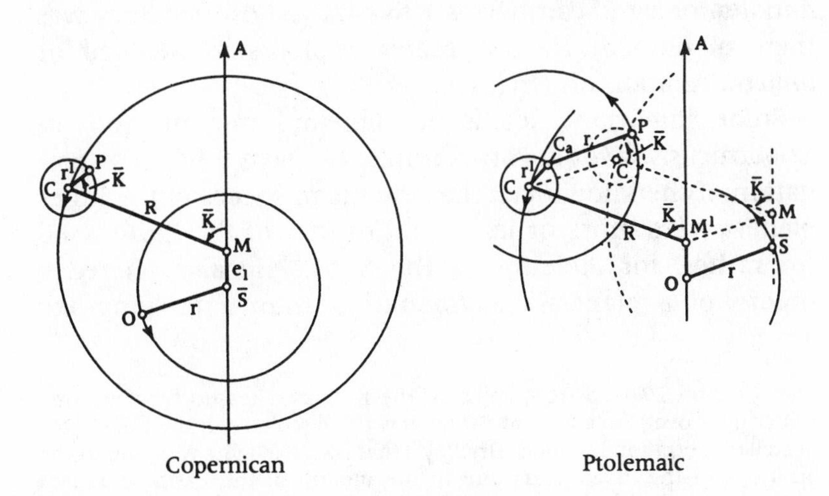

FIG. 4.10.1 *Epicyclic motions {Neugebauer 1968, p. 93}.* The eccentricities SM and OM′ are exaggerated.

around the deferent circle with centre M on the line of apsides SA such that ∠PCM = ∠CMA. The observer sat on O which revolved on a circle with centre at the mean Sun S (an imagined object which traversed the ecliptic at a uniform speed with the same period as that of the Earth's revolution about the actual Sun). In order to switch to a Ptolemaic alternative, the centre of rest is transferred to O by the parallel translations illustrated by the diagram on the right, long known to astronomers: S was now a traveller, and P used an extra epicycle centred on C_2. In a fully Ptolemaic version C_2 would lie on the deferent circle for P.

So Copernicus's theory was less radical a revision of Ptolemy's than is often thought, or indeed than he claimed; indeed, the phrase 'Copernican Revolution' dates only from the 18th century. The assumptions of Copernican astronomy have since been corroborated by observations of various kinds, but at the time observational evidence that would favour it over the Ptolemaic system was not too clear-cut. Copernicus's epicycles were slightly fewer in number, and in general of smaller radius; but the details of the circular and epicyclic motions were neither easier to handle nor significantly closer to data (whose accuracy was then still limited). He laid greater emphasis on the need for *uniform* rotation on circles.

Some important details of planetary motion were in Copernicus's favour. For example, he gave a better explanation of the known fact that the mean speeds of the inner planets were more or less equal to that of the Earth itself (or rather, for Ptolemy, of the Sun).[3] He also offered a theory of a planetary *system* rather than a Ptolemy-like

[3] See {Wilson 1970, p. 109}. Perhaps the first good reason for preferring Copernicus over Ptolemy and Brahe was the discovery in the mid-1720s of stellar aberration by James Bradley – that is, the (slight) variation in the positions of the 'fixed' stars due to the motion of the Earth. Bradley's original record of his discovery is published in {R. Grant 1852, p. 624}. Galileo's discovery in 1610 of the phases of Venus refuted Ptolemy but not Brahe.

guide to the planets. In particular, he not only correctly identified the inner and outer planets and their order from the Sun, but also – and in a valuable innovation whose significance he did not seem fully to appreciate – he used his analysis of the motion of Sun and Earth to find good estimates for the mean distances of each planet from the Sun. He left open the question of the size of the universe, an issue which was to concern many subsequent astronomers and cosmologists.

Copernicus died with his book in his reluctant hands, and his doubts were to be justified by the decades of theological objection and reasoned criticism on unclear points (both from Clavius) that were to follow. Only around the end of the century did significant advance and advocacy arrive, with books by Kepler, Vieta, Stevin, Galileo and others.

4.11 *Kepler on the motion of an individual planet*

{Davis 1992}[4]

The next major theorist was Kepler. As a convinced Copernican, he put this theory into practice in order to establish that each planetary orbit lay in a plane of fixed inclination to the ecliptic, and that all such planes passed through the Sun.

Kepler abandoned not only the Ptolemaic geocentric system but also a compromise model proposed by Tycho Brahe in which the Sun and Moon circled the Earth while the five planets circled the Sun. But Brahe also provided grounds, which Kepler accepted, for rejecting the reality of the celestial spheres. When Brahe moved to Prague, Kepler joined him as his assistant.

After Brahe's death, Kepler set about completing the former's planetary tables. He sought to determine the orbits

[4] In general I follow Davis's interpretation of Kepler's astronomy, which differs from other accounts in emphasizing his reliance on Greek mathematics, especially Euclid and Archimedes. Among more traditional studies, {Wilson 1968} is recommended.

of the planets – which for him meant not only their paths but also their orbital velocities. In this he benefited from the accuracy of Brahe's accumulated observations – usually reliable to within 2′ (literally a hair's-breadth) rather than the then normal tolerance of about 10′. He devised an ingenious way of connecting angular position with angular measure of time to within 2′; but he called it 'the vicarious hypothesis', for the distances it predicted were too inaccurate, so that it could not be used to construct orbits. The main orbit which he tackled was that of Mars, which Brahe's assistant Christian Longomontanus had found impossible to determine. It was very suitable for study since it had a comparatively high eccentricity e (about 0.1), and Mars was relatively near and easily observable.

The main results of Kepler's efforts appeared in the *Astronomia nova* (1609), published in his 38th year. He based his analysis of the orbit of planet P in FIG. 4.11.1 on the eccentric circle with centre B and radius $BQ = a$, for this was equivalent to the simplest possible arrangement of epicycle (centre Z, radius $ZQ = ae$) and deferent circle (not shown in the Figure, centred on the Sun S, radius $SZ = BQ = a$). Rigorously trained in Euclidean geometry at Tübingen University, he naturally used construction by straightedge and compasses to cut off the required length along the extension of SZ. But the position Q on the eccentric circle, determined by the length $SK_1 = SQ$, turned out to be too large; so he tried again with the length SK_2 (where K_2 was the point of intersection of SZ with the eccentric circle), but this was too short. Brahe's data showed that each 'ovoid' (Kepler's word) orbit given in these cases was in error by not more than $\pm 8'$. After four years' struggle, too complicated to describe here, he determined the length SP in two ways:

$$SP = SK = SZ + ZK \quad \text{and} \quad SP = QR = BQ + BR; \qquad (4.11.1)$$

he then found that the orbit obtained by taking the radius

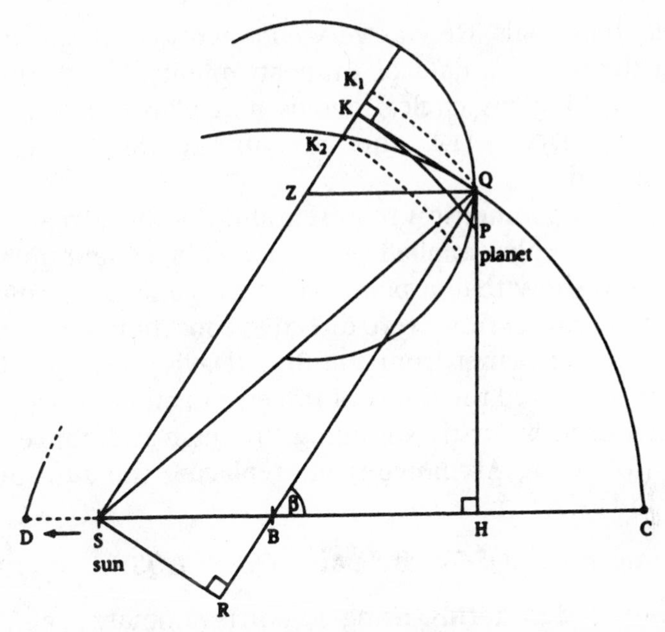

FIG. 4.11.1 *Kepler's Euclidean method for constructing orbits.* The eccentricity SB = ae is exaggerated. BC = BQ = a.

to be SK = SP was right on target (to make a martial pun, in line with his own description of his battle with Mars, 'the God of War'). In terms of the eccentric anomaly $\angle QBC = \beta$ and radius vector SP = r,[5] this defines the curve by

$$\text{Kepler's 'first law':} \quad SP = r = a + ae \cos \beta; \quad (4.11.2)$$

he proved it to be an ellipse from the proportion PH : QH :: $b : a$, where b was the length of the semi-minor axis. Moreover, the ellipse seemed to carry respectable

[5] In {Whiteside 1974} the comparison of Kepler's various orbits is modernized by the use of trigonometric power series. They all take expansions of the forms

$$1 + e \cos \beta + A e^2 \sin^2 \beta + Be^3 \sin^2 \beta \cos \beta \pm O(e^4),$$

and are classified into grades by the magnitude of the first coefficient A. Kepler's approach was mathematically sound, for the shape alone is affected by the term in e^3. It just happens that the ellipse arises when $A = B = \cdots 0$; Kepler did not find it from its own properties, neither did he then specify the Sun at one focus.

Greek credentials. He was very conscientious about reconciling theory with data of Brahean quality, for quantitatively speaking his circles, ovoids and ellipse differed but little (in FIG. 4.11.1 the eccentricity SB is greatly exaggerated).

Seeking a geometrical representation of time, in an early expedient Kepler applied the Ptolemaic equant method (§2.25) to show that at points close to C and D on an orbit small intervals of time were directly proportional to the distances of the planet from the Sun {Davis 1992, pp. 109–14}. He measured the time t of traverse to an arbitrary point P of the orbit by firstly summing 'successive' distances and then (following Archimedes in) replacing the sum by an area; in FIG. 4.11.1,

$$\text{time} \propto \text{area QSC} = \text{circle area QBC} + \Delta\text{QSB}. \qquad (4.11.3)$$

In more modern terms, using eccentric anomaly again, this relation is

$$\text{Kepler's 'second law':} \quad t \propto a^2\beta + a^2 e \sin\beta. \qquad (4.11.4)$$

However, this law was widely ignored until Newton generalized it (§5.10); in the interim, equant arrangements were preferred. In addition, the transcendental form of (4.11.4), which Kepler called 'heterogeneous', led to the separate challenge of solving 'Kepler's equation' (given t, find β). He offered an iterative method of solution; various successors were profitably to take up the question (for example, Lagrange at (6.5.5)).

These two laws indicated the location and the timing of a planet in its orbit, to observationally satisfactory standards. Kepler made an important innovation in realizing the need to provide proofs; he fulfilled this by identifying a cause for each law, finally, in Book 5 (1621) of his *Epitome astronomiae Copernicanae*.

Kepler's reliance upon Aristotle obliged him to regard force as responsible for velocity (that is, both magnitude and direction), and not only for initiating a velocity but also for its continuation {Davis 1992, pp. 165–91}. For him, the only directions permitted by common sense were radial and transverse to that of a force. He turned to the Sun S as the source of both for a planet P – in FIG. 4.11.2, along and perpendicular to the radius vector SP as marked by single arrows. He thought that the transverse component of the motion was caused by the rays of the Sun flailing the planet around its orbit, due to its own axial rotation (a rotation

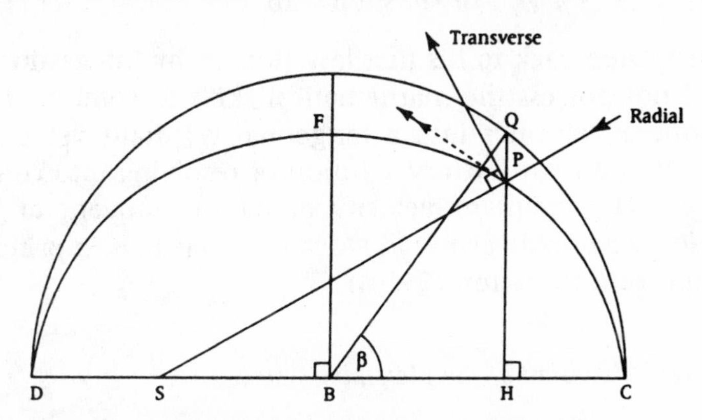

FIG. 4.11.2 *Components of motion in a Keplerian orbit.* BQ = *a*, SB = *ae*, BF = *b*.

which he had predicted in the *Astronomia nova*, and which had soon been corroborated by the observations of sunspots by Christoph Scheiner, Galileo, Harriot and others). He also proposed that the strength of this force decreased as distance increased according to a linear law:[6]

$$\Delta\beta/\Delta t \propto 1/r; \quad \therefore, \text{ from (4.11.2)}, \quad \Delta t \propto (1 + e\cos\beta)\,\Delta\beta. \tag{4.11.5}$$

He sensed that a process akin to integration would get back to the second law, (4.11.4).

Kepler also found mathematically that the radial component of the motion along SP depended upon the position of P in its orbit, so that the Sun could not be directly responsible. Hence, in an extension of the claim made by

[6] In fact, transverse motion is always inversely linear in r for orbital motion; for, taking θ as the polar angle, in Newton's formulation $\Delta\theta/\Delta t \propto 1/r^2$; $\therefore$ for transverse motion $(r\,\Delta\theta/\Delta t) \propto 1/r$.

the physicist William Gilbert in *De magnete* (1600) that the Earth was magnetic, he proposed as cause that the Sun activated magnetic fibres within *any* planet P. This could not be a force, since the fibres did not lie in the direction of the motion. He used this complicated mechanism to argue for the proportionality

$$\Delta r \propto ae \sin \beta \times \Delta\beta, \tag{4.11.6}$$

which related back to his first law (for us, by integration). He did not possess the mathematical skills to combine his component velocities into a tangential resultant velocity, but he found a satisfactory solution of resolving into components. The resultant velocity along the tangent at P, marked by a double arrow in FIG. 4.11.2, was to be a principal concept for Newton (§5.10).

4.12 *Kepler's vision of the planetary system*

There was another and more speculative aspect of Kepler's work which drew upon mathematics. In his first book *Mysterium cosmographicum* (1596), published in his 26th year, he proposed the remarkable idea that the five Platonic solids could explain why there were only six planets (those known to exist at the time). He nested them to produce a model of the spacings (FIG. 4.12.1) as follows:

Saturn – cube – Jupiter – tetrahedron –

Mars – dodecahedron – Earth – icosahedron –

Venus – octahedron – Mercury.

His matching of solid with planet was not arbitrary, but was based upon aesthetic associations. Later, in his *Harmonice mundi* (1619), he used Brahe's data to obtain a better fit; he also showed that the 13 Archimedean semi-regular solids formed a complete list.

The *Harmonice mundi*, with his ontology announced in its title, includes also Kepler's third law of planetary motion, as a proportion: the periods of revolution of any two

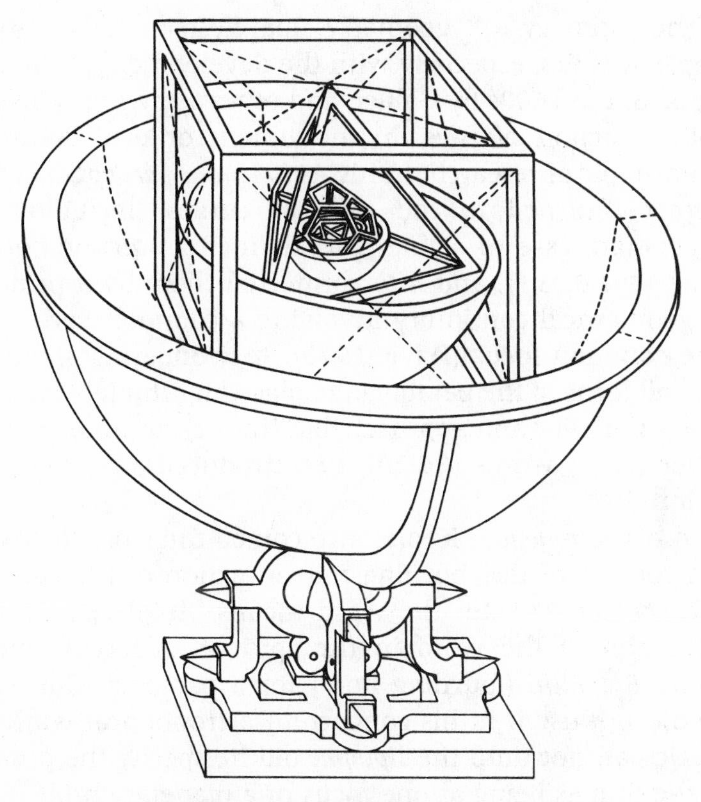

FIG. 4.12.1 *Kepler's model of planetary spacings (*Mysterium cosmographicum, *1596).*

planets stood in the ratio of the $\frac{3}{2}$th powers of the lengths of the semi-major axes of their orbits. It is an exact relation between a body of negligible mass acted upon by a force acting from a fixed centre, and it holds for ellipses of high eccentricity as well as for circles.

To some extent Kepler drew upon mystical metaphysical doctrines: he saw ubiquitous examples of certain shapes as manifestations of God the geometer, who had made man in His image {Field 1988}. In addition, his Platonic convictions led him to use properties of abstract numbers (*numeri numerantes*) numerologically to explain phenomena of the world. These interests related to his concern with astrology, in which he was also skilled.

The primacy of geometry manifested itself also in Kepler's optics, especially with the development of the telescope in the 1600s by Galileo and others. Laws of reflection and refraction of rays, in instrument or eye, became a major topic of research. In his *Astronomiae pars optica* (1604) Kepler returned to the classical curves (including the degenerate case of a line-pair) defined as sections of the cone (FIG. 2.23.1), and interconnected them by a principle of geometrical continuity involving a common vertex and one common focus {Davis 1975}. In wondering about the second focus of the parabola, he placed it infinitely far away along the axis, and approachable from either direction – in other (later) words (§4.20), he introduced a double point of infinity.

In his *Astronomiae* Kepler also coined the word 'focus', in the context of the 'burning-mirror' action of the parabola (§2.22); for in Latin the word means 'fireplace'. He may have chosen the word in the spirit of Albrecht Dürer's name *Brennlini* ('burning line') for a parabola. Curiously, he did not use it in his concurrent astronomical works; in particular, not until the *Epitome* did he specify the position of the Sun as being at one focus of a planetary orbit.

In his *Epitome* Kepler demonstrated to his own satisfaction that his third law also applied to the 'satellites' of Jupiter; he had introduced this name in 1610 for the four moons which Galileo had observed through his telescope. This detail contrasts his theoretical contributions to astronomy with the strong observational approach of Galileo, to whom we now turn.

4.13 *Galileo between mathematics and experiment*

Here is a giant in mechanics, astronomy, physics and optics, a man who laid out basic notions for dynamics, and applied them to problems in hydraulics such as the tides; he also reflected upon mathematical method, and pioneered the use of the telescope in astronomy to make major discoveries about the heavenly bodies. He is perhaps best remembered for his defence of Copernican astronomy, but his arguments were based largely on mechanics (and especially dynamics) in general, which was his main mathematical concern. In this respect he differed from Kepler (and also Copernicus), for whom astronomy was the prime interest. Indeed, it is rather ironic that Galileo was put on trial, in 1632, ostensibly for defending Copernicanism;[7] for Protestant Kepler's *Epitome* had already gained the seal of disapproval in 1619 by being placed by the Catholic Church on their index of forbidden writings.

Apart from accepting the primacy of geometry, there was, as it were, a world of difference between their ways of analysing mechanical phenomena. Kepler put forward his ideas in a theological and metaphysical spirit, whereas Galileo was more reserved on religious matters. Kepler sought for explanations of phenomena, whereas Galileo preferred to remain at the kinematic level, without commitment to notions of (for example) force. Kepler is often notoriously difficult to follow, while Galileo was admired even in his own lifetime for his clarity, including his use of the dialogue form to present his arguments. In these dialogues his mouthpiece, Salviati, was aided by an offstage genius called 'the academician' whom he was too modest to identify.

Galileo associated mathematics with measurement, thereby hoping to unite reason with observation; but the balance he struck between these two areas is a matter of historical debate. For long he was revered for his experimental prowess and the way it help him form dynamical principles; but from the 1930s a generation of historians largely followed Alexandre Koyré {1957} in stressing the theoretical side of his work. ({McMullin 1964} contains an

[7] {Redondi 1987} has argued that Galileo was also prosecuted for advocating atomism, which went against the Catholic understanding of the transsubstantiation of bread in the Eucharist. {Russell 1995} raises the question of whether Galileo was condemned for asserting that the Sun was at the centre of the Universe or of the Earth's orbit.

assemblage of articles from both these traditions.) Over the past 30 years a middle way has been advocated by Thomas Settle, Stillman Drake and others, who have argued that Galileo's manuscripts show that he actually carried out many of the experiments which he discussed. Problems of chronology, however, remain difficult and perhaps intractable, for the dating of many texts is uncertain even after watermark analysis of his sheets of paper. The small selection of results and methods described below come largely from his *Discourses and Mathematical Demonstrations Concerning Two New Sciences* (1638) (to translate his intriguing title *in full* for once), but he had found many of them before; the years around 1610 seem to have been especially fruitful.

In various philosophical respects, Galileo accepted some of the criticisms of Aristotle. One of these concerned the occurrence of continuity (its importance was not doubted). One criticism, due to the Venetian G.B. Benedetti (1530–90) in 1585, goes as follows. Aristotle regarded a motion as split into discontinuous parts if a moment of rest occurred: for example (of a kind important for Galileo), between the upswing and the downswing of a pendulum. However, Benedetti pointed out that if you look straight ahead, and imagine that a point P is describing a closed circular path at the same level as your eyes, what you see is motion in a straight line (FIG. 4.13.1).

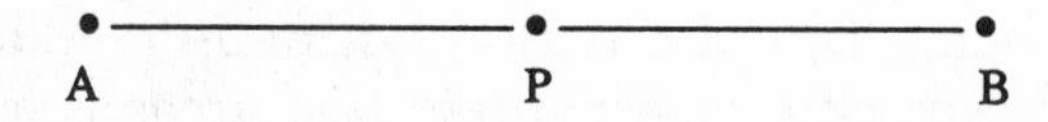

FIG. 4.13.1 *Circular motion viewed at eye-level.*

Now, according to Aristotle, there is a discontinuity each time P reaches A or B and then 'reverses'; however, in fact P is passing through A or B and continuing on its horizontal path, so that there is no reason to single out either point as important {Ariotti 1972, Section 1}. Galileo may have known of arguments of this kind.

4.14 *Galileo on the mysteries of fall*

{Wisan 1974}

This rebuttal of Aristotle was significant for Galileo in his study of the pendulum, which was one of his first interests. In 1602 he declared that the period of (for him, continuous) oscillation of the simple pendulum was independent of its amplitude and of the weight of the bob. Desimplification of this analysis was to become a lasting theme in mechanics (§5.12, §10.6, §14.12); Galileo started off some lines of enquiry by coming to doubt his own first claim.

The pendulum was an example of a falling body, the general laws for which were of prime concern to Galileo. He was encumbered by the gap between thought about fall in a vacuum and experience of fall in resisting air (and in focusing upon the former he displayed a move away from Aristotle). Naturally, he started off concentrating on velocity, but he came to see that acceleration was also a key concept. His experiments with bodies rolling down planes of various inclinations or dropped from a tower (maybe the famous one in Pisa) led him to two important proposals: firstly, that the velocity of a body B at a point P was proportional to time of travel; and secondly, that the differences between distances traversed in successive equal time intervals increased in proportion to 1 : 3 : 5 : . . ., so that the distance to P was proportional to $(\text{time})^2$. In this summary, I shall write '$t(\mathrm{X}, \mathrm{Y})$' to denote the time taken for B to travel from point X to point Y along some curve or line.

Another main result, known as the 'double distance law', stated that if a body fell from A to C along a slope and then moved horizontally to H, then

$$t(\mathrm{A}, \mathrm{C}) = t(\mathrm{C}, \mathrm{H}), \quad \text{where } \mathrm{CH} = 2\mathrm{CA}; \qquad (4.14.1)$$

Galileo knowingly neglected the effect of impact at C and the sudden change of direction there. He recognized the significance of CH in that it involved uniform horizontal motion, which was his most general concept of inertia. This

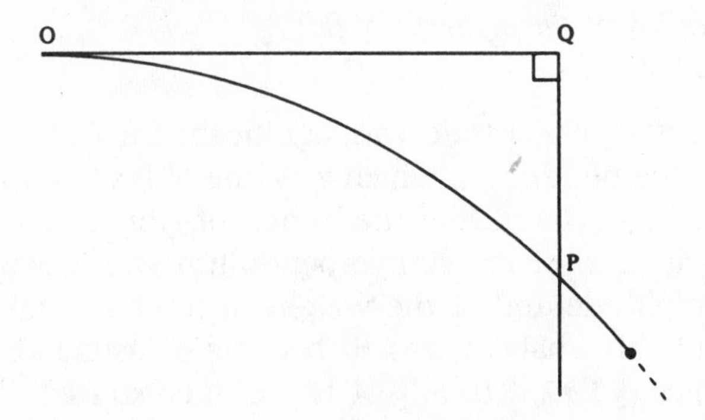

FIG. 4.14.1 *Galileo's parabolic path of a projectile.*

also arose in his work on the path of a projectile P (FIG. 4.14.1), for it suggested that OQ was proportional to $t(\mathrm{O}, \mathrm{P})$ while QP was proportional to $t^2(\mathrm{O}, \mathrm{P})$, and thus that the curve OP was a parabola. One application of this result was to determine the path of a projectile: with a vertical axis, as in the Figure, or maybe inclined to the vertical when allowing for air resistance. Harriot accomplished this also, in a sage note written unfortunately only for himself {CE 8.11}.

This analysis also exhibits another feature of Galileo's thought, in which he was somewhat of a pioneer: not only the composition of motions by the parallelogram law but also their *superposition*. He also came to affirm the importance in dynamics (or motionics) of uniform motion relative to the Earth (hence the later phrase 'Galilean invariance'); but he must have been aware that that direction, though locally horizontal, was circular around the Earth, not 'absolutely' rectilinear and hence reaching out into space.

In such ways Galileo generalized the twofold character of motions on the Earth noted by Copernicus (§4.10), but not always successfully. For example, he regarded the tides (that is, motion of the sea relative to the Earth) to be due only to the revolution of the Earth around the Sun and its rotation about its own axis; unlike Kepler before him, he granted no role to the force exerted by the Moon.

Another important result for Galileo was his law of

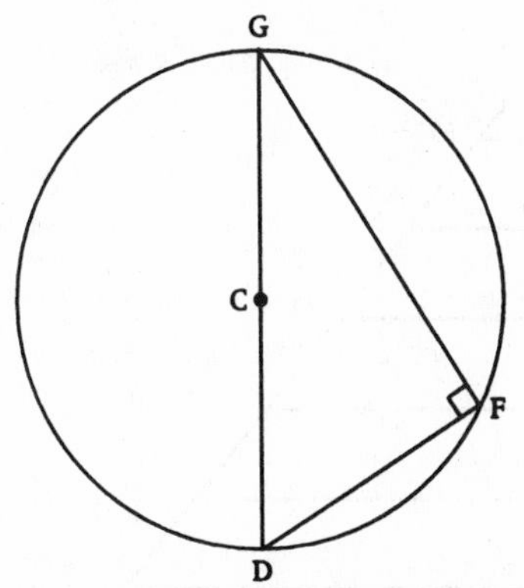

FIG. 4.14.2 *Galileo on the equal time of descent.*

chords, probably found in the early 1600s. A circle was set in a vertical plane (FIG. 4.14.2) and a body fell under gravity along GD, GF and FD (F an arbitrary point on its circumference); then the law stated that

$$t(\mathrm{G}, \mathrm{D}) = t(\mathrm{G}, \mathrm{F}) = t(\mathrm{F}, \mathrm{D}); \qquad (4.14.2)$$

that is, the time of descent along any chord from G or to D was constant. He also thought that the circle was the curve of quickest descent from F to D; but here he was to be corrected (§5.6).

Galileo's proofs often took the form of prosodic reasoning which drew upon diagrams in which lengths represented distances, or velocities or accelerations at an extremity. (On occasion they were muddled together on the same diagram.) Propositions were frequently expressed as proportions and sometimes manipulated algebraically and/or by using properties from Euclidean geometry. On occasion he used sequences of parallel lines to represent successive positions of a body. For example (FIG. 4.14.3), if a body was accelerating along AL in equal parts ab, bc, . . . such that its velocity $v(\mathrm{H})$ at any point H was proportional to AE, and thus also to EP on the intercept to the arbitrary sloping line AX, then

$$v(\mathrm{E}) : v(\mathrm{I}) :: \text{area of AEP} : \text{area of AIX} :: (\mathrm{AE})^2 : (\mathrm{AI})^2, \qquad (4.14.3)$$

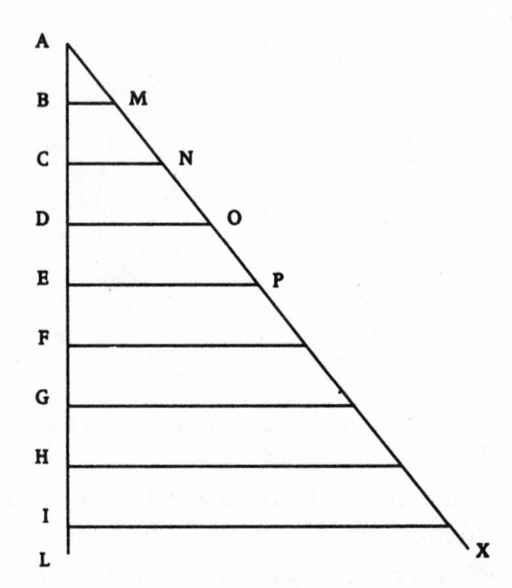

FIG. 4.14.3 *Galileo's proof of the law of descent.* {Wisan 1974, p. 208}

since the triangles were made up of the sum of all the horizontal lines down from A. Thus, as we saw with Kepler, he invoked the idea of an area taken as a sum of lines, which he would later criticize in Cavalieri (§4.19). (In fact, his argument was faulty, though not with respect to the point of view stressed here.) He also drew upon isomorphism between points on different lines, a type of thought which probably was to lead him to his famous 'paradox' in the *Two New Sciences* that the integers 1, 2, 3, . . . were isomorphic with the even ones 2, 4, 6, . . .

In many ways Galileo looks like a modern scientist, and is often presented as one. However, apart from obvious anachronisms his *immediate* influence was much less than might be inferred; in particular, being published in Italian reduced his impact abroad.

4.15 *Calculation by hand and by handle*
{CE 2.5}

Mathematics gained great publicity during this period when Pope Gregory reformed the calendar by taking 10 days out of October 1582, so that it would match up better with observed reality. The Jesuit Clavius played an important role in the preparatory calculations. The additional move of admitting as leap years only those at the end of centuries divisible by 4 brought about a very accurate system – for the Catholic domain, anyway; Britain accepted it only in 1752 (when New Year's Day moved from 25 March to 1 January), and Russia (as the Soviet Union) in 1918 {Aveni 1989}.

The modifications of the calendar can be expressed by continued fractions, and they may have been handled from ancient times by anthyphairesis (§2.13). In modern terms, the mean length of the solar year is 365.2422 days, or 365^d, 5^h, 48^m, 46^s. Now, in the notation of eqn (2.13.1),

$$2{,}422 : 10{,}000 = [0, 4, 7, 1, 3, 4, 1, \ldots], \qquad (4.15.1)$$

and the first four continuants give the following additions to 365 days:

1/4: too great an excess at 6^h; the intercalation of Julius Caesar.

7/29: 7 intercalary days in 29 years; 5^h, 47^m, 35^s.

8/33: 8 intercalary days in 33 years; 5^h, 49^m, 5.45^s; credited to al-Khayyam in the 11th century.

$31/128 = (3\times 8+7)/(3\times 33+29)$, and so combining the previous two continuants: 5^h, 48^m, 45^s; maybe Arabic also.

This desire for numerical accuracy is typical also of many of the developments described in the previous eight sections. Because of the extreme difficulties often involved, methods were sought to speed up and simplify computations. The main mathematical insight, already exemplified in the method (3.27.3) of *prostaphairesis* (partial evaluations) in the addition and subtraction theorems in trigonometry, had been provided by Stifel and others (§4.5); adding or subtracting powers corresponded to multiplying or dividing the corresponding numbers. The task was to extend this method so that any number, not just integers, could be a power (or could take a power). The outcome was logarithms.

Nowadays we think of them primarily in terms of the logarithmic *function;* but this notion was not to come through until Euler in the mid-18th century (§5.24). A prime concern at this time was the compilation of logarithmic tables – numerical rather than classical analysis. Thus a main mathematical problem was to choose sequences of numbers and of difference equations from which values could be computed to several places of decimals. Indeed, the word 'logarithm' was coined in this sense by the Scotsman John Napier (1550–1617), apparently as an abbreviation for the Greek *logos arithmos,* which we might translate as 'number of the expression' (compare §2.12 on *logos*). He was one of the main pioneers, publishing his first collection of tables in 1614 {Carslaw 1914}. Two years later, at the request of the East India Company, an English edition was prepared by Wright (the chart-makers' friend in §4.8). The following year, Napier died.

Unusually among the pioneers, Napier compiled tables for the logarithm (which I denote by y) of the sine (x) of a number, not of the number itself; as was usual at the time, the sine was taken to be a length. In order to obtain his desired accuracy, he chose the length 10^7 to represent $\sin 90°$ (the value used by Regiomontanus at FIG. 3.27.2), and worked to eight significant figures, one of them decimal (in the use of which (§4.5) he was a pioneer). To obtain suitable values for y he formed various geometric progressions, each starting downwards from 10^7. The most pertinent ones used the ratios $1 - 5 \times 10^{-3}$ for 69 terms and $1 - 10^{-2}$ for 21 terms; they were intercalated to give him a sequence S of 69×21 values for y from 10^7 to (about) $10^7/2$, corresponding to the range of numbers from 90° to 30° (that is, $\sin^{-1} \frac{1}{2}$).

Napier adopted a kinematic conception of logarithms. While the sine x 'traversed' a line of infinite length with uniform speed, its logarithm y started along a line L of length 10^7 with the same initial speed, but then slowed up in steps such that each segment was crossed in the same period of time and at a speed proportional to the length of

L which was still to be crossed. Had he been thinking calculus and functions (which of course he was not) then he would have been using this relationship:

$$\text{if } y := \log_N x, \quad \text{then } x = 10^7(1-10^{-7})^y \qquad (4.15.2)$$

(where 'N' is for 'Napier'). To find suitable arithmetic progressions for x, Napier used an inequality which I write as follows. For two nearly equal sines (possibly including 10^7) in the ratio $p:1$, with $p-1$ very small and positive, he showed that

$$10^7(p-1) > \log_N p > 10^7\left(1-\frac{1}{p}\right); \quad \text{also, set } q := 10^7\left(\frac{p^2-1}{2p}\right), \qquad (4.15.3)$$

which gives the average of the bounds of the inequalities. He chose to start out from $\log_N 10^7 = 0$, and selected values of q to provide appropriate differences to generate the arithmetic progressions of values of the logarithm y for the sequences S of values of x; this produced his 'radical table'. Intermediate values of x and y were found via (only) linear interpolation. To obtain the logarithms for values of y between 30° and 15° he formed additional columns of *differentiae* (a rather tedious procedure) from a relation which I write as

$$\log_N \sin 2x + \log_N(10^7/2) = \log_N \sin x + \log_N \sin(90^\circ - x). \qquad (4.15.4)$$

The same formula got him down from 15° to 7.5°, and so on indefinitely.

Given Napier's choice of 10^7 for the antilogarithm of 0, the addition and subtraction properties had to take the forms

$$\log_N x + \log_N z = \log_N(xz/10^7)$$

$$\text{and} \quad \log_N x - \log_N z = \log_N(10^7 x/z). \qquad (4.15.5)$$

Later he saw the advantage of running them from 0 as $\log_N 1$ to 1 as (in fact) $\log_N 10^{10}$; after his death his friend

Henry Briggs (1561–1630) at Oxford worked out a scheme of this kind, and introduced further methods of interpolation {Whiteside 1961, Sections 3–5}. He also presented much of the current theory of the trigonometries.

Napier's contemporaries adopted some strategy of this kind, although usually they produced the logarithm itself (in the case of the German Jost Bürgi (1552–1632), antilogarithms). Other figures concerned with logarithms included Kepler, principally in consequence of his astronomical work. In his *Rudolphine Tables*, published in 1627 after 25 years' work on the orders of his employer, the Holy Roman Emperor Rudolph II, he provided not only the right ascensions (corresponding to longitudes) but also reasonably accurate declinations (latitudes) of the stars, a feature curiously absent from previous tables.

Two of the more immediate problems were not making errors in calculation and getting the printer to set the digits correctly; these were to persist well into the 19th century. An obvious strategy was to mechanize arithmetical operations as far as possible. Napier himself conceived of a 'promptuary' – a box containing strips 'of ivory or any other firm white material' (hence the later name 'Napier's bones'). A strip was divided into ten squares, and each square into a regular grid of which each component was divided into two triangles; each triangle was either left blank or carried an integer from 0 to 9 according to a certain system. The product of two (large) integers $A \times B$ was determined by laying side by side the strips corresponding to each digit of A and then reading off the partial products from the diagonals of the array given by the digits of B. The method became quite popular, often in simpler forms.

A further stage was taken by Kepler's friend the German polymath Wilhelm Schickard (1592–1635), who designed a calculating machine in 1623. Unfortunately his achievement was not recognized until the 1950s {Freytag Löringhoff 1987}, and later pioneers were wrongly given priority – for example, Pascal (1642) and Leibniz (1673). All these machines were digital, based on sets of rotating circles of some form (Schickard used cylinders), with linking mechanisms to transfer integers up the place values {CE 5.11}. Analogue alternatives came around 1620 from the Englishman Edmund Gunter (1581–1626), who pioneered a type of slide-rule which was to be developed as a device known as the 'logarithmic line'; it was well received, especially by engineers and navigators when they needed a quick answer rather than a meticulously accurate one. Mathematics was coming to be aided by mechanical devices, both digital and analogue, at a time when the use of accurate instruments in astronomy and geodesy was growing rapidly.

4.16 *Reconciling the harmonic realm*

{CE 2.10}

Another area of numerical interest at this time was the division of the musical octave. We recall from §2.11 that the Pythagorean use of 2s for octaves did not mesh exactly with (3/2)s for fifths, and that mathematical resolution or compromise was wanting. While musicians had long devised various systems, and harmony retained its place as one of the classical sciences, the amount of attention paid by mathematicians to specifying scales with pleasing effect increased from the late 16th century, partly due to the growing place of numerical methods and partly to the need for a scale flexible enough to accommodate advances in musical composition, especially the use of modulations.

The method of 'just intonation', based on 2 : 3 for the fifth and 4 : 5 for the third, was shown by Benedetti and others to lead to unacceptable discordances. An alternative was 'mean-tone temperament', so called because it placed the tone at $2 : \sqrt{5}$, between 8 : 9 and 9 : 10; but the restricted role which it assigned to notes corresponding to black keys on the keyboard was inhibiting for modulations. 'Equal temperament' democratically assigned the ratio $1 : \sqrt[12]{2}$ to each semitone; it suited Stevin (among a few others) but did not become customary until the 18th century.

For some figures, the question involved issues beyond proportions. Vincenzo Galilei, a fine but forgotten composer, suggested that the tension as well as the length of the vibrating string should be considered – an ironic anticipation of the vibrating string problem in view of the scientific career of his son! Galileo himself followed others in interpreting consonance in terms of coincidence of vibrations so that, for example, a minor sixth at 5 : 8 was coincidental every 40 vibrations. Kepler's brand of geometric realism (§4.11) let him take such an idea further, to the point where a similarity (*not* an analogy) in God's structure was discerned between the consonant intervals of the octave and, for example, the ratios of the smallest and greatest apparent speeds of the planets in their orbits. Curiously, nobody seems to have used logarithms to calculate these values before Huygens in 1691.

A range of systems of intervals was developed during the early 17th century; their encyclopedist was to be Mersenne, who described many in his *Harmonie universelle* (1636–7), published when he was nearly 60 years old. The title refers to the wide range of possibilities, for he proposed no universal winner. He also (thought that he) studied the problem of finding beautiful melodies by calculating the numbers of combinations and permutations of consonant intervals {Knobloch 1979}. More valuably, he analysed sound and acoustics; and on the related topic of the pendulum he showed in 1634, seemingly just before Galileo, that the time of oscillation was proportional to the square root of its length. Again, the next stage was to come from Huygens (§5.12).

4.17 *Some British algebra*

It is appropriate to have noted Mersenne, for he belongs to the new generation of mathematicians and scientists with whom the rest of this chapter is concerned. Firstly, the British need notice. In some ways the role of William Oughtred (1575–1660) was similar to that of Recorde (§4.5), for his textbook *Clavis mathematicae* went though several editions during the century following its début in 1631 (in Latin, an English translation following in 1647). He wrote less clearly than his predecessor, but his treatment of geometry and arithmetic (the only branches tackled) were graced by a large repertoire of symbols for both numbers and operations, including '×' for the multiplication of two numbers.

In the same year, 1631, Harriot appeared in print, but unfortunately after his death and in the editorial hands of unimpressive followers. The material in his *Artis analyticae praxis* had been prepared during the 1600s. As the title suggests, he worked explicitly in Vieta's spirit. The book contained a detailed examination of polynomial equations and some important notations. For equality he chose a Recorde-like sign, that of Gemini, the twins in the Zodiac, 'Ⅱ'; and in the same line of thought he introduced '⦤' and '⦥' for inequalities, which were modified (by the printer of *Praxis*!) into '<' and '>'. Thus his work had some influence, although seemingly only from the 1680s, with his compatriot (and Oughtred's pupil) John Wallis, who regarded Harriot as a victim of plagiarism (§5.14). The supposed culprit was Descartes.

4.18 *Descartes: algebra or geometry?*

{Bos 1981}

Wallis found the alleged evidence in Descartes's *Géométrie* of 1637. In this book, published in his 42nd year, Descartes took up the tradition due to Apollonios and Pappos (§2.27) of determining a curve as the locus of a point satisfying given conditions (including by constructions using mechanical devices). His innovation was to assign letters to (variable and constant) lengths on the resulting diagram (a, b, c, . . . for constants, . . ., x, y, z for variables), and to apply algebra to find the equation of the locus. He also wrote powers above the third without concern for the traditional restriction imposed by the three dimensions of space, thereby bringing an element of arithmetic back into algebra by implicitly using laws such as $x^m/x^n = x^{m-n}$ for integers m and n. By these innovations he enriched but also complicated the relationships between algebra and geometry. Like Galileo on 'demonstrations' (§4.11), Descartes reflected upon methods used in algebra beyond the Cardano-like 'art' of obtaining results; indeed, his *Géométrie* was published as an appendix to his philosophical *Discours sur la méthode* of correct reasoning.

Descartes's methods led to a hierarchy of curves, in that a locus could depend upon another given curve, and so incorporate the corresponding expression in its own equation. To bring some order to this ensemble of curves, he classified some by their algebraic counterparts. Linear and quadratic forms corresponded to the classical constructions by straightedge and ruler; cubics and quartics tied in with curves given by the intersection of (say) a parabola with a circle; and quintics and sextics were given by a circle intersecting with a third-order curve, to be called by Newton 'the Cartesian parabola'; Descartes stated its equation as

$$\text{'}y^3 - 2ayy - aay + 2a^3, \text{ égal [...] } axy\text{'}. \qquad (4.18.1)$$

Normally Descartes used a symbol, '∞', for equality. For his algebra he drew partly upon Vieta (§4.7): letters much more than numbers; superscripts for powers; geometrically oriented terms above three dimensions (though only one collection of them), including 'square of square' for x^4. The corresponding bodies were 'supersolids' (*plusque solides*).

Descartes also handled some aspects of the theory of equations, especially because he associated their roots with points of intersection of curves. He took for granted the fundamental theorem of algebra, in the form that the number of roots of a polynomial equation was equal to its degree; this had been stated explicitly in the *Invention nouvelle d'algèbre* (1629) by his compatriot Albert Girard (1595–1632).[8] In addition to 'imaginary' roots, Descartes split real

[8] Girard also emphasized the use of brackets in expressions. His other work included a new edition of Stevin's *La pratique d'arithmétique* (§4.5), incorporating a translation of parts of Diophantos which led to some contributions to number theory, and spherical trigonometry. His neglect by contemporaries is surprising; residence in Holland cannot be the decisive reason.

ones into the 'true' positive ones and the 'false' negatives. He expressed polynomials with positive or negative coefficients indifferently (and marked absent terms with an asterisk); and in this context he came to a result still known after him. This was his 'rule of signs' – that the number of positive roots of a polynomial was not more than the number of changes of sign in the sequence of its coefficients (for example, +– or –+ in $x^2 - 6x + 8 = 0$); similarly, the number of negative roots was bounded above by the number of preservations (–– or ++). He exemplified the theorem only for equations of low degree; general proofs were to come in the early 19th century (THEOREM 9.6.1). Wallis was to attribute this result to Harriot; however, it is not stated in the *Artis*, and the text does not suggest that he knew it.

In enhancing a geometric tradition, Descartes algebraized it: from the construction of curves to the construction of equations. But, as he clearly indicated in his title, he was using algebra to help geometry, not the other way round {Bos 1984}. His work generated considerable interest, including the construction of curves by mechanical means; but eventually it atrophied under the weight of its own complications (compare §13.5 on invariants). However, he was not trying to invent analytic geometry as we now understand it, for many of the basic features of that subject are absent. For example, he assigned letters *only* to lengths: he did not systematically specify the axes in fixed directions at right angles, or assign scales to them (the metric was the same in all directions, so that diagrams were literally to scale). Further, most of his examples were qualitative in the sense that the coefficients were algebraic rather than numeric. Let us now look at an important type of problem which he studied: the determination of tangents and normals to a curve.

4.19 *Pre-calculus: a variety of methods*

{Whiteside 1961, Sections 8–10; CE 3.1}

Interest in this type of problem, and in its companion of calculating areas under curves, grew rapidly from the

1630s, and with Newton and Leibniz it blossomed into the 'full' calculus, a branch of mathematics which was to rival all others in importance (§5.5). Many methods were developed; there is space here only for Descartes's main contribution, and for one method of finding areas.

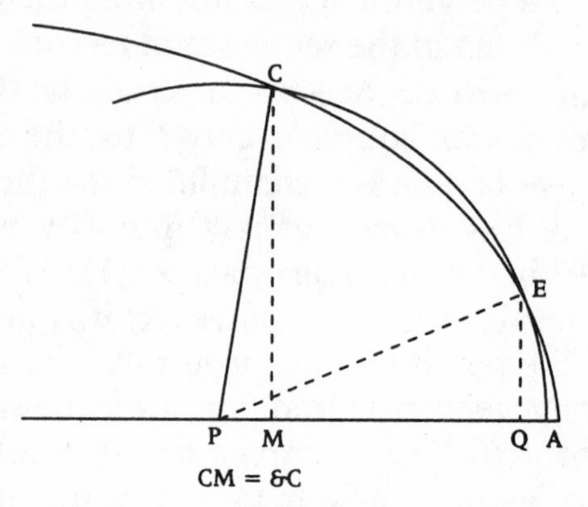

FIG. 4.19.1 *Descartes's diagram for finding the normal to a curve (*Géométrie, *1637).*
CM = x, MA = y, AP = v.

FIG. 4.19.1 shows Descartes's own diagram for finding the normal to a curve AC at an arbitrary point C. (Note the absence of axes, and the directions of x and y about A as origin.) His method was to draw a circle with centre P (an arbitrary point on a *known* fixed line AP) and radius CP, and let it cross AC again at E. Now, as P varied, so did E; and when P lay along the normal at P, E would *coincide* with C. Now bring on the algebra: C and E would be the roots of the quadratic equation in AP = v giving the points of intersection of curve and circle, and so the normal was found from the condition of equal roots. The generality of the method is perhaps less than its author pretended; for example, the curve should be reasonably well-behaved – entirely convex or concave, say. Various other mathematicians produced pre-calculus methods at this time. One such was Fermat, who was developing an analytic geometry of his own, and formulated one similar to Descartes's in that it used a condition of equal roots.

Among the methods for calculating areas, perhaps the greatest publicity was given to that of Bonaventura Cavalieri (1598?–1647), a follower of Galileo, initially after the appearance of his *Geometria indivisibilibus* in 1635, his 38th year. With his notion of 'indivisible' to *fill out* an area by a continuum of parallel lines passing through it (compare Kepler at (4.11.5)), his main principle may be stated as follows. Imagine two closed curves C_1 and C_2 in the plane, and a straight line (the *regula*) cutting them in segments A_1B_1 and A_2B_2, respectively; then, for some constant K,

$$\text{if } A_1B_1 : A_2B_2 = K, \text{ then also (area of } C_1) : (\text{area of } C_2) = K. \qquad (4.19.1)$$

If one area was known, then the other could be calculated.

Cavalieri made good use of his principle, and also extensions of it (for example, involving powers of lengths). Among his results was an analogue to the integral of x^2 from $x = 0$ to $x = a$ being $\frac{1}{3}a^3$. There was some doubt over its generality; as with Descartes, curves should have a friendly shape, such as convex. But the main difficulty was the change in dimension involved in going from lines like A_1B_1 to areas like that of C_1 (we noted this with Kepler's areal law in §4.11). Cavalieri used a concept called 'all the lines of the given figure' to try to justify (4.19.1); but critics such as Galileo remained unconvinced, and after a conceptual rethink by Leibniz in the 1670s it was abandoned (§5.5).

However, before that Cavalieri's compatriot Evangelista Torricelli (1608–48) came up with a valuable extension of pre-integration in 1644. He used a version of (4.19.1) where the *regula* across one closed curve could itself be curved, and showed that the solid or revolution produced by rotating about the x-axis any infinite section of the hyperbola $xy = 1$ when $x \geq a$ (a a positive constant) had an infinite surface area *but a finite volume* (in fact, π/a). This discovery that an object could be finite in one dimension but infinite in another was widely admired, and opened up an intriguing aspect of integration. Torricelli also developed methods to calculate lengths of curves (that is, rectification); Descartes had thought this impossible to effect, but it was to prove important in the development of the full calculus.

The British also made contributions here. Wallis developed a method of 'arithmetic integration' in his *Arithmetica infinitorum* (1655) using the summation of series. In an offshoot to finding the area of a circle he found a formula for π, still known after him, which rivals Vieta's (4.7.1) in its beauty:

$$\frac{4}{\pi}=\frac{3.3.5.5.7.7.9.\ 9.11.11\ldots.}{2.4.4.6.6.8.8.10.10.12\ldots.}; \qquad (4.19.2)$$

Lord Brouncker (1620?–84) then found an alternative expression in terms of continued fractions. Isaac Barrow (1630–77), in his *Lectiones geometricae* (1670), gave a geometrical treatment of many tangent and area problems, including the 'integral' of sec λ for the Mercator projection (4.8.1). But by then we are in the time of greater things from his successor Newton (§5.5).

4.20 *Desargues: sketches of perspective*

Not all developments in geometry were in the direction of algebra. Descartes's compatriot and senior by five years, the engineer and architect Girard Desargues (1591–1661), had taken up both *three*-dimensional Apollonian geometry and mathematical techniques in linear perspective and stonecutting, seeking in both cases projective properties which remained constant while metrical ones changed. Of his various works the most searching was his *Brouillon proiect ... du cone avec un plan* (1639, his 49th year); it is translated and assessed in {Field and Gray 1987}.

The end of Desargues's title shows that conic sections supplied the context. His principal concern was to extend Pappos's idea of the 'involution' (Desargues's own term) of four or more points (§2.27) to the quartet on a line made up of its two points of intersection A and B with the conic, a 'pole' point P chosen outside the conic, and the point of intersection N with the polar line determined by P (FIG. 4.20.1). His main innovation was to see that properties such as involution were true of all conic sections and so constituted a projective property of them. To help him fulfil his aim he used points at infinity (like Kepler before him: §4.11).

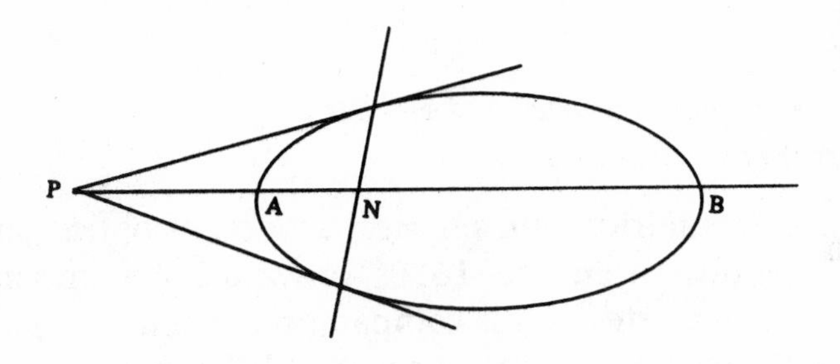

FIG. 4.20.1 *Polar line to an ellipse.*

Desargues's book quickly brought a fine sequel: a little *Essay pour les coniques* (1640) by the youthful Pascal, a fellow member of Mersenne's circle. It included this version of the theorem still known after him: if any hexagon is drawn inside a conic, and the three pairs of opposite sides are extended to intersect (if necessary, at a point at infinity), then the three points of intersection lie on a straight line. Later Desargues himself produced a similar theorem, which has preserved *his* name: if two triangles, lying in the same plane or not, are in perspective relative to a point in space, then corresponding sides intersect in three collinear points.

Such results made both men major figures in the origins of projective geometry; but even Desargues's title – 'rough drafts' – suggests hesitancy, the print-runs were tiny (as a man of some wealth, he did not have to worry about his book sales), and with justice some readers found his presentation to be inferior to Apollonios's. Thus the subject slumbered along until the early 19th century with J.V.

Poncelet (§8.4: the adjective 'projective' seems to be due to him, not to Desargues's *proiect*). Similarly, Desargues had new ideas for engineering geometry, especially a method of 'double turning' a solid into two perpendicular planes to find the sections, which extended the builders' tradition of double projection (§3.22); but these also languished somewhat until the descriptive geometry of the late 18th century founded by Poncelet's future professor, Gaspard Monge (§8.3).

4.21 *Optics with Descartes and Hobbes*

{Shapiro 1973, Sections 2–4}

Both the geometrical and physical aspects of optics gained fresh attention from the 1630s onwards. An important figure was Descartes, in his companion appendix *Dioptrique* (also 1637) to the *Géométrie*. As part of his expansion of the realm of curves in that work, he characterized as 'mechanical' those curves that could not be defined by a polynomial equation, and as examples he constructed four convex curves, called 'ovals', which he used to study reflections and refractions. On the physical side he interpreted light as 'a tendency to motion' caused by the centrifugal pressure of the particles of the aether, like a tennis ball hitting a flat surface; apart from the game in question, it is strikingly similar to a model developed by Ibn al-Haytham, which had been translated into Latin in 1572 as Alhazen's *Opticae*. (It related also to Descartes's vortex model of motion in general, postponed to §5.8 since his principal work on mechanics was published only well after his death in 1650.) Descartes's main work was on refraction, where he found the sine law. In fact it was known already to W. Snel, after whom it is usually known; and before him to Harriot, who rejected the opinion of most contemporaries that optical phenomena occurred only at the surface of a medium, and pioneered the study of effects in its immediate interior.

Other workers included the philosopher Thomas Hobbes, who published an essay in Mersenne's book *Universae geometriae* (1644) which gained attention. For him light was a real motion, caused (but how?) by the vibration of luminous sources; in particular, a ray, conceived as the path traced by the front of a propagated pulse, occupied 'solid space'.

But the physical, including mechanical, bases for all optical theories were shaky. For example, Hobbes argued that light travelled more slowly through water than through air, whereas Descartes asserted the opposite. Judgement on this point was to be important in the early 19th century when A.J. Fresnel's wave model faced a particle theory (compare §10.12); but what was being discussed at *this* time, before Ole Römer's experimental demonstration in 1676 of the finite velocity of light (§5.13), when everybody followed the Aristotelean view that propagation was instantaneous?

4.22 *Fermat: number theory, in integers*

{Weil 1984, Chap. 2}

Fermat also worked on optics: in 1662 he re-proved the sine law of refraction by means of his pre-calculus method (§4.19). But his main concern was with number theory, to which he came by the following circumstance. Bombelli's humanistic conversion to the Diophantine approach to algebra (§4.6) was continued by C.G. Bachet (1581–1632), who published in 1621 the Master's surviving original Greek text for the first time, together with his own Latin translation. Fermat obtained the book, and not only studied it but also wrote so many comments about it – and even in it – that after his death his son included them in a new edition of 1670.

However, in the 1620s Fermat was the only mathematician so preoccupied with this still isolated subject. He aided the seclusion by publishing nothing (in fact, none of his writings saw print before his son made an edition of them in 1679), and even to his various correspondents he rarely disclosed his methods; so it is not surprising that interest was modest.

Nevertheless, Fermat made fundamental contributions. For example, while Bachet's motive had been to extol antiquity, Fermat's reactions exposed its inadequacy for current needs. He decided to restrict solutions to integers and avoid those in rational numbers, which Diophantos had allowed (§2.26). This was a major innovation here, but not won easily. In particular, when in a rare lunge towards publicity he issued a series of 'challenges' to mathematicians in 1657 {Hofmann 1944}, he claimed to have found the complete collection of solutions (X, Y) to the equation

$$X^2 - dY^2 = 1, \quad \text{with } d > 0 \text{ and not a square;} \qquad (4.22.1)$$

but his claim was criticized by Wallis and others for forbidding rational solutions.

Fermat often started out from a Diophantine-like problem. Quite a few involved quadratic forms; not only (4.22.1) but also, for example, representations of prime numbers in the form $a^2 + kb^2$ for some small values of k (positive or negative). He also claimed to have proved a guess due to Diophantos, that every number can be written as the sum of at most four primes; he was mistaken, but a proof did come (§6.13).

Among theorems involving higher powers, an important case is now known as his 'little theorem' (a silly adjective): if p is a prime number not dividing a given number a, then p is a factor in $a^d - 1$ for value of d which is a factor of $p-1$ and for all multiples of d. He also followed (Diophantos and) Book 9 of Euclid's *Elements* in examining the sum $s(a)$ of the proper divisors of a: for example, in seeking perfect numbers, where $s(a) = a$, and amicable pairs (a, b), for which $s(a) = b$ and $s(b) = a$; his first challenge of 1657 was in this area.

Fermat's most widely known contribution was a question, not an answer. In his copy of the Diophantos edition he annotated the section on solutions to Pythagoras's equation with the suggestion that no integral solutions can be obtained for the equations with higher powers:

$$x^n + y^n = z^n, \quad \text{where } xyz \neq 0 \text{ and } n \geq 3, \qquad (4.22.2)$$

and that if only the margin of the book were wider he would write out the proof. This guess, known under the name 'Fermat's last theorem' (another silly one, as the last word assumes it to be true), gradually became recognized as an outstanding problem in mathematics.

The general proof, established in the mid-1990s by Andrew Wiles, draws upon topology and abstract algebra, of which Fermat had no inkling. This suggests that his claim was mistaken, although doubtless it was made sincerely. What kind of proof might he have had in mind? Although he dealt with $n = 4$, he left no trace of a general method. Probably he hoped to deploy a method of his own devising, which he called 'infinite descent', a type of inverse mathematical induction (though without a known first term): assume a solution, then show that there is a numerically smaller one, then . . ., until a contradiction is reached. He had used it successfully on other problems, and it is his major contribution to number theory. The subject still remained in the margins of mathematics, however; we return now to some central colours in its rainbow.

5

The calculus and its consequences, 1660–1750

This chapter is dominated by two of the greatest men of Western culture: Isaac Newton and G.W. Leibniz. Their careers and contributions, and those of their followers in Britain and the Continent very much respectively, are surveyed up to §5.15, with an interlude in §5.8 to introduce the mechanics of Descartes and of Huygens. The rest of the chapter deals largely with the newcomers of around the 1730s who secured the supremacy of the Continent over Britain. Probability theory makes its début here, on both sides of the Channel. As usual I confine the survey largely to publications of the time, although much information did circulate by manuscript or letter. {M. Cantor 1907} is still useful for its 900 pages surveying theories and results, though mostly in pure mathematics and especially the calculus and related topics.

5.1 *Biographies: Newton versus Leibniz*

{Westfall 1980; Aiton 1984}

Isaac Newton (1643–1727; 1643 in the modern calendar of §4.15) went up to Trinity College, Cambridge in 1661. His studies were interrupted by the great plague; the story of all the work resulting from his flight home to Woolsthorpe in Lincolnshire is an oversimplification {Whiteside 1966}, but it was in these years that he developed the fluxional calculus and made his initial insights in mechanics and astronomy. Succeeding Isaac Barrow in 1669 as the second Lucasian Professor of Mathematics at Cambridge during a phase of great mediocrity in the university's development, he immersed himself in mathematics, physics, optics, alchemy and Biblical chronology. In these early years he hoped for publication; but the Great Fire of London in 1666 had devastated the publishing industry in Britain, and scientific works were particularly disadvantaged. These circumstances, together with his antipathy to criticism, led him to retreat into his own world, where he wrote for himself incessantly.

The publication of the *Philosophiae naturalis principia mathematica* in 1687 (his 45th year) owes as much to the intellectual and financial encouragement of Newton's mathematical junior Edmond Halley (1656?–1743) as to any personal drive or public demand; but the book gradually built up an influence in various areas. The next major book was an *Opticks* of 1704, his 62nd year, to which he added various speculations about science and also an appendix containing the first major publication of the fluxional calculus.

In 1696 Newton had left Cambridge to become a Grand Old Man in London Town as Warden of the Mint, and Master of the Mint in 1700. The post required him to be a Member of Parliament, so he was duly elected for Cambridge in 1701 (for the second time). He was knighted in April 1705, for services to the Whig Party, of course; he lost his seat a month later. His scientific work was largely over, but he allowed his follower Roger Cotes to prepare a second edition of the *Principia* in 1713 and Henry Pemberton a third in 1726, just before his death. Another follower, William Whiston, had edited a volume of his algebra lectures as *Arithmetica universalis* in 1707, which also had later editions. Indeed, from 1720 Newton was in print pretty often, with editions and translations of various books (TABLE 5.1.1). I draw attention to the *Arithmetica*; despite its having exercised a considerable influence on British algebra (§5.14), it is often ignored.

During the 1710s Newton indulged in publication of a less desirable kind. President of the Royal Society since 1703 (and until his death), he managed there a priority dispute over the invention of the calculus, with the help of some followers, especially John Keill. Maybe his alchemical activities were getting to his brain (his hair had become saturated in mercury[1]); at all events, the affair split the

[1] Alchemy belonged to the occult sciences in the original sense of the word, 'hidden'; thus its true purpose was not revealed and is now often misrepresented. The associations with astrology and the correspondences with the planets belonged to more ancient lore, and by the 17th century

TABLE 5.1.1 *Principal editions and translations of Newton's scientific books up to the mid 1730s.*
Most editions were published in London or Cambridge; their number is indicated after the short title. They were written in Latin (L) or English (E); various French editions appeared later. An editor or translator is named where appropriate. Some reprints have not been noted.

1687	*Principia* 1 L	1722	*Arithmetica* 2 L
1704	*Opticks* 1 E	1726	*Principia* 3 L (H. Pemberton)
1706	*Optica* 1 L (S. Clarke)	1728	*Arithmetic* 2 E
1707	*Arithmetica* 1 L (W. Whiston)	1728	*De mundo systemate* (J. Conduitt) L E
1713	*Principia* 2 L (R. Cotes)	1728	*Optical lectures* 1 E
1717	*Opticks* 2 E	1729	*Lectiones opticae* 1 L
1719	*Optica* 2 L	1729	*Principles + Systemate* 1 E (A. Motte)
1720	*Arithmetick* 1 E (J. Raphson)	1730	*Opticks* 4 E
1721	*Opticks* 3 E	1732	*Arithmetica* 3 L
1722	*Arithmetica* 2 L		

mathematical world into British and Continental camps, and his own coterie was to lose out.

The target, or victim, was Gottfried Wilhelm Leibniz (1646–1716). At the age of 15 he went to the university of his home town of Leipzig (where his father was a professor) and then Altdorf. In 1667 he entered the service of the Elector of Mainz and soon began his lifelong passion for letter-writing. In 1672 he visited Paris as the companion of the elector's nephew, on a diplomatic mission. His stay of four years was to be fateful for his mathematical and scientific career, for as well as making contacts (that with Christiaan Huygens was especially important), he formulated his version of the calculus there (§5.5). He returned to

the quest for gold served as a front for experimental psychology, with the practitioner inhaling chemical fumes to get a 'high'. It seems to be no coincidence that philosophers of the time, such as John Locke and Thomas Hobbes, were deeply concerned with the philosophy of mind.

Germany, destined for Hanover where he became librarian to the Duke of Brunswick and Lüneberg, and remained in ducal service for the rest of his life. His official duties were historical and diplomatic, and in these capacities he was able to travel around Europe and meet some of his correspondents. But he never met Newton, not even in the spell during his Paris period that he spent in London (when Newton was at Cambridge). His final years were clouded by Newton's attack on him, and his death in 1716 was greeted by a marked outburst of apathy from the Hanoverian Court.

Leibniz's activities were ceaseless: not only letters and official paperwork but also masses of research work of all kinds. As well as encompassing most branches of mathematics, physics and philosophy of his time, he even tried his hand at engineering on occasion, such as a calculating machine (designed in Paris) and pumps driven by windmills in the Harz Mountains.

5.2 *Heritage and editions*

Neither Newton nor Leibniz married, nor did they seem to be 'interested'. After their deaths they both left large quantities of manuscripts, although Newton is known to have been through his pile and burnt a lot. Various editions have appeared since then. The first edition of Newton's works, edited by Samuel Horsley, appeared between 1779 and 1785. The most notable scholarship for Newtonian mathematics is the eight massive volumes of *Mathematical Papers* edited by D.T. Whiteside {Newton 1967–81}. I shall not refer much to the papers here, since most of them never saw print before and thus were not influential, but they do form the main single source for Newton's mathematics and its pre-history.[2]

[2] A companion edition of Newton's optical manuscripts is currently being prepared by A.E. Shapiro for publication by Cambridge University Press. Newton scholarship is surveyed by articles and reviews in *Notes and Records of the Royal Society*.

For Leibniz the situation is more difficult. The edition by C.I. Gerhardt of (among other things) his *Mathematische Schriften* (1849–63) in seven volumes was followed in the 20th century by a plan for a complete edition of all his writings, in six series divided by subject area; but the scale of transcribing so many notoriously difficult folios and then furnishing commentary has caused the edition to be planned out over long periods, with the result that many manuscripts have appeared first elsewhere, in journals or as books. Again, as we concentrate upon the works known at the time, this difficulty of scholarship can be side-stepped; but its existence must be acknowledged.[3]

5.3 *Camp followers*

Table 5.3.1 lists the principal followers who produced work by the mid-1730s. The word 'camp' conveys the idea of tenseness in each school; in old age Newton showed signs of dictatorship. But the atmosphere was not much better among the Leibnizians: for example, often one posed a problem, gave his answer and showed that it was indeed an answer (the 'synthetic solution'), and it was up to the rival to see if he could obtain it also. If he succeeded, then the two might well swap criticisms.

Special mention must be made of the Bernoulli family, whose achievements dwarfed all other followers of that time and eventually led to the domination of the Leibnizian calculus over its Newtonian rival. I confine myself here to the initial pair.

The family, druggists in Basel, seem not to have exhibited any special mathematical ability before the brothers Jacob and Johann, who formed the first, remarkable, mathematical generation. After travelling in France and else-

[3] The Leibniz edition is called *Sämtliche Briefe und Schriften*, and a few volumes in the mathematical series have begun to appear in recent years. The journal *Studia Leibnitiana*, together with a series of book-length supplements, provides authoritative information on Leibniz and his context, and on all aspects of Leibniz scholarship.

TABLE 5.3.1 *The Newtonian and Leibnizian camps: main newcomers up to the mid-1730s.*
Countries of main residence are marked thus: EN England FR France GE Germany SC Scotland SW Switzerland

Newtonian	Leibnizian
Cotes, R. (1681–1716) EN	Bernoulli, D. (1700–1782) SW
Craig, J. (16??–1731) SC	Bernoulli, Jacob (1657–1705) SW
de Moivre, A. (1667–1754) EN	Bernoulli, Johann (1667–1748) SW
Gregory, D. (1659–1708) SC	Bernoulli, Niklaus I (1687–1759) SW
Halley, E. (1656–1743) EN	Euler, L. (1707–1783) GE, Russia
Keill, J. (1671–1721) EN	Fontaine, A. (1704–1771) FR
MacLaurin, C. (1694–1742) SC	Hermann, J. (1678–1733) SW
Raphson, J. (?–1715?) EN	l'Hôpital, G.F.A. (1661–1704) FR
Stirling, J. (1692–1770) SC	Tschirnhaus(en), E.W. (1651–1708) GE
Taylor, B. (1685–1731) EN	Varignon, P. (1654–1722) FR
Oldenburg, H. (1618–1677) EN	Malebranche, N. (1638–1715) FR

where, Jacob obtained the chair of mathematics in his home town in 1687, and settled down to studies in the Leibnizian calculus, mechanics and optics, and probability. The main influences were Leibniz and Huygens, an interesting pairing of the innovative with the conservative. Meanwhile Johann had fallen under the Leibnizian spell, and in 1691 he was contracted to give calculus lectures to l'Hôpital, who published the first textbook on calculus in 1696 (§5.6). Four years later Johann took up a position at Groningen University in Holland, but he returned home in 1705 to succeed his brother in the Basel chair. His work was also focused on the calculus and mechanics, and by then he had instituted rows over priority and method with Jacob.

The most talented student of Jacob in Basel was Jakob Hermann (1678–1733). An early defender of the Leibnizian calculus, he thereby earned the support of its founder, who secured him posts in Italy and Germany. In 1724 he went to the new city of St Petersburg to join the Academy of Sciences that Peter the Great had just founded there (partly upon the advice of Leibniz), but soon grew homesick; finally he secured a post in 1731 in Basel, but died two years later. His main, and important, contribution was to pioneer the use of the Leibnizian calculus in mechanics. Also in St Petersburg from the start was Johann's son Daniel, and Leonhard Euler: they make their bows in §5.16, though a few contributions by Euler are mentioned before then.

Especially from the mid-1720s, each school followed the way of its chosen master, but the quality of the Newtonians' work gradually fell below that of the Leibnizians. The British became content to produce textbooks on the fluxional calculus, mostly stereotypes {Guicciardini 1989, Part 1}; they also gave us some dreary mechanics books. It was a steep decline from the great things that Newton had produced at Cambridge.

At the foot of each column of Table 5.3.1 is a marginal but significant player. Oldenburg was an important international correspondent, like Mersenne in Paris a generation earlier (§4.2). In Paris, Malebranche led a circle of thinkers on science during a period of controversy around 1700 over the quality of the (Leibnizian) calculus, and himself commented on some aspects; a philosopher of Cartesian leanings, he was rather cold to Leibniz's ideas.

5.4 *Major new institutions*

In an act of destiny (or 'pre-established harmony', as Leibniz would have put it), new national societies specifically for science were formed just before the arrival of our two giants. The three most important were the Accademia di Cimento in Florence (1657); the Royal Society of London (1662), where Newton was to be so influential; and the

Académie Royale des Sciences in Paris (1666), a centre favoured by Leibniz. They all launched journals, and two other journals started around this time, covering science although not confined to it: the *Journal des savans* in Paris in 1665, with connections to the French Académie which increased around 1700; and the *Acta eruditorum*, founded by Leibniz and some other scholars in 1682 and published in his home town of Leipzig.

So began a modest new era of communication and professionalization of science. However, for economic and practical reasons publication still remained very precarious; so postmasters such as Oldenburg were essential especially for the rapid spread of news and results, and letter-writing continued to be a major means of contact.

Mathematics occupied a curious place. On the one hand, during the early 17th century John Dee's ideas came to command considerable authority across northern Europe, as the father-figure of 'The Rosicrucian Enlightenment', an important influence on both science and philosophy. His emphasis on mathematics may have served to counteract the stress that Francis Bacon was placing on the importance of experiment in science {Yates 1972}. This division between the Deean and Baconian sciences was to last until the early 19th century, when new developments led to a different distinction (FIG. 7.3.1).

5.5 *Limits or increments? Two different calculi* {CE 3.2}

Both Newton and Leibniz went though the usual crucial errors and partial insights before their theories took final form; for reasons of space only the latter will be given here. Table 5.5.1 indicates the main points; individual rows of this Table are cited below.

Newton developed his calculus initially between 1664 and 1670. At first he was motivated largely by the algebraic treatment of curves inspired by Descartes (§4.18). An early crucial move concerned the binomial series, which he extended to non-integral exponents by an interpolative induction (in the scientific sense) upon the coefficients:

$$(a+x)^n = a^n + na^{n-1}x + [\tfrac{1}{2}n(n-1)]a^{n-2}x^2 + \cdots, \quad (5.5.1)$$

where the dots really do go on for ever. The motive came from finding the area of the semicircle, a problem which he could now generalize to the functions $(1 - x^2)^n$ for any real number n. He published the series first in the *Principia* in a rather offhand way, as part of the analysis of the attraction of a point to an infinite plane {Book 1, Scholium to Proposition 93}.

Recognizing the limited range of functions expressed by polynomials, by 1669 Newton was systematically using power-series expansions to express 'transcendental' functions. He then applied his version of

$$\int x^n \, dx = \frac{x^{n+1}}{n+1} \ (+\text{constant}), \quad n \neq -1, \quad (5.5.2)$$

and differentiated and integrated them term by term to obtain the related functions. His language was now that of variables x flowing with respect to 'time' (abstractly conceived) at rates measured respectively by fluxion $\dot{x}$ and fluent $\grave{x}$ (row 1: these notations date only from around 1691). The inversion principle is expressed in row 6 (though, I believe, not explicitly stated by him); their rates or aggregations of fluxions were expressed by fluxions and fluents of higher orders (row 7). He recognized the element of indeterminacy caused by the arbitrariness of flows, so that only ratios of them needed to be calculated. While this theory was not used in the *Principia*, Newton did speak there of 'first and last ratios of quantities': the adjectives signified that (to us) the right- and left-hand derivatives were equal, but with the latter construed in terms of $0 \to \Delta x$ rather than $\Delta x \to 0$.

TABLE 5.5.1 ***Principal features of the calculi of Newton and Leibniz.***

Topic	*Newton*
1 First-order concepts	fluxion $\dot{x}$; fluent $\grave{x}$
2 Conceptual basis	limits (but used unclearly)
3 Dimensions	changed by limit processes
4 Basic derivative concept	$\dot{y}/\dot{x}, \ddot{y}/(\dot{x})^2, \ldots$
5 Basic integral concept	$\grave{x}$
6 Inversion principle	$\dot{\grave{x}} = \grave{\dot{x}} = x$
7 Higher-order concepts	$\ddot{x}, \dddot{x} \ldots; \grave{\grave{x}}, \grave{\grave{\grave{x}}}, \ldots$
8 Main place of infinitesimals	in moments $o, o^2, \ldots$ of time
9 Increments	$o\dot{x}, o^2\ddot{x}, \ldots$
10 Base variable; criterion	'time'; $\dot{t} = 1$ [so $\ddot{t} = \cdots = 0$]
11 Main mathematical techniques	Term-by-term integration and differentiation of power series

Leibniz's great discoveries occurred in October and November 1675, while he was in Paris {Hofmann 1974}. He started out by making an algebraic version of Cavalieri's method of determining areas, and ran into the same problems of changing dimensions that we noted in §4.19. But a week was a long time for Leibniz, and soon he was back with a remodelled theory in which *preservation of dimension* was a key idea {Bos 1974}. Extrapolating the idea of a forward difference to the infinitesimal case, he denoted by 'dx' that small an increment on x. The operator 'd' (for 'differentia') was dimensionless; for example, if x was a variable line, then dx was a very short line – also variable, and so susceptible to a second-order increment ddx, and so on. The integral was similarly conceived via the operator $\int$ (for 'summa', using the fancy 's' used in legal writings of the

Topic	*Leibniz*
1 First-order concepts	differential increment dx; sum $\int x$
2 Conceptual basis	infinitesimal differences [−] and sums [+]
3 Dimensions	preserved by 'd' and '$\int$'
4 Basic derivative concept	dy/dx [that is, ÷]
5 Basic integral concept	$\int y\,dx$ [that is, sum of $(y \times dx)$]
6 Inversion principle	$d\int x = \int dx = x$
7 Higher-order concepts	ddx, dddx, . . . ; $ddy/(dx)^2$; $\int\int x$, . . .
8 Main place of infinitesimals	in differentials dx, ddx, . . .
9 Increments	dx, ddx, . . .
10 Base variable; criterion	$dx =$ constant [so $ddx = \ldots = 0$]
11 Main mathematical techniques	Characteristic triangle; analogies with −, + and ÷, ×

time); $\int x$ was a very long line, $\int\int x$ a very very long one, and so on. The two operators were inverse, following the inversion of '−' and '+' which underlay each (row 6, which he did state).

Like Newton, Leibniz recognized the degree of indeterminacy and so went for ratios: the slope of the tangent, 'dy/dx' (row 4), really did denote the ratio $dy \div dx$, itself normally finite, of the two infinitesimals in the 'characteristic triangle'. The integral, as an area, was taken to be the sum ($\int$) of a sequence of very thin rectangles, y high and dx wide (row 4), like a loaf of sliced bread; thus a second beautiful structural similarity with arithmetic was obtained, this time with the inverse pair '×' and '÷'.

Naturally many details are analogous to Newton's; for example, the discard rules for higher-order infinitesimals.

To take an elementary example, the product rule, we have respectively

$$\dot{\overline{xy}} = (x + o\dot{x})(y + o\dot{y}) - xy$$
$$= ox\dot{y} + oy\dot{x} + o^2\dot{x}\dot{y} = o(x\dot{y} + y\dot{x}) \qquad (5.5.3)$$

and

$$\mathrm{d}(xy) = (x + \mathrm{d}x)(y + \mathrm{d}y) - xy$$
$$= x\,\mathrm{d}y + y\,\mathrm{d}x + \mathrm{d}x\,\mathrm{d}y = x\,\mathrm{d}y + y\,\mathrm{d}x. \qquad (5.5.4)$$

This example is very instructive, for in a hasty early moment Leibniz had offered $\mathrm{d}x\,\mathrm{d}y$ as the answer (though he corrected himself before publishing), while Newton was so bothered by the mysteries of limits that in the *Principia* he faked a proof by proving that (in the notation of (5.5.3))

$$\dot{\overline{xy}} = (x + \tfrac{1}{2}o\dot{x})(y + \tfrac{1}{2}o\dot{y}) - (x - \tfrac{1}{2}o\dot{x})(y - \tfrac{1}{2}o\dot{y})$$
$$= ox\dot{y} + oy\dot{x}, \qquad (5.5.5)$$

thereby 'losing' the unwelcome extra term {Book 2, Lemma 2}.

Some of the many results and techniques stemming from these theories will be noted in §5.6–§5.7 and elsewhere. Each man is generally regarded as having surpassed his predecessors by recognizing that differentiation and integration were inverse processes. This is quite true, but it is not the whole truth, or even the basis of it. Their advances rested on the insight that, given a function $f(\mathrm{x})$, new *functions* were sought; something corresponding to the derivative function $f'(x)$ for the tangent, and something like the *indefinite* integral $\int f(x)\,\mathrm{d}x$ for the other. Thus *they brought the algebra of variables to the calculus*; as we saw in §4.19, even Descartes had used double-roots conditions for tangent (or normal), and his contemporaries had found specific areas (definite integrals, in our terms). Only when these new concepts were available could inversion be *stated* (in the forms given in row 6).

In addition, their basic concepts for differentiation were linear:

$$\mathrm{d}(a\,f(x) + b\,g(x) + \cdots) = a\ \mathrm{d}f(x) + b\ \mathrm{d}g(x) + \cdots, \qquad (5.5.6)$$

with $a, b, \ldots$ constants for Leibniz, with a corresponding statement for Newton. This property does not obtain in general for the subtangent and subnormal functions;[4] however, our heroes may not have had it in mind as a prior desideratum.

There is another similarity. We saw in §4.18 that Descartes's introduction of symbolism in geometry was not analytic geometry in our sense in that he had not stipulated axes for the basic variable. This practice came in with the calculus, especially with Leibniz; Newton was a pioneer in the use of polar coordinates in the plane. In this connection it is worth noting that the integral comes with a sign, positive or negative; getting used to this, and sorting out the sum of areas with different signs, took some practice.

Despite these points of resemblance, the two methods are profoundly different (so making the priority row a nonsense).[5] Leibniz's 'differential and integral calculus' (the word 'calculus' is his in this context) is often put over as an algebra, but it is really grounded in geometry, because of the central role of preservation of dimension by 'd' and '$\int$' (row 3) and their distinction; it is this that gives the theory its strength. By contrast, as part of his 'method of fluxions and fluents' Newton recognized the importance of power series; indeed, he was the first great mathematician to

[4] The subtangent function (5.5.6) for two terms leads to the differential equation $gf'^2 + fg'^2 = 0$, with solutions in the f, g-plane; for the subnormal we easily find that $fg = K$ (constant). When more terms are taken, the solutions become very complicated.

[5] The same can be said for Isaac Barrow as a source for Newton: his *Lectiones geometricae* (1670) (§4.19) contain geometrical forms of both halves of the fundamental theorem, but not alongside or even with recognition of the link between the two, and no prominence given to either. Some of Leibniz's followers, and certain historians, have maintained that an influence was involved.

assert the central importance of such series in analysis. Two successors appear in §6.5 and §12.7.

5.6 *Leibnizian differential equations*

{Hess and Nagel 1988}

Leibniz first published his calculus in two typographically awful papers on a 'New Method for Maxima and Minima' in the volumes of *Acta eruditorum* for 1684 and 1686; within a few years he and his followers, especially the Bernoullis, were developing new methods rapidly. In particular, Johann Bernoulli was under contract to lecture to the Marquis de l'Hôpital, and so contributed much to the first textbook on the calculus, his lordship's *Analyse des infiniment petits* (1696); in particular, the so-called 'l'Hôpital's rule' on calculating the value of $f'(x)/g'(x)$ if $f(x)/g(x)$ is indeterminate as 0/0 for some value of x.

One context of the calculus was the 'quadrature of curves', the tradition of determining particular areas (although now often conceived as indefinite integrals); another was the inverse tangent problem, where a curve had to be determined from some given property of its tangent. An important problem of this latter type, that of a chain hanging in equilibrium, was the subject of a competition in the *Acta* in 1691. Huygens had already refuted a conjecture of Galileo that it was a parabola, and now he came up with an answer arrived at by conceiving the chain as a collection of tiny weightless cords of equal length and bearing the same weight. By a clever analysis using a complicated diagram involving projections of these lengths onto an axis, making other projections, and assuming as known the rectification of the parabola, he gave a method of constructing the curve – quite correct, but not much fun to do. By contrast, Johann Bernoulli offered FIG. 5.6.1, showing a part of the half-chain, from arbitrary point A of support with coordinates (y, x) to the low point B $(0, 0)$, of length s and unit line density. It was in equilibrium under the constant horizontal force F applied at B, the tension T at A,

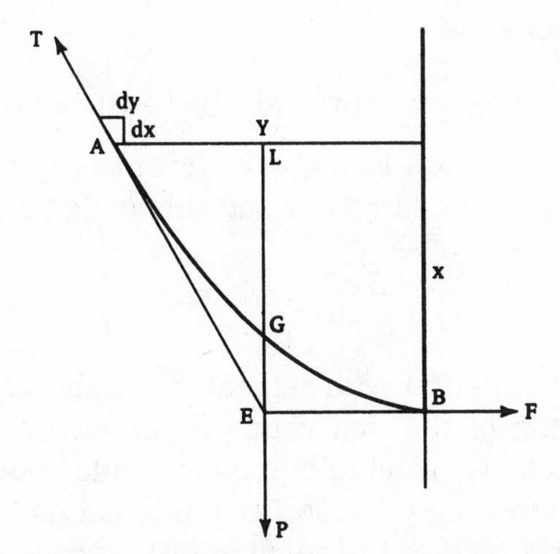

FIG. 5.6.1 *Johann Bernoulli's analysis of the hanging chain, 1692.*

and its weight s acting at the point E below the centre of gravity G of AB; so, by composition of forces, we have literally a differential equation:

$$s : F = \mathrm{d}x : \mathrm{d}y, \text{ or } \mathrm{d}y = F\frac{\mathrm{d}x}{s}. \therefore \mathrm{d}y = F\frac{\mathrm{d}x}{\sqrt{(x^2 - F^2)}} \quad (5.6.1)$$

by various manipulations; the solution was the integral of the last term, which was related to the quadrature of a hyperbola. At this stage Bernoulli did not know the analytical solution in terms of the logarithmic function, but the problem was 'solved', and in a manner much easier to follow than Huygens' procedure.

The construction of curves by mechanical means was very fruitful for inverse tangent problems, especially ones to do with traction {Bos 1988}. The term 'tractrix' for the curve traced on a rough plane by an object pulled by a connecting string is indeed due to Leibniz, who in 1693 formed a differential equation for the purpose. The Bernoullis tackled a desimplified version of this example, in which the length of the string was proportional to the length dragged along the x-axis. Analysis of this problem led to the equation

$$y\sqrt{dx^2 + dy^2} = p(x\,dy - y\,dx), \qquad (5.6.2)$$

with p the proportion factor, and clever changes of variable to the length of the tangent t and subtangent s yielded

$$\frac{dt}{pt} = \frac{dz}{z^2 - 1}, \quad \text{where } z := \frac{s}{t} \qquad (5.6.3)$$

– that is, an equation with separated variables t and z, and moreover one of the first examples of such an equation. The solution by algebraic means could now proceed (though, interestingly, Jacob Bernoulli had not yet learned to allow for a constant of integration). Typically, Huygens found a virtuosic but complicated geometric solution involving a similar curve, and presented it synthetically; but this problem impressed upon him the power of calculus methods.

These cases involved first-order equations, but the importance of higher-order differentials was recognized. In particular, Jacob Bernoulli published in 1694 his 'golden theorem' (as he called it), for the radius of curvature R of a curve:

$$R = \left(\frac{ds}{dx}\right)^3 \div \left(\frac{d^2y}{dx^2}\right), \quad \text{with } ddx = 0. \qquad (5.6.4)$$

The victory of the Leibnizian calculus was sealed with results such as these, which marked the end of a phase of thinking about algebra and geometry which started with Descartes's over-simple identification of algebraic expressions with geometric curves (§4.18). He had rather confused the situation by calling 'algebraic' those curves given by polynomials; and his restriction of legitimacy to such curves became eroded by his successors, who worked with the other, 'mechanical', means of generating curves, such as by traction. The determination of points or curves needed the power of the calculus, and thereby gave it the breakthrough to mathematical stardom.

Quadrature problems also contributed to the success {Greenberg 1995, Chap. 7}. In particular, in 1694 Johann Bernoulli announced his 'universal series' for 'all quadratures and rectifications, and integrals of differentials' for a function $n(z)$:

$$\int n\,\mathrm{d}z = nz - \frac{1}{2!}\left(z^2\frac{\mathrm{d}n}{\mathrm{d}z}\right) + \frac{1}{3!}\left(z^3\frac{\mathrm{d}^2 n}{\mathrm{d}z^2}\right) - \ldots \qquad (5.6.5)$$

a result still known as 'Bernoulli's series' on the Continent. But it belongs more to British specialities in the calculus, to which we now turn.

5.7 *Newtonian series*

{Whiteside 1961, Sections 5–7; CE 4.3}

For some reason Newton kept his calculus to himself for many years, though copies of some manuscripts were circulating. Finding this distasteful, he printed two of them in 1704 as appendices to the first edition of his *Opticks*, under the interesting name 'Treatises of the Species and Magnitude of Curvilinear Figures' on the title page. But by then the Leibnizians were well ahead, as we have just seen.

The Master's emphasis on the manipulation of series steered his followers' thinking much more in that direction than had happened on the Continent, despite such important cases as Bernoulli's (5.6.5). A few results from Newton and his senior Lord Brouncker (1620?–84) involved continued fractions, which provided greater strength than power series by furnishing *rational* approximations to functions ($P(x) \div Q(x)$, not just $P(x)$); but the calculation of coefficients was then often not practicable.

Two significant contributors of series were the friends Abraham de Moivre and James Stirling, the latter notable for having obtained Continental instruction (in Italy, when young). They both considered series with many terms, and means of approximation to the rectification of the ellipse

{Tweddle 1988}; the formula

$$n! \approx \sqrt{2\pi}\left(\frac{2n+1}{2e}\right)^{[(2n+1)/2]} \tag{5.7.1}$$

from Stirling's *Methodus differentialis* (1730) is still known after him, though usually in a slightly different version due to de Moivre, who used it and similar results in probability theory (§5.22).

The best-known series is called 'Taylor's series', after one of Newton's followers, although it had been known already to Newton, Leibniz and others. Brook Taylor's context and procedure is a fine example of normal fluxional mathematics. It goes back to an interpolation method presented by Newton in Book 3 of the *Principia* {Lemma 5, after Proposition 40} which he used to determine the paths of comets: 'To find a curved line of the parabolic kind which shall pass through any given number of points', whether equally spaced or not. Newton's avoidance of the calculus here led to a ponderous presentation; Taylor found virtually the same result in calculus terms in his book *Methodus incrementorum directa et inversa* (1715), which was innovative in stressing the difference mathematics. His notations were rather horrendous, with underdots marking differences, so I state his series as follows: for a function $x(z)$ (his letters) with forward difference Δz,

$$x(z+n\Delta z) = \sum_{r=0}^{\infty}\left[nP_r\Delta^r x(z)\left(\frac{\Delta z}{\Delta x}\right)^r \Big/ r!\right]. \tag{5.7.2}$$

He now slid to the limiting case Δz"→"0 by putting $n\Delta z = v$ and replacing the quotients of differences by the corresponding orders of fluxion to obtain the series which has made his name so widely known:

$$x(z+v) = \sum_{r=0}^{\infty}\frac{x^{(r)}(z)v^r}{r!}. \tag{5.7.3}$$

He raised no questions about what we would call rigour

(limit processes, expandability, convergence, remainder term); he had a great result, which he went on to use in problems {Feigenbaum 1985}. It is of course somewhat akin to eqn (5.6.5), of Johann Bernoulli, who did not miss the opportunity to explode (especially with the priority dispute steaming on all boilers); but it seems that Taylor's work was independent.

Colin Maclaurin gave the series (5.7.3) its name; and later his own name became (unnecessarily) attached to the case given by $z=0$. His *Treatise of Fluxions* (1742) – the most substantial fluxionalist text, and the last major one – was written in response to criticisms of the foundations of the calculus made by Bishop Berkeley (1685–1753) in his *The Analyst* (1734). In humorous but penetrating ways this man of the cloth rebuked the mathematicians for the mysteries of their doctrine (both Newtonian and Leibnizian) and proposed a solution in terms of algebraically compensating errors in the formulation of the basic concepts. He showed its correctness in the case of a parabola; but in fact special properties are involved, which do not apply in general {Grattan-Guinness 1970a}.

5.8 *Slings and bobs: Descartes and Huygens on principles for mechanics*

{Barbour 1989, Chaps. 8–12}

We pick up now from §4.14 the next stages in the development of mechanics. The peak will, of course, be Newton; but before that some note needs to be taken of two intermediate figures.

Descartes published some of his main thoughts on mechanics in his *Principia philosophicae* (1644); but he left others in manuscript, seemingly as a precaution after the trial of Galileo in the 1630s (§4.13). Many of these manuscripts appeared posthumously as the book *Le Monde* (1664). The delay was not a catastrophe, for his ideas belong more to the armchair philosopher than to the real student of mechanics. Nevertheless, his influence was remarkably extensive. Four features are notable, especially for their negative effects on Newton and/or Leibniz.

1. Descartes advocated a relativistic interpretation of space and time. Perhaps in reaction to Galileo's plight, he 'hid' his acceptance of Copernicus's model this way.

2. He offered a string of rather notional laws of collisions between bodies, with some absurdities grounded in his advocacy of the conservation of (mass × speed) rather than momentum (that is, mass × velocity). Indeed, even his understanding of mass was fogged up with volume in that he affirmed extension as the essential property of matter (and thus of a body).

3. He advocated in more detail than hitherto a vortex model of planetary motion in which the planets of a system were swept around their orbits by fluids rotating in concentric circles about the system's sun (FIG. 5.8.1). However, his account faced difficulties: the fact that the outer planets had to move faster than the inner ones, the errant motion of the comet on its path RDI, and the behaviour in the boundary regions between neighbouring solar systems, where their vortices move in opposite directions (as he clearly illustrated in FIG. 5.8.1); thoughts of astronomical instability cannot be avoided. In any case, how could this *two*-dimensional scheme represent spatial processes? A strikingly similar situation arises at FIG. 15.10.1 with Clerk Maxwell's model of the electromagnetic aether.

4. In a valuable insight possibly stimulated by vorticial motion, Descartes emphasized that if a body B were swung around in a sling and then released, it would move along the tangent at its point of release from its previous circular motion, and not along the radius through that point. He gave as an example a ball placed inside a horizontal tube which accelerated outwards when the tube was rotated about a vertical axis.

These last remarks drew attention to properties of inertia and of centrifugal force in rotation (which appeared in his optics: §4.21). For the former concept Descartes had a

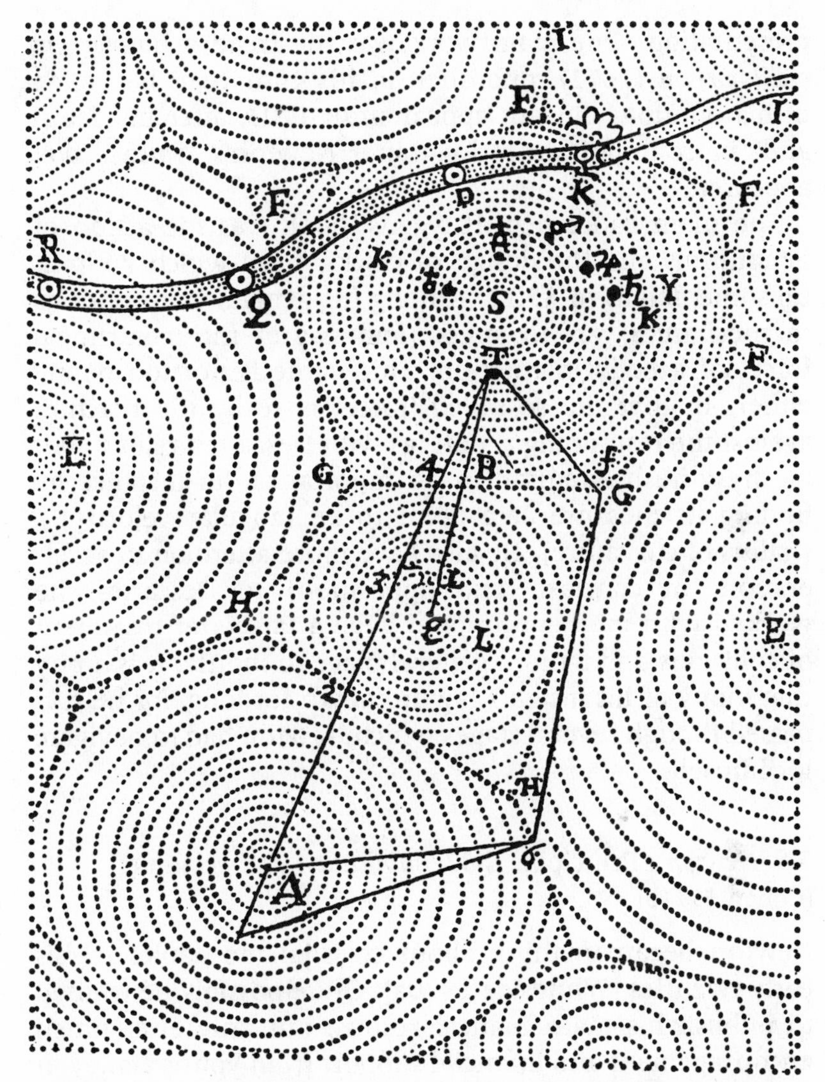

FIG. 5.8.1 *Descartes's vision of vorticial motions (*Le Monde, *1664).*
The Sun is S, and the Earth *(terre)* is T. A comet is passing through on the path RDI. Other vortex centres are A, E, ...

rather unhappy idea, called 'force of rest' – a name which is due to our second figure. Like Descartes, Huygens delayed publication of some of his ideas on mechanics, but apparently only because of dissatisfaction with his presentation. One main work was the treatise *Horologium oscillatorium* (1673), published in his 44th year; but some of his offerings did not appear until 1703, after his death and (more importantly) long after Newton's *Principia* had appeared.

The main concern of Huygens' book was the action of pendulums and the design of clocks, from which we pick up some details in §5.12. Among his general contributions to mechanics was to stress (and name) the 'centrifugal force' of a body in rotation. This name is not too happy either, in that the so-called 'force' is really the reaction against the restraining action (as with Descartes's sling); but he drew fresh attention to its importance, for instance by finding its measure at a point to be (angular velocity)^2/ (radius of curvature). He also gave more attention to acceleration. In contrast (and opposition) to Descartes, he worked with momentum, and gave a far better treatment of collisions and a more general place to inertia. It is a pity that many of his insights lay unknown for long, for he had passed though several stages which Newton had to navigate for himself.

5.9 *Newton's* Principia, *1687: publication and principles* {Cohen 1971b}

Newton began to think about celestial mechanics in his early twenties {Whiteside 1970a}. Quite possibly he wanted to publish his mathematical results; but only after much encouragement and financial support from young Halley in the mid-1680s did he refine and write up his thoughts on mechanics in book form, publishing them in 1687 with the blessing of the Royal Society ('Imprimatur S. Pepys, Reg. Soc. Praeses. Julii 5. 1686'). The *Philosophiae naturalis principia mathematica* ('Mathematical Principles of Natural Philosophy') was presented in three Books: 'the motion of bodies' in free and in resisting media respectively, and then 'the system of the world' (meaning our planetary system).

The print-run was modest (perhaps around 400 copies), and the reception of the book was not rapid: in particular, vortex theories remained sufficiently durable to invoke some counter-propaganda from Cotes in the preface to the second edition of 1713 (750 copies). But that edition, and its successor of 1726 (1,250 copies), gained a better response and by the 1720s the book was becoming widely recognized, with many kinds of positive and negative reaction in place. These editions contained various revisions, the second being arguably the most satisfactory, but the general shape and purpose of the work remained unchanged.

Far too much material in the various editions, and in reactions to them by readers, had been produced by 1750 for even a summary to be given here. In the next three sections I shall examine a few basic features, and mention some others later. For the modern reader the most accessible version – and so quoted here, though not always reliable – is the English translation (1729) of the third edition by Andrew Motte in the revised edition made two centuries later by Florian Cajori {Newton 1934}. A new translation of the third edition is being prepared by I.B. Cohen.

The gradual genesis of Newton's ideas may be charted through a variety of manuscripts {Herivel 1975}. He was not forthcoming on his sources of influence, but it seems that he was respectful about some aspects of Galileo; aware at least of Kepler's laws, although perhaps only at a late stage, and maybe not directly from Kepler's writings; and violently opposed to Descartes on several counts, such as relativity of space and time, conservation of motion, and vortices. Indeed, the title of his book may have been chosen to upstage the ambitions of that Frenchman's *Principia philosophicae*.

During his priority row over the calculus, Newton claimed that he had used his fluxional version to think out many of the results set out in *Principia* but had recast them in geometrical form for publication. No historical evidence has survived to confirm this claim; and, while (as mentioned in §5.2) he made bonfires of many papers in his last years, his claim is regarded as unlikely {Whiteside 1970b} – that is, he was 'remembering selectively'. Certainly a presentation in the new language of the calculus, unknown to any contemporary, would have bewildered most readers: much better to use the universal languages of Latin and geometry. However, there were several techniques similar to those in his calculus. One of them is limit theory; Book 1 began with a section on the 'first and last ratios'.

Another point in common between mechanics and the calculus underlies the second of the three 'axioms, or laws of motion' which followed the opening definitions. We know it as force = mass × acceleration, but that is exactly what Newton did not say; instead, 'The change of motion is proportional to the motive force impressed; and is made in the direction of the right line in which that force is impressed.' Far from speaking of a 'continuous' process in which force leads to acceleration, discrete even if infinitesimally *small increments* of change are invoked, presumably caused by corresponding micro-impulses of acting force {Cohen 1971a}. This line was probably taken to fit in with the manner of modelling continuous phenomena as sequences of infinitesimal effects, like the moments '*o*' of time in his conception of the calculus (as in (5.5.3)). However, in Definition 8 he had already characterized motive force as 'proportional to the motion which it generates in a given time'; this would have helped him to see his formulation as equivalent to the version of the law which we normally adopt.

The intimate association of force here with acceleration is worth stressing. We saw in §4.12 that Kepler and others followed tradition in laying emphasis on velocity; but from now on (change in) velocity was usually associated with impact and collision. Thus Newton also clarified the contrast with 'inertia' (his word, possibly taken over from Kepler), which took prime place in the first law {Gabbey 1971}. In stressing there (from Descartes?) that 'Every body continues in its state of rest, or of uniform motion in a right line, unless it is compelled to change that state by forces impressed upon it', Newton made clear, in this his first law, that both static and dynamic equilibrium (as we now call them) were allowed; but many later figures were less lucid, and often their talk of equilibrium was to be dominated by the static case.

Newton also pioneered the recognition of the importance of centripetal forces, which received early publicity in the opening definitions and in Book 1, Section 2. Indeed, he may have been partly stimulated by the contrast between centripetal and centrifugal to elevate it to the status of a law, the third one, that 'To every action there is always opposed an equal reaction', with a reformulation as 'The mutual actions of two bodies upon each other are always equal, and directed to contrary parts.'

However, such a simple balancing of forces will not deal with all dynamical situations, especially if there is a rotating frame of reference {Bertoloni Meli 1990}. Indeed, in working drafts for the book Newton considered up to six laws, and this trio are wanting in their (non-)handling of angular momentum and frames of reference {Westfall 1971, Chap. 8}. Newton himself provided a good case study, when he described a bucket of water rotated about its axis so that the water rose up the sides and gradually acquired the angular velocity of its container {Scholium to Book 1, Definition 8}; but the understanding of the use of rotating frames in dynamics was to remain rather hazy until the 1830s (§10.7).

Newton used this example as part of his stand against Descartes, when he proclaimed (in the general 'Scholium' between the definitions and the laws) that space and time were absolute. However, it is surely difficult to reconcile this position with dynamic equilibrium in the first law, for if such an invisible artefact does exist, how do we distinguish uniform motion from absolute rest within it? Indeed, how do we know that space and time are 'there' to start with? (§16.25).

5.10 *Newton on the motion of an individual planet*

Seht die Sterne, die da lehren	See the stars, so as to learn
Wie man soll den Meister ehren	How one should honour the master
Jeder folgt nach Newton's Plan	Each one follows after Newton's plan
Ewig schweigend seiner Bahn.	Ever silently its path.

Albert Einstein {Hoffmann 1972, pp. 141–2, with my translation}

Newton's dynamics was dominated by a blend of considerations of gravitational attraction, collision and inertia; for him inertia could obtain in *any* direction, not merely in the (local) horizontal as with Galileo in §4.14. For example, in FIG. 5.10.1 a body (planet) Y was moving under the action of some force f towards a centre of attraction (Sun) S {Book 1, Propositions 1–2}. Time was divided into equal infinitesimal parts (like the base fluxion of Table 5.5.1, row 10), during which Y passed from A to B to C to . . . . (The *curve* ABC . . . is to be ignored for the present.) Inertial force on its own would carry Y from B along AB an equal distance to c; but centripetal attraction to S created a force acting along cC, parallel to BS. By Euclid, Book 1, Proposition 41 (explicitly cited) on equal areas of triangles,

$$\Delta SAB = \Delta SBC = \Delta SBc, \qquad (5.10.1)$$

thus establishing an areal law; further, the resultant of these two forces carried Y from B to C.

Eight aspects of this analysis are noted here {Brackenridge 1995}.

1. To Newton's own satisfaction, (5.10.1) contained Kepler's second law (4.11.4) as a special case; it also helped him to avoid arguments invoking vortices. In addition, in Proposition 31 he gave a construction for approximately determining the position of a planet on its elliptical orbit at any instant, which corresponded to approaching the root of Kepler's equation.[6] Propositions 11–13 contained Newton's

[6] Newton's construction is a geometrical version of the 'Newton–Raphson' iterative approximation to a zero of a function $g(x)$ by passing from the guess $x = a$ to $x = a - g(a)/g'(a)$; but this formulation of the procedure, using the calculus, came only with Thomas Simpson in 1740! {Kollerstrom 1992}.

proof of Kepler's first law (4.11.2) for an orbit executed under 'any' central law of force, and Kepler's third law (§4.12) was proved in Proposition 15. Yet he never mentioned Kepler in any of these passages.

2. In his first work on dynamics, in 1665, Newton had considered the bouncing of a perfectly inelastic ball inside a circle under the combined effect of its inertia and an impulsive force towards the centre; it followed paths such as a square, an octagon, . . .; in the limit, the circle itself. The mature Proposition, embodied in FIG. 5.10.1, is a refinement, with Y 'bouncing' along from A to B to . . . along its orbit.

3. However, Newton's Figure is conceptually misleading in being discrete; for the *Principia* deals with infinitesimals in its analysis of continuous motion, with B 'next' to A, and so on. Thus lengths like QR and QT must be infinitesimal of at least the second order (as at (5.11.1) below).

4. One of his main innovations in dynamics was to consider *both* displacement *and* direction varying continuously: Kepler's circles and Galileo's straight lines each varied in only one of these ways. In FIG. 5.10.1, AB is like a Galilean

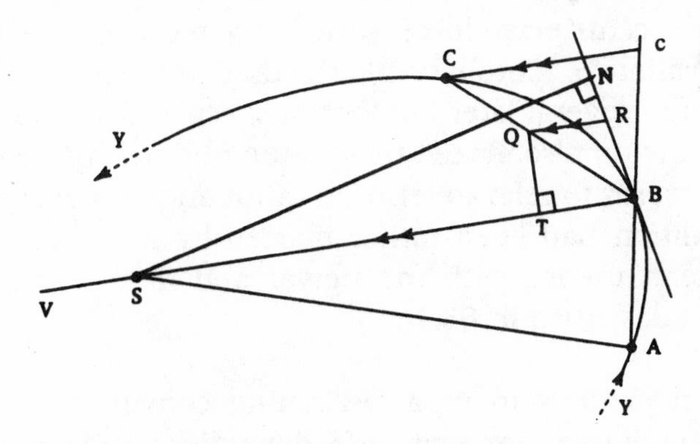

FIG. 5.10.1 *Newton on the planetary orbit, based on* Principia, *Book 1, Propositions 1 and 6.*
The curvature of the part of the orbit shown is exaggerated. A planet Y orbiting the Sun S experiences inertial motion along AB, and a centripetal attraction cC towards S. BR is the tangent to the orbit at B; V lies on the osculating circle passing through B.

horizontal inertia, while cC resembles a Keplerian radial force (although operating on a different mechanical basis from theirs).

5. The force f was *not* stated to be inverse-square. Indeed, it was not even required to decrease as SY increased, and in Proposition 10 Newton showed that if S were the centre of an elliptical orbit, then $f(\mathrm{SY}) = \mathrm{SY}$. But he did assume that the orbit was convex; finding conditions for f which yield convex orbits is a hard mathematical problem – one which he and his successors usually ignored.

6. Newton explicitly allowed for dynamic as well as static equilibrium, for he permitted S, and therefore the whole configuration, to move 'with a uniform rectilinear motion'; but he did not make clear whether, or which, later Propositions could also be construed this way. Maybe he assumed that dynamic equilibrium applied and mentioned uniform motion in order to exclude cases where S was accelerating.

7. In the second and third editions of the *Principia*, Newton sketched out an alternative dynamics based upon the circle of curvature of the orbit at B rather than its tangent; in the notation of FIG. 5.10.1, BV is a chord of that circle, and the centripetal force varied inversely with $\mathrm{SN}^2 \times \mathrm{BV}$ {Proposition 6, Corollary 3}. He used this variant in corollaries to several later Propositions, especially 9–11.

8. Newton also graced these later editions with a clearer use of words to refer to terms in ratios and proportions. The first edition had been rather marred by a mixture of the Euclidean theory with the newer and more arithmetical version described in §3.26.

Newton exhibits to us a fascinating convolution of new and traditional aspects of dynamics and attendant geometry: 'revolutionary or reactionary?', as {Brackenridge 1988} has well put it. On the one hand, he showed some preference for three motions with pedigrees in Galileo and Huygens, among other predecessors: rectilinear with uniform acceleration; parabolic motion under a uniform force;

and uniform circular motion (related also to centripetal force). On the other hand, he concentrated upon the orbits of the heavenly bodies, whereas his predecessors usually gave precedence to the bodies themselves. In particular, with perturbation theory, and also with motion in a resisting medium, he saw that Kepler's method (4.11.3) of measuring time by area could not be used. His prime concern with orbits led him to innovations in refining the differential geometry of curved motion, thinking carefully (if not always with satisfactory rigour) about their infinitesimal components, and admitting orbits which went beyond the normal repertoire of curves.

One class of 'advanced' convex curve was spirals {Book 1, Proposition 9}, which were involved in the remarkable Lemma 28: 'There is no oval figure whose area, cut off by right lines at pleasure, can be universally found by means of equations of any number of finite terms and dimensions.' Newton's term 'oval' probably followed Descartes's use (§4.21) for a convex curve O defined by an algebraic equation; the straight lines issued from some interior point P, and the theorem stated that neither the area nor the perimeter of any sector of O could be expressed by algebraic functions, however many times O was swept out. He deployed the spiral $rq =$ constant as generator, and used the infinity of its points of intersection with any straight line as the basis of his (intuitive, and indeed flawed) proof.

One motivation for this result may have been Newton's search for general properties of convex curves, of which planetary orbits were special cases; another was surely to underline the limitations of polynomial expressions in celestial mechanics (with Kepler's eqn (4.11.4) doubtless an example in mind). A full understanding of his theorem would require the function theory of the 19th century (§8.13).

5.11 *Vacuum or vortices? Two different mechanics*

Although Newton's approach to mechanics came to occupy a very prominent position, it was not the only one. Here I shall contrast some aspects of it with the views of Leibniz; the differences between the two men, and the issue of plagiarism, are admirably handled in {Bertoloni Meli 1993}.

The circumstances of Leibniz's publication were to help fuel the priority dispute mentioned in §5.1. Newton had sent a copy of the *Principia* to Hanover for Leibniz, not knowing that he was away on a diplomatic mission. Leibniz saw a review of the book in the *Acta eruditorum* early in 1689 in Rome, and in response to the *review* he recorded his own ideas in two papers on mechanics and one on optics, which were then published in the *Acta*. He saw the book itself only later in the year, and promptly made sets of notes on it. By priority-row time, Newton thought – justifiably, it seems – that Leibniz's published papers included plagiarisms of his own work.

Some of the differences between the two men are exemplified by their methods of determining the motion of a body moving around a point of central attraction. By contrast with Newton, for Leibniz a planet Y in FIG. 5.11.1 passed from B to C (again in a constant infinitesimal unit of time) due to a circular motion BL caused by the rotation

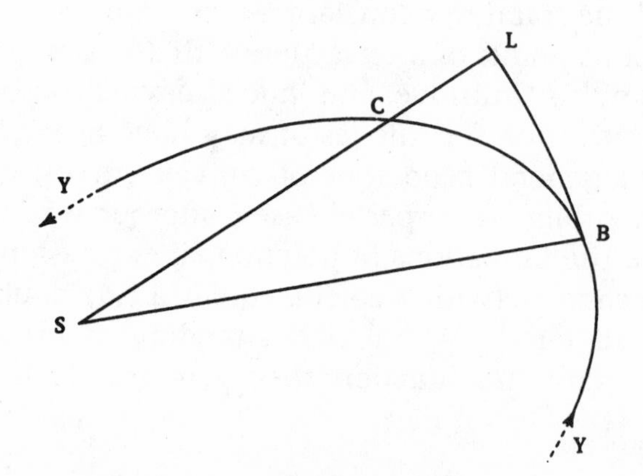

FIG. 5.11.1 *Leibniz on the planetary orbit.*
As in FIG. 5.10.1, the curvature is exaggerated. A planet Y orbits the Sun S; vorticial motion is directed along BL, an arc of the circle centred on S, and centrifugal force acts along LC.

of the vortex, proceeding with an angular velocity inversely proportional to the radius (equivalent to an areal law), together with a force LC from S which combined gravitational attraction with centrifugal force caused by the rotation of Y {Aiton 1988a}. The status of centrifugal force constituted an important point of difference between the two men. Leibniz followed Descartes (§5.8) in taking it to be a genuine force, created by the action of a vortex. Newton began by accepting it as real, but by the time of the *Principia* he had relegated it to a fiction, explicable in terms of the real centripetal gravitational force and the third law of motion. However, he was at pains to stress that his talk of gravity was metaphorical, for he admitted that he did not know its source {van Lunteren 1991, Chap. 2}.

In Book 1 Newton considered the continuous orbit ABC ... of FIG. 5.10.1, in contrast to his earlier polygonal approximations involving (5.10.1), with BR the tangent at B {Proposition 6, Corollary 1}. For ABC he found proportions for both the centripetal force towards S (now assumed to be stationary) and for time:

$$\text{force} \propto 1 \bigg/ \left(\frac{SB^2 \times QT^2}{QR}\right), \quad \text{time} \propto (B \times QT), \qquad (5.11.1)$$

with QR an infinitesimal of at least second order. In Propositions 11–13 he showed that the ellipse, hyperbola and parabola each satisfied the inverse-square law; these determinations have become known as the 'direct problem' (given the orbit, find the force). He stated the converse implication as a corollary in Proposition 17; but this 'inverse problem' is much harder to resolve than the direct one. (Newton did not use the phrases 'inverse' and 'direct problem', and unfortunately the adjectives were switched in the 20th century.) This aspect of his dynamics was noted around 1710 by Jakob Hermann and Johann Bernoulli, members of the Leibniz camp, and Newton himself recognized the need for more ample explanation in the second edition of the *Principia*; but even his account has recently raised doubts among historians about its completeness (see especially {Aiton 1988b}).

In Propositions 39–42 and 57–60 of Book 1, Newton generalized some earlier theorems by determining the planar orbit of a body attracted to a point by 'Any [well, "any"] kind of centripetal force', and some special cases. After these two-body problems he posed in Propositions 65–66 the three-body case (in general, but with Sun, Moon and Earth in mind), and followed a general survey with special cases in 22 corollaries. Thus was launched the study of the three-body problem in various forms, a major challenge for mathematicians in the centuries to follow (§6.8, §10.17, §14.18).

In Book 2, Newton analysed three different laws of resistance: with velocity, $(\text{velocity})^2$, and a linear combination thereof. His examples ranged not only through fluid mechanics but also pendulum motion, where in Section 6 he encapsulated the desimplifications wrought by Huygens (§5.8). His analysis of the propagation of sound drew upon a relationship between the 'elastic force' sending out the sounds and the 'elastic fluid' transmitting them {Propositions 48–50}; but his prediction of velocity was way out, and so in the third (1726) edition he introduced a fudge factor to yield *exactly* the best known current value. In the 1800s, P.S. Laplace was to seek an alternative explanation more in line with Newtonian principles (§10.10).

Leibniz's paper on resistance stressed the cases of velocity and $(\text{velocity})^2$ (respectively, 'absolute' and 'respective'), though his conception of resistance itself was different (to accommodate aethereal resistance, I suspect). He formed and solved first-order differential equations, in his calculus; but he did not publish many, maybe because some solutions were incomplete {Aiton 1972}.

Book 3 of the *Principia*, 'The System of the World' (another swipe at a title of Descartes?), dealt with details of the celestial and planetary mechanics of our Solar System. Here the inverse-square law, which had frequently but not exclusively featured earlier, moved to centre stage. Newton picked out several well-known properties of individual planets, and also analysed the near-parabolic orbits of comets. One main theme was the shape and the motions of the Moon, and its effects on phenomena such as the Earth's tides; but, despite great efforts, the failure to match observed lunar data was to prove a major threat to the veracity of the system (§5.25). A draft version of Book 3 was published posthumously in 1728 in the original Latin (as *De mundo systemate*) and in English translation (TABLE 5.1.1); the latter is available in the modern edition {Newton 1934} of the *Principia*.

The approaches of the two men differed not only in method but also in ontology. Although Newton considered motion in a resisting medium in Book 2, he concluded it with arguments against the existence of an aether, citing its alleged need to be as dense as the bodies it carried, and its supposed failure to obey an areal law when in (necessary) stability. For Leibniz the aether was real, both in its capacity to carry body Y along BL in FIG. 5.11.1 and also to transport Y from layer to layer of the aether in the direction LC. In addition, he understandably objected to Newton's notion of action at a distance. He could have added that vortex theory explained why the planets went the same way around the Sun, a fact which Newton had to adjoin as an additional hypothesis – indeed, Newtonians were always to keep quiet about this lacuna.

On the other hand, the theorems in the *Principia* involving equipotential surfaces (to use the modern term) and/or attractions of bodies to external points were a major point of superiority over Leibniz; indeed, they were to constitute one of Newton's main mathematical influences, and contributed to the understanding of his proofs {Greenberg 1995, Chap. 1}. One of the main results claimed that the attraction experienced at an external point to a collection of bodies was towards their common centre of gravity {Book 1, Proposition 89}; the proof drew repeatedly on the composition of forces. Two Propositions later, Newton found a geometrical expression for the attraction of a spheroid of revolution at an external point P on its axis. Less happy was a corollary, that when P lay inside the spheroid the attraction was proportional to its distance from the centre; the correct result (10.6.2) was to come from S.D. Poisson in 1814.

Their positions relative to Descartes, taken in these and other writings, exemplify some further differences. Newton rejected vortices, while Leibniz worked with them. Newton took space and time to be absolute, whereas for Descartes and Leibniz they were relative. Newton's opening sally in the *Principia* that 'the quantity of matter is the measure of the same, arising from its density and bulk conjointly' {Book 1, Definition 1} is partly consonant with Descartes on matter as extension (§5.8), although Descartes too emphasized density. Leibniz objected to Descartes's position on the grounds that matter had further properties such as resistance (or lack of it) to penetration by other bodies; he also rightly found fault with Descartes's laws of collision. This view led him to emphasize the *vis viva* of a body, roughly corresponding to our notion of its kinetic energy (note the cases of this in Huygens in §5.12), and indeed stimulating an alternative approach to mechanics in which energy considerations were given a prime place (§6.12). (The word 'dynamics' (*dynamica*) was coined by Leibniz, in 1695.) The issue was related to the *range* of bodies to which analysis could apply. For example, in FIGS 5.10.1 and 5.11.1 Y was really a mass-point, but this concept was still to come, from Euler (§5.25). Finally, while Newton worked with velocity and acceleration, Leibniz used *conatus* (internal tendency towards motion) and *sollicitatio* (infinitesimal change in velocity), notions introduced by Thomas Hobbes in rather vaguer forms.

5.12 *The inconstant pendulum*

{Yoder 1988; CE 8.13}

The importance and complexities of this simple instrument continued to concern scientists as its importance grew with its use in, especially, clocks. Huygens was a major figure, with his treatise *Horologium oscillatorium* (1673) (§5.8). Stimulated by the interest taken in time-keeping by the Accademia del Cimento, he contributed to the design of reliable clocks and chronometers, and in his theory of the pendulum he believed that he had achieved his goal. FIG. 5.12.1 shows his design, in which the principal innovation was restricting the movement of the supporting strings to between the cycloidal arcs KV and KI. For in FIG. 5.12.2, in which BA was a cycloid with tangent BI, and FA a semicircle cut by an arbitrary horizontal line GI, he showed that

$$t(\mathrm{B,E}) : ti(\mathrm{BI}) :: \text{length of arc FHA} : \mathrm{FG}\ [:: \pi : 2], \qquad (5.12.1)$$

where I write 't(B, E)' for the time of descent of a particle down the curve BE, and 'ti(BI)' for the time of uniform descent down the straight line BI at the average velocity acquired over Bθ. From arguments such as these, Huygens came to his most durable contribution, building upon Mersenne's discovery (§4.16): the formula

$$T = 2\pi\sqrt{l/g} \qquad (5.12.2)$$

for the period T of oscillation of a simple pendulum of length l. The special allure of the formula was the means it gave for calculating the gravitational force g.

Huygens' proof resembles Galileo's (FIG. 4.14.2) in using a sequence of horizontal lines, and Cavalieri's (§4.19) with GI working like a *regula*; further, the method used exhaustion (§2.21) by examining properties of each thin horizontal slice into which FBIG was divided. The purpose of the result was to reduce the unknown time to known quantities and thereby deduce that the time of descent to the apex A was the same for all such curves. Thus, as the strings followed a cycloid, equal times of oscillation would be performed by the bob down to A and up the other side.

This construction exemplifies an important part of Huygens' mathematics. Given a point P on a curve i, the centre

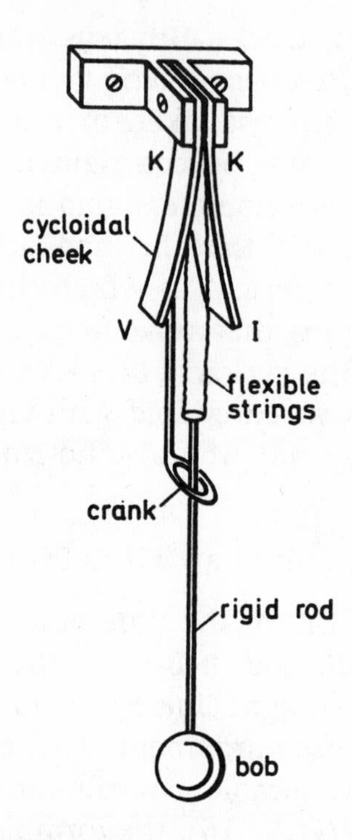

FIG. 5.12.1 *Huygens' cycloidal pendulum (*Horologium oscillatorium, *1673)*
{Ariotti 1972, p. 395}.

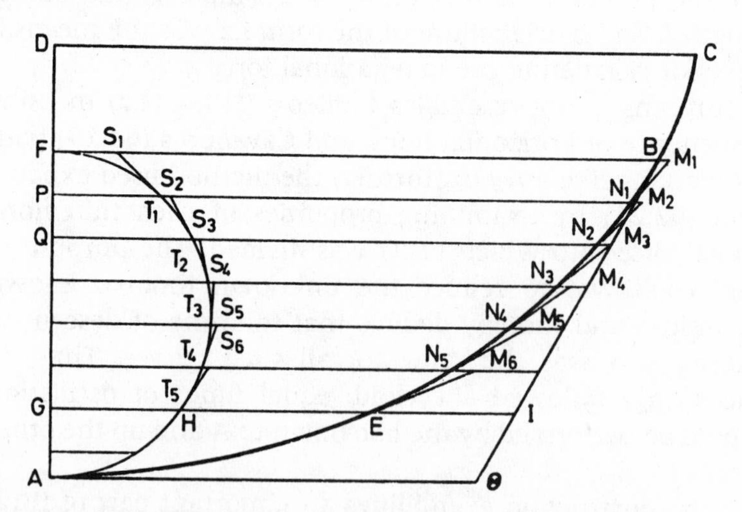

FIG. 5.12.2 *Huygens' proof of the time of equal descent*
{Ariotti 1972, p. 401}.

C of curvature of i at P generated a curve e called the 'evolute' of i. Conversely, i was the 'involute' of e, which he specified this way: lay an inelastic string along i, and gradually peel it out straight; then at any point C its end is at P, so that i is generated. The cycloid was its own involute and evolute. Here, and also in his optics and geometry, he made great use of these pairs of curves, and showed their value for mathematics.

However, while his results were to be of great value to clock-makers, Huygens' hopes for pendulum theory were not to be fulfilled. Part of the trouble lies in the idealizing kinemetic character of his construction, which, while not notional, omitted real effects such as torsion and friction. In addition, current observational work with the instrument had revealed variations in time of oscillation with altitude: the suspicion was growing that the Earth was spheroidal in shape (§6.8), and so such variations should be expected. A strong advocate of this view in the 1680s was Robert Hooke (1635–1702), a fine physicist and experimentalist, supported by Newton in the *Principia* with the formula that $(\text{time})^2$ varied directly with the length of the string and inversely with the distance to the centre of the Earth and 'its absolute force' of attraction {Book 1, Proposition 52}. Newton mentioned Huygens' result as a special consequence, not the general law for which its inventor had hoped.

In general, Hooke and Newton were not particular friends; for Hooke had groped towards the importance of gravitation, the inverse-square law and the analysis of curvilinear orbits before Newton. Hence he saw the *Principia* as plagiarism, and said so in characteristically grumpy style {Nauenberg 1994}. While Newton would have gained some inspiration from these ideas, the mathematical content of the book was undoubtedly his own, and beyond Hooke's ken. Hooke is best remembered for the linear law relating *tensio* to *vis* in elastic bodies (1679), not new with him but given with fresh generality. Further, in demonstrating the importance of the microscope, he contributed notably to optics – our next subject.

5.13 *Optics: bullets or waves?*

Huygens featured also in the development of physical optics in this period, publishing his book *Traité de la lumière* in 1690, when he was in his early sixties {Shapiro 1973}. According to him, the finite velocity *c* of the propagation of light established by Römer in 1676 showed that it *had* to be a wave phenomenon in the aether. Upon this argument, which is actually invalid, he built a beautiful theory. The aether contained tiny elastic particles, which produced light waves when excited into vibration. He did not propose any periodic properties for the aether; on the contrary, the particles were distributed irregularly. A similar model, but incorporating periodicity, would be proposed by A.J. Fresnel in the 1820s (§10.12).

The mechanism was stated in a principle which is known after Huygens: that the motion of any particle led to a wavefront (not his term) which could be taken as a source of secondary spherical waves propagating with velocity *c*. Thus, after time Δt they had radius $c\,\Delta t$, and each point P of a wave was itself such a source, and so on. 'Rays' were just ways of speaking, for a ray through P was just the normal at P to the wavefront. The envelope of rays generated a curve s called a 'caustic', and the centre of curvature of s at P generated the evolute of *c*; hence *c* was the involute of *e*. His most notable contribution to optics was his theory of double refraction, in which the primary rays came from spherical waves but the secondary ones came from spheroidal waves.

Newton's work on optics began to be published in 1672, and some mentions were made in the *Principia*. But the main production was his book *Opticks*, apparently largely ready by 1692 but then lost in a fire in his room, and so rewritten for publication. It was published in 1704; further editions appeared in 1717 and 1721, and a posthumous one in 1730 (TABLE 5.1.1) which is available in a modern annotated version {Newton 1979}. Apart from also being divided into three Books (§17.18), it is so different from the

Principia that one might not guess that they had the same author; indeed, such guesswork is in order, since for some reason he was not named on the title page. The book was written in English, not Latin (a Latin translation appeared two years later); it contained many accounts of experiments, especially his own, whereas the *Principia* was largely theoretical and drew on others' observations; and, in contrast to the earlier reluctance, it entertained hypotheses aplenty (explicitly in a series of 'Queries' at the end, which were expanded from 16 to 23 to 31 in the next two editions), such as possible properties of the aether. Further, apart from deft use of elementary geometry, little mathematics was used. Indeed, rather strangely, he appended the two fluxional texts (§5.7), but they were omitted from all later editions.

Newton explicitly confronted Huygens on two issues. Firstly, he subscribed to the common 'corpuscular' view of matter of his time, claiming that light was composed of tiny particles of some kind, travelling at velocities dependent upon their colour. In fact, the theory of colours was a main concern: in a discussion of the rainbow he plumped for seven colours, adding orange and indigo to the accepted five (red, yellow, green, blue, violet), giving the seven-colour scheme still standard in Anglo-Saxon culture {Book 1, Part 2, Proposition 9}. His argument for this new convention was partly based on an analogy with the ratios r of the seven main intervals of the musical octave, for he proposed the ratios $r^{2/3}$ for the radii of the two interference rings of neighbouring colours that were produced in the space trapped between two slightly convex faces of prisms when pressed together {Book 2, Part 1, Observation 14}. He went once more against Descartes, who had divided the spectrum into six colours; apparently this time his opposition was not consciously motivated.

The second disagreement with Huygens involved periodicity. Newton hypothesized that a ray passing though a refracting medium to produce colours was susceptible to 'fits of easy reflexion' and 'easy transmission', *regular* alternating spatial intervals varying with the colour of the ray and also with the angle of incidence upon the refracting surface. It was the aether that conferred these properties, so his theory was an uneasy mixture of bullets and waves.

In addition, in Query 25 of the 1717 edition of the *Opticks* Newton dismissed Huygens' theory of double refraction without offering any workable alternative. The consequence was that double refraction virtually disappeared from sight (as it were) for the rest of 18th-century optics, along with Huygens' book; in particular, Euler never mentioned the one or analysed the other in his wave theory (§6.14). Further, apart from Euler physical optics remained largely experimental and qualitative throughout the century. The clash between corpuscular and wave models was to arise sharply again only in the 1810s, with more mathematics thrown around on both sides (§10.12).

5.14 *Some English algebras*

{M. Cantor 1907, Chaps. 87 and 98}

In previous sections algebraic symbolism has frequently been evident. Here we consider algebra in general, mainly the theory of equations. John Wallis and Newton are taken as two principal contributors; some successors in later generations appear in §5.26.

In his early *Arithmetica infinitorum* (1655), which contained his formula (4.19.2) for $4/\pi$, Wallis used ': :' for brackets and '∞' for infinity. The subsequent *Mathesis universalis* (1657) includes early examples of Greek letters to stand for unknown quantities. But his most important book was an *Algebra* of 1685. Best remembered today for its very partial judgements on Harriot against that Frenchman Descartes, several more positive features are of note. Wallis popularized, but did not invent, the symbol '÷' for the division of two numbers (it may have been created from a rotation of '%'). He formulated problems in indeterminate analysis, including the alligation of gold (tackled by Recorde; see (4.4.1)). In the context of complex roots of polynomial equations, he interpreted $\sqrt{(-bc)}$ as 'a Mean Proportional between $+b$ and $-c$ or between $-b$ and $+c$', which 'may be thus Exemplified in geometry' as a line in the plane from the origin but off the horizontal line through it where $\pm a$ and $\pm b$ were represented {Wallis 1685, p. 268}. To us this idea smacks of the complex plane, but it had to wait until the 19th century for its (gradual) acceptance (§9.7); he was not offering any *general* theory of complex numbers, nor any axis system.

Wallis's writings include a few unpublished results by Newton; for example, his 'parallelogram', in which the coefficients of a polynomial in two variables x and y were laid out on a rectangular grid in positions corresponding to their powers (for example, that of xy^3 in the second row and fourth column). Newton lectured on algebra for a decade from 1673; the lectures appeared in book form first in 1707 under the editorship of Whiston, carrying the interesting title *Arithmetica universalis*. Although Newton feigned disgust at the publication and editing, later editions appeared in English and Latin in his lifetime (TABLE 5.5.1), and the book enjoyed quite an influence, at least in Britain. He ran through common algebra without making too much fuss about negative numbers; irrationals needed a geometrical basis, while complex roots were dubbed 'impossible'. After a lot of examples, he turned to equations, presenting the usual resolution and estimation of roots of equations up to quartics and transformation of equations, and also his relations between the coefficients of a polynomial and the sums of powers of its zeros {Whiteside 1961, Section 2}.

In this respect Newton concurred with Descartes, but otherwise he was critical of his predecessor as usual, especially over the association of algebra with geometry. For example, for Newton criteria of simplicity in one branch seldom carried over to the other, as he observed in his classification of curves defined by algebraic equations

(§5.7). This topic had been one of his first mathematical interests, and now he gave only some simple cases, drawing on his own ideas and also the methods of transforming polynomials to ones of lower degrees published by the German Ehrenfried von Tschirnhaus(en) (1651–1708) in the *Acta* in the 1680s. He also included some analytic geometry, upon which topic l'Hôpital's *Traité analytique des sections coniques* was posthumously published in the same year.

5.15 *Newton's seven main legacies*

The influence of Newton's various contributions was to become so great that it is helpful to identify seven strands – the fourth and sixth of which are usually overlooked. I follow an order roughly corresponding to his chronology of work; the comments pertain mainly to the period covered by this chapter and the next.

1. His *fluxional calculus* became an infatuation of English mathematicians when they finally came to hear of it, and until the 1740s they kept up pretty well with Leibniz and his followers. However, by 1750 the competition became much too stiff, for reasons indicated in §6.1. But his classification of curves, and other manipulations using his calculus, remained powerful stimuli to analytical and early algebraic geometry.

2. The *Principia* eventually revolutionized the approach to *mechanics* to such an extent that even historians misread his predecessors because of his dominance. The law of gravitation gradually gained supremacy over vortex models, along with his laws; the methods of the calculus came to replace geometrical modelling, although the Leibnizian version was adopted on the Continent. There, however, from the 1740s, his mechanics was to be challenged by alternative principles, competition in the terrestrial branch having become quite fierce (§5.22). Less enthusiastic than Newton's countrymen for the inverse-square law, Continentals tended to start from those Sections (3, 9 and 11 of

Book 1) which assumed central forces independent of the law.

3. In *celestial mechanics* the new chances to analyse perturbation theory and lunar motions were particularly welcome. The Good News was spread with such enthusiasm that the deficiencies of the theory were ignored or glossed over. Apart from the mystery of gravitational action, two major silences relative to Kepler and the vorticists stand out: Newton had no means of explaining why the planets went the same way round the Sun, or why they lie in those orbits rather than in others.

4. Completely different was Newton's practical theory of *the motions of the Moon*, prepared at the time of his transition from Cambridge to London in 1694–5 and first published in 1702. No gravity or inverse-square law operated here – it was all epicycles and kinematics. But his equations gave a good fit, and were used throughout the 18th century for compiling tables of longitude for navigators.

5. His theory of *optics* was the most articulate for the 18th century, a time for bullets and (with less enthusiasm) fits. The mathematical component of all theories remained quite modest, however.

6. His '*universal arithmetic*' set a certain agenda in Britain for the scope of common algebra and the solution of equations for over a century (§9.14). The influence on the Continent was much slighter.

7. Because of the success, a scientifico-philosophical movement known as '*Newtonianism*' or 'Newtonian philosophy' became popular, especially among educated folk wishing to say the Right Thing. It consisted of mechanics and optics done his way, laced with elements of pertinent geometrical methods. Sometimes enthusiasm was also expressed for not hypothesizing, and for the power of contemplative minds – a contradiction which went unquestioned.

Leibniz's influences were somewhat different. The principal one was his version of the calculus, which was to transform many areas of mathematics, especially in applications. His proto-energy mechanics moved more slowly (as it were), but was around as an alternative to the Newtonian regime. In probability theory his lines of thought were modest, but came indirectly through Jacob Bernoulli (§5.22).

5.16 *Five newcomers*

During the 1720s and 1730s, five important new figures appeared on the scene. Four of them emerged on the Continent; I take them in chronological order of birth.

Daniel Bernoulli (1700–82) was the son of Johann. He trained as a doctor but soon exhibited remarkable mathematical gifts. After a spell in Venice, he went with Jakob Hermann to join the new Academy of Sciences in St Petersburg. There he stayed until 1733, when he returned home to Basel to take the chair of anatomy and botany, adding the physics chair to it in 1750.

Son of a Protestant minister in Basel, and pupil of Johann Bernoulli, Leonhard Euler (1707–83) spent two long periods in St Petersburg, 1731–41 and 1766–83; the years between were passed in the Academy of Sciences at Berlin under Frederick the Great of Prussia. He was heavily involved in organizational and academic matters in both institutions (for example, the publication of their respective journals), and his home in St Petersburg became a centre for the discussion of science. Yet for productivity he has few rivals in the history of mathematics, in quantity, quality or range. He lost one eye in 1738 and the other in 1771, but then he became still more productive, walking round and round his working-table while amanuenses took down the results of his thoughts on a slate and wrote them up for him afterwards – one of the great miracles of achievement among the blind.

No branch of mathematics escaped Euler's attention, and most received significant contributions from him; in major ones like the calculus and mechanics he was a principal player, and to the recherché number theory he was long the only regular contemporary contributor. His style was renowned for its clarity and order, although the latter is sometimes deceptive; in particular, in mechanics he was apt to follow the general principles of a theory with an unexplained selection of cases-that-I-find-interesting rather than by a systematic choice.

Euler died in 1783, but the St Petersburg Academy kept publishing new material from him until the 1830 volume of their *Mémoires* (as their journal was called by then), and other items appeared in some posthumous editions. The full edition of his works, however, did not start to be published until 1911, and is still in progress, though most of the 29 + 31 + 12 volumes of its first three series are published. The fourth series, still in its early stages, will contain the known correspondence and notebooks, both also prodigious in quantity. They are being accompanied by a complete edition of the works (including letters) of the Bernoullis, not only the three already discussed but also more obscure but interesting kin such as Daniel's brother Niklaus II (1695–1726). Partial editions of the works of Jacob and Johann were published in the 1740s, the latter self-edited.

Each of the other three newcomers was largely self-taught. Alexis Claude Clairaut (1713–65) was so precocious that he gained election to the French Academy in 1731 with an innovative study of differential geometry (§5.26), although he was under-age. Jean le Rond (1717–83) was named after the Paris church at which he was found after his illegitimate birth; as a young boy he was given for some reason the surname 'Daremburg', which he understandably Frenchified to 'd'Alembert'. Neither man married, though each enjoyed life in all its respects. Both men passed their careers entirely in Paris, although Clairaut took part in an important geodetic expedition to Lapland (§5.25) in the mid-1730s. He also helped to prepare notes for his friend Madame Chatelet's translation of Newton's *Principia* (1756). D'Alembert and Clairaut worked – even competed –

in the calculus and mechanics, and philosophically showed an empirical penchant. However, their mathematical styles differed, for Clairaut worked geometrically (and also contributed to the branch), while d'Alembert showed an algebraic cast (and also worked in probability theory). He is best remembered for the *Encyclopédie* which he edited with the philosopher Denis Diderot, and which appeared in 35 enormous volumes between 1751 and 1780, nearly bankrupting the French publishing industry.

In England, meanwhile, Thomas Simpson (1710–61) had taught himself mathematics, especially fluxions, on which he wrote a textbook (1737). By then installed in London with a wife old enough to be his mother, he made his living as a teacher and textbook writer; but his research, in which he was the main follower of de Moivre in probability theory, was distinguished. He knew Continental methods, and indeed in geometry was influenced by Clairaut; but his talents flourished just as Britain disappeared into its mathematical fog.

5.17 *Complex numbers and their algebra*

Among the discussions between Leibniz and the Bernoullis, an interesting one from the early 1700s concerned the logarithm(s) of a negative number. My '(s)' is deliberate, since the appearances of multiple values when $\mathrm{i}=\sqrt{-1}$ was around were still not well understood. It/they arose in, for example, evaluating this integral by splitting the integrand into its complex factors and using partial fractions to yield

$$\int_y^\infty \frac{1}{1+x^2}\,\mathrm{d}x = \frac{1}{2}\ln\left(\frac{y+\mathrm{i}}{y-\mathrm{i}}\right), \quad y \text{ real.} \qquad (5.17.1)$$

Particular values of y led to objects such as $\ln(-1)$ which, as Johann Bernoulli noted, seemed to possess properties such as

$$\ln(-1) = \ln(1/-1) = \ln 1 - \ln(-1) = -\ln(-1); \quad \therefore \ln(-1) = 0. \tag{5.17.2}$$

Leibniz's way out was to restrict the differentiability of $\ln x$ to positive values of x. The admission of complex values for the function itself was not clearly recognized, although it was known that $\sqrt[n]{1}$ took n values, including complex ones when $n > 2$.

In the 1740s Euler and d'Alembert considered the controversy among their forebears over (5.17.2) and the evaluation of complex numbers of all kinds. D'Alembert showed in 1747 (in a book 'On the General Cause of Winds', of all things) that when c and d were complex numbers, c^d could be reduced to the linear form. In 1743 Euler found from series solutions to the appropriate differential equations that

$$2 \cos x = e^{ix} + e^{-ix} \quad \text{and} \quad 2i \sin x = e^{ix} - e^{-ix}, \tag{5.17.3}$$

which clarified relationships between the trigonometric functions and complex numbers; the canonical equations were given in his *Introductio* of 1748 (§5.27):

$$\cos x \pm i \sin x = e^{\pm ix}. \tag{5.17.4}$$

These equations were not the first of their kind, for Roger Cotes (1714) and de Moivre (1722) had respectively stated theorems corresponding to

$$ix = \ln(\cos x + i \sin x) \text{ and}$$

$$(\cos x + i \sin x)^n = \cos nx + i \sin nx; \tag{5.17.5}$$

however, they had not been cast in this form.

One important context in which complex numbers were deployed is the fundamental theorem of algebra. The result, which dates back to 1629 (§4.19), was then often understood as claiming that a polynomial $y=f(x)$ in x with real coefficients can be decomposed, uniquely, into linear and quadratic factors in x. The importance of the result was clear; proof, however, turned to be very difficult {Gilain 1991}. I note here two efforts from the 1740s. Euler started from (5.17.4) and an analogy between a polynomial of degree n and solutions of an nth-order ordinary differential equation; but he really studied the form of the decomposition after assuming its existence. D'Alembert (in, among other places, his monograph on winds) claimed the existence of complex roots if no real ones could be found, and for such cases he expanded x in a series of (possibly fractional) powers of y and considered the corresponding curve; but, for example, he did not establish the convergence of the series. Both men and some others continued to worry about this question; we shall pick it up again in §9.3 with C.F. Gauss, who raised the level of study around 1800.

5.18 *Important new series and functions, mainly from Euler*

In a book of 1713 on probability theory (§5.22), Jacob Bernoulli furnished an important sequence, still named after him though used in some contexts which he did not envisage: the 'Bernoulli numbers' B_r. They arose for him as coefficients in the summations $\sum_{k=1}^{n} k^c$ (k and c integers) expressed as a power series in n up to n^{c+1}. (Following Leibniz's interpretation of the integral sign, he wrote '$\int n^c$'.) Rather typically, he omitted a proof, but he gave the opening non-zero values:

$$\mathrm{B}_1 = \tfrac{1}{6},\ \mathrm{B}_2 = -\tfrac{1}{30},\ \mathrm{B}_3 = \tfrac{1}{42},\ \mathrm{B}_4 = -\tfrac{1}{30},\ \ldots. \tag{5.18.1}$$

Euler (in a paper published in 1736) and Maclaurin in his 1742 *Treatise* came independently to an important extension of this sequence, which is named after them:

$$\sum_{r=0}^{n} f(r) = \int_0^n f(x)\,\mathrm{d}x + G(n), \tag{5.18.2}$$

where $G(n)$ is a complicated expression involving the B_r. Both men saw it as a method of relating series and function;

as usual, properties of convergence of the former were not considered. Euler also provided this generating function (on this notion see §5.22) for the Bernoulii numbers:

$$x/(e^x - 1) = 1 - \tfrac{1}{2}x + \textstyle\sum_{r-1}^{\infty} B_r x^{2r}/(2r!). \qquad (5.18.3)$$

Much interest was generated, especially from the 1730s with Euler, in the beta and gamma functions (this latter name was proposed by Legendre). They were defined in various forms, such as

$$\beta(x, y) := \int_0^1 t^{x-1}(1-t)^{y-1}\,dt \quad \text{and} \quad \Gamma(x) := \int_0^{\infty} t^{x-1} e^{-t}\,dt: \qquad (5.18.4)$$

the latter arose as a generalization of the factorial function $x!$ when x is not an integer. Euler and others found many results for this function which were to be generalized in the early 19th century with the hypergeometric series (8.14.3).

Another remarkable innovation by Euler was to apply analysis to number theory. In 1737 he defined the 'zeta function' $\zeta(s)$ and showed that it furnished means of considering properties of prime numbers p, as follows:

$$\zeta(s) := \textstyle\sum_{r=1}^{\infty} r^{-s} = 1 \Big/ \prod_p (1 - p^{-s}),\ s > 1 \text{ and real, } p \text{ prime.} \qquad (5.18.5)$$

A further definition and result of 1740 may be expressed thus: with x and y as two real variables,

$$f(x, y) := \textstyle\prod_{h=1}^{\infty} (1 + x^h y) = \sum_{k=0}^{\infty} a_k(x) y^k = \sum_{r,s=0}^{\infty} b_{r,s} x^r y^s. \qquad (5.18.6)$$

The form of the k-expansion revealed two properties.

Firstly, for each r and s the integer $b_{r,s}$ was the number of different ways in which r could be expressed as the sum of s distinct smaller positive integers. This result initiated the study of partitions of integers, which Euler continued around 1750 with this generalization of Fermat's little theorem (§4.22), now named after him: if a and n are relatively prime, then

$$(a^{\phi(n)} - 1) \text{ is a multiple of } n, \text{ where } \phi(n) := \prod_{r=1}^{m} \left(1 - \frac{1}{p_r}\right) \tag{5.18.7}$$

and p_r are the m prime factors of n; $\phi(n)$ counted the number of integers less than n and prime to it.

Secondly, $f(x, y)$ was the generating function of the functions $a_k(x)$. This notion had recently emerged in probability theory, which we now consider.

5.19 *Chance, justice and probability*
{CE 10.1}

Of all the examples of the difference between thinking and theory mentioned in §1.7, that of probability is perhaps the most convoluted and hidden – and like the origins of all earliest mathematics, partly covert. So only now do we note its emergence. Historical study has been commensurably late in arrival, with much of the best work done only in the last two decades. {Hacking 1975, 1990} and {Sheynin 1977} examine the extraordinarily wide-ranging early cultural background, {Hald 1990} treats the technical details.

The genesis of genuine theory is usually attributed to the correspondence between Pascal and Fermat on games of chance that took place in 1654; apart from some earlier forays by Girolamo Cardano, this seems to be a fair appraisal. There was a considerable social stimulus for the development of theory at this time of great political unrest: an infatuation in many countries with gaming and gambling. Here were classes of relatively simple problems that could be analysed, and maybe desimplified versions of the theories achieved could later be formed to deal with harder problems such as demography or the setting of annuities. But probability was still a somewhat recondite branch of mathematics, to which many figures paid little or no attention.

Much of the early theorizing carried a strong legal flavour, either explicitly in cases such as the rectitude of judges, or at least in the context of, for example, dice or coin (not) being manufactured fairly. The notion of equal possibility for potential outcomes sometimes arose, although there was a danger of arguments becoming circular if equal probabilities were involved. Often the basic mathematical concept was expectation rather than probability (the ratio of expectation to total outcome) itself.

While a variety of philosophical positions was adopted, no clear-cut distinction was made between subjective and objective probability; this was to come only around the mid-19th century (§8.10). However, it became understood that testimony for data differed in character from observational evidence.

5.20 *The launch of expectation theory*

{CE 10.2}

The game of chance considered in 1654 by Pascal (then in his thirties) and Fermat, his senior by two decades, was then called the 'problem of points': the winning or losing of some n points is interrupted when the (two) players have scored less than n, so that a fair allotment of their shares has to be made. Expectation was calculated, by Fermat seemingly in terms of listing all possibilities and assigning the respective takings in proportion, and by Pascal by multiplying the probability of each kind by its associated amount of winnings. Their findings enriched knowledge of combinations and permutations, and helped combinatorial probability to become established, under the binomial expansion (5.5.1) for integral exponent n:

$$(p+q)^n = p^n + np^{n-1}q + nC_2p^{n-2}q^2 + \cdots + npq^{n-1} + q^n,$$

$$\text{with } p+q=1. \qquad (5.20.1)$$

Pascal's considerations of the combinatorial aspects were published in his posthumous tract *Traité du triangle arithmétique* (1665), which contained the triangle of integers now named after him {A. Edwards 1987}. We saw it as (3.2.1) with the Chinese; the Indians and the Arabs had also possessed it, and some Europeans before Pascal.

The first advance on the exchange of 1654 was not long in coming: an elaboration in 1657 by Huygens of Pascal's method for any number of players, in *Tractatus de ratiociniis in ludo aleae* ('Treatise on Reasoning in Games of Gambling'). One of their problems has become known as 'the gambler's ruin': in a game in which each player has m counters and gains/loses one to his opponent every time that a throw of a die is favourable/unfavourable (according to some agreed criterion), what is the probability that a player will win all the counters? (If the favourable probability is a/b, then in general the answer is a^m/b^m.) Huygens' analysis included the novelty of constructing trees of possibilities for the various cases that may arise as the game proceeds {Shoesmith 1986}.

Five years later, in 1662, two of Pascal's associates, the lawyers Antoine Arnauld and Pierre Nicole, members of the Jansenist community in the abbey at Port Royal, southwest of Paris, published their *La Logique ou l'art de penser, contenant outre les règles communes plusieurs observations nouvelles propre à former le jugement* {Auroux 1982}. The title is instructive: they sought to include means of estimating the objective probability of evidence from the past and of predictions for the future, both regarded as non-deductive inference. One example of theirs has become well known, usually in later variants: the improbability that 'an infant arranging at hazard the letters from a printing-office, should compose all at once the first twenty lines of Virgil's *Aeneid*'. This widely read work helped to launch a tradition called 'Port Royal logic', and was to be a positive influence on the Enlightenment tradition in France (§6.17).

Enter Leibniz, literally; for in 1666, his 21st year, this young law student made his début in print with a tract called *Dissertatio de arte combinatoria.* Concerned with refining the classification of moods in syllogistic logic, it contained for that purpose contributions to combination theory. Partly from the influence of such work, he began to think about a logic of probability, with legal applications much in mind. Although as usual he published very few of his thoughts, they seem to have influenced colleagues though correspondence and other contacts {Schneider 1981}.

5.21 *British 'political arithmetic'*
{CE 10.3}

In Britain another tradition began, which was to remain a national characteristic for centuries: the mathematics of mortality and assurance. The first major writer was John Graunt (1620–79), with his *Natural and Political Observations ... Made upon the Bills of Mortality* (1662). He not only pioneered theory in this application but also recognized the co-presence of statistical regularities over large numbers of data and wide variations in small numbers. Medical contexts were among his main concerns; in particular, when appraising the various possible causes of plague, he intuitively perceived the notion of inverse probability. In Holland, meanwhile, Jan de Witt enriched annuity studies in the early 1670s in another way, by disavowing equal possibilities and allowing for some correlation between age and life expectancy.

Graunt was a friend of William Petty (1623–87), a doctor by training but an important innovator in economics and economic geography; he introduced the durable term 'political arithmetic' in a book written in the 1670s and published posthumously in 1690. Following earlier work on the 'political anatomy' of Ireland, Petty compared economic policies in England and Ireland. He discussed methods of assigning levels of taxes, and noted the significance of labour in the growth of national wealth. He supported his arguments with masses of statistical data, especially on population (and collected at a time when there was no census or national registration of births and deaths). He worked on the borderline between economics and probability theory, but he did not much discuss the latter (which, we shall note in §14.5–§14.6, was strikingly absent from much of the later development of mathematical economics).

In 1693 Halley enhanced political arithmetic when he presented to the Royal Society an analysis of expectation in demography in the case of Breslau, a German city-state distinguished by its isolation and thus by the stability of its population. He used its recent mortality figures to predict future survival rates on a constant population, and calculated current values of life annuities at an annual rate of interest of 6%, then appraising the corresponding terms offered by the British government. He also drew upon probability theory in that he tried to predict mortality data by using data on births.

As we saw in §5.9, Halley was the financial sponsor of Newton's *Principia*. Newton exemplifies those scientists who exhibit thinking without theory in this area; for while he rarely considered problems in probability, thoughts of that kind are evident elsewhere in his work {Sheynin 1971}. For example, his model of planetary motion showed such 'wonderful uniformity' that 'blind Fate' could not be responsible {*Opticks*, Query 31}; again, when Master of the Royal Mint he was responsible for the 'trial of the pyx', to establish by weight the authenticity of coins.

The success of Newton's astronomy inspired imitation from probabilists. One of these was the fluxionalist John Craig (?–1731), whose volume *Theologiae Christianae principia mathematica* (1699) imitated Newton not only in his title but also in his presentation of laws. The strength of a belief was proportional to its probability, and tradition in asserting it became feebler with the passage of time but could be strengthened by an increase in the number of testimonies. Thus the return of Christ predicted in the Gospels had to occur before the year 3144, by which time the evidence will have fallen to zero.

A more significant Newtonian was de Moivre; but to appreciate his context we must return first to the Continent.

5.22 *Jacob Bernoulli and de Moivre on laws of large numbers*

Leibniz corresponded with Jacob Bernoulli on probability theory as well as the calculus, and until his death in 1705 Bernoulli was working on a treatise. After much delay the publishable material appeared in 1713 as the book *Ars conjectandi* and was soon acknowledged as a major contribution {Bernoulli 1713}; it included the Bernoulli numbers (5.18.1). He continued many of the concerns of Huygens and Port Royal logic, and calculated possible profits in games of chance; Bernoulli also examined conditions under which probabilities could be added together, according to their defining conditions and whether events, testimonies or beliefs were being entertained. His emphasis was less on expectation than on probability, which he usually interpreted as measuring degrees of certainty.

The chief innovation of Bernoulli's book lay in the fifth chapter of Part 4, where he offered a remarkable new *kind* of theorem, concerning probabilistic inference drawn from a large quantity of data obtained from independent trials, each with two outcomes:

> THEOREM 5.22.1 *Let a trial be 'fruitful' or 'unfruitful' according to whether a certain event does or does not arise from it, and assume that there is a probability p of fruitfulness. Then, after (suitably) large numbers r of successes and s of failures 'it will become c times more probable' that p lies between $(r-1)/(r+s)$ and $(r+1)/(r+s)$, where c is another large positive number, dependent upon $r+s$.*

The proof worked by considering the appropriate terms in the binomial series (5.20.1) for a large value of the exponent $n=r+s$.

Bernoulli knew that he had found a major theorem; he called it 'golden', and so on a par with his formula (5.6.4) for the radius of curvature. He saw its novelty as lying not only in its use of large numbers (it has become known as the 'weak law of large numbers') but especially for relating past to future performance; that is, prior to posterior probabilities. He also suggested upper and lower bounds on the probability, which successors would narrow.

De Moivre made many contributions in the general context of this theorem. Having started off with the manipulation of infinite series in Newton's style (§5.7), he came to see their applicability to probability theory, since 'every term of the Series includes some particular Circumstance wherein the Gamesters may be found'. This passage comes from the preface of the first edition of his *Doctrine of Chances*, dedicated to Newton and published in 1718, when in his early fifties. In addition to the Newtonian approach, this French-born but long naturalized Englishman also extended Bernoulli's considerations of large numbers (to his credit, at exactly the time of the priority dispute over the calculus). His efforts continued in editions of 1738 and (posthumously) 1756, and before them in an important book *Miscellanea analytica* (1730) {Schneider 1968}. Drawing upon Graunt and Halley, he applied his methods to calculate *Annuities upon Lives* (1725), using data to give perhaps the first mathematical relationship (an arithmetic progression) between life expectancy and age. His work was popularized in the 1740s by Simpson.

De Moivre interpreted probability itself in terms of frequencies. His studies carried religious as well as mathematical import; the preservation of divine order was exhibited in the regularity with which events occurred over a long period.

As with Bernoulli, de Moivre used the binomial series (5.20.1) with large exponent n, but examined more closely the manner and rate by which the limiting value of the ratio would (or should) be reached as $n \to \infty$ {Archibald 1926}. He thought of approximating to the step function given by the sequence of terms of the formula with $n = 0, 1, \ldots$ by a parabola if n were continuous, and found a close relative of the curve C, then new, which we now call the 'normal curve'. He also used asymptotic formulae of his own, and results like (5.7.1) derived by his friend Stirling, to find that the points of inflexion were of the order of $(p\sqrt{n})$ from the peak, and thereby came essentially to C. He also sought a formula for the probability that the number x of successes in a run of n trials lay within some measure d of its most probable value np, by summing the appropriate terms of the binomial series and then allowing himself to take the sum as an integral. In the simple case of $p=q=\frac{1}{2}$, this gave

$$\text{Prob}(x \text{ lies within } d \text{ of } \tfrac{1}{2}n) \approx 2\sqrt{\left(\frac{2}{\pi n}\right)}\int_0^d \exp(-2x^2/n)\,\mathrm{d}x. \tag{5.22.1}$$

This formulation is modernized; de Moivre had only a series for 'the Hyperbolic logarithm' of e, and not even an integral (or fluent) sign. But he did supply the crucial exponential function, later to be known as the 'error function', safely convergent in series form, and of major importance in asymptotic theory. He realized that it was not directly integrable, so he relied on quadrature methods to obtain estimates. He also used, and thereby highlighted, difference equations. More importantly, he introduced generating functions $(\sum_r a_r x^r)$ to express the various probabilities of the sums exposed after throwing a large number of dice at once, as the coefficients a_r of the appropriate powers of x. We saw them with Euler in (5.18.6), in 1740; Laplace was to name these functions some decades later, and to extend their use (§6.20).

5.23 *The St Petersburg paradox, 1738*

{Dutka 1989}

The publication of Jacob Bernoulli's treatise in 1713 had been effected by his nephew Niklaus I, who made several valuable contributions of his own at this time, when he was in his mid-twenties. He had recently met the Frenchman Pierre de Montmort (1678–1719), and had this poser placed in the second edition (1713) of the latter's *Essay d'analyse sur les jeux de hazard*. I toss a penny and give you 1 coin if it comes down heads, 2 coins if it does so first at the second toss, 4 at the third, and so on. You seem to be on to a great winner, and hence should put up ∞ coins for the privilege; yet you are then bound to lose, no?

This conundrum aroused much interest over the years. It become known as 'the St Petersburg paradox', the noun intended in the Greek sense of 'para-dox', sitting alongside normal knowledge, the adjective alluding to the fact that in 1738 Niklaus's nephew Daniel Bernoulli published and analysed it in the *Commentarii* of the Academy of that city. Daniel's solution was that at any stage your increase in utility was inversely proportional to the wealth already gained (an anticipation of marginal utility theory of §14.6); others proposed to restrict your capital, or keep the game to a finite (though still possibly very large) number of tosses. It remains a striking teaser.

Although probability theory itself had not greatly advanced between the publications of Niklaus's and Daniel's ideas on the paradox, probabilistic thinking and practice had increased substantially, especially with the growth of insurance companies and centres of gambling – often all part of the same organization. One convolution was that the notion of probability, as expectation divided by total outcome, was gaining in importance and even supplanting expectation itself; in many cases it was much less unclear a concept with which to play. Proof-methods often hinged on the binomial series (5.20.1), which delivered probability values for many contexts (and represented compound interest in some others), and on intuitive uses of proportions (from mortality tables, for example) to get probability estimates; but now they were becoming more refined, with the calculus creeping in. At the same time the idea of chance became chancy, for God could not be construed as a dice-roller. Hence the normal interpretation adopted for probability propositions was epistemic, expressing our ignorance about the outcomes and their causes. In such ways the initial era of expectation and uncertainty gradually changed to one of probability rendered unavoidable by human unawareness.

5.24 *Parts of the partial calculus: isoperimetry and parametric differentiation*

{Engelsmann 1984; Greenberg 1995, Chap. 7}

We now catch up with one particular development in the Leibnizian calculus that was to become a major feature from the 1750s onwards: several independent variables. One would assume that the extension was made from curves $y=f(x)$ in the plane to $z=f(x, y)$ in space, and the passage from ordinary to partial differential equations. Such a change did take place, but with a remarkable intermediate stage involving families of curves $y=f(x, a)$, with a as the parameter in the plane. Thus this stage also contains the origins of the calculus of variations. Here are a few principal features.

1. In 1692 Leibniz argued intuitively that the envelope curve of such a family $V(x, y, a)=0$ was given by eliminating a between this equation and $V_a=0$.

2. Isoperimetric problems became very popular. Perhaps the best remembered, from 1697, is: find the curve of quickest descent of a particle under gravity between two fixed points. The answer was the curve known as the 'brachistochrone' (correcting Galileo's guess (§4.14) of the circle), a special case of a cycloid; it was found as the curve between the points which minimized the appropriate integral.

3. Trajectory problems also received much attention, especially involving orthogonality: find the curve(s) which cut at right angles a given family, so that

$$\text{given } \mathrm{d}y=F(x, y, a)\,\mathrm{d}x, \quad \text{solve } \mathrm{d}x=-F(x, y, a)\,\mathrm{d}y. \tag{5.24.1}$$

The work of Huygens (or at least his interests) was much in the background: Leibniz mentioned caustics in optics as an example of an envelope, and the trajectory problems related closely to involutes and evolutes. The Leibniz camp (including Niklaus I. Bernoulli) was largely responsible for formulating and solving these problems, although Newton also found the brachistochrone. A further figure of the 1730s was the Frenchman Alexis Fontaine (1704–71), who developed a superb 'fluxio-differential' partial calculus of his own although in a ghastly repertoire of notations using both d's and Newton's dots; he also studied properties of total and exact differentials of functions of (usually) two variables {Greenberg 1981}.

When Euler came onto the scene in the 1730s, he worked intensively in this area. One question, not new with him, concerned the interchangeability of partial differentials, where he considered not only

$$G_{xy}(x, y) = G_{yx}(x, y) \text{ but especially } G_{xa}(x, a) = G_{ax}(x, a). \tag{5.24.2}$$

He regarded such equations as universally true, and interpreted the second as going by two different routes from the bottom-left to the top-right corners of a near-square bounded by segments of neighbouring curves. Then he found a means of solving trajectory problems by using conditional 'modular equations', in which the 'module' a was treated as a variable on par with x or y; they transformed into ordinary differential equations.

The variational aspect, increasing in prominence, reached an apex in Euler's book *Methodus inveniendi lineas curvas* of 1744 (his 37th year), in which he used such means, often varying one or more isolated points on a curve, to find and indeed classify curves that met some optimal condition. He went beyond modular equations and saw that the problems could be formulated in terms of integrals of the form

$$\int_a^b f(x, y, p)\,\mathrm{d}x \quad \left(\text{where } p := \frac{\mathrm{d}y}{\mathrm{d}x}\right), \text{ with } f_y = \frac{\mathrm{d}(f_p)}{\mathrm{d}x} \tag{5.24.3}$$

as the first-order necessary condition for optimality. (The equations were not written this way, but the letter p is his; the next stage is described at (6.2.1).) In a lengthy addendum he adapted his various results to the conditions of equilibrium of flexible lines, inaugurating in print his concern with elasticity theory, which was to grow in importance from the 1750s onwards (§6.11). By then the era of partial differential equations had dawned, under circumstances described in §6.4. Meanwhile, other aspects of mechanics were developing.

5.25 *Newtonian and non-Newtonian mechanics*

During the 1730s much new writing appeared on mechanics, mostly on the Continent and not necessarily Newtonian in cast. One author was the Spanish-born French engineer Forest de Bélidor, who brought out *La science des ingénieurs* in 1729 and a two-volume *Architecture hydraulique* in 1737–9. While much of the treatment was based upon experiment or practice, he used Newtonian or energy principles, especially for the analysis of water-wheels.

Euler's *Mechanica* (1736), also in two volumes, actually covered dynamics only. Its main conceptual advance was that of the mass-point, whose motion in free and resistant media was analysed in great detail. He applied Newton's laws along the direction of the tangent, principal normal and binormal of the mass-point's path; a major simplification of this treatment was to come to him in the early 1750s (§6.6).

In 1738 Daniel Bernoulli published in Strasburg a major treatise, his *Hydrodynamica*. He introduced this word here – and also 'potential', but in the phrase *ascensus potentialis*, referring to energy potentially causing actual motion of a body. The opposite concept was *descensus actualis*, a formulation which raised the status in mechanics of the equation for the conservation of energy. The context was a well-known problem, the flow of water out of a container; to render the analysis more tractable he used the 'hypotheses of parallel slices' (as it became known), that water moves in differential slices each with its own velocity (like an open vertical stack of sliced bread being dropped on the floor). He also found a relation now known after him (in other forms) between the pressure in the outlet and the height of the container. His other topics included pumps, jets, gases and air-streams. His book contained the best treatment of the subject before Euler's general equations (6.10.1) of the 1750s.

In 1738 Bernoulli also competed for a prize problem on the theory of the tides proposed by the French Academy {CE 8.16}. Assuming the Earth to be a fluid of uniform density in equilibrium, Newton had somehow calculated a value for the excess of the diameter at the equator over that at the poles due to solar and lunar forces {*Principia*, Book 3, the enigmatic Proposition 19}. Maclaurin proved, in geometrical *Principia* style, that these forces would cause a homogeneous sphere to assume in equilibrium the shape of a spheroid flattened at the poles; Euler showed the importance of the horizontal component of these forces for the generation of tides, and found the equilibrate surface by calculus methods; and Daniel Bernoulli tried to predict the tides occurring at a port, and found better values than had Newton for the difference between mean solar and lunar tides.

Maclaurin's finding was an important extension of ideas towards potential theory in Newton; he had predicted that any rotating planet was a spheroid flattened at its poles by the centrifugal effect of rotating about its axis. His claim of polar flattening (for which Huygens also argued in 1690, but using Cartesian principles, and finding an oval profile) was challenged in 1722 by the French astronomer Jacques Cassini, who countered, from somewhat feeble data, that the Earth was elongated in the direction of its poles. An expedition to Lapland in the mid-1730s produced data that vindicated Newton's calculations for the Earth. This was an important finding, not only in itself but also for enhancing the glamour of planetary physics; one could find out something about the Earth *as a whole* {Greenberg 1995; CE 8.14}.

Clairaut, who was a member of the expedition, extended the calculus in his *Théorie de la figure de la terre* (1743) by the use of 'exact differentials' (his phrase) representing equipotential surfaces. A homogeneous incompressible body of fluid was under equilibrium at point (x, y, z) under an impressed force with components (P, Q, R) if and only if there was a variable Z such that

$$P\,\mathrm{d}x + Q\,\mathrm{d}y + R\,\mathrm{d}z = \mathrm{d}Z. \tag{5.25.1}$$

In modern terms a potential Z was admitted (relative to which the components would be represented by its partial derivatives), and so the differential form on the left-hand side of eqn (5.25.1) was exact. From this relation he sought properties of such surfaces (for example, non-intersection). He also applied his results to curves in space, and so helped to inaugurate differential geometry.

In the mid-1740s both Clairaut and then Euler expressed doubts over the capacity of Newton's law to tell all about the motion of the lunar apogee (the point in the Moon's orbit furthest from the Earth); Clairaut had even wondered in print if Newton's law had to be replaced by $(ar^{-2} + br^{-3})$, with constants a and b to be determined, in order to make it work properly. But in his book *Théorie de la lune* (1752) he realized that his earlier approximations had been too gross. Euler was glad of this vindication, but lunar theory still remained as taxing a challenge as any in astronomy. One of its properties resisting full analysis was the Moon's secular acceleration (the gradual increase in its orbital velocity), for which he accepted the possibility of an all-pervading aether causing frictional effects – an extra-Newtonian element of his own.

So not everything was going Newton's way. In 1743 d'Alembert published his *Traité de dynamique*, in which he offered a rival general principle which is known after him {Fraser 1985}. Suppose that a collection of bodies B at rest had impressed upon them motions with velocities v, which their mutual interaction changed to u; then the motions $w := v - u$ would have left the system at rest. Read in reverse order of occurrence, the principle stated that if the w left the bodies at rest, then the impression of v created velocities $v - w$ of motion. The hope for the principle lay in its (alleged) generality, both for the range of bodies covered ('all' of them) and the kinds of motion impressed (impact as well as gradual changes); curiously, d'Alembert restricted it to static equilibrium. While he stated the principle unclearly, he did give a good range of examples, including elastic collision, vibrating systems of points and a body sliding on a friction-free plane. Although he used diagrams, his treatment was often markedly algebraic; indeed, he was thinking (though not writing) vectorially. This tendency to algebraize set a style for mechanics which was to find a strong disciple in Lagrange (§6.11).

5.26 *Geometry and algebra: a perplexing relationship*

Clearly the relationships between algebra and geometry were deepening, with trigonometry still gaining importance from both sides. A good example of the resulting obscurities is evident in the Swiss Gabriel Cramer, whose *Introduction à l'analyse des lignes courbes algébriques* came out in 1750. Two results are now named after him, though both were in Euler and the first also in Maclaurin.

The first was the 'paradox' that a polynomial equation of degree v in two variables has $v(v+3)/2$ independent coefficients, and so requires that many points of specification in the xy plane. It also (usually) intersects a second curve of degree v in v^2 points (a theorem of Maclaurin); however, these points were common to both curves, so that neither of them was uniquely determined. This paradox is not very deep: as Euler had already explained to Cramer in a letter, the second property was not a *defining* one for either curve, so the points were not independent.

Cramer's second result was his 'rule' for solving v linear equations in v unknowns. We know it as the quotient of two determinants of order v; he presented it in the usual longhand way, of course, as the quotient of two sums of products of the coefficients, writing it out in full for $v = 3$. This was the normal manner in all 18th-century mathematics, both in algebraic cases like this and also for Jacobians in the calculus; and the term 'Jacobian' stresses the 19th-century origins of determinant *theory* (§9.10).

With some other figures, geometry could continue free from algebraic appendages. A remarkable example is Clairaut. In 1731, at the age of eighteen, he presented the basic principes of coordinate geometry in three dimensions; and his textbook *Elémens de géométrie* (1741), intended for beginners, so rejoiced in the connection with physical space as to include pictures of trees, and started off with the measurement of land. Unfortunately, more solemn traditions were to prevail (§6.13).

5.27 *Euler's* Introductio, *1748: the state of the analytic art*

In 1748 Euler published at Lausanne a large two-volume treatise entitled *Introductio ad analysin infinitorum*. The fact that it appeared there when he was based in Berlin, and apparently four years after its completion, suggests the fragile state of the market for advanced mathematics (and maybe also a lack of competent typesetters). However, eventually it became widely read, with a second edition and a French translation in the mid-1790s (and an English one in 1990!). It serves well as a milestone to mark the end of this chapter.

The volumes divided clearly by subject matter. The first dealt with 'analysis' proper; that is, functions, series and equations, together with continued fractions and a little number theory. In an opening chapter 'On functions in general' he gave his explanation of generality, which has proved easy to misunderstand. As an expression, $f(x)$ may be 'composed in whatever way'; but of course, he had in mind the ever-growing repertoire of functions with which he was familiar, but *not* our modern notion of a mapping.

Johann Bernoulli had used such phrases in his own lectures a half-century earlier (§5.6); again the range was wide for his time but limited for ours (and Euler's).

Euler's principal taxonomy of functions was based upon his distinction between 'continuous' and 'discontinuous' functions, where the former took one expression but the latter was composed of different expressions specified over adjacent intervals, and usually joined up (say, $4x$ for $0 \leq x \leq 2$ and $x^2 + 4$ over $2 \leq x \leq 7$). He also distinguished single- from multiple-valued functions, and examined some complex-variable ones, using the important equation (5.17.4) relating sine and cosine to the complex exponential.

Among particular functions, Euler made much use of series expansions of 'transcendental' ones such as the trigonometric, exponential and the logarithmic; for example (and not new here),

$$\sin x = x - \frac{x^3}{3!} + \frac{x^5}{5!} - \cdots \text{ and } \cos x = 1 - \frac{x^2}{2!} + \frac{x^4}{4!} - \cdots. \tag{5.27.1}$$

Thereby he gave algebra pride of place over geometry in trigonometry; and from this time onwards the functions were often interpreted as numbers or ratios of sides rather than lengths or lines.

In the second volume, which summarized analytical and coordinate geometry, Euler tried to work with a simpler classification of curves than had Newton (§5.7); he also considered asymptotes and the intersection of curves, and (in an appendix) surfaces. He increased the use of polar coordinates, and also of parametric variables, to represent curves. Unlike the first volume, this one was replete with diagrams, and the pair provide a good example of how branches of mathematics were classified differently from today. In this respect they complement (and the second one overlaps with) Cramer's book, to appear in 1750.

Euler deliberately excluded the calculus from his *Introductio*. He handled that vast branch of mathematics in detail some years afterwards in companion treatises, as we shall see early on in the next chapter.

6

Analysis and mechanics at centre stage, 1750–1800

6.1 *The dominance of the Continent*

From the 1750s British mathematics stagnated for the rest of the century. In particular, for reasons which are not clear, virtually no efforts were made to advance the Newtonian calculus as was done for the Leibnizian form {Guicciardini 1989, Part 2}; at Cambridge Edward Waring

(1736–98) fused it with Newton's algebra, in odd and badly expressed ways which few people read. Mathematical activity was largely the pastime of amateurs – for whom an astonishing variety of journals was published, containing mathematical puzzles or recreations {Archibald 1929}. A modest awakening in the 1800s will be noted in §7.5.

Virtually all major contributions were made on the Continent, especially in France, and also in some German states and Russia. This chapter surveys them up to but excluding the substantial changes effected in France and then elsewhere in the mid-1790s after the Revolution. The development of analysis, and its uses in mechanics, became the prime source of research problems in mathematics. The word 'analysis' covers the calculus and related topics such as functions, series and differential equations; in this chapter we shall see again the somewhat unhappy association of analysis with algebra.

Mechanics was developing into five distinct though interrelated parts {Grattan-Guinness 1990b}, which I characterize thus:

1. 'celestial', where the heavenly bodies were treated as mass-points;
2. 'planetary', where the heavenly bodies were considered as extended objects (and their shapes were also a major area of study);
3. 'corporeal', relating to bodies of 'normal' size, and incorporating basic principles of the whole branch;
4. 'engineering', covering both natural and man-made artefacts; and
5. 'molecular', where the supposedly fine structure of material was analysed.

On this period {Dugas 1957, Part 2} is of some use.

Concerning personnel, Euler, d'Alembert and Daniel Bernoulli remained outstanding until their deaths (Bernoulli in 1782, the other two in 1783), whereupon the leadership passed to the next trio. Joseph Louis Lagrange (1736–1813), born in Turin, was a founder of the Academy in that city in the mid-1750s; he then succeeded Euler (who returned to St Petersburg) at the Berlin Academy in 1766, moving to Paris in 1787 where he remained until his death. From an early stage his principal aim was to convert as much mathematics as possible into algebraic forms (§6.5); among his predecessors d'Alembert was the most profound influence, although Lagrange did not have the same empirical inclination. Pierre Simon Laplace (1749–1827) passed his entire career in Paris, devoting it without pause (or *sans relache*, as the French were then fond of saying) to celestial and planetary mechanics, to methods in the calculus and differential equations, and to some aspects of probability theory and mathematical statistics. Influenced also by d'Alembert, he indulged in some rather unattractive rivalries with Lagrange and with our third figure, Adrien-Marie Legendre (1752–1833). Also resident only in Paris, Legendre was not so careerist. His work was also more varied, in that he contributed not only to mechanics and analysis but also to Euclidean geometry and number theory; the greatest influence on him was Euler.

Another French trio of note flourished at the time, with an especial interest in engineering mathematics. Gaspard Monge (1746–1818) was a geometer in the full sense of the term, working not only in the subject but also using that style in his work on analysis. The work of Charles Augustin Coulomb (1736–1806) on friction is the most relevant here, but his studies of electricity and magnetism are essential background to §10.14. Lazare Carnot (1753–1823), best remembered for his political career under Napoleon, made interesting contributions to mechanics, algebra and the calculus.

Among other figures were two remarkable polymaths who took mathematics within their ambit from time to time. Ruggiero Giuseppe Boscovich (1711–87), born in Dubrovnik, often used this style of name in his many publications, for he worked mostly in Rome and then Pavia {Bursill-Hall 1993}; he is among the most distinguished Jesuits in the history of science. Johann Heinrich Lambert (1728–77) passed his last dozen years at the Berlin

Academy, as a colleague of Euler (briefly) and then of Lagrange {Lambert 1977}.

Professionally, most mathematicians held posts in academies (which, as publishers of the principal scientific journals, brought out most of the papers to be mentioned in this chapter); there were a few posts in universities. Of the newer kinds of institution, the main ones were devoted to engineering, military or civil, especially in France with the Ecole des Mines and the Ecole des Ponts et Chaussées both founded in Paris in 1747, and the Ecole du Génie at Mézières the following year; elsewhere, for example, the Royal Military Academy at Woolwich near London opened in 1741. {Taton 1964} surveys science teaching in France in the 18th century; comparable studies on other countries would be welcome, although they would inevitably be less rich in content.

In this chapter we start out with analysis, followed by comparable studies in mechanics, in its various parts. The rest treats the other main topics, such as the theory of equations and number theory, geometry, and probability theory. The concluding section returns to the dominance of analysis and mechanics. The vast scale of the material is suggested by the fourth volume of Moritz Cantor's history of mathematics, an excellent compendium covering the period 1759–99 produced under his editorship {M. Cantor 1908}; it contains 1,100 pages, even though it is confined almost entirely to pure mathematics. {Kline 1972, Chaps. 19–24} provides a similar, shorter survey. {Grattan-Guinness 1990a, Chaps. 3–6, 8} contains a brief general overview of the calculus and of mechanics, with extensive references to the secondary literature.

6.2 *Euler: clarity to the calculus, and the mysteries of series*

In 1755, in his late forties, Euler published a treatise entitled *Institutiones calculi differentialis*. These 'elementary principles for teaching the differential calculus' included a basic revision to Leibniz's conception, which was to become standard. We saw in §5.5 that there was an indeterminacy over higher-order differentials, with the result that the form of many expressions depended upon which variable took a constant differential. An important example concerned Jacob Bernoulli's formula (5.6.4) for the radius of curvature of a curve, in which $\mathrm{dd}x = 0$.

Euler regularized practice as follows, by introducing the notion of the 'differential coefficient' {Bos 1974, Chap. 5}. If t were a variable with constant differential, then these coefficients for another variable x and its differential could be specified as functions of t:

$$\mathrm{d}x = p\,\mathrm{d}t,\ \mathrm{d}p = q\,\mathrm{d}t,\ \mathrm{d}q = r\,\mathrm{d}t,\ \ldots,$$
$$\text{with } \mathrm{d}t = k, \text{ and so } \mathrm{dd}t = \mathrm{d}^3 t = \cdots = 0. \qquad (6.2.1)$$

But then the higher differentials of x became

$$\mathrm{dd}x = \mathrm{d}(\mathrm{d}x) = \mathrm{d}(p\,\mathrm{d}t) = \mathrm{d}p\,\mathrm{d}t + p\,\mathrm{dd}t = q(\mathrm{d}t)^2,$$
$$\mathrm{d}^3 x = r(\mathrm{d}t)^3, \ldots; \qquad (6.2.2)$$

that is, the indeterminacy was now eliminated, since p, q, r, ... and $\mathrm{d}t$ were all known functions (in principle, anyway). The preservation of dimensions was unaffected, and integration was still the inverse process (now x was the integral of p, p of q, and so on); but the character of the calculus changed in that the differential coefficient (a forerunner of our derivative, of course) was given a place equal in importance to that of differentials. Joseph Fourier's derivation of the equation (10.11.4) for heat diffusion shows the technique.

With this version of the Leibnizian calculus in place, Euler and his followers were set on a massive research programme in analysis and its related topics. The continuing close connection between differential methods and methods of summing infinite series is nicely exhibited by this wild example, taken from a paper of 1760 {Barbeau

1979}. He showed that

$$s(x) = \sum_{r=0}^{\infty} (-1)^r r! x^{r+1} \quad \text{satisfied} \quad s'(x) + \frac{s(x)}{x^2} = \frac{1}{x}, \tag{6.2.3}$$

a first-order differential equation; the solution led after manipulation to 'the' (or a) sum for the series in various integral forms, such as the singular solution

$$s(1) = \int_0^1 \frac{1}{1 - \ln v} dv \tag{6.2.4}$$

as 'the' sum of a series which had been considered by John Wallis.

Since the strictures of A.L. Cauchy in the early 19th century (§8.5), 'divergent series' became a Bad Thing for many decades, and it has become routine to criticize such manoeuvres as totally unrigorous. However, the patronizing inference that Euler and his contemporaries (and predecessors of §5.6–§5.7) were devoid of intelligence in this context should not be drawn. Obviously he knew that (6.2.3) was not the sum of the series in the orthodox sense of term-by-term addition; but nevertheless it expressed some kind of relationship between that series and the sum function, and the purpose of this type of analysis was to find the various kinds {Hofmann 1959}. What was lacking was a clear understanding that a series did not have only one sum, and that different methods of summation could lead to different sums, or to none at all. In modern parlance, absent was a recognized theory of summability and formal power series; we note its origins in the late 19th century in §16.10.

The integral in (6.2.4) is also typical of Euler's and others' interests. One which gained much attention after a 1750 paper by Euler was the 'elliptic integral' (as it came to be called), specified as an integral of the form $\int R[x, \sqrt{(p(x))}]\, dx$, where R was a rational function of its two variables and p a cubic or quartic; it was natural to choose these polynomials to integrate after the quadratics for the inverse trigonometric and hyperbolic functions. The adjective 'elliptic' was chosen, by Legendre, because one type of this integral gave the length of arc of an ellipse. They turned up in many other contexts, however, and Euler, Lagrange and, above all, Legendre devoted much time to their properties: classification and reduction to three basic kinds, addition theorems, and so on. They also made various applications, such as to finding the time T of oscillation of a pendulum of length l: if the amplitude α were not taken to be small, then Huygens' formula (5.12.1) desimplified to

$$T = \sqrt{\left(\frac{l}{g}\right)} \int_0^{\pi/2} \frac{1}{\sqrt{[1 - (\sin^2 \frac{1}{2}\alpha) \sin^2 u]}} \, du, \qquad (6.2.5)$$

the complete integral of the first kind in Legendre's classification.

Among other functions and properties, in 1768 Lambert did something very original: he proved (more or less) that e and π were 'transcendant' (his word) – that is, irrational. He took the series expansions (5.27.1) of the sine and cosine functions, divided them to produce a series for the tangent function, and converted the result into a continued-fraction expansion of $\tan(a/b)$ for a rational exponent a/b. If this were itself rational, then the expansion would have to display properties which it did not have; thus it was irrational. Now, $\tan \frac{1}{4}\pi = 1$; the latter was rational, hence π was not. He argued similarly for e from the continued-fraction expansion of $e^x + 1$; and for a bonus he gave in the 1760s a definitive presentation of the hyperbolic functions cosh and sinh, including series expansions. Avoiding the use of complex numbers in (5.17.4), he used instead geometrical analogies between properties of the rectangular hyperbola and those of the circle for sine and cosine:

$$x^2 - y^2 = 1 \quad \text{and} \quad x^2 + y^2 = 1. \qquad (6.2.6)$$

6.3 *Differential equations, and their solutions* {CE 3.14}

Many of these series and functions were inspired by differential equations, the core of analysis. In general, functional solutions remained the favoured form, but power-series solutions still often furnished the only practical option. Euler provided a remarkably extensive catalogue of these and other methods in his three-volume treatise *Institutiones calculi integralis* (1768–70); the partial differential equations studied were usually linear.

Among topics, the question of general and particular solutions (§5.21) became of great interest, especially when more and more singular ones were found. Euler and Clairaut had come across them in certain equations, and in 1759 Euler presented some 'paradoxes'. One case involved

$$(y-xp)[y+(2a-x)p]=c^2, \tag{6.3.1}$$

where $p:=\frac{\mathrm{d}y}{\mathrm{d}x}$, a and c constants,

which could not be solved by any known means; yet differentiation with respect to x gave a parametric solution

$$x=a-\frac{ap}{\sqrt{c^2+a^2p^2}},\quad y=\frac{c}{\sqrt{c^2+a^2p^2}}, \tag{6.3.2}$$

which could not belong to the complete solution as it contained no constants of integration.

Lagrange made a valuable contribution to this vexing matter, in a paper of 1776. If a first-order equation took as its 'complete solution' $V(x, y, k)=0$, with k as the constant of integration, the differentiation back gave

$$V_x\,\mathrm{d}x+V_y\,\mathrm{d}y+V_K\,\mathrm{d}k=0. \tag{6.3.3}$$

The extra last term led to two possibilities: $\mathrm{d}k = 0$ led back to k being the constant of integration, but eliminating k from $V_k=0$ and $V=0$ led to singular solutions (which he called 'particular integrals'). In later work he affirmed that such solutions were the expected envelope curves of the family of curves given by $V(x, y, k) = 0$ as k varied. In fact, the relationship between singular and other solutions is more complicated: for example, some solutions obtained in this way do not solve the original differential equation. But he had clarified the situation for many cases, and later work on this very difficult topic proceeded with a surer sense of purpose and pitfall {Rothenberg 1910}.

6.4 *The appearance of partial differential equations*

{Truesdell 1960; CE 3.15}

We saw in §5.21 that the extension from ordinary to partial differential equations involved an intermediate stage in the 1730s of isoperimetry, with the second variable a in $y = f(x, a)$ standing for the variation from one member of a family of curves to its neighbours. The extension occurred only in the 1740s, as we shall now see; but this earlier stage left the calculus of variations closely – and beneficially – wedded to the calculus. Later, Lagrange tried to make the connections as close as possible (§6.5).

The innovation of partial differential equations came in connection with a problem that has become known as 'the vibrating string problem', concerning the small horizontal motion of a uniform elastic string of length K held fixed at its ends. Perhaps its long classical background stimulated special interest; at all events, it caused a considerable fuss for some decades. The equation itself, which became known as the 'wave equation', was not doubted: the horizontal displacement y of any point of the string a distance x from one end satisfied

$$y_{xx} = c^2 y_{tt}, \quad \text{with } y(0, t) = y(K, t) = 0 \text{ and } c^2 \text{ constant} \tag{6.4.1}$$

(actually involving physical constants for the string). First d'Alembert and then Euler found it in the mid-1740s; both also favoured the functional solution

$$y = f(x + ct) + g(x - ct), \tag{6.4.2}$$

where f and g (two functions of integration, appropriate for a second-order equation) were to be found from the initial setting of the string and the condition of the fixed ends. However, d'Alembert felt that the presence of y_{xx} required f and g to be twice differentiable, while Euler insisted that release of the string by plucking should permit them to take a corner. The adjectives 'discontinuous' and continuous' were applied to these two cases, in the senses described in §5.27, with 'continuous' given by one expression and 'discontinuous' allowing for different ones over adjoining intervals of x.

This controversy exposed serious issues in the calculus, and neither side 'won'. Indeed, the issue was complicated further in 1753 when Daniel Bernoulli proposed a truly Pythagorean solution,

$$y = \sum_{r=1}^{\infty} a_r \sin rx, \tag{6.4.3}$$

on the grounds of superposition of tones. His idea was an important early step in superposition theory towards the solution of differential equations by structurally similar means, but it was not a promising one; he had no means of calculating the coefficients, and he even failed to include the trigonometric time terms.

In 1759 the young Lagrange chipped in with a suite of long papers which came close to the Fourier-series solution of which Bernoulli's (6.4.3) was a sort of draft; but in fact Euler's reading of (6.4.2) was his preferred solution, which he duly obtained from his method. Thereafter every Young Turk had a go at solving the equation, but no new insight arrived until Fourier's surprising contribution (8.7.1) of the 1800s. However, elasticity theory benefited, with properties found for vibrating thin laminae as well as strings. Companion studies of the propagation of sound are noted in §10.10.

The most important consequence was that partial differential equations were now on the map, and ever since they have been a major component of mathematics, with ordinary differential equations often supplanted in significance. Forms of solution were a principal concern, for, as d'Alembert stressed, they had to satisfy the initial and boundary conditions as well as the equation itself. The next important types of equations and conditions were found largely in response to other problems in mechanics, with time and at least one space variable serving as independent. A few of them will be noted below; next we shall see a new place given to the calculus of variations, as part of the rise in importance of algebra.

6.5 *Lagrange and the algebraization of analysis* {CE 3.2, 3.5}

The collected works of Lagrange, produced between 1867 and 1892, include 12 hefty volumes of his publications, but they contain very few diagrams. (I have noticed only five, three of them from elementary lectures given in 1795 at the Paris Ecole Normale (§7.1).) His correspondence and manuscripts are just the same; pictures are almost always absent. He has few peers in the single-minded way in which he ran one style – the algebraic – throughout his life. While there were no doubt personal or psychological reasons for his choice, points of philosophy were also involved: in particular, the idea that algebraic theories could both be set on a safe footing and formulated with sufficient generality to serve all mathematical needs.

Lagrange's first exercise in the prosecution of this philosophy was his generalization of Euler's methods of solution by modular equations. He outlined his idea in 1754–5 in letters to Euler, who was very impressed and introduced the name 'calculus of variations' for the expanded subject; they both developed it in papers published from the 1760s {Fraser 1985}. Where Euler might disturb no more than one point from its position on the extremal curve given by the function f (§5.21), Lagrange also allowed all points to be displaced, and he used the letter 'δ' to denote, algebraically, the operation of disturbance on a particular variable – an important step in the history of operator symbols in mathematics. In addition, he did not restrict himself to isoperimetrical problems, with their possible side-constraints {EMW IIA8}.

Lagrange's key results, intuitively reasoned rather than rigorously proved, were that (a) the variation on the independent variable x was nil at the optimal situation; and (b) that δ commuted with both d and $\int$ with respect to the dependent orthodox variables $y, \ldots$ of the function Z which was being optimized: in his symbols,

$$\delta x = 0; \quad \text{and} \quad \mathrm{d}\delta y = \delta \mathrm{d}y \text{ and } \delta \int Z = \int \delta Z. \tag{6.5.1}$$

At this stage the multiplier methods for expressing optimization under constraints were not in place.

An important way in which the algorithms worked was by expanding dy and dZ in terms of their partial differential coefficients and using eqns (6.5.1), applied to the variables of the problem and if necessary for higher powers in d and δ. This converted the second and third of eqns (6.5.1) into the form

$$A\,\delta x + B\,\delta y + \cdots = 0; \tag{6.5.2}$$

and since the initial displacements were arbitrary, the coefficients would be zero, yielding equations $A = 0$, $B = 0, \ldots$, often ordinary or partial differential ones, expressing some optimal situation. (This kind of procedure was displaced in the mid-19th century, first by quaternionic and then by vectorial methods: §9.8, §15.9.) Known methods could (hopefully!) then be applied to solve the separate equations; an example, (6.11.2), will be given from mechanics. In this and other contexts such methods became known as 'variation of parameters'.

Lagrange's first attempt to algebraize a major theory was published in 1772. He tackled nothing less than the calculus, wishing to replace Leibniz's and Newton's approaches and avoid their respective use of limits (which had gained a few adherents, such as d'Alembert) and infinitesimals. His basic *assumption* was that a function $f(x+h)$ could normally be expanded in a convergent Taylor series (5.7.3) in the increment variable h about the value x by purely algebraic means, so that its 'derived functions' (his expression) could be defined from the coefficients of h, thus:

$$f(x+h) = f(x) + hf'(x) + \frac{h^2 f''(x)}{2!} + \cdots. \tag{6.5.3}$$

The integrals were defined as the inverse functions: f' *of* f'', f of f', and so on.

This proposal made Lagrange the second major mathematician after Newton (§5.7) to grant a central role for power series. The idea was good, but was it too good to be true? He warned that there would be difficulties at values of x when $f(x) = \pm\infty$. He presented some intuitive arguments to back up his claim, and thought about the 'error term' created by truncating the series after a finite number of terms, although the name 'Lagrange remainder' applied to the series is anachronistic (compare eqn (8.6.6)). However, many contemporaries were less sure that, for example, the power-series form could be obtained without the use of limits and infinitesimals in the first place {Dickstein 1899}.

Nevertheless, Lagrange's claim powerfully reinforced his use of algebra in mathematics, and led to results which were valuable in their own right. One was a restatement of (6.5.3) relating derivatives to forward differences:

$$\Delta f(x) := f(x+h) - f(x) = \sum_{r=1}^{\infty} \frac{h^r f^{(r)}(x)}{r!} = \exp\left(h\frac{\mathrm{d}f(x)}{\mathrm{d}x} - 1\right). \tag{6.5.4}$$

It was effected by interchanging powers and differentiation so that $(f'(x))^r = f^{(r)}(x)$ – a move criticized by some observers. Another result was the 'Lagrange series', the following generalization of Taylor's produced in 1770 to solve Kepler's equation (4.11.4), but soon found to be of great general utility:

$$\text{given } z = x + yf(z), \text{ then } g(z) = g(x) + \sum_{r=1}^{\infty} y^r \frac{[g'(x)(f(x))^r]^{(r-1)}}{r!}. \tag{6.5.5}$$

I conclude with a general point about functions, which could turn up in any of the contexts described above. Quite often they could take multiple values (for example, inverse trigonometric functions, functions with nth roots, and cases of two functions $f(x) = g(x)$ where f and g took different numbers of values). While the property itself was well known, it was not always clear which value(s) pertained to a given case. The matter could affect applications to mechanics, to which we now turn.

6.6 *A third tradition: the dawn of variational mechanics* {CE 8.1}

With these and many other results, the partial differential calculus and the calculus of variations assumed prime places in mathematics, especially as they found applicability in the various branches of mechanics, such as the wave equation (6.4.1). We start here with certain foundational equations: in particular one of major importance to Euler, and another, affirmed by him, which was to be of decisive influence on Lagrange.

Firstly, we have noted that Newton's second law of motion was often applied along directions significant in a problem, such as the direction of motion of a body or along a normal to it. In a paper of 1752, Euler announced 'a new principle of mechanics' in which the second law could apply to motion in any direction (or combination of directions given by a coordinate system, say). He wrote it as

$$\text{'}2M\,\mathrm{dd}x = \pm P\,\mathrm{d}t^2\text{'} \tag{6.6.1}$$

for a mass $2M$ moving under force P; the 'plus or minus' allowed for motion in or counter to the positive direction of x, and the factor of 2 arose from the use of special units. He let '$2M$' denote a mass-point, a concept which he had individuated earlier (§5.25).

While it is not actually a new principle, Euler's enthusiasm for (6.6.1) is understandable, for his move enormously raised the utility of Newton's second law via a sort of coordinate geometry for dynamics, and thereby gave it a form which has always been standard ever since. In the same paper he applied it to find the 'Euler equations' (as we now call them) for the motion of a continuous extended body around a fixed point. He found the 'principal axes' (his term) of rotation, and put moments and products of inertia on a new level of importance. He studied this part of dynamics in detail, especially in a book published in 1765.

The second equation arose around the same time, in connection with principles for mechanics alternative to Newton's laws of motion. In §5.25, we saw energy principles in place, as well as d'Alembert's principle in the 1740s; now we note the emergence of a third tradition closely linked to algebra and stimulated by the following innovation.

A dispute developed at the Berlin Academy in the late 1740s over a principle which Euler was to call 'the principle of least action' {Pulte 1990}. Its content is well illustrated by one of his early presentations of (6.6.1) in a paper published in 1753. Imagine a single mass M moving over time t with velocity u along a path of length s under the effect of forces V_j acting in directions represented by variables v_j. Then, by (6.6.1),

$$M\,\mathrm{d}u = T\,\mathrm{d}t, \text{ where } T = -\sum_j \left(\frac{V_j\,\mathrm{d}v_j}{\mathrm{d}s}\right). \tag{6.6.2}$$

Multiply by $u(=\mathrm{d}s/\mathrm{d}t)$ and integrate:

$$Mu^2 = K - \phi, \quad K \text{ constant and } \phi := -\int T\,\mathrm{d}s, \tag{6.6.3}$$

a form of the energy conservation equation. Multiply by $\mathrm{d}t$ and integrate:

$$\int Mu\,\mathrm{d}s = Kt - \int \phi\,\mathrm{d}t. \tag{6.6.4}$$

Euler followed Pierre Maupertuis in claiming, as an optimal principle, that since ϕ was a maximum or a minimum in situations of equilibrium, then so was $\int \phi\, dt$ in cases of motion, and hence $\int Mu\, ds$ (which he called the 'action') was minimal or maximal.

Contributing to the controversies were issues of priority between Euler's colleagues Maupertuis and Samuel Koenig (the literature is gathered together in {Euler 1955}); for some later commentators the assumption of the existence of ϕ was objectionable (§10.7). But the main issue was the generality of (6.6.4) for mechanics: the range of bodies to which it could apply (collections of mass-points, continuous bodies, and so on) and how it could be reconciled with the nature of matter {Körner 1904}. Euler's position is quite peculiar: affirming that the principle was of great generality, he imparted a religious element in claiming that God had organized things this way. But when out of church and in the market-place of real mechanics, he often forgot his view: for example, as we shall see in §6.10, in his treatment of hydrodynamics he made no use of it at all.

By contrast, Lagrange took this principle very seriously – but in a secular spirit, as part of the foundation of mechanics, including his 'δ' symbol to express optimality. He developed his position first in a paper written in 1764, but it became well known only from the 1780s onwards {Fraser 1983}. By then, however, he was changing his line, as we shall see in §6.11. First we note some uses of these various traditions in mechanics.

6.7 *Newtonian celestial mechanics*

{Wilson 1980; CE 8.8}

Newton's second law was widely used in astronomy, especially in studying the details of the motions of the heavenly bodies when regarded as mass-points; for it was interplanetary attraction that was held to cause the 'perturbations' of the bodies from their elliptic orbits around the Sun. Here are a few cases.

One difficult problem was analysing the precession of the equinoxes, which Newton had examined in the *Principia* {Book 3, Proposition 39} in terms of the conservation of momentum of the matter contained in the Earth's equatorial bulge, but he had not obtained a sufficiently accurate value for the Earth's angular velocity. In a book on the subject published in 1749, d'Alembert proposed, among other things, that angular momentum should be treated instead, and he came up with much better values {Wilson 1987}.

Two aspects of this work bore upon Euler. Firstly, while his analysis of the problem was very difficult to follow, d'Alembert had produced equations for angular momentum in three axis directions, and probably with some justice he claimed priority for Euler's (far clearer) 'new principle of mechanics' (6.6.1). Secondly, the status of angular momentum was not clear in mechanics; Euler correctly saw it as independent of Newton's laws {Truesdell 1968, Chap. 5}.

At that time, in 1747, Euler introduced a major mathematical technique into celestial mechanics. Imagine the ecliptic to be the plane of this paper, with two planets A and B in motion upon it at respective distances a and b from the Sun S. Then, in ΔASB,

$$\mathrm{AB} = \sqrt{(a^2 + b^2 - 2ab\cos\theta)}, \quad \text{where } \angle\mathrm{ASB} = \theta, \tag{6.7.1}$$

an expression which would be used in various powers (-1 for Newton's second law itself). Each pair of planets considered in an analysis required its own version of (6.7.1), and the passage-work became very complicated. To make it more tractable, Euler proposed that inverse powers of AB be expanded by the binomial theorem in power series in $\cos\theta$, and each power could be converted by (5.17.5) into terms in multiple angles $\cos n\theta$, which were easier to integrate and indeed to handle in general.

These moves brought trigonometric series into celestial mechanics. They were to became a staple technique of

French astronomy, especially with Lagrange and Laplace. The coefficients of the terms were calculated by various means, including difference equations and various numerical procedures; in addition, expansions of other powers of AB were studied in great detail {EMW VI2,13}. For Laplace these series also helped buttress a favoured physical principle: periodic causes lead to periodic effects.

Relative to the Sun, the masses of even the largest planets are small, and so are some of their orbital parameters, such as (with a few exceptions) their eccentricities, and the normal effects of perturbations. Thus methods of approximation were introduced in order to produce simplified but still useful mimics of the real situation, which could be effectively desimplified. This last possibility led Euler to another major innovation. Six parameters, or 'elements', are required to determine completely the motion of an *un*perturbed body on its elliptical orbit around the Sun: for example, two to fix the plane of its orbit relative to the plane of the ecliptic, two to define the ellipse on it, and two to specify the location and velocity of the body at any instant of time. In a paper on lunar theory written in 1756 but delayed by 15 years in appearance, Euler showed that perturbations of these elements were small enough for the body to be regarded at any moment as going along some elliptical path which was gradually changing, and that a method of 'variation of parameters' could be used to calculate the new 'temporary' orbit.

Euler's progress on this course was limited, but the potential for algebraization inspired Lagrange and Laplace. Among later methods of this kind, Laplace introduced in 1776 one for solving differential equations, which he called a 'method of successive approximations'. Let n functions $f_r(x)$ satisfy n mth-order ordinary equations

$$f_r^{(m)}(x) + P_r + \alpha Q_r = 0, \quad 1 \leq r \leq n, \ \alpha \text{ small,} \tag{6.7.2}$$

where P_r and Q_r are functions of both x and of the f_r up to their $(m - 1)$th-order differential coefficients. For the case in which they could be solved when all $Q_r = 0$, Euler laid down means for generating the solutions of the full set. In astronomical terms this allowed terms of successive powers (and thus smallness) in the basic equations of motion to be accommodated.

6.8 *Stability and the three-body problem*
{Gautier 1817; CE 8.9}

One of Lagrange's greatest innovations was to introduce new independent variables (s, u) which are still known after him. Usually they took the form

$$s = \lambda \sin \chi \quad \text{and} \quad u = \lambda \cos \chi, \tag{6.8.1}$$

where λ was the longitude of the node of the orbit and χ the tangent of its inclination to the ecliptic. When the equations were recast in these variables, various symmetries arose in their form which greatly aided analysis.

These variables were applied to a major problem in celestial mechanics: the stability of the Solar System {Hawkins 1975}. Figures such as Newton and Euler were content to allow the Solar System to be susceptible to instability, as God could effect the rescue; but Lagrange hoped that stability could be *proved* from Newton's laws – and the assumption that the planets orbited in the same direction around the Sun. Using these new variables with λ this time as eccentricity of the orbit and χ the longitude of the perihelion (the nearest point of the planet's orbit to the Sun), the equations of motion took a very beautiful form: in matrix/vector form (which of course is anachronistic), they read

$$\frac{d\mathbf{s}}{dt} = \mathbf{Cu} \text{ and } \frac{d\mathbf{u}}{dt} = -\mathbf{Cs}; \quad \therefore \quad \frac{d^2\mathbf{p}}{dt^2} = -\mathbf{C}^2\mathbf{p} \tag{6.8.2}$$

for each vectorial variable (whose elements are the scalar variables for the planets), written here as '$\mathbf{p}$'. The task was to show that all variables always took bounded values, so that the system would not fly apart (unbounded eccentricity) or flip over (unbounded inclination). The analysis corresponded to finding the latent roots λ_r of the determinant $|\mathbf{C}^2 - \lambda\mathbf{I}|$, proving that they were real, and wondering what happened if any of them repeated; but matrix algebra was far away in the future, and Lagrange relied on clever manipulations of the corresponding quadratic form $\mathbf{p}^T\mathbf{C}^2\mathbf{p}$. The argument was not, of course, watertight, but still very brilliant, in Lagrange's and then later in Laplace's refining but purloining hands.

Another great challenge to the accuracy of Newton's laws was the irregularities in the motions of Jupiter and Saturn {Wilson 1985}. Perturbation theory was applied to them with great imagination by Euler, Lambert and Lagrange, but discrepancies were still evident. However, in the mid-1780s Laplace found a means of explaining them in terms of gravitational theory, thus vindicating Newton. His argument rested as usual on expanding the differential equation in powers of the eccentricities and inclinations of the orbits, and by noting that the mean velocities n_J and n_S of these planets were related by the relatively very small expression $E := (5n_J - 2n_S)t$, with t as time. Thus, although E was associated with the third degree in these variables (and therefore very small) for each orbit, its coefficient could lead to large values for terms when landing up in the denominator after integration, and moreover could provide corrections of the required order. Again, Laplace seems to have drawn upon key ideas in Lagrange's recent work; for the purpose of calculation he used a method called 'equations of condition', which is described in §6.20 in connection with his work on statistics.

Another spectacular challenge was the three-body problem, which seemed to be bearing out Newton's suspicion (§5.11) that it lay beyond human capacity for solution. But it had become clear that simplified versions could be soluble, and Euler was particularly fond of them, writing several papers over the years. For example, in the 1760s he took the Moon to be negligible in mass and examined its motion as attracted to the Sun and Earth, finding solutions involving elliptic integrals; he then repeated the study with Sun and Earth moving around each other. (Henri Poincaré was to call this the 'restricted' problem: §14.18.) A few years later Lagrange tackled the general problem, starting out from the energy and angular momentum equations, but he was able only to reduce it to a seventh-order equation.

6.9 *The shape of the Earth, and its potential*
{Todhunter 1873, Chaps. 14–27}

We turn now to planetary mechanics, with some studies of the Earth not treated as a mass-point. The vindication of Newton's prediction of the flattening of the poles (§5.25) now posed the problem of desimplifying the theory of gravitational attraction, knowing the Earth to be a spheroid rather than a sphere. In this and other contexts there arose the differential equation representing the shape in equilibrium: in a rectangular coordinate system the equation $V=0$ of the surface satisfied

$$V_{xx} + V_{yy} + V_{zz} = 0, \tag{6.9.1}$$

with (V_x, V_y, V_z) the components of attraction. In an important remark Lagrange noted in 1782 that V was given by $\int_M \mathrm{d}M/r$, where r was the polar distance in the coordinate system; this increased the role of potentials (for which he provided no name) in these contexts.

The two principal students of attractions were Laplace (whose name is usually attached to (6.9.1)) and Legendre, and an unlovely relationship developed between them from the 1780s onwards, when they were both in their thirties. Both men used Legendre functions (as we now call them; for much of the 19th century they were known as 'Laplace functions'!), and related functions such as surface harmonics. The analyses usually proceeded from taking Laplace's equation in spherical polar coordinates

(r, ω, θ):

$$((1-\mu^2)V_\mu)_\mu + V_{\omega\omega}/(1-\mu^2) + r(rV)_{rr} = 0, \quad \text{where } \mu := \cos\theta \tag{6.9.2}$$

and solving by separation of variables, from which the functions would be terms in μ. (With no ω, separation would lead to elliptic integrals in r, which interested Legendre greatly and – therefore? – Laplace not at all.) The theory was concerned especially with finding an expression for attraction at points not lying on the Earth's axis. Both men found and used the generating function

$$\frac{1}{\sqrt{(1 - 2u\mu + u^2)}} = \sum_{j=0}^{\infty} P_j(\mu)u^j, \tag{6.9.3}$$

which was to become a major tool for Laplace. But, like Newton (§5.25), nobody noticed that Laplace's equation itself obtained only if the point was exterior to the Earth; Poisson was to make the important correction (10.6.2) in 1814.

The surface of the Earth itself provided further difficult problems. Cartography benefited from a lovely paper by Lambert in 1772 in which he laid down basic principles for the drawing of both maps and charts by defining a point on the surface of the (spherical) Earth by arc-length coordinates v and w, respectively along and perpendicular to the equator, and projecting it onto a point in the equatorial plane with coordinates x and y. Some of his and others' procedures involved conformal mappings and (following d'Alembert) complex-variable solutions to Laplace's equation (6.9.1) in the plane, with v and w as independent variables: for example (by Euler), to preserve areas and keep the latitudinal and longitudinal lines perpendicular, so that

$$\frac{y_v}{x_v} = -\frac{x_w}{y_w}; \quad \text{thus } x_v x_w + y_v y_w = 0. \tag{6.9.4}$$

This puts us within a finger's-reach of the 'Cauchy–Riemann equations'; but their importance was not to be recognized until the next century (§8.11).

Lambert also extended the trigonometry into the 'tetragonometry' and 'polygonometry' of rectilinear figures with more sides. At the time, navigators stuck largely to planar and spherical trigonometry in working out a variety of methods of 'sailings': that is, means of representing on paper the path of a ship at sea {CE 8.18}. Their determinations, of longitude especially, were helped by tables constructed from analyses of the motions of the Moon made by Euler, d'Alembert and Clairaut, partly in response to lucrative prizes offered by the Royal Greenwich Observatory in Britain.

But what conditions might the sailors find at sea? The next major step in tidal theory after the prize essays of 1742 (§5.25) was taken by Laplace, drawing upon fluid mechanics, which we now consider.

6.10 *Fluid mechanics: pressure or history?*

Celestial and planetary mechanics were not the only branches where Newtonian principles were examined: within the terrestrial branch, the behaviour of fluids was still rather mysterious. D'Alembert contributed to both areas, including the notion of a velocity potential, defined as the function whose partial differential coefficients gave the components of the velocity of a particle. Euler introduced the general equations in a quintet of papers in the 1750s; a decade later he wrote four more of a rather didactic character, including applications to acoustics. They are gathered together, with editorial description by C.A. Truesdell III, in {Euler 1954–5}.

Rather than using the universal principle of least action (§6.6), Euler followed his own method of differential modelling to find the equations of hydrodynamics by examining the passage of fluid through an infinitesimal cuboid in a space with axes $Cxyz$ under the action of 'force' (P, Q, R)

per unit mass. His basic notions were velocity (u, v, w), density q, and pressure p (the central importance of which emerged here from his treatment of hydrostatics). In the Cx direction,

$$P - \frac{p_x}{q} = u_t + u_x u + u_y v + u_z w + \mathrm{O}(\mathrm{d}t^2), \qquad (6.10.1)$$

with a similar story in the other two directions. The equations were too difficult to solve, but Euler made progress on some special cases and conditions, particularly incompressibility and constancy of mass: respectively,

$$u_x + v_y + w_z = 0 \quad \text{and} \quad q_t + q_x u + q_y v + q_z w = 0. \qquad (6.10.2)$$

Euler also often assumed that both force and velocity took potentials:

$$\mathrm{d}S = P\,\mathrm{d}x + Q\,\mathrm{d}y + R\,\mathrm{d}z \text{ and } \mathrm{d}W = u\,\mathrm{d}x + v\,\mathrm{d}y + w\,\mathrm{d}z + \Pi\,\mathrm{d}t. \qquad (6.10.3)$$

But he knew of the limitations of such assumptions, and gave an example of vortex flow where the latter equation did not obtain. In 1761 he also noticed that W satisfied Laplace's equation (6.9.1); this seems to be its first appearance.

From these bases Euler produced valuable studies on a wide range of phenomena, including the propagation of sound in that elastic fluid, the air. He began with the wave equation (6.4.1) or (more often) more complicated versions suitable for the problem at hand, and the functional solution (6.4.2) or variants such as the functions under some integral, although series solutions had to be used sometimes. In addition to simple aerial propagation, he examined planar waves.

Laplace made an important application to the theory of tides, published in a paper in 1775 {CE 8.16}. He put forward three main equations: for continuity of water-mass; horizontal motions (which Euler had perceived as essential); and the effect of the Sun and the Moon. Solving them proved difficult, even with gross assumptions such as a constant depth for the sea; but he showed that stability would obtain if the mean density of the sea were less than that of the solid Earth.

When Lagrange turned to fluid mechanics in the 1760s, the theory had to be algebraized. So, instead of starting out from cuboids and pressure, he analysed the 'history' (as others were to call it) of a particle of fluid from its initial location (a, b, c) to point (x, y, z) at time t: for incompressible fluids, along Ox

$$q[(x_{tt}+X)x_a+(y_{tt}+Y)y_a+(z_{tt}+Z)z_a]=p_a. \qquad (6.10.4)$$

I have given here his definitive form of the 1780s, when he completed his book on mechanics, to which we now turn.

6.11 *Lagrange's* Méchanique analitique, *1788*

Over the years Lagrange built up his algebraic empire. An important advance came in a paper of 1764 on lunar theory, which he prefaced with his own current preference for the foundation of mechanics: d'Alembert's principle (§5.25), together with a principle of 'virtual velocities', described below. He developed this approach in later papers and gave it a definitive statement in his book *Méchanique analitique*.

Lagrange seems to have finished the manuscript of this work in 1782 (his 47th year), during his Berlin period, but there seemed to be no chance of having it published there – an interesting comment on the market, or rather the lack of it (compare Euler in §5.27). So he wrote it out again and sent it to Paris, where Legendre saw it through the press; it appeared in 1788, a year after Lagrange had moved there himself. Its 512 pages contained no diagrams, as he stressed in his preface: rigour and generality were to be drawn from algebra uncontaminated by geometry. However, the proofs themselves were usually *synthetic*, in the sense of §2.27, starting from assumptions or results already proved and ending with the theorem required. The book would have been better entitled *'Méchanique algébrique'*.

Lagrange had switched allegiance away from the principle of least action, perhaps for philosophical reasons such as its strongly teleological character (that the path, or whatever was determined by (6.6.4), seemed to be inevitable). His new preference stated that if forces $P, Q, \ldots$ acted along lines $p, q, \ldots$ at certain fixed positions, with 'variations, or differences of the lines' $\mathrm{d}p, \mathrm{d}q, \ldots$ caused by disturbance from equilibrium, then

$$\sum_p P\,\mathrm{d}p = 0\ (= \mathrm{d}\phi), \qquad (6.11.1)$$

where the term in parentheses shows his readiness to assume that the summation term admitted a potential ϕ. He probably chose the name 'virtual velocities' under the influence of d'Alembert's suspicion of notions such as force and work. However, he gave no proof of his principle nor justification of his assumption, and both matters excited comment (§10.2). If constraints given by equations $A = 0$ applied to the system, then extra terms of the form $\alpha\,\delta A$ were added into (6.11.1) (with the sum set to zero). We now call the coefficients α 'multipliers'; this was one of their early appearances.

With such equations, and d'Alembert's principle on hand to reduce dynamics to statics, Lagrange started out with statics, obtaining proofs of the composition of forces and of moments. Then he obtained Newton's second law by casting (dynamic) equilibrium in the form (6.10.1) and setting the variations $\delta x = 0$. He also obtained the energy equation, but with an assumption of a potential. But his main aim was his own equations, obtained in 1764: using general variables (which I write 's_r'), he took of a system of particles and/or bodies, considered their *'forces vives'* T and potential (energy) V as given by the first of eqns (6.6.3), and the constraints given by $L_j(s_r) = 0$. Then, stating Newton's second law for these variables and manipulating the algebra so as to collect together the terms in each ds_r, he found

$$\sum_r \left[d\left(\frac{\delta T}{\delta ds_r}\right) - \frac{\delta T}{\delta s_r} + \frac{\delta V}{\delta s_r} + \sum_j \lambda_j \frac{\delta L_j}{\delta s_r} \right] \delta s_r = 0. \tag{6.11.2}$$

As in (6.5.2), the individual equations could be obtained by setting the coefficients of each δs_r to zero. The contrast between this equation and, for example, Euler's (6.6.1) is very striking; while remarkably general, it has little 'feel' of mechanics.

The rest of the book ran through some standard problems, including the motion of a mass-point attracted to more than one centre, a version of the stability argument (6.8.2) for a system of mass-points, and Euler's equations for the rotation of a rigid body; he finished off with his version of fluid mechanics. The coverage was less wide than the opening pages suggested or promised: in particular, absent was the kind of mechanics that engineers had been working on for generations.

6.12 *Engineers' formulae*

Among our major figures, Euler once again tops the poll for his interests, including several studies of the flow of fluids and of the action of pumps and turbines, a volume on gunnery, and some writings on machine theory. Unfortunately, although they appeared as books or in the usual journals most of them gained very little attention, despite many clever things. One item for which he is remembered is a formula of 1759, obtained as an offshoot of his study of elastic curves (§5.24), for the maximum load a stone column can bear without buckling. He also worked on various other aspects of elastic and flexible bodies {Truesdell 1960, Part 4}.

The theory of ships also gained from Euler's attention. He used, for example, the calculus of variations to find the best profile of a ship given the law of water resistance, and the composition of forces to find kinematically the direction of best steerage. Daniel Bernoulli contributed by using the principle of the conservation of energy to show that water pressure dropped when it was swept beneath the hull of a moving vessel, causing it to sink somewhat in the water. This effect is known as the 'Bernoulli squat'.

Bernoulli's result arose as part of his answer to a prize problem posed for 1757 by the French Academy of Sciences. Indeed, the most influential studies were coming from French figures, among whom the mathematical aspect of engineering was very evident. Hydraulics made substantial progress with the rapid development of canal-building. Empirical laws were put forward concerning, for example, the rates of flow of a large body of water at its surface, at various depths, and at the bed of a river or canal {Mouret 1921}. Flow out of orifices was especially tricky because of its great dependence upon the size, shape and position of the orifice, and cavitation (the turbulent behaviour of water immediately in front of it) and contraction (the reduction in the effective width of the outflow due to surface tension). Jean Borda made a valuable contribution to these studies with his introduction in 1766 of a 'mouthpiece' into the orifice to smooth the flow, though response to his idea was slow. In the 1790s Coulomb carried out experiments with his torsion balance to argue that the resistance to motion through a fluid was of the form $au + bu^2$, where of the two constants b was very small.

Coulomb had used this instrument to great effect in the 1770s to study friction and torsion. Indeed, his more substantial contributions lay in the stability of structures such as embankments and arches, and the rupture of beams, many of them initiated in a classic paper which won a French Academy prize in 1776 {Heyman 1972}. His work was particularly useful for clarifying the location and properties of the neutral line in a structure under stress or load. His methods were usually straightforward uses of moments and resolution of forces; the novelty lay in his insights about materials and their constitutive properties.

The work of Coulomb inspired several sections of an

engineering treatise published in 1790: the huge first volume of Gaspard de Prony's *Nouvelle architecture hydraulique,* which went far beyond the remit of its title. He ran through the main parts of mechanics (statics, dynamics, hydrostatics and hydrodynamics) but 'for Artists in general', with examples such as orifice theory, earth pressure on dams, the 'forces of men and animals' (to us, ergonomics, which Coulomb had pioneered), and motion in machines; he also included many tables of data. On principles he was eclectic, but he gave the energy equation some prominence. The second volume, of 1796, largely applied these methods to the design of steam engines, and also to linkages.

Coulomb's approach also excited a strong challenge to Lagrange's from Lazare Carnot. He worked over the principles of mechanics with machines in mind, publishing in 1783 an *Essai sur les machines en général* in which, without naming anybody, Lagrange's principles were opposed. Dynamics should come *first,* with statics as a special case; and potentials should *not* be routinely admitted, as they are in (6.11.1), for in percussion and impact the force functions were discontinuous and so could not appear in Lagrangian exact differentials (5.25.1). His book itself did not make much impact (as it were), but he persisted in the 1800s and inspired a remarkable tradition in engineering mechanics, which we shall record in §10.8–§10.9.

So we complete our survey of the calculus and mechanics. The rest of the chapter treats other branches of mathematics, which proceeded in more modest ways.

6.13 *The conundrum of Euclid's parallel postulate*

{J.J. Gray 1989, Chaps. 4–6}

Six options are available. Three of them arise usually under the assumption that Euclidean geometry is the only one true, or possible: (1) to accept the postulate as it is, or (2) to prove or (3) to replace it by an assumption which seems to be more evident. For the more adventurous, there is the option (4) to envision that other postulates are possible, and even (5) to develop theories based upon them in some detail. We saw in §2.16 that Euclid may have chosen (1) within the philosophical context (4) posed by Aristotle; many successors such as the Arabs (§3.7) had attempted (2) or (3). The option (6) to assert that some non-Euclidean geometry is true seems not have been pursued at that time.

In a remarkable foray, the Italian Giovanni Saccheri (1667–1733) set out for (2) but came close to (5). In a book on the 'vindication of Euclid' published at the end of his life, he developed a line of thought due to Nasir al-Tusi in a work that had appeared in Latin in 1594. Assuming that three angles of a rectangle were right angles, he argued that only the assumption that the fourth was also a right angle could be reconciled with Euclid's other four postulates. That is, he sought a contradiction between these postulates and each of these assumptions (with corollaries):

>. There is no line parallel to a given line through a point in the plane not lying on it (when the angles of a planar triangle add up to more than two right angles);

<. There is more than one such line (when the sum is less than two right angles).

However, he succeeded only with case >; his argument for case < assumed as true at infinity properties of figures that hold only at finite distances, which begged the question.

Saccheri was one of the very few mathematicians who also studied logic. Another was Lambert, who in the 1760s took the same approach to the postulate but saw that case > led to a geometry similar to that of spherical trigonometry, but hopefully based upon the other four postulates. Befitting such a fine student of maps and charts (§6.9), he was close to the revolutionary option (5), but he shied away, for his analyses of both cases led to consequences which struck him as dubious: he characterized the sphere that arose in case > as 'imaginary' . . . . He did not publish his paper, which came out posthumously in 1786.

Lambert's doubts contrast sharply with the attitude of the Scottish philosopher Thomas Reid (1710–96), also in the

1760s. As part of an examination of perception, he found the same geometry on the sphere, which he called a 'geometry of visibles', to describe the images formed upon the retina {Daniels 1989} – and he saw it as an example of realism, in the sense of the Scottish 'common-sense' philosophy of which he was an important founder. In the next century it was to influence compatriot scientists such as William Thomson (§10.19) and Clerk Maxwell (§15.10); unfortunately, his geometry was ignored until non-Euclidean geometries began to be explored again, also in the 19th century (§13.8).

Among other Scottish geometers, Robert Simson (1687–1768) had tried to reconstruct lost works by Pappos and Euclid, and produced in 1756 a highly influential edition of the *Elements*. He was followed by John Playfair, who himself produced a widely used edition of the first six Books of the *Elements*, which first came out in 1795. On the parallel postulate he chose option (3), that of replacement, stating that 'two straight lines [in a plane] which intersect one another, cannot both be parallel to the same straight line', an axiom which is still named after him. Other possible alternatives were explored by Legendre in his *Elémens de géométrie*, a particularly influential textbook in Euclidean geometry which first appeared in 1794 and was into its 12th edition by 1823.

A general query of a different kind about geometry had been raised in 1754 by Boscovich: why do straight lines have priority over curved ones anyway? Maybe some other intelligence 'would not seek, as our geometers do, to *rectify* the parabola, they would endeavour, if one may coin the expression, to *parabolify* the straight line'.[1] His position was stimulated by his affirmation of the continuity of all things, including mathematical ones (such as lines changing continuously from straight to curved ones). This emphasis was

[1] Boscovich's remark, made in his *Elementa universae matheseos* (1754), implies that notions other than dy/dx might ground the differential calculus; it has been taken up only in recent times {Grossmann and Katz 1972}.

also to underlie the development of projective geometry, as we shall see in §8.4.

6.14 *Geometrical optics and differential geometry*

Physical properties of light remained under study, and Newton's corpuscular theory (§5.13) did not reign alone. Euler, for one, invoked his aether (§5.25) to construe rays of light as representing the directions of oscillation of the aetherian particles. But the mathematical aspects of optics were dominated by its geometrical side, especially concerning the properties of lenses for use in microscopes and telescopes. Euler weighed in with three hefty tomes, the *Dioptrica* (1769–1771), as well as papers from the 1740s. One interesting part of his analysis, which used differential equations and forms, concerned chromatic and spherical aberration (when rays, respectively of different colours or of different distances from the axis of a lens, would focus at slightly different points); he thought that the former could be completely eliminated, but John Dollond, the distinguished English instrument-maker and designer of the achromatic telescope, regarded such hopes as excessive.

A problem which required mathematical attention as a result of the refinement of instruments was atmospheric refraction, for which correction factors needed to be calculated. Euler, of course, . . .; but perhaps the best study was Lambert's on the *Propriétés remarquables de la route de la lumière*, his first book, published in 1758, his 31st year. Taking Snel's law of refraction as his experimental base, he laid out a string of theorems and corollaries, trigonometry spiced with a few first-order differential equations leading to some differential geometry, all laid out in markedly Euclidean form. The main theoretical difficulty was not knowing how the density of the atmosphere varied with altitude; he tried polygonal approximations for the path as the ray passed through concentric spherical layers, each of constant density. His work contrasts interestingly with the molecular approach to be proposed by Laplace (§10.10).

Lambert also enriched perspective theory with a book of that title published in the next year, 1759. In addition to studying vanishing points and providing many recipes for constructing perspectival figures, he gave new emphasis to the properties of the projections themselves rather than the original configurations. Refinement of the approach was to lead to descriptive geometry, which we shall note in §8.3 in the hands of Monge.

The companion subject for Monge was differential geometry, then usually called 'geometry of curves and surfaces'. Indeed, the study of three dimensions received much necessary attention, perhaps because of the rise of partial differential equations (§6.4) {Boyer 1956, Chap. 8}. Lagrange and Monge found formulae for the perpendicular distance from a point to a surface, and the equations of the tangent plane and of the normal at any point. Various properties of the curvature of a surface were discovered; some will be noted at (8.3.1), in Monge's definitive presentation.

6.15 *Algebra and the resolution of equations*

Euler was very much to the fore, once again, with a two-volume *Vollständige Anleitung zur Algebra*, in 1770 (in German, indeed, and in fact first out in Russian two years earlier, when he was in his early sixties). This 'Complete Introduction to Algebra' covered not only basic algebraic theories but also certain series expansions, elementary functions such as the logarithmic, and some number theory including quadratic forms. To the French translation of 1773 Lagrange added an extensive set of notes, mainly on continued fractions and number theory.

As usual, Euler gave a reliable presentation; but he gaffed in his algebraic handling of complex numbers, by misapplying the product rule for square roots

$$\sqrt{(ab)} = \sqrt{a}\sqrt{b}, \text{ to write } \sqrt{-2}\sqrt{-3} = \sqrt{6} \text{ instead of } -\sqrt{6}. \tag{6.15.1}$$

{Euler 1770, Articles 148–9}. The error was systematic: for example, he committed it for division as well, and it confused some later writers on the subject.

From this time on, algebra was quite widely taught; long (and often rather tedious) textbooks, more elementary than Euler's, were written on it. Especially influential was the volume by Etienne Bézout (1730–83), first published in the mid-1760s as part of his multi-volumed *Cours de mathématiques*, which reappeared in posthumous editions for a century. These books usually advanced as far as trigonometry and the roots of equations; they also included analytic geometry, largely following Euler's *Introductio* (§5.27). However, the calculus was usually avoided.

At the research level, Bézout himself studied in the 1760s the question of how to eliminate variables from a system of polynomials in several of them. His main result stated that the degree of the final equation (in one variable) was the product of the degrees of the original equations; but his methods for finding (any) lower degrees and his ancillary ideas for reducing the problem to linear equations – coming close to determinant theory – were very striking. A decade later, his compatriot A.T. Vandermonde came even closer to determinants when resolving equations (although not to the determinant now named after him).

Lagrange looked closely at factorizing polynomials {Hamburg 1976}. He reviewed the known methods (§4.6–§4.7, §5.17) of resolving cubics and quartics, and the known 'reduced equations' of lower degree used to determine roots. He proposed such an equation himself, but of higher degree: the 'equation of differences' of a given equation E, defined as having as roots the squared differences $(p_r - p_s)^2$ of all the roots p_r of E; it gave him means of approximating to roots and interpreting complex ones (algebraically). He also considered permutations of roots of equations, and found that some rational functions of them took fewer values than one might imagine. Invariants were again in the air, but in this case it was to be the group-theoretic properties of the permutations themselves that would have the greatest influence upon his successors (§9.4).

6.16 *Quadratic forms in number theory with Euler and Lagrange*

{Weil 1984, Chaps. 3–4}

After the lonely Fermat (§4.22), Euler was the next major contributor, although Lagrange also became heavily involved from the 1760s, and Legendre a decade later. For algebraist Lagrange one attraction was the refinement of proofs, which at the time continued to be rather rudimentary, depending on factoring and cancelling, checking out all combinations of odd and even values for integers, and finding special theorems for particular forms of integer.

The influence of Diophantos is noticeable in the stress on quadratic forms. In 1770 Euler produced a result which Legendre was to name 'the quadratic reciprocity law':

> THEOREM 6.16.1 *Let p and q be two primes, each $\neq 2$, and consider the equations*
>
> $$A := x^2 - qy^2 = ps \quad \text{and} \quad B := x^2 \pm py^2 = qt. \tag{6.16.1}$$
>
> *If* $p = 4m + 3$ or $p = 4m + 1$ *for some integer m, then A has integer solutions for x, y and s if and only if B is similarly so solvable for x, y and t.*

An interesting aspect of this result is the *duality* between p and q, which we shall see elsewhere in mathematics. Among special forms, Euler showed that

$$x^2 - Ny^2 = 1, \text{ with } N \text{ a given positive integer,} \tag{6.16.2}$$

took infinitely many solutions for x and y, which could be found by using continued fractions.[2] This rather shadowy branch of mathematics {CE 6.3} gained some attention because of this theorem and its consequences.

[2] Euler called (6.16.2) 'Pell's equation', after a minor British mathematician; but it is in Diophantos, and the name is absurd. Many results in number theory are now stated in terms of congruences, but this is a later form due to C.F. Gauss (§9.2).

Lagrange's continued study of quadratic forms led to notable results. In particular, he showed, first in his additions of 1773 to Euler's *Algebra*, the following theorem:

THEOREM 6.16.2 *Let variables x and y be transformed linearly by*

$$x = pu + qv \text{ and } y = ru + sv, \text{ and set } G := pr - qs; \tag{6.16.3}$$

all quantities take only positive integral values. If $G = \pm 1$, *and*

$$ax^2 + bxy + cy^2 \text{ is transformed into } Au^2 + Buv + Cv^2, \tag{6.16.4}$$

$$\text{then } (b^2 - 4ac) = (B^2 - 4AC). \tag{6.16.5}$$

His emphasis on the discriminant in (6.16.5), while not new, was to be most fruitful in stressing the role of invariants under transformation.

Concerning forms involving prime numbers, Lagrange proved in 1770 Diophantos's guess that any integer could be expressed as the sum of at most four squares. Then, in 1785, Legendre showed that if the integer did not take the form $4^r(8m + 7)$ (with r and m positive integers), then three squares would suffice.

For forms above the quadratic, Euler proved Fermat's Last Theorem (4.22.2) for cubes and quartics. His methods were so different from one another that he despaired of finding any general proof for all powers – an excellent intuition, in the context of techniques then available or conceivable.

6.17 *The probabilistic convolution*

{Daston 1988}

The application of algebra to probability continued to build slowly upon the innovations described in §5.19–§5.23. However, while most of the major figures of this chapter considered probability at some time, only Laplace was a principal contributor.

This branch of mathematics was the one most clearly influenced by French Enlightenment philosophy, especially for providing techniques which led to its objective of knowledge gained through reason as well as through experience.[3] One of its founding figures was d'Alembert, who debated during the 1760s with Daniel Bernoulli on the manner of appraising the effectiveness of inoculation against smallpox. Bernoulli's training as a doctor attracted him to the problem, and his practice as a probabilist led him to estimate life expectancy after treatment, from areas under mortality curves. He admitted that these curves were very incomplete, which for d'Alembert meant that the study was insufficiently empirical and therefore faulty in principle; he also upheld that area alone was inadequate to characterize mortality anyway. Later he also attacked Bernoulli's solution of the St Petersburg paradox (§5.23) for offending experience by allowing for the possibility of very long runs in the calculation of expectation.

D'Alembert's influence on the Marquis de Condorcet (1743–94) was more positive. Condorcet was not a major mathematician, and his fetish for seeking closed-form solutions for differential equations found few adherents among the informed, and attracted contempt from Lagrange. He did, however, embody Enlightenment ideals in various ways, in particular in his work on probability theory and 'political arithmetic', the British name (§5.21) which he adopted. His contributions have recently been gathered together in {Condorcet 1994}, an excellent edition.

Condorcet envisioned the 'enlightened man', whose presumed capacity to reason out his judgements would form a basis for a new 'social mathematics' (including economics) based upon probability theory {Baker 1975}; but in sad

[3] Many Enlightenment views (especially d'Alembert's opinions) were expressed in articles in the famous *Encyclopédie* which d'Alembert and Denis Diderot edited and published between 1751 and 1780. Some of them were rehearsed again, for example by Condorcet, in the mathematical volumes of its successor, the *Encyclopédie méthodique* (1781 to 1832, when it stopped). However, apart from probability theory I am not convinced that that encyclopaedia greatly influenced the contemporary practice of the mathematical sciences.

irony, he committed suicide in 1794 because of the French Revolution, after which life and society could not be the same. In particular, his intellectual ideal was to be replaced by a much more proletarian reification: Adolphe Quetelet's 'average man' (§8.10). Perhaps Condorcet's most arresting single contribution was a calculation of the return on investment over a period possibly exhibiting different rates of interest, for in his analysis of each stage as dependent upon its predecessors there is, to the modern mathematician, a whiff of the Markov chain (§16.21). Alas, at the time of its publication in 1785, it did not receive much attention {Crépel 1988}. After his death some associates, especially the Swiss-born E.-E. Duvillard de Durand (1775–1832), tried to launch mathematical economics on a probabilistic basis {Israel 1993}; but the products did not sell.

6.18 *Bayes' theorem,* 1764
{Dale 1991, Chaps. 1–6}

While most of the significant work proceeded on the Continent as usual, one of the best-remembered events in probability theory during these decades occurred in England. In 1764 Richard Price, clergyman of the Dissenting persuasion and part-time economist, published with the Royal Society a manuscript by a fellow member of the cloth, Thomas Bayes, deceased two years previously, together with his own introductory letter. These papers highlighted inverse probability in a new way, and generated an approach to the subject now called 'Bayesian' which has always excited adherence and dissent in large measures. Perhaps inspired by de Moivre (the title of the paper mentioned the 'doctrine of chances' of §5.22), Bayes had found a formula for guessing that the probability of the single occurrence of a given event lay between 'any two degrees of probability' (specifically, the location of the area of rest of a billiard ball on a table after its arbitrary launch) in terms of its previous occurrence r times in n trials. In other words, he found a rule for estimating the probability of future performance within certain bounds, given knowledge of the past record.

Bayes understood the breadth of his result, for in a 'scholium' he rethought it in quite general terms. He argued (1) that the event was equally likely to happen on some $s \neq r$ occasions, where $1 \leq s \leq n$ trials of any event, and (2) that the probability of r occurrences was exemplified by the case of the billiard ball, so that his formula could be applied in general. Claim (1) is now called that of 'uniform priors' in the modern Bayesian jargon, and its validity and generality has been much debated {Gillies 1987}. So also has the relationship between the theorem and Jacob Bernoulli's Theorem 5.22.1; it looks like an inverse, but there is the important difference that with Bayes the specification of the range of estimation of the probability (in the original case, the area where the billiard ball came to rest) is independent of r and n, but not so with Bernoulli {Dale 1988}.

One motive for Price's introductory letter seems to have been David Hume's doubts about the efficacy of inductive inference, as stated in his *Enquiry Concerning Human Understanding* (1748); for Price used Bayes' theorem to present a reason (whose existence Hume denied) for the repetition of regular events. But this was not the only source of his interest in probability theory: since the late 1750s he had acted as a mathematical consultant to the Society for Equitable Assurance on Lives and Survivorships (note its legalistic name), especially on calculating the 'reversionary payments' (his name, I believe) from mortality tables.

Bayes' formula for the probability, a ratio of areas,[4] was itself part of his legacy. The denominator corresponded to the beta function (5.18.4), arising from the binomial character of the problem; the numerator was like the 'incomplete' function (to use the modern name), the same integrand defined over a variable range of values within [0, 1] corresponding to the area of rest of the billiard ball.

[4] Interestingly, Bayes did not present his calculation of the probability as the ratio of two fluents, as would have befitted the (anonymous) author of a rather good fluxionalist text of 1736, written in reaction to Berkeley's criticisms (§5.7).

This was the début of a function of considerable future significance for probability theory {Dutka 1981}.

6.19 *Probability theory in science: data and their errors* {Eisenhart 1983; CE 10.5}

Parallelling these social and religious contexts, probability was playing a greater role in science. One fine piece of thinking came from the English astronomer John Michell: in a discussion of 'probable parallax and magnitude of the fixed stars' published by the Royal Society in 1767 (three years after Bayes' piece), he found that the incidence of apparently close pairings of stars was too great for them *all* to be effects of line of sight, and that 'next to a certainty' (as he put it in a passage redolent of Bayesian concern with prior belief) such observed pairs of stars must actually be very close together, perhaps moving under mutual gravitation. His speculation waited 40 years before corroboration by the observations of William Herschel.

Michell's insight arose from analysing data. The most substantial question concerned the 'laws of error' to which data were susceptible by the frailties of observation and equipment. A major innovation was made in the mid-1750s, by Thomas Simpson, the popularizer of de Moivre's work on annuities (§5.22). Pondering upon the rationale of the normal practice of taking the mean of a set of observations as the most probable value, he assumed various discrete and continuous distributions of error from the mean, and used generating functions to estimate the probability distribution of the sum of a collection of independent errors, and thereby of their mean. Twenty years later, Lagrange extended this analysis in a paper suspiciously similar in content, which Euler summarized soon afterwards.

Meanwhile, Lambert had done still better, especially in his book *Photometria* (1760, his 33rd year), which named and indeed innovated the study of optical luminosity. Here errors would arise in the observations; let the corresponding measurements be y_r, each one occurring m_r times. He also questioned the assumption underlying the use of the mean, that errors were equally likely to occur below the true value T as above it, and sought instead a method of estimating a value Y that differed from T with maximum probability. For this purpose he assumed some (unspecified) continuous distribution of error $f(y)$, and proposed that Y be determined by maximizing the product $\Pi_r [f(y_r - Y)]^{m_r}$. This was his second innovation – a notion we now call 'maximum likelihood'.

6.20 *Laplace's first work on 'probabilities'*

Lambert later coined the term 'the law of error' in these contexts, and applied it to the accuracy of astronomical observations. This was very much Laplace's territory, and it motivated many of Laplace's important contributions from the mid-1770s. Although he used only the word 'probabilities' (*probabilités*) to refer to his work, he was also a main founder of mathematical statistics.

Firstly, in a paper on 'the causes of events' published in 1774, Laplace proved a theorem of a Bayesian kind, possibly independently. In a companion paper he also thought about the law of error, showing that the arithmetical average might well not minimize the cumulative error. He defined the error function $f(y)$ (to continue with the above notation) in general terms, requiring it to take (positive) values over all values of y and fall to zero as $y \to \infty$, and to enclose a unit area. Among the particular functions he considered were, with the median as desired value, the distribution $\exp[-|y - Y|/k]$ (k a parameter), now named after him; and, with the mean as target, the distribution $\exp[-(y - Y)^2/k]$.

We saw this latter function with de Moivre at (5.22.1); following Francis Galton (§12.11), we call the associated distribution 'normal'. Especially in a 'Memoir on Series' (1782), Laplace also grasped from de Moivre the utility of generating functions (an expression which he introduced at this time). Writing them in the form $\sum_{r=-\infty}^{\infty} a_r x^r$,[5] he emphasized de Moivre's perception that they functioned as 'carriers': use them to characterize a problem, calculate a value for x^n to represent some desired feature of it, and then read off a_n as (say) the required corresponding probability. The later stage of these studies in probability theory, in the 1810s, would produce from him central limit theorems (§8.8); but he used the functions also in analysis, such as (6.9.3) for the Legendre functions.

In the mid-1780s Laplace correlated theory with data in two other important ways. Firstly, in connection with the problem of the orbits of Jupiter and Saturn (§6.8), he drew upon another technique due to Euler and especially to the German astronomer Tobias Mayer, which was to become standard in astronomy: forming 'equations of condition'. In this case Laplace had to calculate four unknowns for each planet from four linear equations, but he had 24 sets of data to draw upon; so he inserted all sets, gathered the 24 equations together into four groups and added up the equations in each group to form four equations of condition. These he solved in 'longhand' manner – which was still hard work, as determinants were not available.

The second context was geodesy. Given that the length $l(\theta)$ of the arc of a meridian was related to the angle of latitude θ by the equation

$$c(\theta) := l(\theta) - A\sin^2\theta + B = 0, \text{ with } A \text{ and } B \text{ constants,} \quad (6.20.1)$$

together with some measurements or calculations of arc $l(u_s)$ made at latitudes θ_s, how were the 'best' values of A

[5] I commit an anachronism here, as the summation sign became frequent only from the 1820s (for example, with Fourier at (8.7.1)). In the late 18th century series were usually written out term by term; and an interesting feature is the difference between writing '. . .' and '&c.' after a few of them. Sometimes a distinction was made between continuation by means of an algorithm and by preservation of form, or between finite series ('+ · · · +') and infinite ones (' + &c'). The practices were not uniform, and vanished with the introduction of the summation sign.

and B to be found from these data, in order to determine the spheroidicity of the Earth? In the 1750s, Boscovich had taken this as a problem in linear regression using the criterion of absolute deviation (to use our terms); that is,

$$\text{require that } \sum_s c(\theta_s) = 0 \text{ and } \sum_s |c(\theta_s)| \text{ be minimal} \qquad (6.20.2)$$

(for normally $c(\theta_s) \neq 0$). Laplace took over the approach, and extended it to cases where the θ_s were replaced by $n_s\theta_s$, where the numerical factors n_s represented estimates of the apparent levels of accuracy of the observations corresponding to each one of the data. This work contains the seeds of linear programming; but they were to stay in the ground for a long time (§16.22).

So Laplace was advancing his work in various ways, both probabilistic and analytical. He was well aware of the social uses of probability theory, and from the mid-1780s he applied his developing methods to demography, estimating the current population of France at around 25 million – among whom, from certain points of view, he was Number One.

6.21 *Competing traditions in mathematics*

While the number of jobs available in mathematics remained very limited, we see a great increase of work produced by the few employed figures. However, little atmosphere of cooperation attended the research – on the contrary, there were accusations of plagiarism (especially from d'Alembert) and personal disclosure of results by letter was discouraged. Thus, early on in their contact Lagrange told the young Laplace that he preferred to see discoveries only 'in printed sheets'. Similarly, apart from prize problems posed by Academies, the motivations for researches were often unclear; for example, Euler did not state why he was led to develop the equations for dynamics in the early 1750s and not before.

The central branches of the calculus and mechanics grew not only in quantity but also in competition, with two trios of traditions acting across all branches. In the calculus, the dominant Eulerian style was being challenged by Lagrange's faith in Taylor's series, with limit theory around also but not popular even when not in Newtonian clothing. In mechanics, Newton's laws (and angular momentum) were particularly good in the celestial and planetary branches. In contrast, the engineers were attracted by practical utility to use principles of energy conservation and the loss or conversion of kinetic energy into work (not yet the technical terms used); and some more algebraic principles, such as least action and virtual velocities, were intended by their adherents to cover all parts of mechanics {Grattan-Guinness 1990b}.

This last tradition accentuated a related competition of styles. For the algebraic aspects of d'Alembert and the general algebraic style of Lagrange challenged the geometrically guided approaches adopted by Euler and others; Laplace (and Daniel Bernoulli before him) were somewhere between the two positions. The difference of styles reflected contrary aims, often based on knowing the answer beforehand or not: Euler was seeking new results, and needed a geometric style to capture the properties of the physical phenomena, while Lagrange re-proved and reordered them algebraically from general principles (the finding of which was, of course, a novelty in its own right). Thus, for example, when Euler sought the (unknown) equations of rotation of a rigid body (§6.6), he began by visualizing a body rotating; Lagrange later rederived them solely by algebraic means by imposing upon his general equations (6.11.2) the particular features of the problem, such as one point being fixed. Again, the concept of pressure came into Lagrange's equation (6.10.4) for the motion of incompressible fluids, not as a means of saying something about a fluid but as the multiplier attached to the constraint stating that it was incompressible!

We shall see similar differences later, especially in connection with axiomatic and non-axiomatic versions of a theory. We shall also find the tension between algebraic and geometric styles, and competition with mathematical analysis, as the mathematics of the 19th century developed at speed after the French Revolution.

7

Institutions and the profession after the French Revolution

7.1 *A new* époque: *the Ecole Polytechnique*

{Grattan-Guinness 1990a; CE 11.1}

After the Revolution of 1789, everything in France had to be changed, but it took a little time. Most of the educational institutions met their guillotine around 1793; the score of existing universities were abolished, and 15 years were to pass before a new structure was fully in place. However, France was already a major country for the mathematical sciences, and remained so in the new century, and until the 1820s it was undoubtedly the most important one by far. The first two sections of this chapter are devoted to its main institutions and figures; I start with the most important institution.

In the early 1790s the country was threatened by hostile or suspicious neighbours, so it needed to improve its military arm, including educational aspects. The functions of the engineering schools already in existence, civil and military (§6.1), were not basically altered. But a new general engineering school was created in 1794 in Paris: the Ecole Polytechnique. In its definitive programme, which was established around 1799, this school provided engineering students with a general (that is, 'poly-technical') two-year course which included a lot of mathematics. Graduates of the Ecole Polytechnique normally studied for a further three years in one of the specialist 'application schools' (as they were called) for military or civil engineering. Several of the great scientists of Paris were appointed as professors or examiners (a strict division of post), and the Ecole's reputation grew enormously. Underneath the glory, however, some unclarities and disagreements of purpose can be detected.

Firstly, the need to balance civil and military requirements led to political controversies at times, indeed throughout the system of engineering schools. Late in 1804, Emperor Napoleon (as he was about to crown himself) converted the Ecole into a purely military school and imposed a much stricter regime; but civil engineering remained a prime requirement for the country.

Secondly, the staff were eminent and the flow of graduates who became major scientists was remarkably high until the late 1810s; but all this was contrary to the purpose of a preparatory (that is, *elementary*) institution. The eminence of the teachers there forms a striking contrast with the frequent mediocrity of their colleagues at the specialist so-called 'application schools', where *higher*-level teaching was the aim.

Thirdly – and most importantly from our point of view – there were considerable differences and disputes about the amount of mathematics taught and balance across the branches. This is exemplified well by the two founding professors of analysis: pure algebraist Lagrange, and engineer *par excellence* de Prony (we saw his treatise of the 1790s in §6.12). During these early years Gaspard Monge was very important, and his beloved subject, descriptive geometry, gained a large airing; but by the mid-1800s its teaching had been reduced considerably to make more time for analysis. The major influence here was Laplace, never a teacher at the school but very influential as examiner and on the governing council, which he created in 1799 during a brief six weeks as Minister of the Interior. The orientation towards engineering applications was weakened in favour of more general applied mathematics, and even the pure sides.

This tendency increased after the (second) fall of Napoleon in 1815 and the Restoration of the Bourbons. Onto the staff came graduate and Bourbon fanatic A.L. Cauchy, with his innovations in mathematical analysis of the 1820s (§8.5–§8.6) – which, however, met with great hostility from the students and even from his colleagues on the staff. Yet his approach continued to be taught by others after he fled from France with the Bourbons after their fall from power in the 1830 Revolution. Then a counter-reform of the syllabus which brought applications to engineering back into favour was initiated in 1850 by the astronomer Urban Leverrier, whereupon several professors resigned in protest . . .

7.2 *Other French institutions*

The application schools enjoyed a more tranquil atmosphere, and attracted much less publicity. For mathematics, one civil and one military school stand out.

The Ecole des Ponts et Chaussées had de Prony as director for 40 years from 1798, simultaneously with his obligations at the Ecole Polytechnique. Many of its most distinguished graduates went there to study. However, the quality of teaching was indifferent until 1819, when Claude Navier was appointed to teach elasticity theory and applied mechanics; he inaugurated a tradition there (§10.16) which was continued by G.G. Coriolis and Barré de Saint-Venant (§14.8) for the next 60 years.

At the military school in Metz in the Alsace, Colonel J.V. Poncelet began a course in energy mechanics in 1824 which stimulated a tradition both there and, later, in Parisian institutions for a comparable period (§10.8–§10.9). However, the school closed in 1871: it had been moved there in 1795 from Mézières to be further away from the

British, but it was shut down when the region was annexed by the Prussians in the Franco-Prussian war.

A second part of the French system of higher education was the Université, as it was called upon its foundation in 1808. The name is not a happy one, for it was (meant to be) a state monopoly of school teaching, with the country divided into *académies* (another unfortunate term), each run by a Rector. Some higher-level instruction was delivered in the *facultés* of several *académies*, including a few for science. The one in Paris was the most important, of course, but even it was secondary in quality and prestige to the Ecole Polytechnique.

Within this Université a school for clever pupils, the Ecole Normale, was set up in Paris; but it became significant in mathematics only from the 1870s (§11.4). It took its name from a previous school which had opened in 1795 as a national institution for teacher training. It had closed after a few months; but the likes of Lagrange, Laplace and Monge had given relatively elementary courses there.

Finally, there was a unique institution dating back to the 16th century. Called the Collège de France after 1793, it was allowed to continue its functions largely unaltered. Teaching and research were conducted there for their own sake, without enrolment conditions or examinations. Unusually, the standards for mathematics were low until the 1840s, when Joseph Liouville was appointed. However, one of the two chairs of physics was held for the first 60 years of the century by J.B. Biot, a point of significance for mathematical physics (§10), especially when Ampère took the other chair in 1824.

Among research institutions, the Académie des Sciences remained in form and functions more or else the same, although it came gradually to be more active in mathematics than before. From 1795 to 1815 it was the 'class for mathematical and physical sciences' of a new umbrella institution of academies, the Institut de France, which is still in place. I shall refer to the 'French Academy' or '(class of the) Institute of France' as chronology demands. In 1803, two posts of 'perpetual secretary' were set up, the holders to be elected from among the members.

Some other institutions provided both a forum and employment for specific sciences: for example, the Bureau des Longitudes for celestial and planetary mechanics, and the Dépôt Générale de la Guerre for cartography. In addition, a private Société Philomatique met in Paris and rapidly published short papers in its *Bulletin*, thus becoming perhaps the best journal in the world for The Latest News in science.

This ensemble of institutions for research and teaching, and the opportunities thus provided for employment in both, gave a scientist the novel chance to be a truly professional person, with jobs paid for by the state, often in a *cumul* (as the French referred to it) of incomes from a variety of posts (such as with de Prony above). However, while the institutions themselves had no competitors in their particular specialities, there was plenty of competition between their *savants* for patronage, influence and appointment. We see here the origins of many aspects of the modern scientific establishment, with dark as well as bright sides in evidence.

Innovation and convolution in French institutions; excellent examples of each are evident here. The Ecole Polytechnique was the chief novelty, not for its (unexpected) and rapid rise to world eminence but for the new role it played as the general preparatory institution for engineering. Also novel was the Université, although less happy an innovation; it would be modified and eventually dismantled from the 1870s onwards. But many other institutions continued to operate as before, although in different contexts.

7.3 *A new classification of applied sciences*

Table 7.3.1 indicates the principal French mathematicians who worked in the period 1800–40, together with some older figures from the late 18th century met already, and a few younger ones who continued into the 1860s and

TABLE 7.3.1 *Principal French mathematicians, 1800–1840, divided by research interests.*

s t e student/teacher/graduation examiner at the Ecole Polytechnique
ch student at the Ecole des Ponts et Chaussées
m passed military career

Calculus/Mechanics/Engineering	*Mathematical Analysis/Physics*
Bossut, C.S.J. (1730–1814) e	Ampère, A.M. (1775–1836) t
Carnot, L. (1753–1823)	Binet, J.P.M. (1786–1856) s t ch
Coriolis, G.G. (1792–1843) s t ch	Biot, J.B. (1774–1862) s
Delambre, J.B.J. (1749–1822)	Cauchy, A.L. (1789–1857) s t ch
Dupin, F.P.C. (1784–1873) s	Duhamel, J.M.C. (1797–1872) s t e
Francoeur, L.B. (1773–1849) s	Fourier, J.B.J. (1768–1830) t
Girard, P.S. (1765–1836)	Fresnel, A.J. (1788–1827) s e ch
Hachette, J.N.P. (1769–1834) t	Germain, S. (1776–1831)
Malus, E.L. (1775–1812) s e m	Lagrange, J.L. (1736–1813) t
Monge, G. (1746–1818) t	Lamé, G. (1795–1870) s t e
Morin, A.J. (1795–1880) s m	Laplace, P.S. (1749–1827) e
Navier, C.L.M.H. (1785–1836) s t ch	Legendre, A.M. (1785–1833) e
Poncelet, J.V. (1788–1867) s m	Liouville, J. (1809–1882) s t ch
Riche de Prony, G. (1755–1839) t e	Poinsot, L. (1777–1859) s t ch
Puissant, L. (1769–1843) m	Poisson, S.D. (1781–1840) s t e
	Sturm, C. (1803–1855) t

beyond. The last in the list, Sturm, was Swiss by birth but made France his home from 1825. The perpetual secretary for mathematics, physics and astronomy was Delambre, followed in 1822 by Fourier and in 1830 by the astronomer and physicist D.F.J. Arago (1786–1853).

These figures are divided into two groups by research interests, in a manner which requires explanation. Analysis and mechanics formed the bulk of the so-called 'classical sciences'. They stood in marked contrast to the 'Baconian' sciences, in which experiment and observation were in the driving position and mathematics was modestly placed or even absent (§4.3). Among these sciences physics had the closest contacts with mathematics, yet even there nothing more than simple algebra and geometry was normally called upon. Partly for that reason, physics did not have a very high reputation in either research or education in the late 18th century.

This situation changed radically between 1800 and 1830, as major areas of physics came to be mathematicized: in chronological order, heat, physical optics, electrostatics (then called 'electricity') and magnetism, and the union of the last two in electromagnetism. Several of the French mathematicians listed in the left column of Table 7.3.1 were largely or entirely responsible for the initial changes (as we shall learn in §10), which eliminated the sharp division between the two categories of science. What was the relationship then?

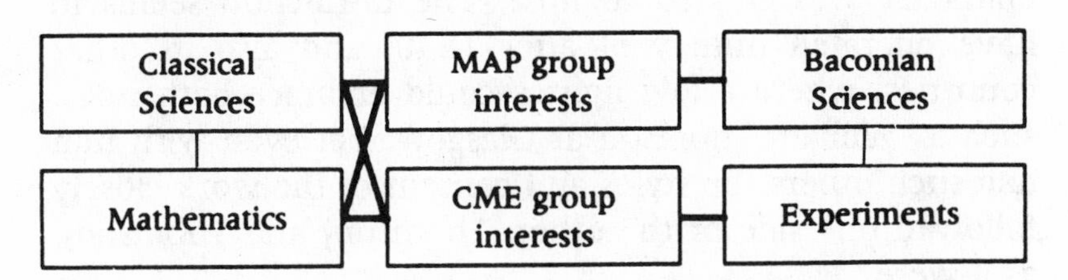

FIG. 7.3.1 *Traditional and new connections between mathematics, physics and engineering before and after 1800.*
The two periods are represented by thick and thin lines, respectively.

FIG. 7.3.1 shows my suggestion, reflecting the two columns of Table 7.3.1. Some mathematicians functioned in mechanics and mathematical physics, with little attention paid to experiments (or at least the execution of experiments); often they were the more able mathematicians as such, and some contributed heavily to pure mathematics (for example, Cauchy will be a major figure in §8). The others worked largely in engineering mathematics, with experiments very much to the fore but mathematical physics ignored; they also used the calculus, but were less drawn to mathematical analysis in general. Geometry belongs to both sides, although with some distinction stemming from the differences in basic concerns: those in analysis worked on (or with) differential and algebraic geometries, while the engineers thought more about descriptive geometry. Projective geometry crossed the boundary line, partly for the links with descriptive geometry and partly because of the major contributions made by military engineer Poncelet.

Such a scenario fits the French remarkably well, with few modifications (for example, Fresnel was also a superb experimenter, while Lamé made fine contributions to engineering mechanics in his early years, which were spent in Russia with Benoit Clapeyron). Part of the reason lies in the prominent place given to engineering as a profession, with the Ecole Polytechnique and the other institutions mentioned above, though it is somewhat surprising that the boundary was crossed so little. The distinction seems to have operated until well after 1840, and also in other countries, where a few figures could embrace both sides, such as William Thomson at Glasgow. Yet even with him and such others the styles and content of the work closely followed one side or the other: Thomson_1 and Thomson_2, as it were.

7.4 *The first sage of Göttingen*

Outside this French *gloire*, one light was burning strongly from afar, as the Parisians knew: Carl Friedrich Gauss (1777–1855). His range was wide (although curiously patchy in analysis): he made quite fundamental contributions to astronomy, potential theory, magnetism, number theory, algebra and aspects of geometry. He was the archetypal ivory tower mathematician, except that it was an observatory he occupied, from 1807, as Director and Professor of Astronomy at Göttingen; the mathematics chair was held in his time by luminaries such as Bernhard

Thibaut (§8.17). Hardly ever leaving the town, he made his contacts with the outside world largely by correspondence – but in this mode he excelled, with extensive exchanges of ideas enriched with some gossip.[1]

Yet Gauss's influence could have been greater had he chosen to alter two strategies. Firstly, he refined his ideas for publication to such an extent that sometimes he got deeply into a problem area which others had barely perceived, so that the reception was quiet – for example, a paper of 1813 on infinite series (§8.14), which slept for 20 years. Secondly, he had many wonderful insights which he kept to himself: not only the now famous diary he kept in his youth and early manhood, but also all kinds of notes, such as a 'beautiful theorem of probability', from perhaps the 1800s, which states the Fourier integral theorem (8.7.6) with complete clarity (an interesting title, therefore) but then stops. He said of his own writings, *'pauca sed matura'* ('little but mature'), but in fact he published many pages, though maybe not always the right ones or in the most readable form.

Apart from him, the only other German mathematician of top rank was Wilhelm Friedrich Bessel, much of whose work in analysis and mechanics was stimulated by astronomy and geodesy, sciences in which Germans excelled. Among the minor figures, Georg Klügel and Gottfried Rosenthal concentrated on producing large-scale dictionaries of mathematics.

There was also a school of combinatorial algebraists active for a generation from around 1780, led by C.F. Hindenburg. It even sported a mathematical journal (then a novelty), the *Magazin für reine und angewandte Mathematik* (1785–9), in which appeared Lambert's posthumous paper

[1] Although several biographies of Gauss have been written, none has either the length or the quality to match their subject. The edition of his *Werke* (1863–1933), therefore, is still essential, including in Volumes 10 and 11 a suite of important articles on most major parts of his scientific work. Quite a lot of letters have been published, but many remain among his manuscripts (which are kept in Göttingen University Library) and those of his correspondents (where known).

on Euclidean geometry (§6.13). The productions of this school were devoted largely to binomial and other series, with applications to the calculus and to probability theory; they were pretty baroque, and perhaps its best consequences came from L.F.A. Arbogast and his followers of post-Lagrangian mathematics, the brothers J.F. and F.J. Français, and F.J. Servois. Minor figures in the French coterie, this quartet exhibit a striking case of cultural influences, for these French algebraists with a German flavour came from the region of Alsace-Lorraine.

7.5 *The 1820s: mathematics goes international*

Reactions to French dominance in science in general began to come from other countries in the early 1810s, as reforms or at least as individual or group initiatives. Imitations of the Ecole Polytechnique were attempted in a restructuring of West Point in the USA, and in new institutions, especially in Vienna and Prague. In Prague, from the 1800s, the mathematician and philosopher Bernhard Bolzano (1781–1848) floated some remarkable ideas on the foundations of mathematical analysis and geometry, but they attracted little international attention until the 1880s (§12.5).

An attempt to launch such a school in Berlin during the 1820s failed, but Prussia had begun a successful revolution of its university system, reducing its literally dozens of institutions, many tinpot, to a small number of more professional establishments {Turner 1971}. Berlin University was the leader, although mathematics was not one of its successes. In the late 1820s, Königsberg saw innovations in various procedures, such as the seminar where research work was presented by its creator, possibly soon after birth; from such strategies came the German love of *Bildung*, of which 'culture' is only a pastel translation. Mathematics and physics were subjects thus encouraged there; Bessel had been resident since 1813, and he was joined in the mid-1820s by youngsters such as Carl Jacobi and Franz Neumann.

Meanwhile, the first two volumes of Laplace's *Mécanique céleste* (1799) at last began to wake up some of the British, who set aside the fluxional tradition {Guicciardini 1989, Part 3}. The Eulerian calculus was learnt by the professors at the English military schools, by John Playfair and a few other Scots, and by Bartholomew Lloyd and his colleagues at Trinity College, Dublin. At Cambridge, Lagrange's algebraic approach found favour with Robert Woodhouse; then in the early 1810s undergraduates Charles Babbage, John Herschel and George Peacock founded the 'Analytical Society', the aim of which was the replacement of fluxions by Lagrangian algebras in research and by differentials in teaching.

Soon a good group of mathematicians emerged, including Augustus De Morgan, who became founder Professor of Mathematics in 1828, his 23rd year, at the new London University. (It was renamed 'University College, London' in 1836, when the University of London was established as an examining federation for it and King's College London, founded in 1831.) They and others contributed remarkably fine survey articles of many parts of mathematics to the supplementary volumes of the *Encyclopaedia Britannica* (1820s), the *Encyclopaedia metropolitana* (1817–45), the *Reports* of the British Association for the Advancement of Science (1831–), and elsewhere.

In Italy a few respectable figures such as Paolo Ruffini and Vincenzo Brunacci maintained a good tradition of work in algebra and analysis, and various others contributed to mechanics. Over and above individuals, a Società Italiana had functioned since 1782 to promote the sciences, with a good representation of mathematics {Grattan-Guinness 1986}. The very existence of such a society is remarkable, for in a country which saw unification only in the 1860s, here was an association with members from all parts of the peninsula and the Mediterranean islands.

As a result of these initiatives, good work from outside

France (indeed, outside Paris) began to appear during the 1820s to join the contributions of Gauss, Bessel and a few others. But the change was rapid: before the end of the decade several other countries could boast of figures to match the *savants*. Table 7.5.1 lists the principal new boys, who will be prominent in the following three chapters. By 1829 nine of them had published significant work. In fact, 1829 itself was a 'good year' for mathematics in both France and abroad: for example, Dirichlet on the convergence of Fourier series (§8.7); major textbooks by Coriolis and Poncelet on energy mechanics (§10.8); Sturm's algorithm for finding the number of roots in a polynomial (§9.6); early work on heat diffusion by Duhamel, Lamé and Liouville, which led to pioneering results in, respectively, potential theory (§10.15), differential geometry and function expansions; Jacobi's book on elliptic functions and Abel's theorem on Abelian functions (§8.12–§8.13); Steiner's presentation of analytic geometry his way (§8.4);

TABLE 7.5.1 *Principal non-French newcomers, 1820–1840.*

English and Irish	*Germans*
Airy, G.B. (1801–1892)	Dirichlet, J.P.G. Lejeune- (1805–1859)
De Morgan, A. (1806–1871)	Jacobi, C.G.J. (1804–1851)
Green, G. (1793–1841)	Kummer, E.E. (1810–1893)
Whewell, W. (1794–1866)	Möbius, A.F. (1790–1868)
Hamilton, W. R. (1805–1865)	Ohm, G.S. (1789–1854)
Lloyd, H. (1800–1881)	Plücker, J. (1801–1868)
Murphy, R. (1806–1843)	Neumann, F.E. (1798–1895)
MacCullagh, J. (1809–1847)	Weber, W.E. (1804–1891)
Others	*Others*
Ostrogradsky, M. (1801–1862), Russia	Abel, N.H. (1802–1829), Norway
Quetelet, L.A.J. (1796–1874), Belgium	Steiner, J. (1796–1863), Switzerland

Gauss on inequalities in the foundations of mechanics (§14.14); some of Cauchy's papers on elasticity theory, and one of Poisson's (§10.16). Not all pieces were winners, though. For example, three amazing works made little impact even on their own authors' later work: namely, Cauchy and Sturm on the latent roots of matrices (§9.10), and Lamé and Clapeyron on locational equilibrium (§10.9).

I give this list (which could be extended) to convey the measure of durable work produced at the time. The years 1828 and 1830 were pretty good vintages, too: Hamilton's astounding first paper on variational optics, and Green's book on electricity and magnetism; Fourier's book on equations, and Peacock's textbook on algebra ...

After 1840 the mathematics was still more international in all its branches, though most countries were stronger in some branches than in others; Table 7.5.2 lists the main

TABLE 7.5.2 *Principal new figures, 1840–1860.*

French	*German*
Bertrand, J.L.F. (1822–1900)	Clausius, R. (1822–1888)
Bienaymé, I.-J. (1796–1878)	Eisenstein, F.G.M. (1823–1852)
Chasles, M. (1793–1880)	Grassmann, H. (1809–1877)
Delauney, C.E. (1816–1872)	Helmholtz, H. von (1821–1894)
Hermite, C. (1822–1901)	Kirchhoff, G.R. (1824–1887)
Leverrier, U.J.J. (1811–1877)	Kronecker, L. (1823–1891)
Puiseux, V. (1820–1883)	Neumann, C. (1832–1925)
Saint-Venant, B. de (1797–1886)	Staudt, K.G.C. von (1798–1867)
British	***Others***
Boole, G. (1815–1864)	Betti, E. (1823–1892), Italy
Cayley, A. (1824–1895)	Bolyai, J. (1802–1860), Hungary
Rankine, W.J.M. (1820–1872)	Brioschi, F. (1824–1897), Italy
Stokes, G.G. (1819–1903)	Chebyshev, P.L. (1821–1894), Russia
Sylvester, J.J. (1814–1897)	Lobachevsky, N.I. (1792–1856), Russia
Thomson, W. (1824–1907)	

newcomers for all countries. The French lost their dominance, though they were still arguably the strongest single nation.[2] Germany continued to rise in importance; Jacobi is worth emphasizing, not only for his splendid work but also as the first Jewish mathematician of note in Europe. The German list excludes Karl Weierstrass and Bernhard Riemann, whose work made most impact after 1860, as we shall see in §11.

Other countries to boast major figures included Russia, with a tardy international recognition of Mikhail Ostrogradsky and Nicolai Lobachevsky; however, with very few exceptions the posthumous papers of Euler, which the Saint Petersburg Academy published until 1830, made very little impact. Until mid-century Italy could boast only Enrico Betti as a first-ranker in the period 1800–60; but it maintained a good cohort of mathematicians in many branches, and as a country is under-represented in these tables.

7.6 *Three mathematical journals*

Mathematics was being published in quite a wide variety of journals: for the French, in those of the the Ecole Polytechnique and the Academy, some military journals, and several commercial enterprises. France also boasted the first international abstracting serial for all sciences, and also engineering and economics: the *Bulletin Universel des Sciences et de l'Industrie*, run under the direction of the naturalist the

[2] A general question in 19th-century history concerns the supposed sharp decline of French science after the 1830s. While some decline is evident, one should note: (1) the level achieved earlier, not only in mathematics, was so high that the only way forward was down; (2) the appearance of decline is enhanced by the rise of other countries (and their mathematics); (3) 'historians', even of science, rarely try to understand mathematics anyway, in contrast to its high place in French science then (and at other times too); and (4) engineering of all kinds tends to be marginalized – but this is indefensible for France, where, for example, Poncelet inspired improvements from the mid-1820s with his own work and that of followers such as Morin (§10.10).

Baron de Férrusac. The coverage in its eight parallel series was quite extraordinary; its closure in 1832 after eight years (because of the termination of governmental support) was a great misfortune. The mathematical sciences had their own series {Taton 1947}, and for several of its later years the young Sturm was its editor. In other countries there were fewer suitable journals, although academies took papers when required (Gauss in the Göttingen Academy, for example). None of the publications mentioned so far was devoted to mathematics alone, or even primarily; but three journals of that kind did start.

Joseph Diez Gergonne (1771–1859), professor of mathematics at Nîmes in the French Université system, launched in 1810 his *Annales de Mathématiques Pures et Appliquées*. He intended it to have an educational purpose, and some papers with teaching potential did appear there. But educational policy as such was not debated, and to a large extent it became a home for Unusual Mathematics, the most prolific contributor of which was one J.D. Gergonne (sometimes as 'a subscriber'). Until the 1820s the majority of authors were provincial French teachers like himself; after that the Paris *savants* sometimes honoured it with their products. It ceased in 1832, the year Gergonne was made Rector of the *académie* of the Université at Montpellier (whither he had moved in 1816), and the workload became too great for him to continue.

Meanwhile, the *Journal für die reine und angewandte Mathematik* had been founded in Berlin in 1826 by August Leopold Crelle (1780–1856) {Eccarius 1976}. Although he imitated the titles of Hindenberg's *Magazin* and Gergonne's *Annales*, his own was immediately seen as of advanced calibre. He took articles from all countries, and with the young Abel he even effected the publishing revolution of translating his French texts into German before printing. Mathematician and (principally) engineer, he was also a voice in the Germany governmental ear; upon his advice some young Germans were able to travel abroad, and research mathematics was encouraged in general {Eccarius 1977}.

Finally, back in France in 1836, Liouville, only 28 years old, began his *Journal de Mathématiques Pures et Appliquées*. It was issued by the important publishing house of Bachelier; a son of the owner belonged to the same year's intake at the Ecole Polytechnique as Liouville. The title almost exactly imitated that of Gergonne's recently terminated serial, and Liouville referred to it in his opening editorial; but in level it compares much more with Crelle's journal, and from the start it enjoyed comparable success.

Thus the mathematical research journal was becoming a species. Indeed, in the early 1840s two more were launched, with a focus on educational needs: in Germany the *Archiv der Mathematik und Physik*, and in France the *Nouvelles Annales des Mathématiques*. Thereafter other journals started (and in some cases finished) in various countries.

The growth of scientific publishing in the early years of the 19th century is quite extraordinary: for example (although an exception in degree), French mathematicians were publishing in over 30 journals, magazines and even newspapers. The production of books, whether textbooks or research treatises, was also vast. The paper varied in quality from moderate to rubbish (and was usually expensive); but the typesetting varied from adequate to superb in both appearance and accuracy, especially given the demands of notation in books presenting or using the calculus. For example, in those days before linear algebra, systems of differential or linear equations were printed out in full, usually each one on its own line, so that reading was easy and patterns could be spotted. The publishing industry constitutes a major facet of the history of science of the time, yet hardly anything is known of that history.

7.7 *States of the arts around 1800*

Finally we return to 1800, where the last chapter left us and the next three will start. Partly because of the opportunities afforded by the Revolution, some French mathematicians, and also Gauss, published major works around

TABLE 7.7.1 *Major books published around 1800.*

Author	*Date*	*(Short) title, volumes*	*Ed.*	*Ref.*
Lagrange	1797	*Théorie des fonctions analytiques*	1	§8.2
Lacroix	1797–1800	*Traité du calcul différentiel et intégral*, 3 volumes	1	§8.2
Lagrange	1798	*Résolution des équations numériques ...*	1	§9.6
Legendre	1798	*Essai sur la théorie des nombres*	1	§9.2
Laplace	1799	*Exposition du système du monde*	2	§10.4
Laplace	1799	*Traité de mécanique céleste*, volumes 1–2		§10.4–6
Legendre	1799	*Elémens de géométrie*	2	§6.14
de Prony	1800	*Mécanique philosophique*		§10.2
Lagrange	1801	*Leçons sur le calcul des fonctions*	1	§8.2
Gauss	1801	*Disquisitiones arithmeticae*		§9.2

that year in various branches of mathematics. Some authors summed up the immediate past, and all thought that they were peering ahead towards new ideas. Table 7.7.1 lists these works, with indications of the sections where they are discussed or noted here. There was also a major history of mathematics, by Jean Etienne Montucla (1725–99), completed after his death in four volumes by the astronomer J.J. Lalande.

One of these authors, not mentioned so far, is Sylvestre François Lacroix (1765–1843). He produced no research-level mathematics, but his output of textbooks, especially in pure mathematics, was remarkable – the best after Bézout (§6.15), and one of the greatest of all time. For example, the young Analytics at Cambridge translated his short textbook on the calculus into English in 1816, as part of their effort to oust the fluxions. Lacroix also stands out from his genre by his philosophical sensitivities (perhaps imbibed from his mentor Condorcet) and his deep historical knowledge (for example, he helped Lalande with Montucla). It is fitting to end this chapter with him; for he starts the main story of the next one.

8

Mathematical analysis and geometries, 1800–1860

8.1 *Preliminaries*

Before meeting Lacroix, some features of this and the next two chapters need to be explained. For the most part, Chapters 8, 9 and 10 cover the first 60 years of the 19th century, and deal respectively with the calculus and geometries, algebras and number theory, and mechanics and mathematical physics. This order best reflects the dependency among branches: for example, the calculus is prominent in the next two chapters.

So also are the French, for reasons just described: in all three chapters the mathematicians listed in Table 7.3.1 dominate the scene until the late 1820s. The various young foreigners listed in Table 7.5.1, especially the Germans, then begin to make their appearance. Among their older compatriots, Gauss haunts the pages, often as the distant sage portrayed in §7.4, with a somewhat shadowy presence.

A few books cover substantial amounts of all 19th-century mathematics. Felix Klein's survey {1926, 1927} of the century is still worth reading for its general range, although it is rather slight on French mathematics; and {Kline 1972} contains a great deal of information about results and theorems, mostly in pure mathematics. Among more recent books, {Dieudonné 1978} and {Rowe and McCleary 1989} are compendia which contain some valuable articles.

Apart from an excursion into probability and statistics in §8.8–§8.10, this chapter takes up various aspects of geometries and mathematical analysis in tandem. Cauchy's innovation of the latter, for both real and complex variables, is the single largest topic; the former is largely the work of Monge and Gauss, and certain of their associates. Both branches concerned Lacroix.

8.2 *Lacroix and the varieties of the calculus*

Lacroix exhibited his great educational concerns, vast historical knowledge and philosophical preferences in the three large volumes (1,880 quarto pages) of his *Traité du calcul différentiel et du calcul intégral* (1797, 1798, 1800). At the time they constituted an innovation in mathematical writing: for example, the tables of contents included extensive references to the original texts. In the first two volumes he catalogued (and occasionally commented upon) the state of knowledge of all parts of the calculus and related topics; solutions of differential equations made up the largest single topic, but he also covered differential geometry and the calculus of variations. The third volume was a remarkable vade-mecum of difference mathematics, including methods of interpolation, solutions of difference and difference–differential equations, summation of series, and generating functions.

As a record of methods and results the volumes are surpassed only by the 2,360 updating pages of his second edition (1810, 1814, 1819). One change that was not made was the correction of a mistake in the calculus of variations which occurs in the second volume of both editions. It concerns the variation of the repeated integral

$$\delta \iint F(x, y, z(x, y), z_x, z_y)\, dx\, dy \tag{8.2.1}$$

with respect to the dependent variable z. He forgot that the limits on the y-integral were normally functions of x, which would have to be accommodated in the ensuing procedures. Several of his successors also worked with such integrals: Gauss in 1829 (§8.16), Poisson in 1833 and the latter's compatriot P.E. Sarrus (1798–1861) in 1846 {Todhunter 1861}. These integrals are important when studying area, volume and weight, and constitute an underrated part of this subject.

The *Traité* was the apex of Lacroix's aspirations. Most of his other writings were directed towards an octet of more elementary textbooks; covering arithmetic, algebra, trigonometry, geometry, analytic geometry and the calculus, they constituted a comprehensive *Cours* in the tradition of Bézout (§6.15). Like the *Traité*, they began to appear in the late 1790s, but continued in numerous editions, often with print-runs of 3,000 copies or more, beyond his death in 1843.

The *Traité élémentaire* on the calculus (editions from 1802 onwards) shows Lacroix's encyclopedist attitude very nicely, for over the years he shifted from Lagrange's approach towards limits. By contrast, in his widely read *Réflexions* on the calculus (1797, 1813), Lazare Carnot moved away from limits and towards differentials. These shifts typify the rather confused situation of the time. It was reflected at the Ecole Polytechnique, where a teacher was permitted to offer his own choice – or even mixture of methods – within the curriculum.

Thus Lagrange, as a founder professor of analysis there, told his algebraic story in detail in two high-level textbooks: *Traité des fonctions analytiques* (1797, 1813) and *Leçons sur le calcul des fonctions* (1801 onwards). He treated in more detail his faith in the Taylor series (5.7.3) and the extensions such as his own series (6.5.5), and his approach to the general and singular solutions of differential equations. (The consequences for algebras themselves are recorded in §9.11–§9.12.) By contrast, the other founder professor of analysis was Gaspard de Prony, who, mindful that it was an engineering school, taught the Eulerian calculus and also many numerical techniques. Lagrange was succeeded in 1799 by Lacroix, who used his own books in a melange of methods which was continued by his own successor of 1808, Poisson (a former student). This variety was to be challenged by another student, Cauchy; but he met with considerable opposition (§8.6).

8.3 *Monge: geometry, differential and descriptive*
{CE 3.4–3.5}

Lacroix also produced an influential textbook on 'application of algebra to geometry' (editions from 1797 onwards) for his *Cours*, following a volume on trigonometry. He pioneered writing at this level on both analytic and coordinate geometry, with especial attention to the conic sections. His volume was soon joined by textbooks from colleagues such as Biot, who used 'analytic geometry' in his title (1802 onwards).

Usually such books avoided the calculus; it belonged to a companion topic, which also featured at the Ecole Polytechnique. We call it 'differential geometry'; its principal advocate at the school, Monge, entitled his textbook 'Application of analysis to geometry' (1795 onwards: the source of Lacroix's title above?). He had come to the subject partly via partial differential equations, whose solutions he interpreted geometrically, in opposition to the algebraic regime of Lagrange.

Another name used was 'geometry of curves and surfaces', for the range of functions treated was much greater than the second-degree regime of analytic geometry. Monge and students such as Dupin found many important results, and simplified others. They included the differential equation for minimal surfaces (a form of which had first been found by Lagrange), families of surfaces, and this pretty version of a theorem of Euler: if the principal radii of curvature at a point P on a surface are R_1 and R_2, then the radius R_u through the section making angles u and $90° - u$ with the principal sections is given by

$$\frac{1}{R_u} = \frac{\cos^2 u}{R_1} + \frac{\sin^2 u}{R_2}. \tag{8.3.1}$$

Among other achievements, M.A. Lancret picked up an insight due to his friend Joseph Fourier, and used it to make fundamental contributions to the study of curves in space by studying the rates of change with arc length of the angles between neighbouring normals an tangents; these notions later became known as 'torsion' and 'contingence'. His early death in 1807, his 34th year, deprived Monge's group of one of its finest members.

The work by Monge and his followers in this area was perhaps his most substantial contribution to mathematics teaching and research; but it was not his main concern at the school. Influential in its founding, he larded the curriculum with masses of his beloved 'descriptive geometry'. The name is his, referring to the analysis of an object or shape in space by examining the projections of it onto planes; it extended the study of projective properties by figures such as Desargues (§4.20) which by then had been largely forgotten. He had developed his ideas initially in the 1780s (his thirties), while teaching at the military school at Mézières, and applied them to the design of cannons. However, this made it Secret Mathematics, and he was not able to proclaim the details until he began teaching at the the Ecole Polytechnique. His textbook *Géométrie descriptive* appeared from 1795 onwards, in editions by his followers, of whom the most ardent was Hachette.

In Monge's scheme, the plane was divided into four quadrants by two lines intersecting at right angles. Following the builders' tradition of double projection (§3.22), he projected a spatial body into two perpendicular half-planes, which were then folded out into one plane; properties of these two planar projections could then be studied (adjusted, maybe, and the modified body reconstituted). This sounds good sensible stuff, and when applied, for example, to stone-cutting or surveying, it enriched traditional engineering practices (especially at the Ecole Polytechnique) with great effect. It soon found an unexpected but perfect application in Bonaparte's expedition to Egypt in 1798 (§2.2). Several graduates of the Ecole went, and were able to make such good drawings of the carvings on the temples that their contributions to the vast ensuing report, the *Description de l'Egypte* (1809–30), have been of permanent historical value.

Unfortunately, Monge wanted to elevate his subject to something like a branch of mathematics; so he filled his book with posh pure theorems about, say, the number of different spheres that can touch a given quartet of ellipsoids. But such questions cannot give the subject the weight that he hoped it would; at the same time, they contaminated the practical aims upon which the subject had been nourished. Thus, while versions of the subject were taught and practised in many countries {Booker 1979}, it became marginalized by the mathematical community, for good reasons: in particular, Laplace had its place in the Ecole Polytechnique's curriculum considerably reduced. The best fruits were some nice applications made by followers such as Dupin, and the individuation of projective properties by Poncelet.

8.4 *Poncelet: geometry, projective and dual* {EMW IIIAB4a; CE 7.6}

In Monge's theory, figures were projected onto planes by collections of parallel lines. His former student at the Ecole Polytechnique, Poncelet, generalized this process by projecting along pencils of intersecting lines – thereby recovering essential features of Desargues's work. Although by profession a military officer and by occupation a principle figure in engineering mechanics (§10.9), Poncelet devoted much of his time to this topic from the 1810s, publishing at his own expense a *Traité des propriétés projectives des figures* in 1822. Gergonne also came to these considerations around the same time, so of course there was a priority dispute, largely on Poncelet's part.

Poncelet's theory concentrated upon lines and conic sections in the plane, expressed in a rich way in his book as dual theorems printed together as double columns on the page; for example, a result about points, lines and planes had its dual in a similar result about planes, lines and points. In such ways the logical relationships between these algebraical objects were stressed more than the objects themselves {Nagel 1939}. Several theorems related poles of a conic to their polar lines. FIG. 8.4.1 shows one of his most beautiful (and also characteristic) results, now known as his 'closure theorem'. If a polygon can be inscribed as shown between two conics C and D, then another one can be constructed starting from *any* point P on C, and in the same number of steps {Bos and others 1987}.

One bold feature of Poncelet's theory concerned points of intersection of lines with lines and with conics. For the former case he reinvented Kepler's point of infinity (§4.12) in order to assume that intersection always happened. He was influenced by Lazare Carnot's ideas on continuity in geometry, especially as expressed in *Géométrie de position* (1803). Carnot had considered properties such as this: in an acute-angled triangle ABC, C lay between A and B, but as ∠ABC gradually – that is, continually – passed to and

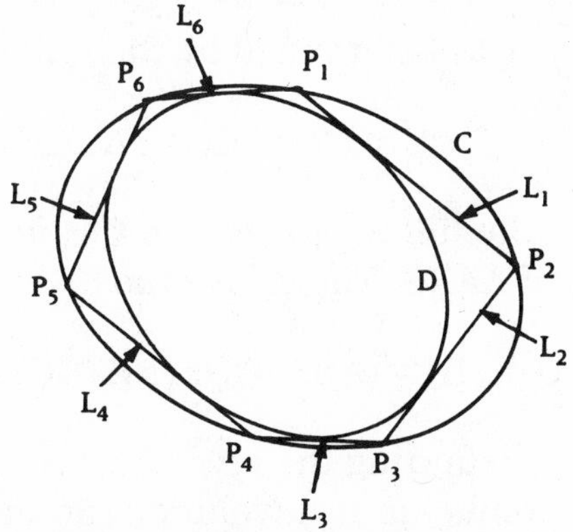

FIG. 8.4.1 *Poncelet's closure theorem, 1822.* {Bos and others 1987, p. 290}

then beyond 90°, C suddenly fell outside AB. Poncelet elevated this insight into a *general* 'principle of continuity' (his phrase), according to which properties would normally be maintained under gradual change.

This proposal soon met a bad reception from Cauchy, on the general grounds that geometry could not provide a grounding for continuity (§8.5); but even among enthusiasts for projective geometry after 1830 there was some unease {E. Kötter 1898}. One group, including Chasles (who also wrote extensively on the history of geometry), Steiner and von Staudt, called themselves 'synthetic geometers' because they extended notions such as cross-ratio connected with four collinear points in involution on a conic (FIG. 4.20.1) to cases where the line did not cross the conic. The other approach, favoured by Möbius and Plücker among others, brought up the geometry/algebra question again by deploying complex numbers to discuss such cases {EMW IIIC1–2}.

Möbius also highlighted cross-ratio in his book *Der barycentrische Calcul* (1827). The word 'barycentric' relates to the centre of mass of a system of masses, and the way in which its position depends upon the comparative values of the given masses. FIG. 8.4.2 has two readings here, both of which he saw. Firstly, look only at C inside ∠AFG: its position as the centre of gravity of masses *a*, *f* and *g* at A, F and

G, respectively, was determined by the ratios

$$a:f:g = \Delta CFG : \Delta CGA : \Delta CAF. \qquad (8.4.1)$$

Secondly, take the full Figure, where C is now an arbitrary interior point of ΔAFG. The cross-ratio

$$(A, G, B, H) := (AB/BG) \div (AH/HG) = -1 \qquad (8.4.2)$$

(the minus sign reflecting the reversed direction of HG), so that the four points are in involution. So are A, E, D and F; hence the ratio served as an invariant under projective transformations, like ordinary distances in Euclidean geometry.

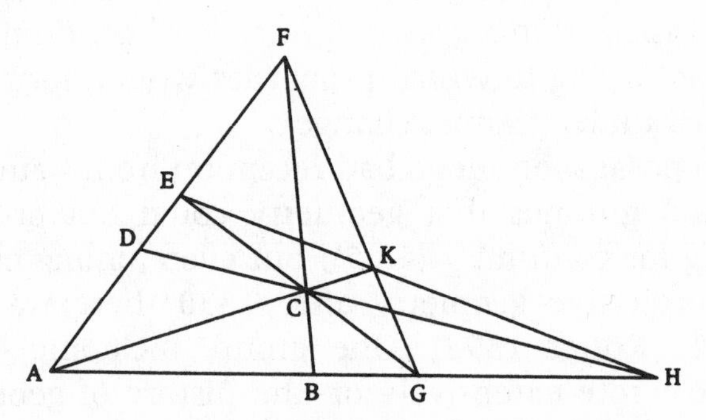

FIG. 8.4.2 *Möbius's barycentric coordinates and cross-ratio, 1827.*

Steiner followed the philosophy of his compatriot mentor, the Swiss educational reformer J.H. Pestalozzi, on the need to bolster observation with conscious reflection; he was remembered for lecturing in a darkened room so that the students had to visualize projections in their mind's eye. Thus his work relied upon intuition and mental conceptions more than on formal proofs. For example, in 1826 he stated without proof that, given three circles M_r, three more could be found, m_r, touching as shown in FIG. 8.4.3. If the M_r are of infinite radius, the theorem becomes that of his Italian contemporary Giovanni Malfatti (1802), about inscribing three circles in a triangle.

Many such theorems about curves in contact came out of Steiner's meditations; if any proof acompanied them, they were often based upon methods of inverting figures, of which he was a pioneer. In this way some of Monge's theoretical ambitions for descriptive geometry were fulfilled. But he and, especially, Plücker went beyond French concern with conic sections to consider cubic and quartic

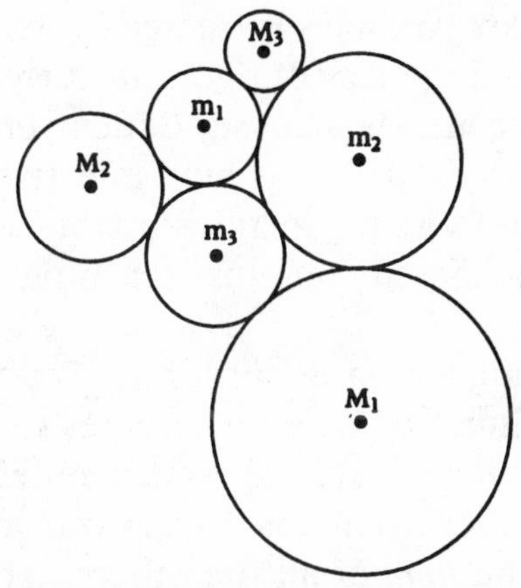

FIG. 8.4.3 *Steiner's theorem on circles in contact, 1826.*

curves {EMW IIIC4}. This helped to individuate algebraic geometry and also to lead to the line geometry of Plücker's distinguished student Felix Klein (§13.9), who also described the history of the invasion of this French topic in the 1820s by German-speakers, all then in their thirties {C.F. Klein 1926, Chaps. 3–4}. They published many of their papers in the early volumes of Crelle's journal, and some in Gergonne's (§7.6).

8.5 *Cauchy's 'mathematical analysis': the* Cours d'analyse, *1821* {Bottazzini 1986; CE 3.3}

Cauchy's criticism of Poncelet's geometrical conception of continuity was typical of his style: 'mathematical analysis' (his term), referring to a general doctrine of limits, based on an articulated *theory* and not just thinking. He wished to avoid both the looseness of geometrical thought (of which Poncelet's principle was to him a good example) and the inductions and mysterious operator objects of Lagrange's approach. However, in two respects he was in accord with Lagrange: no diagrams, and usually synthetic (sic) proofs starting out from assumptions and definitions. This and the next section discusses his innovations in real variables; his theory of complex variables is treated in §8.11.

Dissatisfied with the lack of rigour in current conventions in both real and complex analysis, Cauchy began to develop his ideas in the mid-1810s, his own mid-twenties. The first main presentation was made in his textbook *Cours d'analyse* {Cauchy 1821}, based on his teaching at the Ecole Polytechnique.

In seeking a higher status for limits, Cauchy began with this general definition: 'When the values successively attributed to the same variable approach indefinitely a fixed value, so as to differ from it as little as one might wish, this latter is called the *limit* of all the others.' This formulation covered passage both over integers ($n \to \infty$), and over a continuum of values ($x \to a$) by zigzagging to and from both sides as well as solely from below (↑) or solely from above (↓). I call this approach 'limit-avoiding'.

An important application of limits was to replace geometrical intuition about continuity, and also Euler's specification of a continuous function in terms of one algebraic expression (§5.27), by this *definition*: for Cauchy a function $f(x)$ 'will remain continuous with respect to the given limits, if, between these limits, an infinitely small increase of the variable always produces an infinitely small increase of the function itself'. He also re-expressed it for continuity 'in the vicinity of a particular value of the variable x', and proved {Chap. 2} various theorems for continuous functions, of both one and several variables.

Another major innovation concerned the convergence of the infinite series $\sum_j u_j$, defined by the property that the nth remainder term lim $r_n \to 0$ as n increased indefinitely; the

nth partial sum s_n of the series approached the sum s ({Chap. 6}; Cauchy popularized the use of these notations, and also 'lim'). The exegesis of convergence included the first batch of tests for the convergence of infinite series (ratio, root, condensation; and in 1826 the integral test); indeed, they constituted some of the first responses to Cauchy's book by other mathematicians, especially outside France {Grattan-Guinness 1970b, Appendix}. Cauchy himself held that non-convergent series should be avoided.

One of Cauchy's main applications was to the binomial series in the form

$$f(\mu, x) := \sum_{j=0}^{\infty} {}_\mu C_j x^j, \quad \mu \text{ real, } x \text{ real or complex;} \qquad (8.5.1)$$

he proved that it converged when $|x| < 1$ and was continuous in x, and to sum it he showed that it satisfied the functional equation

$$f(\mu + v, x) = f(\mu, x) f(v, x), \qquad (8.5.2)$$
$$\text{with solution } f(\mu, x) = (f(1, x))^\mu = (1 + x)^\mu.$$

This was a showpiece; not only did he deploy his new views on continuity and convergence, but he also avoided using the calculus. The growing place of functional equations at this time is described in §9.11.

Among other theorems proved by Cauchy was:

THEOREM 8.5.1 *The sum function of an infinite series of continuous functions of x over a finite range of values of x is itself continuous.*

Doubt was soon cast upon the proof by Abel, in a paper of 1826 in Crelle's journal translated into German by the editor. He studied convergence of eqn (8.5.2) for real and complex values of μ as well as of x (hence 'Abel's theorem' on the convergence of power series comes from this paper). For him the trouble with this theorem was that Fourier series (8.7.1) were composed of such functions over a finite range, but that the sum could be discontinuous. While some aspects of Cauchy's definitions can vindicate his proof

{Laugwitz 1989}, his analysis was not fully in command of *two or more variables varying together*. This was to be a major talking-point among the Weierstrassians (§12.3), with regard not only to functions but also to the calculus – for example, the integration of an infinite series of functions term by term. To this branch we now turn.

8.6 *Cauchy's* Résumé *of the calculus, 1823*

Cauchy's *Résumé des leçons données à l'Ecole Royale Polytechnique sur le calcul infinitésimal* (1823) contained 20 lectures on the differential calculus followed by 20 on the integral calculus. Moreover, each lecture was printed on *exactly* four pages – the theory of limits applied to printing, doubtless intentionally.

Cauchy's principal innovation here has become so normal that he is not usually given credit for achieving it. He took over Lagrange's words and notation for the derivative, but he defined both derivative and indefinite integral of a function *independently*; thus the fundamental theorem of the calculus became a genuine *theorem*, requiring some sufficient conditions of the function, instead of the (more or less) automatic reversal that we saw with Leibniz, Newton and Lagrange. His formulation runs thus:

$$f'(x) := \lim\,[(f(x+h)-f(x))/h] \quad \text{as } h \to 0; \tag{8.6.1}$$

$$\int f(x)\,\mathrm{d}x := \lim \textstyle\sum_r\,[(x_r - x_{r-1})f(x_{r-1})], \quad \text{with } x_0 \leq x \leq X, \tag{8.6.2}$$

$$\text{as } \max_r |x_r - x_{r-1}| \to 0;$$

THEOREM 8.6.1 *If f is continuous, then*

$$\frac{\mathrm{d}}{\mathrm{d}x}\int f(x)\,\mathrm{d}x = f(x). \tag{8.6.3}$$

He also allowed that the limits in (8.6.1) and (8.6.2) might not exist, and used limit theory to extend the definition of the integral to infinite intervals and to infinite-valued functions; but he forgot to state the other half of the theorem, that

$$\int f'(x)\,dx = f(x) + K, \text{ where } K \text{ is a constant.} \tag{8.6.4}$$

The proof of the theorem was effected via the first mean-value theorem of the integral calculus. It dates from the early days of the branch, and Cauchy added an excellent extension: if two functions f and F and their derivatives were continuous and F monotonic over $[x_0, X]$, then

$$\frac{f(X) - f(x_0)}{F(X) - F(x_0)} = \frac{f'[x_0 + \theta(X - x_0)]}{F'[x_0 + \theta(X - x_0)]}, \text{ where } 0 < \theta < 1. \tag{8.6.5}$$

An important result in the book was Cauchy's counter-examples (published first in 1820), such as $\exp(-1/x^2)$ around $x = 0$, to Lagrange's belief (6.5.4) that every function took a convergent Taylor expansion. Thus the series was relegated from principle to theorem, where its convergence had to be proved. By repeated integration by parts he found an integral expression for the remainder R_n after n terms, which a mean-value theorem converted to a differential expression; so instead of Lagrange's (6.5.3), he had

$$f(x + h) = \sum_{r=0}^{n-1} \frac{h^r f^{(r)}(x)}{r!} + \frac{h^n f^{(n)}(x + \theta h)}{n!}, \quad \text{where } 0 < \theta < 1. \tag{8.6.6}$$

While this was not a completely original result, his grasp of the issue was superior to his predecessors' in precisely the ways he advocated. He also initiated a new way of defining the differential $df(x)$ {Taylor 1974}.

How were these major innovations in mathematics received at the Ecole Polytechnique? Both staff and students at this military engineering school *hated* them, and tried to have them stopped, or at least modified. Matters came to a head in March 1821, just before the *Cours* appeared, when the students walked out of the 65th of his prescribed 50 lectures on analysis. They were confined to barracks for their action, but the ensuing correspondence between the directorate and the War Ministry shows that they regarded Cauchy's kind of highbrow meditation as quite inappropriate for the aims of this engineering school {Grattan-Guinness 1990a, Chap. 20}. Nevertheless, he continued with only one rebuff, which is noted in §8.15.

Social factors seem to have been involved. While not a political animal, the Professor was a fanatic for the Catholic Bourbon cause at a time of great social tension in France, especially after the assassination of the heir to the crown (the Duc de Berry) in 1820; and the Ecole Polytechnique had been created during the Revolutionary period. Presumably the government was not anxious to take sides against its prominent fan. The only thing to stop Cauchy was the Revolution of July 1830, when the Bourbons fell and went into exile, accompanied by ex-Professor Cauchy as tutor in mathematics to the pretender to the throne.

8.7 *Fourier's Fourier analysis*

{Grattan-Guinness 1970b, Chap. 5; EMW IIA12}

In addition to educational tribulations, Cauchy faced mathematical problems with his analysis, especially the one to which Abel had pointed: Fourier series. We take here only the 'pure' aspects of Fourier's innovation, which was made in the mid-1800s; his context was the mathematical analysis of heat diffusion, which is considered in §10.11.

The series arose from solving the partial differential equation for heat diffusion by separating the variables and inserting the initial heat distribution $f(x)$ for the body in question; this led to equations such as

$$f(x) = \sum_{r=0}^{\infty} a_r \cos rx + b_r \sin rx, \quad 0 \leq x \leq 2\pi. \qquad (8.7.1)$$

(The use here of 'Σ' is faithful to Fourier, for he was a pioneer in its deployment.) The most obvious question is the determination of the coefficients, which he duly effected by multiplying through (8.7.1) by $\cos rx$ and integrating over $[0, 2\pi]$ to obtain the integral formulae for a_r:

$$2\pi a_0 = \int_0^{2\pi} f(x)\,\mathrm{d}x \quad \text{for } r \geq 1. \qquad (8.7.2)$$

$$\text{and} \quad \pi a_r \text{ or } \pi b_r = \int_0^{2\pi} f(x) {}^{\cos}_{\sin}\, rx\,\mathrm{d}x \quad \text{for } r \geq 1.$$

But deeper issues were involved. Euler had in fact found these formulae in a paper which Fourier had not seen; but he had not followed on with the consequences which Fourier was to initiate.

In my view,[1] Fourier's principal insight concerning (8.7.1) centred on '='. One of the reservations about Daniel Bernoulli's advocacy (6.4.3) of trigonometric series was that, while the $f(x)$ on the left-hand side of (8.7.1) was general, the series on the right was periodic; but Fourier realized that the specification in (8.7.1) of a range of values for x answered the point, for periodicity together with oddness *or* evenness of the sine *or* cosine series determined the manner of representation over the rest of the x-axis – in general different from $f(x)$, *and also different from each other*. He gave examples, with diagrams, of full, cosine and sine series for the function $\frac{1}{2}x$ over $[0, \pi]$ (FIG. 8.7.1).

Nevertheless, Lagrange remained unconvinced – dare one guess, uncomprehending. One difficulty for him was that f, as a function present in the initial conditions of diffusion, was not shown *explicitly* in the series, unlike the functions in the preferred functional solution (6.4.2) of differential equations.

Among the many properties that Fourier found, one stands out for its different readings over the decades: Parseval's theorem. M.A. Parseval (1755–1836) was a talented

[1] My interpretation of the significance of Fourier's understanding of the periodicity of the series is not held by other historians who have studied the matter; for example, it is not mentioned in the extensive study {Yushkevich 1976a} of the development of the notion of function. However, I remain unreformed.

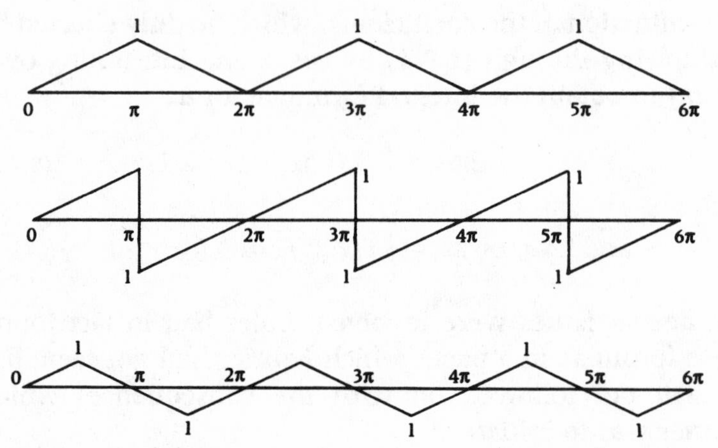

FIG. 8.7.1 *Fourier's three series for the function $\frac{1}{2}x$, 1807.* {Grattan-Guinness 1990a, p. 599}

eccentric contemporary who published this way of summing power series in 1806:

THEOREM 8.7.1 *If*

$$f(x)=\textstyle\sum_{r=0}^{\infty} a_r x^r \quad \text{and} \quad g(x)=\sum_{r=0}^{\infty} b_r x^{-r}, \tag{8.7.3}$$

then

$$\textstyle\sum_{r=0}^{\infty} a_r b_r = \dfrac{1}{2\pi}\displaystyle\int_0^{\pi} [f(e^{iu})g(e^{iu}) + f(e^{-iu})g(e^{-iu})]\, du. \tag{8.7.4}$$

He also offered the version to which we are accustomed, using sines and cosines instead of exponentials, but again as a summation. Fourier also found this version for himself, apparently independently, regarding it as one among many properties of the series. Neither man perceived the deeper consequences of the relation that were to accompany and aid the birth of functional analysis (§16.12), although Fourier also stumbled upon the completeness aspects of the functions.

For reasons to be described in §10.11, Fourier did not get his work into print until the early 1820s – by which time his heresies had become orthodoxies. So when the young

Dirichlet arrived in Paris in the mid-1820s, he found both man and work congenial and significant, and placed with Crelle in 1829 a classic paper on the convergence of the series. While Fourier had possessed a clear understanding of convergence, he had tried to prove it for 'any' function, as had his rivals Poisson and (strangely) Cauchy. By contrast, Dirichlet absorbed Cauchy-style rigour and sought sufficient conditions of f for convergence to be established, and he found them in 1829 as:

> THEOREM 8.7.2 *If $f(x)$ has a finite number of turning-points and discontinuities over $[0, 2\pi]$ of x, then its Fourier series converges to $f(x)$, or to the average of its values at points x of discontinuity.*

In a remarkable act of prophecy he also gave an example of a function lying outside these conditions, 'the characteristic function of the rationals' (as it is now called): $f(x) := A$ *or* $B \neq A$ when x is rational *or* irrational. He clearly thought it to be pathological; for example, it could not have an integral (compare §16.9 on measure theory).

In a revision of 1837 Dirichlet introduced the notations '$f(x \pm 0)$' for the right- and left-hand values of f at a point of discontinuity, and also extended the conditions to cover a finite number of values of x where $f(x) \to \pm\infty$. The quest for still weaker sufficient conditions would lead to new ideas from various mathematicians; seeking necessary ones stimulated Riemann (§12.1).

Fourier's other major mathematical innovation came in 1811, with some help from Laplace. All his series solutions were for finite bodies of various shapes; diffusion in infinite bodies needed a different form. Laplace's clue, found in 1809, was the following integral solution to the diffusion equation in temperature $z(x, t)$ over an *infinite* interval of x:

$$\sqrt{\pi}z = \int_0^\infty f(x + 2s\sqrt{t}) \exp(-s^2)\, ds, \text{ where } f(x) := z(x, 0). \tag{8.7.5}$$

This is not the so-called 'Laplace transform' (16.13.3), but was important in its own right. In particular, it stimulated Fourier to trick-cycle his way through infinitesimals and find forms such as

$$2\pi f(x) = \int_{-\infty}^{\infty} f(u)\,\mathrm{d}u \int_{-\infty}^{\infty} \cos[q(x-u)]\,\mathrm{d}q \qquad (8.7.6)$$

(I follow his way of writing the double integral), with companion double-integral solutions of the diffusion equation. With the initial function f shown explicitly, neither it nor the associated solution caused controversy; the 'Fourier integral transform' (the modern name) soon became a major tool in the new mathematical analysis, especially in the hands of Cauchy (§8.11).

8.8 *Laplace on the state of 'probabilities'*

Laplace had two reasons of his own for taking up Fourier integrals. One related to heat theory itself (§10.11); the other to probability and statistics, which had risen high on his agenda since his early forays (§6.20). He encapsulated his knowledge in a *Traité analytique des probabilités* (1812, 1814, 1820). The second edition was prefaced by a chatty *Essai philosophique* on the subject(s); it was published separately in editions to 1825 and has reappeared from time to time ever since, including recently a fine one by B. Bru, and one in English by A.I. Dale. The *Essai* was more widely read, and is remembered for his advocacy of determinism. The *Traité* was written in his more usual opaque manner; in it he used generating functions and Bayesian means of evaluating posterior probabilities, and presented many results oriented towards large-number theory of various kinds.

Among the latter the most noteworthy result is what is today called 'the central limit theorem'. This name was introduced in the 20th century by Georg Polya, the 'central' having the sense of 'central importance'; Laplace gave it no name, but he was certainly aware of that importance. Taking a continuous distribution function $f(x)$ of errors in independent observations within some range of values of x and with mean X, he found an integral incorporating the function $\exp(-x^2)$ which approximated to the probability that the mean of a large number n of these observations lay only of the order of $1/\sqrt{n}$ distant from X. The thrust of this amazing result lay in the fact that f had 'disappeared', allowing the mean of the observations to be appraised from the integral alone. Hence the normal distribution based upon $\int \exp(-x^2)\,dx$ gained in significance. Laplace's use of $\exp(-x^2)$, evident also in his concurrent integral (8.7.6) for heat diffusion, highlighted a line of thought born in de Moivre's analysis of large numbers of trials (§5.22) nearly a century earlier, and his own first work in the 1770s on 'probabilities' (§6.20).

8.9 *Much ado about 'least squares'*

{Eisenhart 1964; CE 10.4}

This is a well-known story, partly for a fracas after the method's launch, but mainly because of the great importance that was assigned to the method. It was proposed originally by Legendre in 1805, in an appendix to a short book on the motion of comets. He introduced the name, but not the probabilistic interpretations that were later to be attached to it, for his technique was purely a criterion of optimization. He determined the 'best' line that would fit a collection of data by minimizing the sum of the squares of their distances from it when they were plotted as points in the plane. Imitating the presentation of Boscovich's absolute deviation in (6.20.2), I write the equation of the line as

$$c(x, y) = 0, \text{ where } c(x, y) := y - Ax - B; \text{ and put } c_s := c(x_s, y_s). \tag{8.9.1}$$

Then A and B were determined from the requirements that

$$\textstyle\sum_s c_s = 0 \quad \text{and} \quad \sum_s (c_s - c)^2 \text{ be minimal.} \tag{8.9.2}$$

The usual first-order optimal conditions of the calculus were then applied to the second condition.

Probability theory came into the story when Gauss published the method in his book *Theoria motus* (1809) on the motion of heavenly bodies (§10.5), for he adjoined an assumed normal distribution of the errors of observations. (He had known the method himself since about 1795.) Soon afterwards Laplace derived it in various probabilistic ways in his *Traité* and elsewhere, including a proof that the optimal values could also be obtained from the maximum likelihood function (§6.19) of the data. Gauss also claimed priority for it over Legendre, and Laplace was not exactly a friend anyway; so the chattering started.

The method spread quite rapidly. Astronomers were very appreciative, especially after much cajoling by Gauss's former student J.F. Encke from the late 1820s. But it remained controversial, on issues such as generality (where figures such as Fourier made great claims); on the error from true or from presumed mean values (the case of residuals); on the error relative to maximum likelihood or just by the least-square property itself; and on the association with nature or with belief {Knobloch 1992}. Curiously, objections of the kind that Boscovich had already raised were not often voiced: namely, that the squares of outlying data contributed disproportionately much to the sum (for example, if the mean is 9, then datum 13 contributes 16 but datum 22 lends 169). The mathematical advantage of least squares – that its algebra is much easier to handle than that of absolute values – was crucial in its rise to prominence.

8.10 *Judgements on probability theory*

With Laplace's *Traité* serving as a statement of position, and least squares gaining in popularity, probability and (mathematical) statistics became more evident in science and society. Organizations were founded in various countries to collect data and maybe theorize on methods for their analysis. An example was the Statistical Society of London, in 1834; one of its founders was Benjamin Gompertz, who in

1825 had suggested the law that the tendency of humans towards mortality increased with age at some exponential rate. Such initiatives not only gave a greater presence to statistics at this time, but also relaunched the word, now with the meaning of the numerical analysis of society; it had been introduced in the late 17th century to refer to the description of states (by 'statists'), especially in Germany where the many small *Staaten* were amenable to such assessment {Hacking 1975, Chap. 2}.

One of the most prominent figures was the Belgian Adolphe Quetelet (1796–1874), astronomer and meteorologist by profession but social statistician by inclination {CE 10.8}. In the new spirit of the rights of man in an industrializing society, he replaced Condorcet's enlightened intellectual aristocrat (§6.17) with the 'average man' (*homme moyen*), an artefact around which actual people and their attributes were distributed, for him like the centre of mass of a body, or observations (with their inevitable errors) of a star with respect to its true position. His new subject of 'social physics'[2] treated laws of error around this desideratum, and also laws of the large-number type about the (near-)constant values in a society (such as its annual tally of deaths or marriages).

In Paris, Laplace's faithful follower in all things, Poisson, began to work in statistics from the mid-1820s (his mid-forties). His contributions included the distribution now known after him (and also the one named after Rayleigh, mentioned in §1.1). He also announced a general but rather vague law, to the effect that as one observed an ever greater number of events of two types (for example, the births of boys and girls), the ratios between them approached a fixed limiting value. Successors would sharpen this result into a law of large numbers (§12.11).

[2] Quetelet's Parisian contemporary Auguste Comte (1798–1857) also used the phrase 'social physics', which he changed to 'sociology'. He accused Quetelet of plagiarism, and was in general hostile to social statistics despite the fact that it had attracted Fourier, who exercised much influence on his philosophy in general (§10.11).

In contrast to the democratic flavour of Quetelet's approach, Poisson was conservative in his choice of preferred applications: his own treatise, published in 1837 (his 56th year), carried the title *Recherches sur la probabilité en matière criminelle et en matière civile*. But just when he was bringing his Laplacian approach to a climax by sorting out things for the judges, his *confrères* in the French Academy suddenly broke with it. He presented the preamble of this book there in December 1835 and in the following April, and on this second occasion Poinsot and Dupin were moved to condemn the enterprise in principle {Daston 1988, Chap. 6}. In Poinsot's case one can imagine that dislike of Poisson (§10.3) played a role, but there were also substantive issues, reflecting a change in the sensibilities of the time (but why then and not earlier?) by which this kind of applied mathematics was now regarded as illegitimate. From then on (but not necessarily because of this exchange), we see probability and mathematical statistics focusing much more upon its other provinces, such as the characteristics of the average man, or heaps of error-strewn scientific data. The distinction between subjective and objective interpretations of probability became more explicit, with the latter gaining somewhat greater favour.

On the technical level, it is very surprising that, despite the popularity of least-squares methods, most emphasis in this period was laid upon measures of location such as mean and median (and the 'average man'), and that little attention was paid to dispersional parameters such as standard deviation and quartiles. They came through only much later in the century, as we shall note in §16.21.

8.11 *The slow rise of Cauchy's complex analysis*

{Bottazzini 1986; CE 3.12}

From time to time, the mathematical statisticians came across integrals with complex integrands. For example, in his *Traité* Laplace converted a generating function into a Fourier-type integral of this kind; he also used characteristic functions, which use series in e^{ix}. Often these integrals could not be evaluated, but ruses such as using Euler's equation (5.17.4) to replace $\cos x$ or $\sin x$ by e^{ix}, and taking the real or imaginary parts of the complex evaluation, could yield answers.[3] These results were usually correct, but conceptually the methods were mysterious.

This was the situation which the 25-year-old Cauchy addressed in 1814, in a long paper which launched one of the greatest innovations in the entire history of mathematics: a comprehensive theory of complex-variable integrals. He proposed a condition for appraising but also extending current practices which I express as follows:

$$\text{let } I(x,y) := \int f(z)\,dz, \text{ where } z := g(x,y), \text{ and set } P := I_{xy} - I_{yx} \tag{8.11.1}$$

(z real or complex, x and y real); then if $P = 0$, the known methods of evaluation are rendered valid. He re-evaluated some recent integrations due to Legendre, and a few by Laplace involving $\exp(-x^2)$ (§8.8), by integrating $f(z)$ over appropriate specified intervals of integration of x and y. However, in the 'singular' event (his adjective) that $f(z) \to \pm\infty$ or took no determinate value for certain values Z of z, then maybe $P \neq 0$; and he evaluated the converse repeated integrals as an integral over values of x and y close to Z. For example (one of his),

$$\text{given } f(x,y) = \frac{y}{x^2 + y^2} \text{ with } Z = 0 + 0i, \quad \text{then } \int_0^1 \int_0^1 P\,dx\,dy = -\pi/2. \tag{8.11.2}$$

From these seeds Cauchy launched complex-variable analysis. At this early stage the limits on integrals were real; the poles of these functions were always simple; and conditions were presented after equating the real and imaginary parts of complex expressions. He did not actually write

[3] Often these functions were discontinuous relative to one of their parameters, a feature which encouraged Cauchy to think about (dis)continuity when working out his *Cours d'analyse* (§8.5).

down the best known of these, the 'Cauchy–Riemann equations' (as we now call them):

$$\text{if } f(x+\mathrm{i}y) = u(x,y) + \mathrm{i}v(x,y), \text{ then } u_x = v_y \text{ and } v_x = -u_y. \tag{8.11.3}$$

In addition, there was no talk of the complex plane, or of contours within it; on the contrary, he was already on the track of his no-diagrams mathematical analysis. Indeed, in the *Cours d'analyse* of 1821 all the basic real-variable definitions were to be imitated for complex variables.

A major advance in evaluations came in 1817, when Cauchy discovered the Fourier integral theorem (8.7.6) for himself, and used it with complex exponentials as well as with sines and cosines in the integrand. The next principal step of his general theory appeared in a short book of 1825, where he defined the complex integral as the limit of a sum like (8.6.2), but also allowed the limits to be complex. His main theorem was that the value between these limits did not change if the function was always continuous over the sequence of intermediate values involved (he tacitly assumed that the derivative $f'(x)$ was continuous also); however, across a point z_0 of singularity with residue f_0, $\pm\pi \mathrm{i} f_0$ had to be taken into account, and $\pm 2\pi \mathrm{i} f_0$ when including it. (It is remarkable that he allowed *both* cases here; normally the first is excluded.) He now had an expression for the 'residue' (his word, from 1826) of a multiple pole (never his word); and he still tried to remain free from geometrical considerations. By 1830 he had the idea of an integral over a contour C, but with the incomprehensible notation '$\int_0^{C}$', and still without invoking the complex plane.

Only in a paper of 1846, after 30 years, did Cauchy grandly announce that 'nothing prevents us' from admitting geometry and setting up integral theory in terms of the residues from poles within or on a contour. The canonical result of this kind states that

$$\int_C f(z)\,\mathrm{d}z = \sum_u 2\pi \mathrm{i} f_0 \tag{8.11.4}$$

for residues of poles u of f lying within C.

A subtle but notable advance from 1825 is involved: between obtaining the same value for an integral by taking two routes between complex values z_0 and z_1, and evaluating as zero the integral along a contour from z_0 to z_1 and back again. The difference concerns the need for *topological thinking* about the complex plane, in particular whether poles lie inside, on or outside a given contour. Such thinking was to become theory with Riemann and his successors (§13.12).

Another reason for Cauchy's change of style seems to have been the rise in prominence of power-series expansions of $f(z)$. He had made advances of this kind around 1830, with a 'calculus of limits', and in the 1840s used eqns (8.11.3) to specify the continuity of a function. (The word 'analytic' came then to be used in this context.) In addition, in 1843 his compatriot P.A. Laurent had submitted to the French Academy a theorem which is still known after him (although he never had his paper published). It established a result similar in form to Laplace's generating functions (§6.20) in real variables, and of comparable importance in handling functions which cannot take a Taylor expansion. This is a somewhat developed version:

THEOREM 8.11.1 *If $f(z)$ be continuous within given limits of $|z|$, then*

$$f(z) = \sum_{r=-\infty}^{\infty} a_r z^r, \qquad (8.11.5)$$

with the a_r specified by certain known contour integrals.

Cauchy reported on the paper and found many corollaries, such as the circle of convergence about a given point z_0 extending up to the nearest singularity of f: a result which unifies power series with the complex plane.

Although he never wrote a comprehensive single account, Cauchy now had a pretty general theory at last – and the only one, for alternative approaches from Weierstrass and Riemann were still to come. It was given an influential treatment by C.A. Briot and J.C. Bouquet in their *Théorie des fonctions doublement périodiques* (1859), which drew also upon lectures given by Joseph Liouville.

Their title referred to a principal application of complex analysis, to which we now turn.

8.12 *Foreign analysis: elliptic functions*
{Houzel 1978; EMW IIB3; CE 4.5}

Just as the endeavours of Legendre and others with complex integrands were eclipsed by young Cauchy's integration (8.11.1) of complex functions in 1814, another of Legendre's concerns was overshadowed a dozen years later by two more youngsters, but this time Norwegian and German. Abel and Jacobi came to essentially the same insight around 1827: to replace elliptic integrals, Legendre's lonely preoccupation for 40 years, with their inverses – elliptic functions (as Jacobi was definitively to name them). As so often, Gauss knew a thing or two, but he published only a few remarks about the lemniscate (8.13.1) below.

Between 1825 and 1828, when in his seventies, Legendre published his *magnum opus* on his integrals, the three-volumed *Traité des fonctions elliptiques*. He started from the basic types of integral which he had found (including beta and gamma functions, eqns (5.18.2) and (5.18.3)), and ran through many transformations and other relationships, and also tables of values. His 'simplest' form was to be expressed by Abel in the form

$$y := \int_0^x \frac{1}{\sqrt{[(1-c^2u^2)(1+b^2u^2)]}}\,du, \qquad (8.12.1)$$

with c (for which he wrote 'e') and b real; but Abel saw that theory would progress much more fruitfully if $x=f(y)$ (the converse of Legendre's manner), and also if y could take complex as well as real values. In addition, he defined two further functions of y, corresponding to Legendre's second and third kinds of elliptic integral.

Soon afterwards, and seemingly independently, Jacobi came to essentially the same insight. He took 1 and $-k^2$ for the values of the constants in his version of (8.12.1), and wrote the inversion relationship as

$$\sin^{-1}x = \mathrm{am}\, y, \quad \text{so that } x = \sin \mathrm{am}\, y \tag{8.12.2}$$

(where 'am' abbreviated 'amplitude'); he also defined two further functions. The sine function was used because of an analogy between (8.12.1) and the integral

$$x = \int_0^{\sin x} \frac{1}{\sqrt{(1 - u^2)}} du. \tag{8.12.3}$$

The emphases of the two men differed somewhat: Abel studied kinds of function, while Jacobi examined transformations and the effect on c/ib (which became known as the 'modulus' of the function). But both men reworked in similar ways the properties that Legendre had found, and added new ones: for example, points of discontinuity, expansions in infinite series and products, and addition theorems (that is, the analogues of equations such as those for $\sin(A \pm B)$ in trigonometry). Of fundamental importance was the double periodicity in y, as real multiples and/or complex sums of $f^{-1}(1/c)$ and $f^{-1}(1/ib)$. Complex values for y, on which Legendre and his contemporaries had somewhat floundered, were now seen to hold the key to the theory, especially periods in complex ratio. Jacobi was helped in his understanding of their properties when he introduced 'theta functions', his name for certain infinite-series expansions in complex exponentials; he even came to redefine elliptic functions as quotients of products of them {EMW IIB7}.

They both published in early volumes of Crelle's journal, and Jacobi also produced a book in Latin in 1829, his 25th year (and also Abel's 27th and last). It carried the phrase *functionum ellipticarum* in its title, and marked the definitive transition from Legendre's use of it for his integrals. In 1832 he introduced a powerful method for inverting integral functions which became known as the 'Jacobi inversion problem' when it stimulated many developments in function theory and complex analysis (§13.13).

The whole area soon generated a massive industry, particularly in the German states, where important figures included A. Göpel (1812–47), J.G. Rosenhain (1816–87) and F.G.M. Eisenstein (1823–52). It also provided attractive opportunities for German schoolteachers to show off *Bildung* (§7.5) by knocking out some properties for a particular kind of function.

In France, Poisson publicized the functions by reporting on Jacobi's book to the French Academy in 1830. Among successors, Liouville found in 1844 that if such a function had no poles, then it was a constant function; at the same time Cauchy found a form of this result. In such ways these functions helped to stimulate complex function theory before Cauchy had given it a general grounding. Conversely, the functions often came to be *defined* in terms of a given pair of complex periods.

In addition, all the applications known to Legendre were rehearsed and developed: for example, the dynamics of rotating bodies (an especially nice piece of work by Jacobi in 1849), pendulum theory as expressed in the integral (6.2.5), and geometry. Jacobi made a striking use in geometry in 1828, when he proved Poncelet's closure theorem (FIG. 8.4.1) by noticing that the points P_0, P_1, ... were given by an arithmetic progression of values of am y of (8.12.2).

8.13 *Abel's Abelian functions*

{EMW IIB7; CE 4.6}

'What a head on him has this young Norwegian', said the aged Legendre of Abel's work on elliptic functions. Yet perhaps his richest gift came in a paper on a class of integrals to which his name is now attached.

Abel submitted a long paper to the French Academy in 1826, and published the main result, pretty baldly, in Crelle's journal in 1829. How wise of him: Cauchy produced his report for the Academy only that year (and three months after Abel's death), and the Academy rushed the big paper into print in 1841.

The background can be explained by this simple but influential example. Imagine two points, A and B, 2 units apart, with mid-point O. The equation in polar coordinates of points P(r, θ) in a plane through AB for which PA × PB = 1 is

$$r^2 = \cos 2\theta. \qquad (8.13.1)$$

The corresponding curve is a pair of symmetrical loops through O, looking like a bow-tie; indeed, its inventor, James Bernoulli in 1694, called it the *lemniscus* (Latin for 'ribbon'), and it is still known as the 'lemniscate'. Now, the length of arc from O is given by

$$s(r) = \int_0^r \frac{1}{\sqrt{(1-u^4)}}\, du, \qquad (8.13.2)$$

a kind of elliptic integral which had fascinated mathematicians from the 1720s. In particular, a result of Euler which Legendre highlighted was this addition theorem: if

$$s(p) = s(q) + s(t), \qquad (8.13.3)$$

then p was a rational function of q and t, and also symmetric (that is, expressible as a ratio of power series, and also $p(q, t) = p(t, q)$).

Abel's theorem was a vast generalization of such addition theorems:

THEOREM 8.13.1 *Let R be a rational function in x and y, variables which themselves are related by a polynomial*

$$f(x, y) = 0; \text{ and let } I(a, x) := \int_a^x R(u, y)\, du, \qquad (8.13.4)$$

with a constant. Now, if

$$\sum_{r=1}^n I(a_r, x_r) = 0, \text{ then } G(x_1, x_2, x_3, \ldots) = 0, \qquad (8.13.5)$$

where the a_r are constants, the x_r are variables and G is a polynomial.

Thus a transcendental relation – the first of (8.13.5) – between integrals, the Abelian functions, was reduced to a 'geometric relation' (as we now say) G between the variables. Furthermore, $f(x, y) = 0$ determined an index number M such that any number $m > M$ of integrals could be reduced to M of them, where their M upper limits v_s were algebraic functions (that is, the form of (8.13.4) rearranged as $y = h(x)$) of the original m limits.

With this amazing result, the young Abel ushered in another massive study of elliptic and many other functions that would last for the rest of the century. Complex analysis was again essential: only partly present in Abel's formulation, it led to restatements where both x and y were complex variables, and I in eqn (8.13.4) was expressed as a contour integral. The idea of the index was to be deepened by Riemann (§13.13).

One early use of eqns (8.13.5) concerned a mystery previously noted by the pioneers of the calculus, especially in Newton's remarkable theorem on the area and perimeter of a convex curve (§5.10): whether a function does or does not have its indefinite integral in closed form (a phrase often used then was 'in finite terms'). Some strange things go on here, since functions very similar in form differ in their integrals: for example, $\exp(-x)$ causes no trouble, but the de Moivre–Laplace error function $\exp(-x^2)$ has no finite form, and tables of its values had to be produced. Liouville sought general criteria in the mid-1830s, and found Abel-like theorems such as this one: if an algebraic function $f(x)$ has an integral in closed form, then it can also be expressed in the form

$$\int f(x)\, dx = g_0(x) + \sum_{r=1}^{n} a_r \ln g_r(x), \qquad (8.13.6)$$

where the a_r are constants and the $g_r(x)$ algebraic functions. He also gave various example of functions, including elliptic ones, which did (or did not) have such integrals {Lützen 1990, Chap. 9}. In the 1840s he extended his results to such finite-form solutions for differential equations, an issue which had been a fetish for Condorcet (§6.17). Before we come to them, however, some other functions await us.

8.14 *Special functions*

{EMW IIA3 and 10; CE 4.4}

This name is attached to a collection of functions separate from the trigonometric, logarithmic, elliptic and Abelian functions, although various relationships hold between them. Among other properties, they (can) satisfy an ordinary differential equation in an infinite power-series expansion, and possess a generating function. Special functions often arose out of physical applications from the mid-18th century, but usually only with a few properties; the systematic theory belongs to this period. Some of the concurrent applications are recorded in §10.12–§10.15. For a long time real-variable properties dominated, but especially from the 1840s forms involving complex variables were added. Textbooks on one or several functions began to appear from the 1870s; {Whittaker and Watson 1927} is a comprehensive classic, graced by many historical references and remarks.

Perhaps the two most significant functions, as well as the earliest, are known today under the names of Legendre and Bessel. The former were introduced by their generating function in (6.9.3), in the unlovely *pas de deux* of Legendre and Laplace; successors such as Poisson used them in mathematical astronomy and elsewhere. New results included 'Rodrigues' formula' to express the associated functions in a closed form: it was found in 1816 by Olinde Rodrigues, one of the early Jewish mathematicians in the modern era, as part of his doctoral thesis in the French Université system.

The Bessel functions have a scrappier background in the 18th century, though Euler and Daniel Bernoulli had found series for some of them. They come in orders, $J_n(x)$ (in the modern notation) distinguished by a parameter n; and Bessel's own chief contribution (1824) was to find some of the recurrence formulae relating it to its neighbours at $n-1$ and $n+1$. Before that, however, Fourier had individuated $J_0(x)$ with the first catalogue of basic (real-variable) properties (not all new with him): the differential equation

$$xJ_0''(x) + J_0'(x) + xJ_0(x) = 0, \qquad (8.14.1)$$

its series expansion, its integral form, the reality of its zeros n_r, and especially the expansion of 'any' function in a series:

$$f(x) = \sum_{r=1}^{\infty} a_r J_0(n_r x), \qquad (8.14.2)$$

where the formulae for the a_r were known, in integral form.

This last result was particularly important for Fourier (and for many other practitioners afterwards); for he came to it when tackling the diffusion equation for heat in a cylinder, and so went into cylindrical polar coordinates. He hoped that the function in the radius variable x would have the same kinds of property as the trigonometric functions in (8.7.1), especially real zeros and the expansion of 'any' function to produce the general solution to the differential equation. His success here was epoch-making for the applicability of the Bessel function of other orders (in fact, quite a few expansions like (8.14.2) were found for them). When complex variables became more widely used, many further properties emerged: for example, contour integral forms, and reinterpretation of expansions in complex variables underpinned by Theorem 8.11.1.

Various other special functions were found during the century, and became known after a discoverer, though maybe not the first one: Jacobi, Chebyshev (§14.7), Hermite (already in Laplace's *Traité* on probability), Laguerre (partly known to Euler), and so on. The hypergeometric series came to be recognized as an umbrella for many of them, particularly in the complex domain. The inspiration here came from Gauss, who had produced a beautiful paper in 1813 on it, defined in the form (including the '*F*' notation):

$$F(a,b,c,x) := 1 + \sum_{r=0}^{\infty} \left(\frac{\Pi_{s=0}^{r}(a+s)\, \Pi_{s=0}^{r}(b+s)}{(r+1)!\, \Pi_{s=0}^{r}(c+s)} \right) x^{r+1}, \qquad (8.14.3)$$

where a, b and c were real but x could be complex. He found not only conditions for its convergence (before Cauchy's doctrines on this topic) but also various relationships arrived at by transformations; he kept to himself the differential equation that it satisfied. Eventually notice was taken, in the mid-1830s, when the young Ernst Kummer found that equation and 24 solutions to it; once again, further steps were to be inspired by the work of Riemann (§13.14).

8.15 *Differential equations: solutions and existence*

Among the major properties of all the functions considered in the last three sections is the differential equation(s) that they satisfied. The general theory of these equations was also in progress; a few features are recorded here.

Firstly, an important innovation was again due to Cauchy. Regarding $\int f(x)\,dx$ as the solution to a differential equation led him via this gorgeous analogy to the novel problem of establishing the existence of solutions:

$$\text{from solving } dy = f(x)\,dx \text{ to solving } dy = F(x,y)\,dx. \qquad (8.15.1)$$

Moreover, his method of solution was the same, in that constructing the integral as a sequence (8.6.2) of partition sums was imitated in sums of the form

$$\textstyle\sum_r [(x_r - x_{r-1})F(t_r, u_r)], \text{ where } x_{r-1} \le t_r \le x_r \text{ and } y_{r-1} \le u_r \le y_r \qquad (8.15.2)$$

for continuous F, and seeking conditions for the existence of a limiting value as the partition became finer.

Cauchy served up this beautiful and original mathematics to the hapless students at the Ecole Polytechnique in 1824, and even had many of his notes printed for their delectation. But for once the objections of the school authorities prevailed, and the notes vanished completely until 1976 {Gilain 1981}. Furthermore, and most untypically, Cauchy himself published little on his methods in later work, so that the origins of this method remained obscure until the recent rediscovery of the notes. This is why it is known as the 'Cauchy–Lipschitz method', honouring also its independent development in 1868 by Rudolf Lipschitz (§12.9).

Cauchy's juniors Charles Sturm and Joseph Liouville collaborated closely (which was then unusual among mathematicians) from the mid-1830s in many areas of mathematical analysis, publishing joint papers in the early volumes of Liouville's journal. One was in the geometrical representation of residues, which Cauchy was still resisting (§8.11); another lay in solving an important class of second-order differential equations, of which Fourier's solutions of the Bessel equation (8.14.1) had been a special case, and maybe an inspiration. They produced some fine work, which is now still being developed; but it made surprisingly little impact at the time, and they soon dropped it {Lützen 1990, Chap. 10}.

8.16 *Cauchy and Gauss on differential geometry in the mid-1820s*

A textbook of Cauchy's on differential geometry published in 1826 included a beautiful definition of the order of contact between two curves C_1 and C_2 at a point P, independent of the calculus but using limits. It depended upon the magnitude of $\angle P_1PR_2$, where R_1 and R_2 were the points of intersection of C_1 and C_2 with a tiny circle with centre at P whose radius decreased to zero.

At this time Gauss was also thinking deeply about differential geometry. Two papers of his were published in 1825 (in German, responding to a prize problem posed by the Copenhagen Society of Sciences) and 1829 (in Latin). Aware that many geometrical or spatial properties such as length and volume are invariant under transformations, he wondered if curvature could be handled in this way. Taking the coordinates (x, y, z) of a point P on a surface S as functions of two associated parametric variables p and q, but independent of metric properties, he expressed the differential element $\mathrm{d}s$ of a line drawn upon the surface through P as

$$\mathrm{d}s^2 = E\,\mathrm{d}p^2 + 2F\,\mathrm{d}p\,\mathrm{d}q + G\,\mathrm{d}q^2, \tag{8.16.1}$$

and found a (complicated) differential expression for the measure k of curvature at P which, as a function of p and q, depended only upon S *and not upon the space in which it was situated.* He then obtained relationships satisfying the property that lines upon S would not change in length if S were distorted or bent continuously into another one.

The most 'remarkable theorem' – as Gauss rightly called it – stated that k would remain unchanged in value (*invariata* was his word: hence, presumably, the choice of 'k', from *konstant*). With deft use of the calculus of variations he then studied properties of geodesic lines on S (that is, lines of shortest path on S between any two of its points) {Bolza 1916}. He also integrated k over a triangle with geodetic sides on S, to calculate the measure by which the sum of its three angles fell above or below two right angles. His finding continued, perhaps consciously, the line of thought opened up by Lambert and others (§6.13).

The significance of Gauss's insights was not quickly recognized, nor was it developed by Gauss himself; but they took wing in Liouville's own contributions {Lützen 1990, Chap. 17}, which included translating the 1828 paper from Latin into French for his 1850 edition of Monge's textbook on differential geometry (§8.3). Soon after that the next major steps forward were to be taken in Gauss's Göttingen, when Riemann used the invariance of expressions such as (8.16.1) to consider relationships between geometry and space (§13.7).

8.17 *Fringe mathematics in fringe towns: non-Euclidean geometry*

{J.J. Gray 1989, Chaps. 9–11; CE 7.4}

This last consequence involves the foundation of geometry. One apparent proof of Euclid's postulate (option 2 in §6.13) had been proposed by his mathematician colleague at Göttingen, Bernhard Thibaut, in the 1800s. Draw a triangle on the floor and walk one circuit around its sides. Your turn through its three external angles totals four right angles; but then its three internal angles add up to two right angles, Q.E.D. The unintended assumption is not obvious; but Gauss spotted it. Walk the circuit again with your arms outstretched on either side, and you should get it also.

In fact Gauss had gone much further; in considering option 5, the possibility of non-Euclidean geometries. He never published, however, and thus the question was left to others to resolve independently. It would be ranked among the most famous achievements of the entire century, but up to 1860 the interest was rather slight, so the account here will be brief.

When pondering the matter around 1800, Gauss corresponded with his Hungarian friend Farkas Bolyai (1775–1856) about proving the parallel postulate. Soon afterwards, Bolyai was appointed to the college in a provincial town in Hungary (now Tîrgu Mereş in Rumania). One of his brightest pupils there was his own son János (1802–60), who went on to extend his father's studies in what spare time he had from his military career. In the mid-1820s, while based in the town now known as Timişoara in Rumania, he thought out many of his ideas. He published a version in 1832, in an appendix to the first volume of a textbook on elementary mathematics written in Latin by his father and published in Tîrgu Mereş. János had not been encouraged by his father (perhaps this helped him form his contrary point of view in the first place), and now he gained no publicity, not even on the title page of the book. Gauss was most complimentary to the young man on receiving a copy, but announced his own priority – whereupon János became so upset that he never published in mathematics again.

At the time, and seemingly independently, similar ideas were occupying Nikolai Ivanovich Lobachevsky (1792–1856), professor of mathematics in Kazan in the Volga basin. He was much more *au fait* than the Bolyais with mathematical developments, publishing papers in analysis; but, as with them, his results were usually published in the local town, another mathematical non-centre. He did present his geometry in a 25-page paper in French for Crelle's journal in 1837, but blew his chance by writing unintelligibly. A short 61-page book of 1840 in German from a Berlin house might have fared better, but in fact did not do so; and his university duties (between 1825 and 1846 he was Rector and/or Librarian) precluded him from conducting a better marketing campaign. {Bonola 1955} contains English translations of the main works of Bolyai and Lobachevsky; {Rosenfeld 1988, Chaps. 6–7} is particularly rich on the Russian context for Lobachevsky.

János Bolyai and Lobachevsky differed somewhat in their approaches. Bolyai's title referred to 'the absolutely true science of space' (where 'absolute' was his term for any proposition provable without using Euclid's parallel postulate), while Lobachevsky's mentioned geometry. But the principles and most of the technical details were similar. Each man conceived of a geometry, initially on a continuous surface of revolution, in which two lines, m and n, which were asymptotic to each other could be regarded as parallel. The 'angles' of a triangle made up of such lines added up to less than 180°.

Both men realized that such a non-Euclidean geometry was not unique, and specified it by a function involving the angle of parallelism (our term, more or less Lobachevsky's). They defined a metric for measuring 'distances', fleshing out the insight implicit in Lambert's contributions (§6.13) that the formulae of spherical trigonometry were absolute. Consider in FIG. 8.17.1 the continuum of 'lines' m passing

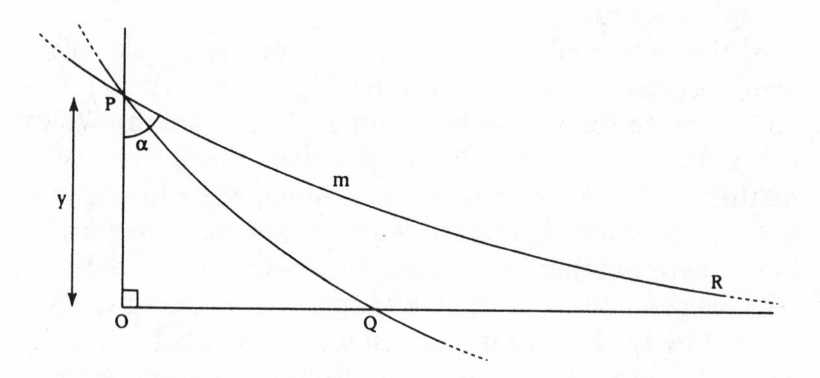

FIG. 8.17.1 *Lobachevsky's angle of parallelism in non-Euclidean geometry, 1840.*

through P (rather unhelpfully, Lobachevsky drew them as straight lines). Some of them, such as PQ, meet the axis of revolution OX, but others do not; let the asymptote PR, the 'line' which never approaches OX, make an angle α with the y-axis at P. In Euclidean geometry $\alpha = 90°$, but here it is less, and depends upon OP ($=y$); but which function $f(y)$ is it? Drawing upon the absolute trigonometry of triangles and solving a resulting functional equation, Lobachevsky found that

$$\alpha = f(y) = 2\tan^{-1}(k^{-y}), \text{ where constant } k > 1. \qquad (8.17.1)$$

This formula was the key to defining a particular geometry of this type as characterized by k; in particular, it could be used to measure the *amount* by which the 'angles' of a triangle added up to less than 180°. Further, using (5.17.4) to relate $\cos y$ to e^{iy} led to similar corollaries such as

$$\cos f(y) = \tanh y \text{ if } k = e. \qquad (8.17.2)$$

In this way complex numbers came into the story; Lobachevsky called his paper 'imaginary geometry', and Bolyai (who also knew (8.17.1)) was more liberal in using i to derive such relations. Both men also showed how to interpret some well-known definite integrals in their geometries.

Bolyai produced only a little appendix in a textbook published in a backwater of Hungary, while Lobachevsky supplied an unfortunate paper, a rather better book, and also lots of Russian that hardly anybody could read; moreover, the consequences for mathematics were profound. So it is not surprising that the reception was slow to arrive[4] – not until the 1860s, in fact, with the preparation of the works of Gauss as one source.

8.18 *The geometry of place*

This aspect of geometry was to make as momentous an impact on mathematics as the attempts at non-Euclidean geometry just recorded, and this time in the form of topology. However, it took still longer to come to fruition (§16.15–§16.16), and the steps about to be described were very preliminary. Again the influence of Gauss is evident, although its degree is hard to appraise.

Von Staudt worked more slowly than his compatriot geometers (§8.4), developing his ideas only in books such as *Geometrie der Lage* of 1847, his 50th year. He followed Poncelet and Steiner in studying projective properties such as cross-ratio and stating theorems in dual pairs, but in this 'geometry of place' he went beyond them in excluding metrical properties entirely. Intersection and incidence were among his staple concepts; for example, a conic section was understood as a path of points together with their tangents. Indeed, there might be no real points, in which case the notion of involution was extended to complex spaces (without, however, using complex *numbers*).

[4] For example, Martin Bartels (1768–1836) was teacher and later friend of the young Gauss; and then he moved to Kazan in 1807 as founder professor of mathematics, where he taught the young Lobachevsky; he was also interested in geometry. Yet he played no part in the introduction of the non-Euclidean brands {Biermann 1978}.

This concern with relationships between points in continuous space rather than with (say) the distances between them was taken further by J.B. Listing (1808–82), in a short book entitled *Vorstudien über Topologie* published in the same year as von Staudt's. Listing had studied under Gauss in the late 1820s (like von Staudt 15 years earlier), and worked mainly in physics, but this book contained some elementary thoughts about positional properties of objects – and also the début of the word 'topology' (from the Greek *topio*, meaning 'site'). The traditional name, due to Leibniz, was *analysis situs* ('analysis of place'); it continued to be used.

A more significant work in this direction was Listing's paper of 1861 published by the Göttingen Academy containing a 'census of spatial complexes'. He focused upon a result found by Euler in 1758 relating the numbers of vertices (V), edges (E) and faces (F) of a convex polyhedron:

$$V-E+F=2. \tag{8.18.1}$$

Modifying (and even refuting) this theorem for more complicated figures had occupied some mathematicians on occasion, including von Staudt {Lakatos 1976, Chap. 1}, and Listing felt it timely to make a list. In the process he created this exotic surface for which it did not hold: take a rectangle, turn one edge through 180° and join it up to its opposite edge. One might call it 'the roll of Listing', but it is known today after a fellow constructor (and student of Gauss, in 1813): 'the strip of Möbius', for A.F. Möbius had written a paper on 'the contents of polyhedra', in response to a prize problem set by the French Academy, but published in 1865 by the Leipzig Academy. The strip does not satisfy Euler's theorem, since it has only one face, only one edge (an important feature, rarely mentioned) and four vertices (or none, if one ignores the join). Beyond that, however, it was a curiosity; for example, each man made his construction in 1858, and did not rush into print with it. General theories about its consequences were to wait until the end of the century (§16.15).

The same is true of an earlier consideration by Möbius, of 1840: dividing a kingdom into five regions such that each one bordered upon the other four. The most famous question of this type is called the 'four-colour problem', on the possibility of colouring unambiguously 'any' planar map with only four colours. The profundity of this guess was not perceived for decades; and for once it came not from a student of Gauss but from one of Augustus De Morgan's students in England: De Morgan published it himself in 1852. He contributes also to our final remarks in this chapter.

8.19 *Cauchy's influence: its scope and limits*

The influence of Gauss has been much apparent in these last sections, and earlier; but many of his insights were not published before his death in 1855, and those of his associates were often poorly advertised. So Cauchy is the major figure of influence in this chapter, and it is appropriate to end with him. Although he left the Ecole Polytechnique in 1830 (§8.6), his doctrine of mathematical analysis stayed behind, and became the normal approach to real-variable analysis, as we shall see in §12. However, the process was gradual: even his successors at the school, Navier and Sturm, did not adopt all his principles.

Outside France, a good example is provided by De Morgan. In his widely read textbook on *The Differential and Integral Calculus* {De Morgan 1842}, he began with an outline of the theory of limits and defined derivative and integral as in eqns (8.6.1) and (8.6.2); but he did not mention Cauchy, and even gave Euler's name 'differential coefficient' (§6.2) to the derivative! He also treated Arbogast's 'calculus of derivations' (§9.11) and manner of handling infinite series (§6.2), which Cauchy would not tolerate. He did not even rehearse there the treatment of continuity of functions which he had given in 1835 in an algebra textbook: '"let me make x as small as *I* please, and I can make $7+x$ as near to 7 as *you* please"' {De Morgan 1835, pp. 154–5} –

the first occurrence of the usual form of continuity which is used today and called the '(ε, δ)' form. Cauchy had introduced these Greek letters into analysis in a few of his proofs, but we saw in §8.5 that his definition of continuity was sequential.

The scope (+) and (as it were) limits (−) of Cauchy's formulation of mathematical analysis may be summarized under the following sextet of pairs:

1+. Cauchy placed much emphasis on the logic of necessary and/or sufficient conditions (N and S, say) required to establish the truth of a theorem T. As he claimed in the introduction to the *Cours d'analyse*, this feature gave his approach 'all the rigour that one requires in geometry', especially Euclid's. Future work would involve strengthening the necessary conditions and weakening the sufficient ones, and maybe even finding necessary and sufficient ones, T′, in the scheme

$$S_1 \to S_2 \to \cdots \to T(\equiv T' \equiv \cdots) \to \cdots \to N_2 \to N_1. \qquad (8.19.1)$$

This feature of theorems has become so standard that even historians have not recognized its novelty – the ultimate compliment, of course.

1−. However, in common with almost all mathematicians, he did not examine logic as such (for example, (8.19.1) was never given in this explicit form), least of all to mathematicize it. This was to be left to De Morgan and George Boole, with origins in algebras (§9.13); and to Giuseppe Peano and his followers (§12.13), in consequence of the levels of rigour achieved in mathematical analysis by the Weierstrassians.

2+. Cauchy exposed the power of the limit concept itself, especially in formally uniting theories of functions, series and integrals instead of leaving them only loosely connected.

2−. However, while his treatment of single limits was basically clear, the specification of upper and lower limits was rather vague, and they were not properly distinguished from least upper and greatest lower bounds. Furthermore, his handling of multiple limits was shaky. Riemann and Dedekind, and then the Weierstrassians (including Cantor), were to chime in here, loudly.

3+. Cauchy brought out the virtue of broad definitions of basic concepts. The theory of (dis)continuous functions benefited especially, for a plethora of terms was swept away.

3–. However, the depth of his rigour went only so far: the real line was still left unstructured, and collections of points (or whatever) were handled informally. Similarly, he often used symbols quite sparingly, really too little in multiple-variable situations.

4+. As an important example of 1+, Cauchy offered the fundamental theorem of the calculus as a *genuine theorem, complete with discussible conditions,* instead of the naive assumptions of his predecessors that differentiations and integrations could be automatically reversed.

4–. However, in both the fundamental theorem and elsewhere (such as Theorem 8.5.1) he put perhaps too much emphasis on continuous functions as a 'safe haven' for analysis. Furthermore, he did not seek necessary and sufficient conditions for integrability, multiple integrals were not well distinguished from repeated ones, and line and surface integral were not defined at all.

5+. Cauchy revolutionized the use of complex variables, developing a general theory in close analogy with real-variable analysis.

5–. However, his avoidance of geometrical styles restricted his development of function theory. Riemann and Weierstrass were to suggest powerful alternatives.

6+. Cauchy brought new levels of order into mathematical theories. He even pioneered (from the mid-1820s) the systematic numbering of equations in a paper or book.

6–. However, the chronology of his voluminous writings exhibits extraordinary disorder: often a string of shortish papers strewn with repetitions, instead of one comprehensive account – of his complex analysis, for example. (His follower the Abbé Moigno performed a great service from the 1840s by writing book-length versions of various of his theories, which were widely read.) Impatience could triumph over the 'rigour' of this strange insider-outsider who so affected many parts of mathematics.

9

The expanding world of algebras, 1800–1860

9.1 *Common and uncommon algebras*

From mathematical analysis and geometries, we turn to concurrent and partly related developments in algebra. Two new features are the rise of number theory as a branch of mathematics, and the emergence of British (especially English) mathematicians into the international community; but the dominant theme is expressed in the title – that algebra was definitely becoming algebras, with a range of new ones appearing in a short time. They enter here chiefly in §9.8–§9.14: quaternions, as extensions of complex numbers; differential operators such as d/dx taken as the objects of the algebra; functional equations, with the function f

(not its values) itself similarly treated; and algebras in logic.

The novelty of these algebras somewhat retarded their growth and recognition; indeed, in contrast to the neighbouring chapters, the story told here records a goodly proportion of (partly) missed opportunities. The French provide many major figures, as usual, but the dominance is not as marked as in the last and the next chapters, largely because of the activities of Gauss. Among the general historical literature, {van der Waerden 1985} and {Scholz 1990} are valuable, though both sadly overlook the concerns of §9.11–§9.13.

9.2 *Gauss on the algebraic properties of numbers, 1801*

{Bachmann 1911}

The century opened wonderfully with Gauss's book *Disquisitiones arithmeticae* {Gauss 1801}, which elevated the intellectual level of algebraic number theory at a stroke, and so greatly contributed to its rise in importance in mathematics. Although he was only 24 years old at the time, many of its results were already old news to him; as he explained in the preface, much of the book had been ready for four years and wanted only for financial support. This came eventually from his employer, the Duke of Brunswick and Lüneberg (a descendant of Leibniz's employer a century before) and published in Leibniz's home town of Leipzig. A French translation rapidly appeared, in 1807; but German and English ones followed only in 1889 and 1966 respectively.

The contrast with Legendre's *Essai sur la théorie des nombres* (1798), the first comprehensive treatise on number theory, is quite stark. Legendre had summarized very capably the main findings to date, with lists of results on prime numbers, quadratic forms, continued fractions, and the like, providing tables of integers possessing various properties. Gauss also tackled such topics, but sought theorems which exposed underlying structures; for example, in some places he implicitly deployed group theory. The book is significant above all for the depth of its concepts and the proofs inspired by them; a few main examples are noted here.

Gauss began with a new notation for congruences which has remained standard ever since, and which carries a structural feel in its very formulation. For example (his own),

$$\text{`}-7 \equiv 15 \pmod{11}\text{'} \quad \text{because} \quad -7 = (-2) \times 11 + 15; \tag{9.2.1}$$

−7 was the 'residue' relative to the 'modulus' 11. Sections 1–3 of the book were taken up with properties and solutions of congruence equations of various degrees, and also included the first general proof of the unique factorization of an integer into primes.

The fifth section, by far the largest, was concerned with quadratic forms. Gauss's focused upon the 'determinant'

$$D := (b^2 - ac) \text{ of } ax^2 + 2bxy + cy^2 \tag{9.2.2}$$

(all letters denoting integers), and the corresponding expressions for forms of more than two variables. He transformed (x, y) into variables (X, Y) by linear relations, and sought necessary conditions that the new determinant would equal the old one in value (contrast Lagrange's sufficient conditions in (6.16.4)). Equation (9.2.2) is an example of a determinant as we now use the word (§9.10), and it has the same character of an invariant under transformation.

The seventh and last section examined the consequences of the fact, known from Euler's equation (5.17.4) for e^{ix}, that the equation

$$E := x^m = 1 \ (m \text{ an integer}) \quad \text{had } m \text{ roots} \tag{9.2.3}$$
$$x = \exp(2r\pi i/m), \ 1 \le r \le m.$$

These values corresponded to m points placed uniformly round a unit circle centred on the origin; so, thanks to J.J. Sylvester in 1880, the equation has become known as 'cyclotomic', meaning 'divide the circle'. The question, of Greek origin, about constructing the corresponding regular

m-gon with straightedge and compass was now expressed in terms of decomposing E into linear and quadratic factors. Some values for m had been dealt with; Gauss gave a *general* criterion based on the theorem that the full solution of E can be found by solving equation of degrees equal to the factors of $m-1$. In particular, for $m=17$, the factors are all 2's, which yield quadratic equations; hence the classical construction was possible for the 17-gon.

9.3 *Gauss's concern for a rather fundamental theorem*

Elsewhere, Gauss proved that any polynomial in real coefficients,

$$x^m + Ax^{m-1} + Bx^{m-2} + \cdots + M = 0, \quad (9.3.1)$$

can be rendered as the product of linear and quadratic real factors; there followed the corollary that a polynomial in complex coefficients breaks into linear complex factors. He highlighted the importance of the theorem by offering one proof in 1799 (his dissertation, defended at the University of Helmstedt), two more in 1816, and a variant of the first in 1850. The modern name, 'the fundamental theorem of algebra', seems to have come in late in the 19th century; Gauss himself never used it.

This theorem is a beautiful example of the status of a proof: how general, and based upon which presuppositions? Three of Gauss's quartet drew upon setting $x = re^{i\theta}$ and using de Moivre's theorem (5.17.5) to connect $\genfrac{}{}{0pt}{}{\sin^n}{\cos^n}\theta$ with $\genfrac{}{}{0pt}{}{\sin}{\cos} n\theta$ (he actually used an equivalent procedure avoiding i); he then converted the resulting expression into two polynomial equations of degree m in r and coefficients in $\genfrac{}{}{0pt}{}{\sin}{\cos} n\theta$, $n \le m$. Plotting these equations as algebraic curves in the plane, he claimed that they had to intersect in at least one point; the theorem then followed easily. But he construed the continuity of the curves in the traditional way which Poncelet would later exhibit also (§8.4), bolstered by some intuitive use of topology. His second proof was purely algebraic, but assumed that a polynomial equation of odd degree possessed a real root, and also that a quadratic equation with complex coefficients had a pair of complex roots: reasonable enough, but assumptions nevertheless. Criticizing predecessors such as Lagrange for assuming the existence of roots in some way, he deepened the foundations, but he reached geometrical and topological depths which were still mysterious in his time (§8.18).

9.4 *Irresolving the quintic and permuting the roots: Abel and Galois*

{Kiernan 1971; CE 6.1}

Lagrange's studies of the factorization of polynomials (§6.15) was the inspiration behind the advances described here. Considering rational functions f of the roots $a, b, \ldots$ of a polynomial, he had found results about the number of different functions that f could represent under all possible permutations of its variables. (The rather loose talk of the time, which I shall follow, spoke of the number of 'values' that f could take; to present-day mathematicians f is a sort of functor.) Should the quintic, like the cubic and quartic, be resolvable 'by radicals' – that is, should all its roots be expressible in a finite number of arithmetical operations and extractions of roots, then such a function should take no more than four values; but he could not find one.

The Italian mathematician Paolo Ruffini (1765–1822) became obsessed with the question, and wrote at great length for his Società Italiana (§7.5) on the impossibility of finding such a function. A few mathematicians, such as Cauchy, thought that Ruffini had succeeded, but doubts remained. It seems that he had (more or less) achieved his aim for methods based upon Lagrange's methods; however, he had not shown that no others existed {Bryce 1986}.

Cauchy himself had contributed a paper in 1815 on the number of values that a non-symmetric function can take: his argument embodied various properties of group theory in terms of products of 'substitutions' on 'permutations' of the letters (in modern terminology, permutations on arrangements). Unlike his predecessors, he individuated the notion of rearrangement by introducing the notation '$\begin{pmatrix} A_1 \\ A_2 \end{pmatrix}$' for the substitution of A_2 for A_1. Another new algebra, of groups, was now within reach, but Cauchy did not grasp it here.

Enter Abel again, with a decisive paper in Crelle's journal in 1826. By a brilliant new argument independent of any of Lagrange's procedures, he showed that each root of the quintic could be expressed as a rational function of any one of the other four. Then he considered possible numbers of values, and combined Lagrange's result with Cauchy's extension to show that no such function could be found; thus the resolution by radicals could not be achieved, neither for the quintic nor for any higher-order permutation. Neither Abel's proof nor his *a fortiori* corollary just mentioned was entirely watertight; however, he had made the breakthrough, ending the three-centuries long restriction of the theory of equations to radicals and related geometrical constructions. Tools from analysis, such as the elliptic functions of §8.12, could now be used in resolving polynomials of the fifth and higher degrees.

Abel also highlighted algebraic features of roots and functions of them. So had Lagrange before him, in a way which may be illustrated by the simple function ab/c of three quantities. The permutation of the quantities that tranforms the function to ba/c leaves the function unchanged, unlike the one that yields bc/a. Lagrange had shown that those permutations of the zeros of a polynomial that preserved a rational function of them formed a group, but that this was not necessarily true of the permutations that preserved the values of the function.

My formulation above uses a form of words and theory which Lagrange's successors were to introduce. A major figure was Evariste Galois (1810–32), famously tragic for writing out in a letter many of his mathematical findings on what he knew would be his last night on Earth before being shot dead in a duel the next morning. The letter was soon published, but in a general Paris journal of rather republican reputation, and so failed to reach the right audience. Previously he had placed two notes with Gergonne, and three (two dealing with equations) in Ferrusac's *Bulletin*, thanks to editor Sturm.

Galois built upon the work of Lagrange, Cauchy and Gauss (from whom he adopted the '≡' notation (9.2.1), presumably having read the *Disquisitiones arithmeticae* in French). Like them, he looked for conditions for equations to be solvable by radicals; and he found such conditions in properties of 'substitutions' on 'permutations' (Cauchy's words). But he treated substitutions as objects of an algebra of combination in an explicit way which Lagrange had not sought and Cauchy had not attained. He divided them into 'groups' (his own word[1]), and recognized closure – that a combination of substitutions was itself a substitution. The part of his letter dealing with equations announced a few decomposition theorems and consequences about the permissible degrees of solvable equations, but expressed in terms of the group itself (for example, involving what are now called 'normal subgroups'), not the parent equation. To the modern mathematician this is strikingly familiar; in September 1832 it must have seemed weird, even to the trained mathematician.

Neither young man made the impact that his work deserved during his short life. Galois was known around town as a loud-mouthed and opinionated republican, not a good reputation to have either before or after the 1830

[1] Galois's use of the word 'group' conflated notions which we would now distinguish as subgroups and cosets. A little-known historical detail attends its use. Poinsot reviewed the second edition of Lagrange's book on equations (§9.6) after its publication in 1808. The great man praised his piece; so when a posthumous printing was made in 1826, it was placed at the head. Now at one point Poinsot considered Lagrange's linear function L of the roots of a quintic, and divided the 24 different values of L^5 into six 'groups' by various criteria involving roots of unity. While he did not formalize this line of thought, there is evident similarity with that of Galois, who had surely seen this important recent book on algebra.

revolution; thus, while he had gained Cauchy's attention by 1829, he lost it soon afterwards. We speak now of 'Galois theory', but Galois himself had only sketched out some outlines. Abel was not well known to most contemporaries (§8.13); for example, he sent his 1826 paper to Gauss, who did not read it (but it was Abel's fault for not slitting the pages of the copy before posting it). So the next stages in the evolution of algebra had to wait until the mid-1840s. Then, with suspiciously close timing, Cauchy launched himself afresh into a long series of articles on the permutative properties of the group of substitutions, while Liouville (re)published Galois's papers, letter and manuscripts in his journal {Wussing 1984, Part 2}. These considerations were deepened in the 1850s by the Italian mathematician Enrico Betti, as Galois theory began to become more international.

9.5 *The consolidation of number theory*

{Ellison and Ellison 1978; H. Edwards 1977}

The next two major students of number theory after Gauss were younger compatriots. In 1823, during his time in Paris (§8.7), J.P.G. Dirichlet made an impressive début in print with a discussion of equations of the form

$$x^5 \pm y^5 = Az^5 \text{ (all integers)} \qquad (9.5.1)$$

for various values of A; it soon led Legendre to prove Fermat's Last Theorem for the case $n = 5$. His main concern was with quadratic forms, including finding a formula for the number of different classes of forms with the same given determinant (9.2.2). His lectures at Göttingen over many years on these forms raised considerable interest among followers {EMW IC1–2}, especially the students Riemann and Dedekind (§13.3).

Some of Dirichlet's later work enhanced analytical number theory in various ways. In 1837 he introduced an expansion of Euler's zeta function (5.18.5), now known as the 'L series', to prove a guess by Legendre that an arithmetic progression whose first term and difference are co-prime contains an infinity of primes (2, 5, 8, . . ., for example, but not 2, 4, 6, . . .). This lovely result was related to an old question on the distribution of prime numbers among the integers (§12.12).

The other main figure, Carl Jacobi, also worked in both analytic and algebraic number theory. To the former part he applied elliptic functions to establish Fermat's guess that any integer can be expressed as the sum of at most four squares. In the latter part he, and later also his compatriot Gotthold Eisenstein, followed Gauss in studying higher-order laws of reciprocity (Theorem 6.16.1). They were phrased in the Gaussian form (9.2.1): let $n > 2$ and maybe prime, and be given

$$x^n \equiv p \pmod q \text{ and } x^n \equiv q \pmod p,\ p \text{ and } q \text{ prime}; \qquad (9.5.2)$$

then if one equation had a *or* no solution for x, so *or* neither did the other equation.

The results were extended by Ernst Kummer in 1847 to a certain kind of prime number which he called 'regular' and for which he proved Fermat's Last Theorem. His proof was motivated by news from Dirichlet that the field of complex numbers did not admit unique factorization. A simple example, taken from a different context but noticed at the time, is given by adjoining $\sqrt{-5}$ to the integers; for

$$9 = 3 \times 3 \quad \text{but also} \quad 9 = (2 + \sqrt{-5}) \times (2 - \sqrt{-5}). \qquad (9.5.3)$$

It had profound consequences for algebra; in particular, it sunk a proof of Fermat's Last Theorem proposed, also in 1847, by Gabriel Lamé.

Kummer drew upon cyclotomic numbers (which he called 'the most general form of complex numbers'), polynomials P of the form $\sum_{r=1}^{\lambda-1} a_r\alpha^r$ in a complex root α, given by eqn (9.2.3), of unity, where λ was a prime number and the a_r were integers. He found means of factorizing them uniquely into ideal prime factors (as we would now say: his rather unhappy term was 'ideal complex numbers'). The discovery of such numbers, and of their attendant properties and relationships to P, came largely from studying congruences; the results and methods were to enrich not only algebraic number theory but also abstract algebras (§13.3).

9.6 *Roots of equations: detection and approximation* {CE 4.10}

Lagrange's *Traité de la résolution des équations numériques* (1798) was the chief inspiration for the advances to be reported here. It contained the reprint of two papers of the 1760s on real and complex roots of polynomial ('numerical') equations, together with a dozen notes; a second edition ten years later contained two more notes. He devoted much space to factorizing the equations, and considering properties of their roots such as symmetric functions of them, and Newton's and other relationships between these functions and the coefficients of the polynomial. Lagrange also treated at length methods of detecting real and complex roots, and approximating to them by continued fractions, functions of their roots, and other means.

Some durable advances were made, with Fourier as a stimulus. In his teens in the mid-1780s he had proved Descartes's rule of signs (§4.18) by a nice inductive proof on the degree n of the polynomial $f(x)$, and also generalized the theorem to

THEOREM 9.6.1 *An upper bound on the number of positive roots of* $f(x) = 0$ *in an interval* $[a, b]$ *of* x *is given by the number of changes of sign (that is,* $+-$ *or* $-+$*) between the sequences of signs in*

$$\begin{aligned} &f(a),\ f'(a),\ f''(a),\ \ldots,\ f^{(n)}(a) \\ \text{and}\quad &f(b),\ f'(b),\ f''(b),\ \ldots,\ f^{(n)}(b). \end{aligned} \tag{9.6.1}$$

Similarly, a bound on the number of negative roots was given by the number of preservations ($--$ *or* $++$).

Fourier taught these results in the late 1790s at the Ecole Polytechnique but published them only in 1820, his 53rd year. His definitive presentation came in his book *Analyse des équations déterminées,* which appeared incomplete and posthumously in 1831 under the editorship of Navier. He also considered there methods of approximating to roots, including approaching them from both smaller and larger values. Another method of the time, for obtaining the successive decimal places of the root, is known today after the English mathematician W.G. Horner (1819), although he was not the only or even the first discoverer; predecessors can be found as far back as the Arabs.

The weakness of Fourier's theorem is that it does not detect complex roots. The young Charles Sturm pondered this point, and came up with the answer in 1829: the same procedure as Fourier's but replacing the derivatives after the first with this sequence:

$$f_0(x) := f(x),\ f_1(x) := f'(x),$$

$$f_{j-2}(x) := f_{j-1}(x)q_{j-1}(x) - f_j(x), \quad \text{with } 2 \leq j \leq m; \qquad (9.6.2)$$

that is, the Euclidean algorithm (2.13.1) but with a reversal of sign on the remainder functions $f_j(x)$. This algorithm, still known after him, caused an unusual measure of interest at the time {Sinaceur 1991, Part 1}, although, as with Fourier's, the computations are pretty tiresome. Cauchy was to grumble that a relevant formula of his published in 1831, which used a contour integral (8.11.4) for $f'(x)/f(x)$, was ignored.

In his *Traité,* Lagrange had also presented a formula for the expansion in a continued fraction of a zero R of a polynomial. This result led to another valuable finding: proofs of the existence of transcendental numbers – those that cannot be expressed as a zero of a polynomial with integer coefficients. From Lagrange's formula, Liouville in 1844 derived an inequality for R, and choosing coefficients so that the inequality was infringed led him to transcendental numbers, such as $\sum_{r=0}^{\infty} N^{-r!}$ for N an integer. He had hoped to prove e and π to be transcendental, but he could not find the right expansions {Lützen 1990, Chap. 16}. Another 30 years were to pass before e fell (to Charles Hermite in 1873, following Liouville's ideas) and another decade for π (Ferdinand Lindemann, 1882).

9.7 *Complex numbers: algebraic or geometrical?*
{Argand 1874}

Many of the considerations noted so far involved complex numbers; so it is not surprising that their status was reappraised. Over the centuries the algebraists had heaped opprobrium upon them as they struggled to live with their necessity: 'sophistic', 'impossible', monstrous', 'ridiculous', 'tortures', 'fictitious', 'chimeras', 'false' and 'useless' were among the nouns and adjectives used, although complimentary terms such as 'true' and 'useful' were occasionally bestowed. Even our standard words 'imaginary' and 'complex' carry an aroma of disapproval. They are due respectively to J.R. Argand (1768–1822) and to Gauss: contrasts in reputation, the latter being one of the greatest figures, the former a bookseller in Paris.

However, despite his non-fame in his lifetime, Argand is a familiar name in textbooks today for his geometric representation of 'imaginary quantities', which he proposed (after mentioning some of the other terms) in a short book of 1806. Following Wallis (§5.14), though maybe not consciously, he defined i as the mean proportional between -1 and 1; but in addition, unaware of its evaluation by Euler as $e^{-\pi/2}$ (see §6.15 for the context), he thought that i^i needed a third dimension. Nobody reacted to Argand until a discussion took place in Gergonne's *Annales* in the early 1810s between him, Legendre and J.F. Français; they straightened him out on i^i.

Apart from his obscurity, the general lack of interest was probably caused by the fact that Argand was answering a question which mathematicians were not then asking. We noticed that the roots of the cyclotomic equation (9.2.3) could be represented by points around a unit circle; but the algebraic manner of handling complex numbers *in general* seemed to be satisfactory, notwithstanding Euler's blunder (6.15.1) over the product of two such numbers. Furthermore, we saw in §8.11 that, in his approach to complex analysis, for a long time Cauchy deliberately avoided geometrical resources. Forms of the complex plane were proposed from time to time, but only with Gauss's suggestion of 1831, partly in connection with the fundamental theorem of algebra, did it start to gain favour. Even then, W.R. Hamilton was soon to suggest a new algebraic version in 1833, in which the complex number was understood as an ordered pair (x, y) of real numbers satisfying the required algebraic properties.

9.8 *From complex pairs to quaternions*

{Crowe 1967, Chap. 2; CE 6.2}

A decade later, Hamilton was to make a more important, and famous, extension to algebra. His work on algebraizing mechanics after Lagrange (§10.17) led him to seek an algebraic means of saying in three dimensions what $a + \mathrm{i}b$ said very nicely in two. The target was easy: let

$$z := a + \mathrm{i}b + \mathrm{j}c, \quad \text{with 'length' } |z| := \sqrt{(a^2 + b^2 + c^2)}; \tag{9.8.1}$$

find the rules for i and j which would let z and $|z|$ satisfy the required properties. Now, + and − were straightforward, but × proved impossible: $|zw| \neq |z|\,|w|$, no matter what Hamilton tried. Then an 'aha' moment occurred on a walk into Dublin on 16 October 1843, in his 39th year: try *four* units, 1, i, j and k, and impose laws upon the 'quaternion' q defined as a linear combination of them:

$$q := a + \mathrm{i}b + \mathrm{j}c + \mathrm{k}d, \quad \text{where } \mathrm{i}^2 = \mathrm{j}^2 = \mathrm{k}^2 = \mathrm{ijk} = -1; \tag{9.8.2}$$

$$\text{and } \mathrm{ij} = \mathrm{k} \text{ and } \mathrm{ji} = -\mathrm{k},$$

with permutations among i, j and k. Commutativity was lost, but associativity was preserved.

This unexpected venture into four algebraic dimensions (though with no corresponding geometrical reading) gave Hamilton the breakthrough, for now q and $|q|$ could do all that was required of them. Working out the details, with numerous applications, was to occupy him for much of the rest of his life. His *Lectures on Quaternions* (1853) told the story to date in 753 pages, and was succeeded by a more introductory *Elements* which, appearing after his death in 1865, was longer still. Since application to mechanics and mathematical physics was one of his main concerns, he adapted the definition of q in eqn (9.8.2) to define the operator of partial differentials

$$\triangleright := \mathrm{i}\frac{\mathrm{d}}{\mathrm{d}x} + \mathrm{j}\frac{\mathrm{d}}{\mathrm{d}y} + \mathrm{k}\frac{\mathrm{d}}{\mathrm{d}z}; \text{ thus } -\triangleleft^2 = \left(\frac{\mathrm{d}}{\mathrm{d}x}\right)^2 + \left(\frac{\mathrm{d}}{\mathrm{d}y}\right)^2 + \left(\frac{\mathrm{d}}{\mathrm{d}z}\right)^2 \tag{9.8.3}$$

represented the Laplacian operator. Furthermore, given the quaternion function

$$V := \mathrm{i}X + \mathrm{j}Y + \mathrm{k}Z, \text{ then} \tag{9.8.4}$$

$$\triangleright V = -[X_x + Y_y + Z_z] + [\mathrm{i}(Y_z - Z_y) + \mathrm{j}(Z_x - X_z) + \mathrm{k}(X_y - Y_z)]. \tag{9.8.5}$$

Nowadays we split expressions such as this into their vector and scalar components (both words due to Hamilton, incidentally, 'vector' being the Latin for 'carrier'). The rise of vector algebra and analysis was long and slow (§15.9), and quaternions gained many adherents before the turn of the century. Indeed, together with various extensions such as octonions to eight elements, they gained a following strong enough for an International Association for Promoting the Study of Quaternions and Allied Mathematics to run from late in the century until 1913, defending the faith against vectorial heresy {Hankins 1980, Chap. 7}. Such institutional support is, I believe, unique for a subject in mathematics. Curiously, nobody seems then to have tried to extend complex-variable analysis by developing a theory of functions of a quaternion.

9.9 *Grassmann on the algebra of geometrical 'extensions', 1844* {Schubring 1996}

Contemporary with Hamilton's insight, and partly stimulated by similar concerns, were the researches of Hermann Grassmann (1809–77). Thanks to his remarkable father, Justus (1779–1852), he gained great insight into the algebraic methods in mathematical physics (and also questions of mathematical education). His main work is entitled *Die lineale Ausdehnungslehre* (1844), 'The Linear Doctrine of Extension'. Like Hamilton's theory, it suffered eclipse by vector analysis even though it has some advantages (for example, the possibility of defining an inverse magnitude).

In his formulation Grassmann was influenced mathematically by A.F. Möbius and philosophically by the *Dialektik* (1839) of the neo-Kantian Friedrich Schleiermacher, whose lectures he had heard while a student in Berlin {Lewis 1977}. An important aspect is pairs of opposites, known among German philosophers as *Polarität;* for example, pure mathematics (or mathematics of forms) and its applications, discrete and continuous, space and time, analysis and synthesis. This latter distinction underlay Grassmann's algebra of 'extensive magnitudes': two of them, a and b, could be combined in a 'synthetic connection' to form '$(a\frown b)$', where his parentheses indicated that a new object had been formed by the connection; conversely, an 'analytic connection' $(a\smile b)$, involving decomposition, sought an a, given b and $a\frown b$. That is,

$$\text{given } c := a\frown b, \quad \text{then } a = c\smile b; \qquad (9.9.1)$$
$$\therefore\ a = (a\frown b)\smile b = a\frown b\smile b = a\smile b\frown b,$$

granting the shedding of brackets (he was innovative on this procedure also) and admitting the basic laws satisfied (or not) by '$\frown$' and '$\smile$', such as 'exchangeability' (commutativity) and distributivity.

Grassmann's use of 'analytic' and 'synthetic' corresponds more closely to the traditional uses of these terms (§2.27) than was normally the case! He was also unusually elaborate in examining basic properties, including linear combinations and expansions of a magnitude relative to a basis (in other words, implicit use of a vector space): he may have chosen the unusual word *lineale* for his title to refer to geometrical straightedge constructions, and perhaps also to allude to *Linie-alle* – 'all linear'.

Of the various interpretations given to these connections, the pair which attracted a fair amount of interest were the 'outer and inner products' of a and b, of which the vector ($\cap$) and scalar ($\cup$) products of two vectors a and b are respective examples. Grassmann was quite explicit about the increase of dimension of the former; he also spoke of the *signed* area of the parallelogram which a and b defined, and used it to explain the law of non-exchangeability:

$$(a \cap b) = -(b \cap a), \text{ in contrast to } (a \cup b) = (b \cup a). \qquad (9.9.2)$$

He stressed that his definitions and constructions could be continued up to any finite number of dimensions, and that they were independent of any coordinate system.

Like Bolyai and Lobachevsky with their non-Euclidean geometries shortly before him (§8.17), Grassmann spoilt the communication of his innovations by obscure exposition, and especially by being born and pursuing a career in the mathematical non-centre of Pomerania (now Szczecin, in Poland). A revised edition of 1862, containing less philosophy and following a more rigorous presentation, fared little better. His principal followers, from the 1870s onwards, usually adopted the technical features and the aim for generality, but set aside most of the philosophical context. An exception was his younger brother Robert (1815–1901), who applied his system to logic, and gained a little recognition there (§13.21).

9.10 *Some traces of determinants and matrices*

This talk of forerunners of vector analysis naturally raises similar questions about the emergence of linear algebra.

Again, historical discontinuities and lost opportunities dominate our hindsight, but the 5th chapter of Gauss's *Disquisitiones arithmeticae* marks a first step. There he wrote quadratic forms in two and three variables in an array form of the coefficients, and gave formulae equivalent to making a linear transformation of the variables, finding an inverse matrix, and reducing to triangular forms; and we saw at (9.2.2) that his 'determinant' of a quadratic form served as an invariant quantity under transformation. But he did not invent the theories that *we* can see were within his reach; for example, he wrote the quadratic form

$$ax^2 + a'y^2 + a''z^2 + 2bxy + 2b'xz + 2b''yz \qquad (9.10.1)$$

in three variables (x, y, z) as

$$\text{the array } `\begin{pmatrix} a & a', & a'', \\ b, & b', & b'', \end{pmatrix}' \qquad (9.10.2)$$

$$\text{rather than the matrix } \begin{pmatrix} a & b & b' \\ b & a' & b'' \\ b' & b'' & a'' \end{pmatrix}.$$

However, this work seems to have played a role in the first of two stages involving remarkable coincidences for Cauchy. The first concerns a second paper published in 1815 in which he worked out some details of the theory of permutations noted in §9.4. He considered functions which could take only one or two 'values', and for the latter he handled the product of two-by two transpositions of letters a_i and a_j by expanding the product of the differences $a_i - a_j$, replacing each power a_i^m of a letter by the new element $a_{i,m}$, and then considering the 'determinant' array of these elements, with the sign of each term assigned as + or − according to whether the corresponding substitution was even or odd. He wrote down the ensemble of new elements *as* an array, although he did not make great use of it; however, he proved the product theorem for determinants.

The coincidence was that Cauchy submitted this paper (and its predecessor) to the Institute of France on the same day, 30 November 1812, as his friend J.M.P. Binet (1786–1856) sent in one on the same product theorem. Neither paper made any impact, unfortunately, so that there was little advance before the second coincidence occurred on 27 July 1829, when the French Academy received papers from Cauchy and Sturm.

In recent years Cauchy had deployed quadratic forms in a variety of contexts, including differential geometry (§8.16), the kernel of some second-order differential equations (§9.11) and elasticity theory (§10.16); reduction to a sum of squares was a major concern. This paper went beyond these special cases to give a beautiful outline of the 'spectral theory' (as it is now called) of matrices and the associated quadratic forms, with the terms displayed in a rectangular or square array, which he called a 'tableau'. He showed the reality of the latent roots of a symmetric matrix and the orthonormality of the latent vectors, the interlacing of the roots with those for leading submatrices, and the conversion to sums of squares. The paper is startling {Hawkins 1975}; equally so is Cauchy's failure to recognize its significance, for he did not use it to rework any of those concurrent cases, and in later years he largely forget its contents! His title referred to 'the equation with the help of which one determines the secular inequalities of the motions of planets'. He must have had in mind the Lagrange/Laplace attempt to prove that the Solar System was stable (§6.8), but he did not make any such application in the paper.

The title may have been suggested by Sturm, whose paper dealt with an even more remarkable case; beyond Cauchy's

$$\mathbf{Ax}=k\mathbf{x} \quad \text{to} \quad \mathbf{Ax}=k\mathbf{Bx}, \tag{9.10.3}$$

which we now call 'the generalized latent root problem'. Sturm used no such name, and wrote out no arrays, but he clearly realized the novelty of his treatment, which arose from solving a system of (in his case, five) ordinary differential equations ($\mathbf{A}\dot{\mathbf{x}}+\mathbf{Bx}=\mathbf{0}$) by linear sums of special solutions, like Fourier's trigonometric series (8.7.1), for example.

The mathematician who most clearly appreciated the strength of these ideas was Jacobi. In the early 1830s he developed some of Cauchy's recent ideas; then in the early 1840s he worked out many basic properties of determinants, following Cauchy's rules for assigning + or − to products corresponding to even or odd 'substitutions' of the letters. He individuated them as objects in their own right rather than consequences of other theories: for example, the coefficients did not even have to be ordinary quantities. We owe the modern use of the word 'determinant' chiefly to this work; the first of several textbooks was produced in 1851, by William Spottiswoode.

Spottiswoode – an English mathematician! What had the British been up to? First, some French background.

9.11 *'Separating the symbols': algebraists after Lagrange* {CE 4.7, 4.9}

In his book *Calcul des dérivations* (1800), the Alsatian mathematician L.F.A. Arbogast (1759–1803) extended Lagrange's algebraic approach to the calculus by 'separating the symbols' (his phrase) and detaching 'd/d*x*' (written 'D') from '*y*' in the derivative 'd*y*/d*x*', and '*f*' from '*x*' in '*f*(*x*)'. For an important example of the first, Taylor's theorem (8.6.6) was converted from

$$\Delta f(x) := f(x+h) - f(x) = \sum_{r=1}^{\infty} \frac{h^r f^{(r)}(x)}{r!} \qquad (9.11.1)$$

$$= \exp\left(h\frac{\mathrm{d}f(x)}{\mathrm{d}x}\right) - 1 \text{ to } \Delta = \mathrm{e}^{hD} - 1, .$$

where '1' was the identity operator, and each order $\mathrm{d}^n/\mathrm{d}x^n$ of differentiation was identified with its power $(\mathrm{d}/\mathrm{d}x)^n$ (a sinful move to critics!). This equation linked difference and differential calculi; it also led to summations, via

$$\sum = \Delta^{-1} = [1 + (\Delta - 1)]^{-1} = 1 - (\Delta - 1) + (\Delta - 1)^2 - \cdots, \tag{9.11.2}$$

and all sorts of related magic such as $\sum^n$ for any positive integer n. Arbogast and his followers not only enlarged the 18th-century tradition of manipulating series (§6.2); as with permutations but much more explicitly, they also extended the realm of algebra in a fundamental way by developing these two new ones *in which the objects were neither numbers nor geometrical magnitudes*. Today these algebras are called 'differential operators' and 'functional equations'; then the more descriptive names 'calculus of operations' and 'calculus of functions' were used.

Laws and rules of manipulation for these new algebras had to be found. A notable contribution was made in 1814 by Arbogast's follower F.J. Servois (1767–1847): seeking the fundamental properties of both algebras, especially functions, he characterized f as 'distributive' and f and g as 'commutative with each other' if, respectively,

$$f(x + y + \cdots) = f(x) + f(y) + \cdots \text{ and } f(g(x)) = g(f(x)). \tag{9.11.3}$$

This is the origin of these now standard algebraic terms.

If we ignore the purpose of these equations as definitions of properties, we can consider them as two examples of functional equations, where the *function* itself is the unknown; one seeks one or many functions (or maybe there are none) which satisfy the equation for some given ranges of values of the variables. In the first example f is construed as a function of its *several* variables $x, y, \ldots$; with the second the task could be to find f, given g. These equations had appeared from time to time in the 18th century {Dhombres 1986} but now they gained more prominence, especially in foundational contexts; we saw an example at (8.5.2) in Cauchy's derivation of the binomial series. Attempts to elucidate Lagrange's approach to the calculus focused much attention on them; indeed, eqns (9.11.3) arose here.

Servois's laws were applied also to functions of differential operators. Some French mathematicians used such methods to solve linear partial differential equations, difference equations (which may be treated as special kinds of functional equation) and mixed equations; but Cauchy saw them transgressing his aspirations for rigour in analysis. So in 1825 he proffered a new foundation for them in the Fourier integral solution (8.7.6) taken with a *complex*-variable kernel, $\exp[iq(x-u)]$, and where the form of the f imitated that of the operators of the equation: f_y was mapped to iu and Δ to $\exp(ihu)-1$ (following (9.11.1)), and so on up to any degree or number of independent variables. Second-order equations led him once again to use quadratic forms.

9.12 *Young English operators: Babbage and Herschel*

{Panteki 1992, Chap. 2}

When the British reformed mathematics in the 1800s and 1810s, they decided not to update Newton's fluxional calculus but to go for some Continental version. The principal reform was led at Cambridge by undergraduates Charles Babbage and John Herschel, who formed with others an 'Analytical Society' in 1813. For teaching they preferred Euler's form of the calculus; but for research purposes they adopted Lagrange's approach to propagate the algebraic faith (note the linking of 'analysis' with algebra again). Babbage concentrated on the calculus of functions, seeking solutions in ingenious but rather unrigorous displays of symbol-spinning; he manipulated functions and series, used self-substitutions of functions into equations ($f(f(x))$, for example) and deployed cunning changes of variable. He pioneered the use of the form $F^{-1}GF(x)$ as a solution, where F and G were to be found from the given function(s); it is now called 'conjugate', and turns up in other branches of mathematics (§13.12, §16.19).

The notation 'f^{-1}' was popularized by Herschel to denote the/an inverse function of a function f, together with '$\sin^{-1}$' and kindred notations for the inverse trigonometric functions. Some of his results involved continued fractions, for which he proposed the notation which has become standard:

$$'\frac{c_1}{a_1+}\frac{c_2}{a_2+}\frac{c_3}{a_3+}\cdots\frac{c_x}{a_x}'. \tag{9.12.1}$$

His interests overlapped with Babbage's, but he gave more attention to the calculus of operators and difference equations, especially generalizations of (9.11.1).

During the 1820s Babbage and Herschel turned to other things, in Babbage's case showing the same concern with algebra, algorithms and signs {Grattan-Guinness 1992a}: his 'Difference Engines', which make him the principal father-figure of computing. For the remaining 40 years of his life he worked hard on the project, but he failed to appreciate the problem of planning the construction of machines containing so many more parts than any other artefact of his time. It was in this respect – not because of any supposed limitation of Victorian technology – that he eventually failed, though his efforts helped later figures to gain success {Lindgren 1990}. From the scientific point of view the most remarkable machine was the 'Analytical Engine' of the 1830s (note the adjective again), which included not only mechanical calculation but also aspects of stored programming {CE 5.11}.

9.13 *The new generation of English algebraists: Boole and De Morgan*

{Panteki 1992, Chaps. 3–9; CE 5.1}

The mathematical efforts of these two Young Turks lingered awhile after they turned to other matters in the early 1820s; but the late 1830s saw a renaissance with the arrival of four important members of the next generation: the Irishman Robert Murphy, Augustus De Morgan, Duncan Gregory (a descendant of Newton's follower in Table 5.3.1, who died prematurely in 1844) and George Boole. A host of more minor followers (including Spottiswoode: §9.10) were influenced by Boole to study operators, but the most striking innovations lay in the applications of algebras to logic.

In 1836 De Morgan wrote the first extensive account of functional equations in an article for the *Encyclopaedia metropolitana* (§7.5). Although it gained little attention, it influenced his own ideas on logic. In the 1830s he linked syllogistic logic with Euclidean geometry by trying to expose the logical structure of Euclid's *Elements* – and he ran into great difficulties, since many other procedures are occurring, such as forming definitions and making constructions. Nevertheless, he pursued the search in his book *Formal Logic* (1847), where he also rehearsed an algebraic approach to logic which he continued in a series of articles. He used brackets and dots (these signifying negation) to represent the logical form of a proposition involving classes *X* and *Y*, as in

'*X*).(*Y*' for 'No *X* is *Y*'

and '*X*[.[*Y*' for '*Y* sometimes not in *X*'. (9.13.1)

De Morgan's main contribution to logic was a paper of 1860, in which he finally broke free of the limitations of syllogisms (while retaining the methods used for their deduction) to outline a logic of two-place relations. It is strange than nobody before him had pondered the fact that, say, 'John is older than Mary' cannot be stated *as* a relation within syllogistic logic. De Morgan now did so, with an extensive catalogue of relations and their converses and contraries. Although he did not emphasize it, the structural similarity with functional equations was strong: between fg and (son of friend of), say, or between f and f^{-1} and 'father of' and 'son of'. The logic of relations remained a central feature of algebraic logic, and was extended to relations with more than two places (§13.4).

In another coincidence to join Cauchy's pair (§9.10), De Morgan's book appeared on the same day in 1847 as did Boole's short book containing *A Mathematical Analysis of Logic*. The background to Boole's study lay principally in an analogy with differential operators. Closely following earlier work by Gregory, in 1844 he had laid out three principal laws for differential operators π and ρ (his symbols) on functions q and r; in his book he now presented a very similar trio of laws, which he wished 'elective symbols' (explained in a moment) u, v, x and y to obey. He carried the similarity of structure even to the same names, two taken from Servois's (9.11.3):

$$\pi\rho q = \rho\pi q \quad \text{commutative law} \quad xy = yx \tag{9.13.2}$$

$$\pi(q+r) = \pi q + \pi r \quad \text{distributive law} \quad x(u+v) = xu + xv \tag{9.13.3}$$

$$\pi^m \pi^n q = \pi^{m+n} q \quad \text{index law} \quad x^2 = x \tag{9.13.4}$$

Boole saw logic as mathematical psychology, an application of mathematics to the mind, giving *The Laws of Thought*, as he called his second book, of 1854. An elective symbol x denoted the mental act of picking, say, dogs from a universe of animals. Equation (9.13.2) asserted that electing dogs and English animals also yielded English dogs or, equally, dogs English; eqn (9.13.4) stated that to pick dogs twice made no difference from a single election. By 1854 his interpretation of x, as the class of dogs, had largely been preferred, and I shall use it from now on. These classes lay within a universe 1, and to (9.13.4) were adjoined these important corollaries:

$$x + (1 - x) = 1 \quad \text{and} \quad x(1 - x) = 0, \tag{9.13.5}$$

where 0 was a (somewhat unclear) notion of 'nothing'.

Boole's algebraic treatment of logic was much more radical than De Morgan's, though he did not seem to realize it until writing the second book, where syllogistic logic occupied only the 15th of the 15 chapters on logic itself. (It also included a long discussion of probability theory, much of it based upon treating compound events as combinations of simple ones in the sense of his algebra.) Representing a proposition as an algebraic equation in terms such as x, y, z, . . ., his aim was to state a collection of propositional premises in this way, and then use 'expansion theorems' to express a chosen class as a proposition comprising a logical function of the others, maybe with side conditions attached; such propositions would then constitute consequences of the premises. He recognized that solutions were not always unique, and so used 'v' to denote an indeterminate class which could be combined with known ones. For example (one of his), the consequences of the single premise

$$x(1-z)-y+z=0 \quad \text{were}$$
$$z=vxy+(1-x)y, \text{ with } x(1-y)=0; \qquad (9.13.6)$$

that is, z, taken as the unknown class in the premise, comprised none/some/all of y's which were x's and also those x's which were not z's, together with the side-condition that all x's were y's.

One of the cornerstones of Boole's approach concerned disjunction; that the class $x+y$ was interpretable only if x and y were disjoint. In *The Laws of Thought* and in manuscripts he tried to show that this law of mathematical psychology fitted in with the workings of the mind. But the argument is weak (why is it not possible to form the class of, say, Europeans or women?), and it was rejected by his first principal commentator, Stanley Jevons (1835–82), who argued in his *Pure Logic* (1864) that $x+y$ was admissible for all x and y. Jevons corresponded at the time with Boole over the example $x+x$, which for Boole could not be formed in the first place; his view, that $x+x=x$, was to prevail in the further development of the algebra of logic (§13.4).

Soon after Jevons another Englishman took up Boole's algebra, for a different purpose. The chemist Benjamin Brodie proposed that '$x+y$' denote the weight of two substances with weights x and y, and 'xy' the weight after their (assumed) chemical combination, and he tried to develop an algebra based upon the law

$$xy = x + y. \tag{9.13.7}$$

He was unsuccessful, but his correspondence at the time with De Morgan, Jevons and others reveals well how Boole's algebra of logic had been received {Brock 1967}.

9.14 *Peacock and the principles of forms*

{CE 6.9}

The main feature of this chapter has been the growing range of algebras, with their laws similar but also sometimes different from those of common algebra. It is appropriate to end, however, by recording that this common algebra itself came under scrutiny, especially in the *Treatise on Algebra* (1830) by George Peacock (1791–1858), the principal co-reformer at Cambridge with Babbage and Herschel (§9.12). Following in the English tradition in algebra established by Newton's 'universal arithmetic' (§5.14), he asserted that in 'universal arithmetic' negative numbers could not be granted status in isolation; this was permissible only in 'symbolical algebra', where 'arbitrary signs and symbols' operated 'by means of defined though arbitrary laws'. This latter stance is reminiscent of abstract algebra as it is now understood, but such a view is illusory – algebraists then still hoped that an algebra handled objects of some kind or other, over and above the structure that its laws exhibited. This was as true for the new algebras as it was for the orthodox one(s): $x + y$ in Boole's algebra of logic, or the failure of Hamilton to interpret triads, are cases of *lack* of content, not avoidance of it.

Of the figures discussed above, De Morgan worked most significantly in both common and new algebras, and vacillated quite considerably in his views over the years, trying to make contrasts between 'the art and the science' of algebra {Pycior 1983}. To him and to others, especially in

England, the expanding world of algebras was also a perplexing one.

One feature common to most of these algebras was that linear combinations were prominent. This form occurred also in many of the applications of mathematics of the time, to which we now turn.

10

Mechanics and mathematical physics, 1800–1860

10.1 *Making physics out of mechanics*

This chapter describes the introduction of mathematics to areas of physics: in rough chronological order they are heat theory, optics, electrostatics (then called 'electricity'), magnetism and then electromagnetism. Some of the mathematical methods and techniques were new, but many had been tried and tested in mechanics, which itself continued to develop in all its parts and traditions (§6.1, §6.6) and to interact with these new fields. Up to 1830 almost all the principal findings were made by the French mathematicians of Table 7.3.1, together with the astronomer and physicist François Arago (1786–1853). By then all topics were established, and thereafter all were studied internationally. Bernhard Riemann is again a point of cut-off, at around 1860, between this chapter and §14, in which work during the rest of the century is reviewed; other such figures include James Clerk Maxwell.

{Burkhardt 1908} is an unsurpassable 1,800-page survey of the pertaining mathematical methods; the transition to the more international scene is especially well done. Later he prepared a 536-page snippet {EMW IIA12}, largely on trigonometric series to around 1850. {Whittaker 1951} provides a fine overview of the physics for the whole century, although his treatment of the mathematical aspects can be rather cursory. {Grattan-Guinness 1990a, Chaps. 7–18} gives a global overview of the French work.

10.2 *Foundations and principles of mechanics*

Mathematicians of the 18th and early 19th centuries are often regarded as cavalier in their attitudes to rigour. As we saw in §6.2, this is unfair and indeed inaccurate in connection with series; and mechanics shows clearly that attitudes were sharp *when the need was perceived*. J.L. Lagrange's *Méchanique analitique* (1788) provoked several studies, especially on the assumption of the principle of virtual velocities (6.11.1): could it be proved, and if so, in terms of what?

Various successors offered proofs: some algebraic, as Lagrange liked; others, geometric {Lindt 1904}.

A paper of 1798 by Joseph Fourier is an important example of the latter kind: he even tried to prove the lever principle by means of a functional equation (§9.11). In FIG. 10.2.1 the force P at A balances the weight at B on the other

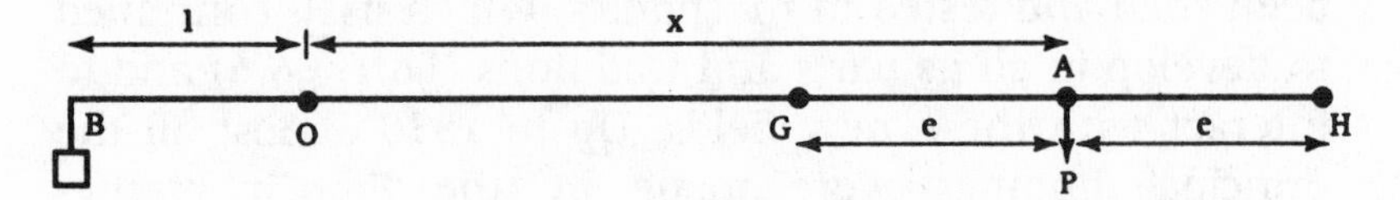

FIG. 10.2.1 *Fourier's proof of the lever principle, 1798.*

side of the pivot O; what function of P and x (= OA) is the moment Q? By replacing the force P by two forces $\frac{1}{2}P$, one at G and one at H, where GH = $2e$ (an arbitrary measure), he found that Q was linear in P (that is, $Q = Pf(x)$) and that f satisfied

$$2f(x) = f(x+e) + f(x-e), \text{ with } f(0) = 0; \therefore f(x) = x \tag{10.2.1}$$

after a bit of calculus, giving $Q = Px$ as required. Among other considerations in this paper was the insight that the principle of virtual velocities did not have to be expressed as an equality of moment of the displaced forces – the inequality $\leq$ would do. This opened up an intermittent but valuable line of thought about inequalities in mechanics (§14.4).

A heuristic illustration of the principle was offered, also in 1798, by Gaspard de Prony. He reworked it in his textbook *Mécanique philosophique* (1800) of his courses at the Ecole Polytechnique. The philosophy of his title was the *encyclopédiste* enthusiasm for classifying: in his case, notations, definitions, theorems and problems, which he laid out in separate columns on his right-hand pages – a most unusual feature.

Soon afterwards, the law of composition of forces received another valuable reappraisal, from one of de Prony and Fourier's former students at the Ecole Polytechnique. The first publication by Louis Poinsot was a textbook, *Elémens de statique*, published in 1803, his 27th year. He filled a significant gap in basic statics in such a clear and comprehensive way that one can only wonder that nobody before him had noticed the need. As was well known, but was regarded as merely an anomaly, the resultant of forces *P* and *Q*, as given by the parallelogram law, broke down if *P* and *Q* were equal in magnitude but opposite in sign and not collinear (↑ ↓) – because there was no parallelogram. The practice since at least the Renaissance had been to treat the force-pair as a torque and accommodate it appropriately in the given mechanical situation. But Poinsot realized that *a fundamental issue in mechanics* was involved here; to resolve it, he named the force-pair a 'couple', and showed that it obeyed the same laws of composition as did forces. He also showed that any system of forces in space could be reduced, though not uniquely, to a force and a couple. His treatment provides an example of a theory done first also done best; far superior to any textbook account of my acquaintance, it lacks only vector algebra to abbreviate the statement of properties.

This book appeared in editions up to the 12th (1877); Poinsot began to include later papers from the third edition (1821) onwards. One of them is an essay, first published as a booklet in 1834, in which he brilliantly applied his theory of the couple to dynamics. A rectilinear motion was construed as a couple of equal rotations in the plane normal to its direction; this led to a simplification of the theory of rotation of a continuous body B about a point P (§6.6) as the rotation of a cone with summit P about another cone at P but fixed in space, the line of contact at any instant being the axis of rotation. Similarly, he construed the motion as the rolling of the momental ellipsoid (the quadratic form with moments and products of inertia as coefficients) along a plane set parallel to the couple which initiated the motion. The point of contact would generate a curve on each surface: the 'polhode' on the ellipsoid, and on the plane and so fixed in space the 'herpolhode' (originally 'serpolhode', after *serpenter,* 'to wind'). All these notions soon became standard fare in dynamics, and contributed to a rich tradition of geometrical thinking in mechanics {Ziegler 1985}.

The algebraic side gained some attention from A.F. Möbius. His *Lehrbuch der Statik* (1837) contained some lovely results concerning the maximum and minimum values that the moment of a system of forces in space could take with respect to all the lines through a given point M; his barycentric calculus (§8.4) was an aid in defining coordinates for these lines. In some respects his theory helped to inspire Hermann Grassmann's algebra (§9.9).

10.3 *Prize new solutions for hydrodynamics,* 1815

{Burkhardt 1912}

Poinsot's colleague S.D. Poisson produced his own textbook based upon his teaching at the Ecole Polytechnique, in editions of 1811 and 1833. Claiming to cover all basic mechanics, he showed his firm non-friendship for Poinsot by failing to present couple theory at all; but otherwise he gave a good coverage of most terrestrial parts (the engineering side was rather light, and astronomy was deliberately excluded).

One of these parts was hydrodynamics, where Poisson largely followed Lagrange; but an advance was made in 1815, when both he and his former Ecole Polytechnique student Cauchy found the differential equations required to analyse motions in a *deep* fluid body, beyond the shallow bodies covered by existing theory (§6.9). The occasion was a prize problem put up by the Institute of France. Poisson was one of the judges, but he still sent in his own piece, which was published pretty promptly. The young contestant Cauchy won the prize, but his paper appeared only after a 12-year delay; typically, during the wait he added important appendices.

Given a rectangular coordinate system, the basic problem was to calculate the height z of the surface above (or the depth below) the horizontal equilibrium xy-plane of a particle of fluid at (x, y, z) when the fluid body was disturbed in some way; the prime concern was to determine the peaks and troughs of waves. Both men provided important new solutions, by means which exhibit striking similarities and differences. They worked with a velocity potential and used Laplace's equation (6.9.1) among their basic assumptions; but Poisson started out from the Eulerian model (6.10.1) using pressure, while Cauchy adopted Lagrange's 'history' model (6.10.4) Given their general inclinations, one would have expected Poisson to follow Lagrange, and Cauchy to take after Euler.

Assuming the fluid body to be of constant depth h, Poisson used Fourier series (8.7.1) to solve the equations, and then converted the solution into a double integral which included within it the expression

$$\int_0^\infty \mathrm{ch}[q(h-z)]\,\mathrm{sech}(qh)\,\mathrm{d}q, \qquad (10.3.1)$$

an important new form which we now call the 'convolution integral'. Allowing the fluid body to have variable depth, Cauchy deployed Fourier integrals (8.7.6). Neither man mentioned Fourier, whose work was available though little published (§10.11). For Cauchy the connection increased when, as announced in a short note of 1817, he discovered the Fourier integral theorem, apparently for himself. At all events, it was to become one of his staple techniques in mathematical analysis, especially for complex variables (§8.11).

Presumably independently, both men came up with an important refutation of a guess by Lagrange. In an analogy with his own work on surface waves on shallow fluid bodies, Lagrange had thought that internal waves would be propagated with uniform velocity; but Poisson and Cauchy found instead that uniform acceleration obtained. In addition, Poisson groped towards an idea which Fourier grasped much more clearly in an 1818 survey of his integral solutions: the modern term is 'group velocity' (Fourier's word was 'furrows'), the speed at which a group of waves moves, slower that those of its component waves.

10.4 *Celestial mechanics: stability and the brackets*

P.S. Laplace's *Exposition du système du monde* presented a lay account of his conception of astronomy, together with some historical accounts and speculations on cosmology. The first edition came out in 1796; three years later there appeared both the second edition and the first two volumes of his *Mécanique céleste,* followed by two more volumes in 1802 and 1805. This new work provided a detailed coverage of celestial and planetary mechanics, with especial emphasis upon perturbations and potential theory.

Among the general results, Laplace presented one by Euler which had appeared posthumously only the year before: if a given system of forces had torques P, Q and R relative to the axes of a rectangular coordinate system, then the torque M relative to the axis through the origin with direction cosines (l, m, n) was given by

$$M = Pl + Qm + Rn; \qquad (10.4.1)$$

thus M took its maximum value for an axis A with direction ratios (P, Q, R). Laplace saw that, in the context of the angular momentum of the planets of the Solar System, A was the normal to the 'invariable plane' (the plane through the centre of mass of the Solar System and perpendicular to the direction of its total angular momentum vector). This notion provided a valuable base reference for studies of the Solar System, although Laplace held back from positing A as an 'absolute' direction for celestial mechanics. Shortly (perhaps deliberately so) after Laplace's death in 1827, Poinsot made a profound criticism of the derivation: Laplace had excluded from the total angular momentum the rotations of the planets, especially that of the Sun.

Effects of the Sun underlay the last major contribution

to mathematics from the aged Lagrange and one of the first from the young Poisson, made conjointly between 1808 and 1810. We saw in §6.7 that the Lagrange/Laplace analysis of the stability of the planetary system had been made to the first power of the masses of the planets, on the assumption that higher-order terms could be neglected relative to that of the Sun. Laplace had considered such terms in his *Mécanique céleste* (1802–5), but Poisson checked them out systematically, using the expansion (6.7.1) of expressions into trigonometric series. In the course of his work he came across coefficients of the form

$$P\int Q\,\mathrm{d}t - Q\int P\,\mathrm{d}t, \qquad (10.4.2)$$

where P and Q included products of trigonometric terms. The algebraic form attracted his attention, and led him and Lagrange to start out from Lagrange's equations (6.11.2) to express the motions of a system of masses and to use the methods of variational mechanics (§6.6) to study the canonical forms of the solutions to these equations. Both men ended up with expressions to which their names are still attached. Here I give them in the notations now normally used, which were introduced by Lagrange in the second edition of his *Mécanique analytique* (1811): the 'Lagrange brackets' $[b, a]$, a sum of terms in form like (10.4.2), where in effect the potential and kinetic energy functions T and V were differentiated partially with respect to the constants of integration a and b; and their inverses, 'Poisson brackets' (b, a), where a and b were differentiated partially with respect to the independent variables in T and V. They noted the algebraic properties (for each bracket)

$$[a, a] = 0 \text{ and } [b, a] = -[a, b]; \qquad (10.4.3)$$

later Carl Jacobi added the beautiful and powerful relation

$$[a, [b, c]] + [b, [c, a]] + [c, [a, b]] = 0 \qquad (10.4.4)$$

between three variables, when he advanced variational mechanics in the 1830s (§10.17). Expressions of this kind, and the notation, remain in the vocabulary of differential equations (compare §13.9).

In 1834 Jacobi made another contribution to stability theory, of a different kind, when he showed the surprising fact that an ellipsoidal fluid body with axes $a < b < c$ of *different* lengths could rotate uniformly in equilibrium about the shortest of these axes. The requirement is this: that at every point on the surface, the resultant of the gravitational and centrifugal forces must lie along the normal there. He formulated this condition as a transcendental equation, from one of the roots of which he found the case of the spheroid where $b = c$ and rotation occurs about a, which had been shown already (by different means) by Colin Maclaurin in 1742; but other roots led to cases where $b < c$. His result was extended by Joseph Liouville {Lutzen 1990, Chap. 11}, a contribution which was to lead in turn to profound ideas on equilibrium and stability (§14.14).

10.5 *Celestial mechanics: French exactness versus German compactness*

While the French were largely supreme in mathematics, the merit of their preferred strategy in celestial mechanics of using expansions in trigonometric series was questioned. Their view, evident especially in Laplace, was that such expressions would give sufficiently 'exact' (Laplace's frequent word) an analysis if they were continued to include enough terms; but of what *use* are expressions of such lengths? When Charles Delaunay {1860, 1867} applied this principle to lunar theory to produce *chapter-length* series for the perturbation function and for the latitude and longitude of the Moon, with expansions into around 450 collections of terms, followed by 57 methods of estimating terms or checking their smallness, the doubts are understandable – notional science is in the air.

A movement against such approaches had developed among various German mathematical astronomers by the 1800s. Life is too short for such aspirations, they said, so develop *compact* methods which will give workable approximations much more quickly (especially criteria for taking only the early terms of power series); also, seek methods for checking and (if possible) improving accuracy and analysing errors statistically {EMW VI2,2}. Friedrich Bessel is notable for systematizing the methods of 'reducing observations' to calculate astronomical parameters; other concerns in the discipline led him to the 'Bessel functions' of (8.14.2).

These studies in celestial mechanics helped to raise the status of probability and statistics. C.F. Gauss's use of the method of least squares in his *Theoria motus* (1809) (§8.9) also belongs to this movement; indeed, in the next year he determined the orbit of the minor planet Pallas by using the method of solving a resulting system of linear equations by successively reducing their number by one (a technique known now as 'Gaussian elimination'). Pallas had been discovered in 1802 by Heinrich Olbers, five years after he had introduced an efficient method of determining the orbits of comets which became popular.

A rather less successful mathematical story attended the prediction of Neptune in 1846. Motivated by discrepancies between the observed and theoretical positions of the planet Uranus, Urban Leverrier and the Englishman John Couch Adams posited an exterior planet that was perturbing it gravitationally. They predicted different orbits for it, neither of which matched the real one; luckily the two predicted positions were pretty close to the actual position when a search was mounted, so that the new wanderer was spotted.

The story of celestial mechanics in the 19th century is very much one of competition between French and German methods, often to the advantage of the latter. But the nationalities interacted fruitfully over a discrepancy between theory (mainly Laplace's) and observation of the Moon concerning its secular acceleration (§5.25). In 1865 Delaunay came up with an explanation which had been given already in a general way in Immanuel Kant's

Physische Monadologie (1756): that friction between the fluid seas and the solid seabeds could affect the Earth's angular velocity and thereby calculation of motions of other bodies. After Delaunay's calculations, 'tidal friction' became a component of lunar theory.

10:6 *On planet Earth: pendulums and plans*

{Wolf 1889, 1891}

In a paper of 1813 on the attraction of spheroids S, Gauss made an important contribution to potential theory when he considered internal as well as external points P and proved a result which I express in vectors as

$$\int_S \frac{\mathbf{ds} \cdot \mathbf{r}}{r^3} = 0 \textit{ or } -4\pi \quad \text{if P is outside } \textit{or} \text{ inside S.} \tag{10.6.1}$$

It corrected the assumption that Laplace's equation (6.9.1) obtained *always*, a matter clarified, maybe coincidentally, by Poisson in the following year, with the differential equation which is still known after him:

$$V_{xx} + V_{yy} + V_{zz} = -4\pi\rho, \tag{10.6.2}$$

where ρ is the density of material at the interior point P of S. Strangely, neither man dealt with the case of surface points, where $-2\pi\rho$ obtains; Poisson did mention it in 1826, in connection with magnetism (§10.14).

Laplace's *Mécanique céleste* contained lengthy analyses of the shapes of planets that made especial use of Legendre functions. The shape of the Earth was the principal concern, and another approach for its *exact* determination became normal: the measurement of gravity by the simple pendulum. The adjective 'simple' is a misnomer, for this activity led to decimal-point science *par excellence*; this pendulum turned out to be a very complicated instrument when all the possible small effects were taken into consideration – flexure of the pendulum mount, internal friction, possible torsion effects of the suspending wire, errors arising from the motion of the wire on its pivot, equality of the upswing with the downswing, air resistance to the swings, and so on. These and other effects had been studied in the 1790s by the Frenchman Charles Borda, who also made a fine design of the instrument which became standard; his successors included Laplace, Poisson, Bessel, G.G. Stokes, G.B. Airy and George Green.

Borda was also involved in the determination of the 'metre' (his word) in the 1790s by the calculation of the length of the longitude from Paris to Barcelona, which was reported principally in the three volumes of J.B.J. Delambre's *Base du système métrique décimal* (1806–10) {Bigourdan 1901}. From the calculations, a quadrant of the Earth's circumference was determined, and the metre defined as its 10^{-7}th part. Trigonometry was to the fore, but now requiring new levels of accuracy. Legendre contributed this simple but crucial result:

> THEOREM 10.6.1 *To any small triangle T on a sphere there corresponds one on the plane with sides of the same lengths and each angle reduced from the originals of T by the amount* $\frac{1}{3}\omega$, *where* ω *is the excess of the angles of T over* 180°.

More down-to-earth was map-making, not only for producing maps and charts but also for detailed surveys for military campaigns. In France it led to the profession of *ingénieurs géographes*, with its own school in Paris and army corps. The principal *savant* of this calling was Louis Puissant, the foremost cartographer of the early 19th century, and director of the Paris school from 1809 to 1833. His *Traité de géodésie* (1805) and *Traité de topographie* (1807), with later editions around 1820, became standard reference works. He also conceived the projection for a new map of France, choosing a conical projection which gave good areal preservation over the range of latitudes covering France. He was heavily involved in its direction, which continued for decades, with dozens of *ingénieurs* to work out the fine details. Spheroidal trigonometry was employed, helped by some more elegant theorems from Legendre. The significance of

Puissant's contributions was recognized by the French Academy in 1828, when they elected him to succeed Laplace.

10.7 *Near planet Earth: meteorites and projectiles*

{EMW IV18}

In the 1800s Laplace was attracted to another planetary question: the origins of meteorites, whose extra-terrestrial nature had recently been established by several spectacular showers. His idea was that they were stones ejected from the Moon, perhaps by volcanoes. In 1803 his young follower Poisson analysed the idea mathematically as a pair of two-body problems (Moon–meteorite, then Earth–meteorite), in one of his first papers.

A subject of both scientific and military interest was the study of projectiles. Borda had contributed a clever approximate solution of 1772 which, however, ignored the rotation of the Earth. This was too simple for Laplace, who produced a treatment in 1803 which allowed for rotation by deploying both fixed and moving frames of reference in a lovely way. Poisson desimplified the analysis still further in 1838 by considering also the rotation of the projectile; in 1851 his work inspired Jean Foucault to his famous design of a long pendulum to demonstrate the rotation of the Earth {Acloque 1981}.

Laplace's analysis of projectiles included the appropriate components of the 'Coriolis force'. The name is attached to his junior contemporary, who studied the general consequences for mechanics of the use of moving axes in papers of 1832 and 1835. His contribution belonged to a tradition of engineering mechanics, to which we now turn.

10.8 *Engineers' energy mechanics and physicists' energy conservation*

{Haas 1909}

In §6.12 we saw that in the 1780s Lazare Carnot was formulating doubts about the generality of Lagrange's treatment of mechanics, especially relative to the needs of technology. The doubts increased after 1800 among some teachers of mechanics: J.N.P. Hachette and then G.G. Coriolis at the Ecole Polytechnique, Navier at the Ecole des Ponts et Chaussées and J.V. Poncelet at the military school at Metz. The climax was reached in 1829, when both Coriolis and Poncelet published textbooks in which Carnot's views were vindicated {Grattan-Guinness 1990a, Chap. 16}. Here energy principles were given the prime place that he had envisioned, but construed in a general enough form that *forces vives* (the rendering in French of Leibniz's term *vis viva* (§5.11) to express kinetic energy) could be converted into 'work' (Coriolis's term, *travail*) *when impact occurred*. Over a given interval of time, masses m moved under 'moving forces' P and resistant ones P', traversing distances ds and ds' along their respective curves of action, and changing velocities from v_0 to v, expressed as an equation with the form

$$\text{`}\sum P\,\mathrm{d}s - \sum P'\,\mathrm{d}s' = \tfrac{1}{2}\sum mv^2 - \tfrac{1}{2}\sum mv_0^2\text{'}. \qquad (10.8.1)$$

Indeed, the factor $\frac{1}{2}$ in the *forces vives* was proposed by Coriolis, and soon became standard; he also modified the equation to allow for 'his' force and for moving axes of reference.

More significant for theory, however, was the fact that the work terms were *not* assumed to be exact, and so did not take a potential (dQ, say); for in impact and percussion, P and/or P' would be discontinuous functions over time, and so the required equalities of mixed partial derivatives would not obtain. The consequences of this stance were very profound. From a mathematical point of view, it blocked any easy integrations of (10.8.1), since the work terms could not be integrated to give a tidy Q (contrast Lagrange's principle (6.11.1) of virtual velocities). It also dispelled Lagrange's claim that his approach to mechanics was general, exactly as Carnot had envisioned. However, its own generality was limited, in that physical effects (such as loss of heat at impact) were left out.

The assumption of potentials was to be reaffirmed in the 1840s in the espousal by physicists of the 'conservation of energy'; but Q now included 'all' factors – mechanical, physical, chemical and thermic. The mathematical side was considered in most detail in the version by the young Hermann von Helmholtz in an essay with the above title {Helmholtz 1847}. The 'tension forces' $\phi(r)$ acting at a distance r that 'endeavour to move' mass m with velocity $v(r)$ to a position given by $r = R$ was given by an integral, so that (10.8.1) became

$$\int_r^R \phi(r)\,\mathrm{d}r = \tfrac{1}{2}mv(R)^2 - \tfrac{1}{2}mv(r)^2. \qquad (10.8.2)$$

For a collection of such masses, each term in (10.8.2) was summed over them. The integral granted $\phi(r)$ a potential; in a curious slip, Helmholtz interpreted it as a sum of *lines* of length $\phi(r)$.

Heat theory was fundamental to the rise of thermodynamics in the 1850s in, for example, 'the dynamical theory of heat' proposed by William Thomson {C. Smith 1976}. The French engineering background was again evident, but this time in a short book of 1824 on 'the motive power of machines' by Carnot's son Sadi. Apart from a mathematical treatment by Benoit Clapeyron in 1834, his French colleagues had not paid it much attention {Redondi 1980}; but now its idea of cycles was to bear great fruit for energy physics, which enjoyed an extraordinary range of influences in science, and even outside it, throughout the century (§15.2).

10.9 *The performance of man and machine*

The engineers, especially those influenced by Poncelet, applied (10.8.1) to their concerns. Hydraulics benefited rapidly, with a design of a waterwheel by Poncelet himself in which impact was reduced by the use of curved rather than straight blades (1825), and a similar scheme for a turbine by Benoit Fourneyron (1834). A principal cause of energy loss was friction of various kinds; their effects in military equipment were analysed in remarkable detail by Poncelet's colleague at Metz, Arthur Morin.

Another topic in which the equation played a role, together with basic ideas in statics such as the couple, was governors on machines {CE 8.4}. In Britain, Astronomer Royal G.B. Airy, motivated by the problem of steering large telescopes by machine to track the stars, made an important contribution in 1840 by seeking conditions under which the governor would be *prevented* from oscillating between two different locally stable positions and thereby frequently switching the machine on and off.

A further application of energy mechanics was to ergonomics, where the work-rate of men and animals was appraised, especially by Charles Dupin in his teaching and public lectures at the Paris Conservatoire des Arts et Métiers from the 1820s onwards. Mechanics was exerting an influence here upon sociology, to coin the word proposed at this time by the philosopher Auguste Comte. Another influence upon him will appear in §10.11.

Engineers also became concerned with the optimization of economical systems, especially involving transport: were these new railways, for example, a better investment for investors than canals or roads? Usually some function $f(x)$ of the one or several variables involved in the problem was formed, and an answer obtained by solving $f'(x) = 0$. Gabriel Lamé and Clapeyron, young graduates of the Ecole Polytechnique in the mid-1810s, had socialist inclinations and were therefore instructed by the French Restoration government to work with the Russian Corps of Ways and Communications in St Petersburg. There they produced in 1829 a marvellous generalization which could have launched operational research and locational equilibrium theory. They formed the 'warehouse problem', as it is now known, where the optimal location is sought for a depot to supply a system of given locations with known quantities of goods, and they solved it by using a mechanical analogue in which weights were suspended on strings all fastened together to a little ring {Franksen and Grattan-Guinness

1989}. However, although both men followed successful careers in Paris after returning in 1831, neither publicized his innovation; so another mathematical topic was to sleep for a very long time (§16.22).

One absentee from engineering mathematics during this period was the statistical control of production. One would expect the Industrial Revolution to have inspired its growth, but evidently it did not; we find it rising only a century later (§16.22). Perhaps manufacturers held the same view as did Charles Babbage on his calculating 'engines' (§9.12): that parts could be made with such accuracy as to be essentially identical. But the case of Babbage is ironic, as he had helped to encourage the founding of some statistical organizations in Britain during the 1830s (§8.10)!

10.10 *Extending mechanical principles: Laplace's molecular programme*

{Fox 1974; Grattan-Guinness 1990a, Chap. 7}

During the early 1800s, Laplace became more interested in 'molecular' physics (his term). For example, he considered Newton's fudged 'calculation' of the velocity of sound (§5.11) in 1803, and suggested the explanation that the compression and expansion of particles of air caused the sound to pass; and in the 1810s he and Poisson refined the analysis with a study of gases which stressed the difference between the specific heats of a gas at constant pressure and at constant volume.

A next step was presented in the fourth volume of Laplace's *Mécanique céleste* (1805), in which he analysed the curved path of a ray of light as it passed through the atmosphére. He advocated a Newtonian model (§5.13): a ray of light is a tiny molecule, moving fast like a bullet but subject to central forces of attraction and repulsion with the molecules of the atmosphere. From this assumption he formed a differential equation to describe the path. His motive was to correct astronomical observations for atmospheric refraction, because instruments had become accurate enough to allow for it.

Laplace soon extended this molecular conception to cover 'all' physical phenomena. With his friend the chemist Claude Berthollet, who adhered to a similar prevailing molecular model of chemical substances, he formed a school of Bright Young Men – Ecole Polytechnique graduates especially – to apply the idea to as many areas as possible. They formed a school in the strict sense of the term; there was even a name, the Société d'Arcueil (after the district, just outside Paris, where the two leaders lived in neighbouring mansions), and an associated *Mémoires* in which to publish papers.

The programme was dominated by its physical and chemical sides; but for Laplace the mathematical aspect was equally important, and in this respect Poisson was his chief lieutenant, with occasional contributions from Biot. It followed these principles:

1. The force $f(r)$ between two molecules was known only to decline rapidly with the distance r between them; but an inverse-square law could not be assumed, since the shapes of the molecules were unknown.
2. The kinds of forces involved in $f(r)$ were either attractive, like gravitation, or repulsive, due to heat.
3. Each molecule was subject to forces from all other molecules in its vicinity; their effect was assessed where necessary by the usual rules, such as the cosine law of decomposition into components.
4. The aggregation of such forces upon a given molecule was to be expressed by an integral; that is, the structure of the supposed physical actions was reflected in the mathematics, and the actions could cause the molecule to rotate or oscillate as well as translate.
5. Fluids of various kinds (caloric, electric, magnetic) were assumed to exist.

This last principle suited the physics of its day; the fluids were not analysed mathematically.

Laplace's first detailed application was to capillary theory {CE 9.9}, where he produced a remarkable and successful analysis (1806–7, published as supplements to the *Mécanique céleste*). But the greatest success came in optics, where he and his followers assumed that the 'bullets' oscillated and rotated in various ways in order to explain the various phenomena of light. One class of phenomena was called 'polarization' by his most important optician, Etienne Malus, in 1811; the name expressed the belief that the bullet oscillated to and fro like a mounted magnetic needle. The word has remained to this day, even though the theory which inspired it has long gone.

Malus was also responsible for a fine mathematical achievement, in a prize paper for the Institute of France in 1809. One of Huygens' contributions to optics (§5.13) had been a construction using an ellipse to locate the paths of both rays of light produced when an incident ray was doubly refracted. Malus used the construction in a mechanical explanation of the phenomenon by deploying the principle of least action (6.6.4) to determine equations for the velocities of the rays. He won the prize; but Monsieur le Chef Laplace won the praise by appropriating and refining the method and publishing it first.

The programme made progress in other areas of physics, following the principles outlined above. I note a few examples in the following five sections; entering the picture will be the other innovators of mathematical physics – for most of whom Laplace's programme was uncongenial.

10.11 *Outside mechanics: Fourier on the diffusion of heat*

{Grattan-Guinness and Ravetz 1972; CE 9.4}

The first major mathematical breakthrough outside mechanics was made between 1804 and 1807 by Fourier, then in his late thirties. His work was done in his spare time from employment as Prefect of the Department of Isère based at Grenoble, and from his hobby of Egyptology (§2.2).

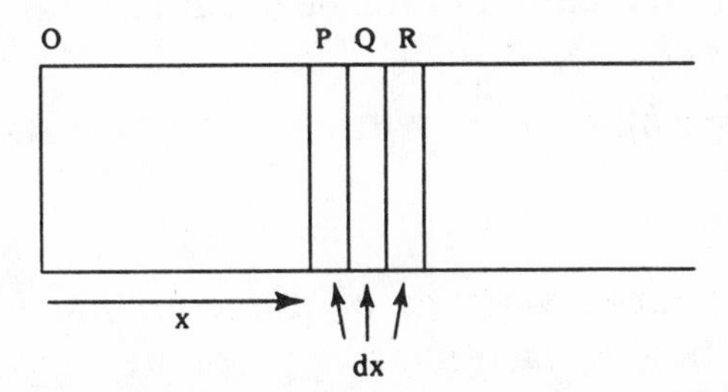

FIG. 10.11.1 *Fourier's derivation of the heat diffusion equation, 1807.*

To find the equation of heat diffusion, Fourier followed the normal practice of the Eulerian differential calculus, assigning constant differentials (6.2.2) to length x and time t. FIG. 10.11.1 shows his model for a thin, straight bar, which diffused heat within itself with coefficient of conductivity K, and also into the environment (set at temperature 0) with coefficient h. He considered the balance of heat inside the differential slice Q of width $\mathrm{d}x$ and x along from one end. Newton's law of cooling, that the heat loss was proportional to temperature difference, gave him the following expressions for the internal flow into Q from slice P (with temperature y at the interface with Q) and out to slice R (with temperature y') over a constant time-interval $\mathrm{d}t$:

$$(-K \div \mathrm{d}x)\,\mathrm{d}y\,\mathrm{d}t \text{ and } (-K \div \mathrm{d}x)\,\mathrm{d}y'\,\mathrm{d}t, \tag{10.11.1}$$

$$\text{with difference } K\,\mathrm{d}(\mathrm{d}y \div \mathrm{d}x)$$

across the slice. The equation of heat balance,

$$(\text{internal heat-flow}) - (\text{external heat-flow}) = \text{net heat gain}, \tag{10.11.2}$$

therefore took the form

$$K\,\mathrm{d}(\mathrm{d}y \div \mathrm{d}x)\,\mathrm{d}t - h(y-0)\,\mathrm{d}x\,\mathrm{d}t = CD\,\mathrm{d}x\,\mathrm{d}y, \tag{10.11.3}$$

where C was the specific heat of the bar and D its density, both assumed to be constant. Tidying up gave the literally partial differential equation for heat diffusion as

$$K\frac{\mathrm{dd}y}{(\mathrm{d}x)^2}-hy=CD\frac{\mathrm{d}y}{\mathrm{d}x}, \text{ with } \mathrm{d}x \text{ and } \mathrm{d}t \text{ constants.} \tag{10.11.4}$$

Fourier produced versions of this equation for various other simple bodies, each in an appropriate coordinate system. For solid bodies such as the sphere, (10.11.4) lost its h-term and described only internal diffusion; the external effect was expressed in boundary conditions of the form

$$k\frac{\mathrm{d}y}{\mathrm{d}x}+hy=0 \text{ when } x=X \text{ (the radius, or whatever).} \tag{10.11.5}$$

For solutions of (10.11.4), Fourier turned to trigonometric series; in particular, when $h=0$,

$$y=a_0+\sum_{r=1}^{\infty}[(a_r\cos rx+b_r\sin rx)\exp(-Kr^2t)], \tag{10.11.6}$$

with $0\leq x\leq 2\pi$.

As was described in §8.7, this was *not* then a normal solution of differential equations. His interest seemed to stem from an older type of model that he had used already, in which a continuous body was replaced by n separate ones (such as n equal blocks along a line instead of a straight bar), n ordinary differential equations were formed and solved, and then $n\to\infty$ to get the required answer. Now, an error made in the specification of K, which he did not detect, had given a false final solution; however, the n-body solution, a *finite* trigonometric series, looked promising enough to be worth imitating in the direct approach using differentials. For the cylinder these series did not work; so he turned to the Bessel function, with excellent consequences noted at (8.14.2).

While the series provided lovely problems for the foundations of mathematical analysis, they were clumsy to use in applications, especially for comparison with empirical data. Fourier himself had to reduce them to the first term when comparing them with his experimental findings, and so he found only weak corroboration.

Fourier submitted a very long paper on all these things to the class of the French Institute in December 1807. There were two disputes: with Lagrange, on the legitimacy of Fourier series (§8.7); and with Laplace, on the fact that (10.11.4) had not been derived using his molecular principles involving the force function f. Laplace himself made such a derivation in a paper of 1810, where internal conduction – just 'K' to Fourier – involved the integral $\int_0^\infty s^2 f(s)\,ds$ of heat interaction between molecules of the body at distance s apart. He also found the integral solution (8.7.5) of the diffusion equation for an infinite body (when the surface condition (10.11.5) did not obtain, of course, since there was no surface); it inspired Fourier to his own integral solution based upon (8.7.6).

Perhaps these success led Laplace to view Fourier favourably; at all events, he was behind the proposal of a prize problem on the subject for 1812, which Fourier won with a second, extended paper. But Institute approval was still ambiguous, and the world at large learned of Fourier's various results only when his book *Théorie analytique de la chaleur* appeared in 1822. By then Poisson was also getting into print with Laplacian reworkings of the principles using molecular modelling and integrals; while he solved the diffusion equation for some further bodies, the work was pretty routine. But also in 1820, Laplace expressed his own *approval* of Fourier's efforts, which had made him sensitive to the importance of heat in physics: they both analysed the cooling of the Earth by solving Fourier's diffusion equation for a large sphere, and used their solutions (Fourier with his series and integrals, Laplace with surface harmonics) to estimate its age. This problem became an important multidisciplinary question later in the century.

Yet Fourier remained philosophically distant from Laplace. He never extended his enthusiasm for trigonometric series to a structurally similar interpretation of heat itself as a wave phenomenon, which *was* a theory advocated at the time; but neither did he affirm the rival 'caloric' heat-as-fluid theory favoured by the Laplace school. For him heat was heat, with cold as its opposite; there was no need to ponder its intimate structure. In this respect he was a positivist – and I use that word deliberately, for Comte was to popularize it in his *Cours de philosophie positive*, of which Fourier was a dedicatee and where his work on heat was hailed as an apotheosis of scientific method.

Comte's book began to appear in 1830, the last year of Fourier's life. In his final years, Fourier saw his heat theory transformed from the heresy of 1807 to normal mathematical physics, and many of the new generation took it up. One consequence is noted at (10.15.2).

10.12 *Inside the aether: Fresnel and the waves of light*

{CE 9.1}

While the Laplacian opticians were worrying in the mid-1810s about the oscillations and rotations of their bullets, a recent graduate from the Ecole Polytechnique and the Ecole des Ponts et Chaussées was spending his spare time from designing a road near Aix in southern France wondering whether optical phenomena could be modelled in terms of motions of the all-pervading aether. These wonderings turned into an obsession that dominated the rest of the short scientific career (1815–27) of Augustin Jean Fresnel; with some help from Arago and others he secured engineering appointments (including secretary to the Commission for Lighthouses) in or around Paris, giving him better contacts and some laboratory facilities. Eventually he won over most of his colleagues to seeing light his way rather than that of the Laplacians; indeed, Laplace is reported to have been greatly impressed by his treatment of double refraction.

Fresnel's initial mathematical contribution occurred in 1816, his 29th year, as part of his first extended study, of diffraction. He found that the 'Fresnel integrals' (as they are now known)

$$\int_{-c}^{\infty} \frac{\sin}{\cos} az^2 \, dz, \text{ with } a \text{ and } c \text{ positive constants,} \qquad (10.12.1)$$

expressed the sum of the interactions of wavefronts near a diffracting source. They lack a closed-form integral; so he compiled a table of their values.

After diffraction, Fresnel passed on to reflection and (double) refraction, and polarization. His main principles were these:

1. The aether is composed of tiny particles of its own, and it is their periodic vibration about their positions of equilibrium that generate wavefronts. With Huygens (§5.13) but unlike the Laplacians, the word 'ray' was just a way of speaking.

2. His first idea was that the vibrations were longitudinal, along the direction of propagation of the ray —←•→—; (10.12.1) arose in this way. But in 1821 he changed to the idea, due to Ampère, that the vibrations moved *transverse* to the direction of propagation. He also assumed that the aetheric particles are arranged in a regular lattice of equilateral triangles:

```
        • • • • •
 . . .              . . .      (10.12.2)
        • • P• • •
```

When particle P vibrated in simple harmonic motion about its position of equilibrium, the changes in the forces exterted on P by the other particles produced a resultant force acting up or down the page at P.

3. Waves possess properties of interference; that is, of interacting so as to support each other in certain circumstances and to oppose each other in different ones. Fresnel found that Thomas Young in England had recently

come to the same idea, although without the mathematical trimmings.

4. Huygens' principle (§5.13): the motion of a particle caused a wavefront which acted as a source of secondary waves.

Fresnel's achievements show not only great experimental skill, but also deft handling of mathematical methods {Kipnis 1991}; as usual, I concentrate on the latter. He did not talk in the holy Parisian language of differential equations, but instead deployed analytic geometry and analogies from mechanics, such as a cosine law of decomposition of the intensity of light in a given direction. The utility of the model (10.12.2) is not clear, since it works in the plane whereas rays are propagated in space (compare Maxwell's model of the electromagnetic aether in FIG. 15.10.1). He also found a new reading of double refraction using the principle of the conservation of energy, though not in the very general conception (10.8.1) favoured by his engineer colleagues: since the intensities were respectively $I \sin^2\alpha$ and $I \cos^2\alpha$ (where I and α were the intensity and the angle to the normal of the incident ray), then the optical 'energy' was conserved during the process of double refraction.

This proposal exemplifies well a major aspect of Fresnel's superiority over the Laplacians: he could *deduce* some results, especially concerning intensity, which they could put forward only as empirical findings. In this respect, therefore, mathematical methods were central to his physical purposes. His superior experimental ability was also crucial. Nevertheless, despite his successes and their special concerns with optics, his work did not gain the same reception among his *confrères* after his death in 1827 as did Fourier's work on heat diffusion: his only immediate successor on theory was to be Cauchy (§10.16).

10.13 *Line and surface integrals: Ampère's electrical electrodynamics*

{Grattan-Guinness 1990a, Chap. 14}

The idea of transverse vibrations in the aether was not Fresnel's: it was proposed by a man who later became his close friend and landlord in Paris. Ampère was a professor of mathematics at the Ecole Polytechnique, and lived round the back; his other interests included psychology and Kantian philosophy. But Hans Christian Oersted's discovery of 'electromagnetism' (his word) in 1820, that a wire carrying current caused a mounted magnetic needle to oscillate, changed Ampère's life that September when Arago repeated the experiment at the Academy of Sciences. The next week he could demonstrate there 'electrodynamics' (his word), the effect of wire upon wire.

This 45-year-old mathematician then devoted the next seven years to physics, in which he had no training. The inspiration may have been religious: in the history of electricity and magnetism there are several figures before, during and after Ampère's time who saw those phenomena to be based in the aether, and interpreted them as Pantheistic evidence of the existence of God. At all events, he produced some substantial work, of which the mathematical parts are the chief concern here. Like his lodger, influences from mechanics were quite profound. Concentrating upon electrodynamics, he decided in a brilliant move to seek 'cases of equilibrium', where *no* action between wires took place; analysis of the manner of motion would be too difficult. He assumed that central forces acted between infinitesimal elements of the wires, thanks to the aether (which, unlike his lodger, he did not attempt to structure), according to the law r^n, r being the distance between them. In imitation of Coulomb's work of the late 18th century on electricity and magnetism, he hoped to prove that n took the Newtonian value of -2. He set to zero the expressions representing either the forces or moments of forces in his cases of equilibrium, and found that, indeed, $n = -2$ by solving the resulting equations.

Mathematically speaking, Ampère's sequence of cases forms an excellent example of desimplification: from a trigonometric expression for the action between two straight elements through a differential expression for curved ones, and then to a line integral for an element acting upon a wire in space; he also used contour integrals for current passing round closed circuits. While the mathematics itself was not 'hard', the use of line and surface integrals was rather innovative: the best example is reported at FIG 10.15.1. He appreciated its importance, for he told himself in a manuscript note: 'My three great things, the attractions, the projections, the areas.'

10.14 *Poisson's isolated contributions to electricity and magnetism*

Poisson could not let all this non-Laplacian mathematics pass unchallenged. His first contribution came in two papers around 1812, his 32nd year, when he suddenly provided an extensive analysis of electrostatics in which he adapted potential theory and Legendre functions from planetary mechanics in order to calculate the net depth of difference between the two electrical fluids deposited upon a spheroid, or upon two spheres in interaction, when in equilibrium. The high mathematical point was a brilliant solution found in the second case to a most complicated functional equation; the physical interest lay in the good correlation found between his calculations of attractive or repulsive forces and the data obtained experimentally 30 years before by Coulomb. Nevertheless, although his colleagues praised his work, little was done to build upon it.

Twelve years later, in 1824, Poisson took up magnetism (but not electromagnetism). The problem he tackled was to determine, in terms of the quantities of boreal and austral magnetic fluids, the 'action' of a magnetic body A upon an isolated pole set at an exterior point P. His model may be conveyed thus. Imagine a flat plate containing a scattering of dry rice seeds, and a pea laid outside it at P; then the plate is a plane section through A with its discrete tiny but not infinitesimal seed-like magnetic dipoles ('magnetic elements') D acting upon the pea at P. Drawing again upon the known potential theory, he imposed a rectangular axis system and calculated the components of the action it exerted upon P as surface integrals over a seed; the total action of A at P was then given by volume integrals over A. One feature of his analysis warrants special discussion.

10.15 *A new stage in potential theory: new roles for surface integrals*

{CE 3.17}

We see line and, especially, surface integrals coming to the fore. The point is worth stressing since, while not completely new at this time, such integrals were still very occasional fare for mathematicians – Cauchy, for example, said nothing about them in the reform of mathematical analysis that he was effecting (§8.6). Three examples are noted here.

1. The most remarkable instance occurred in Poisson's analysis of magnetism; for he 'simplified' (his word) the volume integral expressing the magnetic potential of A at P by converting it into an integral over the surface S of A, with outward direction angles (l, m, n) at the surface, by integrating by parts successively in the three axial directions (x, y, z). Let $\mathrm{DP} = r$, and let k be the 'magnetic density' at D; his result was

$$\iiint_A \left[\left(\frac{\alpha k}{r}\right)_x + \left(\frac{\beta k}{r}\right)_y + \left(\frac{\gamma k}{r}\right)_z\right] \mathrm{d}x\,\mathrm{d}y\,\mathrm{d}z \qquad (10.15.1)$$

$$= \iint_S (\alpha \cos l + \beta \cos m + \gamma \cos n)(k/r)\,\mathrm{d}s,$$

where α, β and γ were certain surface integrals defined over D. Poisson realized that his result was not restricted to magnetism itself; he also extended it to D's which were not convex by breaking up the integral into appropriate intervals of values of x, and to interior locations of P by admitting the local density term from 'his' equation (10.6.2). But the significance of the gold in his hands escaped him. The simplification resulting from eliminating one integration is a minor point; the main one, reflected in our name 'divergence theorem', was to be grasped by Green and his successors (§10.18).

2. Poisson's treatment of magnetism posed a severe challenge to Ampère; for one basic question posed by Oersted's discovery (§10.13) concerned the relationship between electricity and magnetism themselves. Many figures, including probably Poisson, took them to be distinct phenomena although susceptible to this interaction; but for various reasons, especially the electrodynamic effect, Ampère had decided that magnetism was a special kind of electricity. Thus the dipole D had to be replaced by a tiny 'solenoid' (his word, from the Greek word for 'canal'), a tiny coil of electric current lying along the axis of D.

Ampère overcame the challenge of Poisson's work in 1826 by means of a remarkably original theorem. In his FIG. 10.15.1, the right-hand shape was in space, its surface σ bounded by contour s, with an arbitrary element $d^2\sigma$ sitting at m; a second surface σ_1 (not shown), with the same contour, lay close to it. Similarly, the element $d^2\sigma'$ of a surface σ' (also not shown) was located at m'. Ampère took the Poissonian expression (E, say) of magnetic action between m and m' and mapped the surface σ across to the plane Oyz, giving the shape on the left side of the Figure. By transforming the variables appropriately, he found that E was transformed into a version of *his own* expression for the action between an element and an electrical solenoid; therefore the basic link between Poisson's theory and his own electrical conception of magnetism was established. Various integrations were effected to develop the connections between the two

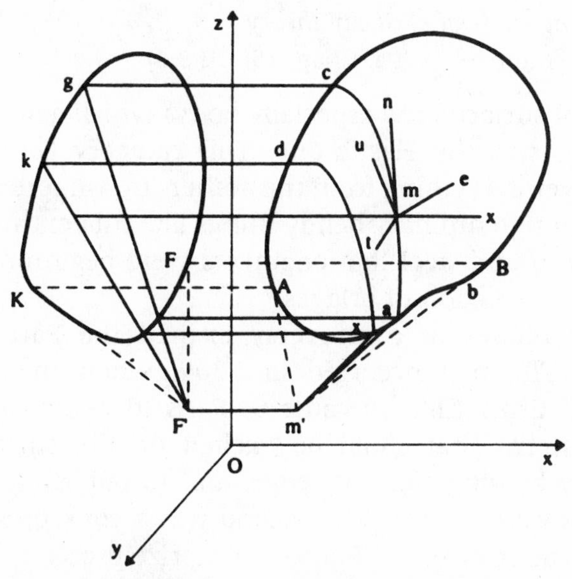

FIG. 10.15.1 *Ampère's translation of Poisson's theory of magnetism, 1826.*

theories: surface to the neighbouring surface, surfaces to element, and finally surfaces to surfaces.

3. At this time Fourier's heat theory drew such integrals from the new generation: Jean Duhamel and Mikhail Ostrogradsky, the latter a Russian visitor in Paris. Their problem was: what mathematical properties underlie the use of trigonometric series (10.11.6)? Both men took two of his elementary solutions u_r and u_s for a solid body B, and by using integration by parts and the surface condition (10.11.5) they showed via (10.15.1) that

$$\iiint_B u_r(x, y, z) u_s(x, y, z) \, dx \, dy \, dz = 0 \text{ when } r \neq s. \qquad (10.15.2)$$

But the sine and cosine functions in Fourier's series had this property, as required. Of course, we can see that this does not truly establish their orthogonality (compare §16.12); but it is interesting to see the way that surface integrals and the divergence theorem appeared again.

10.16 *The stress of elasticity theory*
{Grattan-Guinness 1990a, Chap. 15; CE 8.6}

A type of surface (and especially body) which gained much attention was the elastic one. This category could – and did – cover everything from the aethers in Ampère's household, through ordinary bendy things like rubbers and twigs, to the iron and steel that engineers were beginning to use in the construction of bridges.

Three phases of controversy excited the Paris coffeehouses. The first occurred in 1809, when the German physicist Ernst Chladni came to Paris to demonstrate the sand-patterns that could be excited on the surface of a drum by bowing along its edge, and to publish in French his ideas on acoustics. The second was a consequence: the class of the Institute of France set a prize problem for 1811 on the analysis of the elastic lamina. Poisson was doubtless the intended winner, but he gained election in 1812 anyway following the premature death of Malus. His own paper, of 1814, naturally followed Laplacian principles of molecular modelling, although the vile fourth-order differential equation which he (unintelligibly) produced was insoluble. The prize eventually went, in 1815, to Sophie Germain, the one woman mathematician of note at the time. She followed and extended the geometrical style of Euler; but the best parts of the derivation of the differential equation were due to Lagrange and Legendre, while her solutions, intended to predict Chladni's experimental patterns, were marred by elementary errors in passage work.

The third phase was much more significant. In 1819 engineer Claude Navier, newly appointed professor at the Ecole des Ponts et Chaussées in his 35th year, began a series of papers on elastic lines, surfaces and solids, stimulated by his current responsibility for investigating the feasibility of building suspension bridges. His methods also followed Euler's, but in a more original way. From an analogy between elastic solids and viscous fluids, he also produced in 1822 a form of the equations now known as 'Navier–Stokes' for reasons explained in §10.19.

One feature of Navier's approach was to assume that resultant forces at a point on a line or surface would always act along the normal there. This assumption was criticized in 1823 by Cauchy, in one of his greatest ideas: the resultant might be *oblique* to the surface. (He had been stimulated to find this result partly by Fresnel's analysis of the properties of the aether in double refraction.) This simple change led him to an intense programme of work over seven years and 500 printed pages between 1827 and 1830. His type of model was to be called 'stress–strain' by the Scottish engineer W.J.M. Rankine some decades later (§14.8): isolate any part of an elastic body, and consider how the forces acting at its imagined surface, quite possibly obliquely, affect its expansion and contraction. He always assumed that the relations between stress and strain were linear, with constant coefficients expressing the proportions. Another major innovation was to assume that, even for an isotropic body (that is, one with the same measure of elastic strain at every point and in every direction from it), *two* coefficients were needed: k, a function of location, and K, a function of the extendability of the body there. Thus if A were (say) the x-component of the stress at $P(x, y, z)$ which disturbed it by the measure (p, q, r), then the linear relationship with the corresponding component of strain was given by

$$A = kp_x + K(p_x + q_y + r_z). \qquad (10.16.1)$$

Cauchy produced a wide range of models, of varying degrees of generality. In those for which a molecular structure was assumed, symmetries like Fresnel's (10.12.2) had to be imposed upon them. The most comprehensive models were therefore 'purely' continuous, especially one using 60 coefficients which allowed for the torsion of the elastic body. In these cases Cauchy rejected Laplace's assumption that cumulative action could be represented by integrals (§10.10) and used sums instead, on the grounds that the action of nearby molecules was too discrete to allow an integral to be taken safely. In his last papers, from 1830 on, he applied appropriate models to Fresnel's wave optics, paying especial attention to the difficult topic of optical dispersion (§10.21).

Meanwhile, Poisson soldiered on with even more pages (about 700) of Laplacian elasticity theory. After Cauchy it does not make exciting reading, and in places there is rather a sense of *déjà vu* (where, for example, he replaced integrals by sums, though justified by an incomprehensible argument). Perhaps the most valuable part was an extensive reworking in 1831 of Laplace's theory of capillarity using sums rather than integrals and modifying some other assumptions. But mathematical molecularism was losing some of its prestige; among his younger colleagues, only Lamé was working with it, and then with the particular aim of generalizing coordinate systems (§12.6). In capillary theory itself, Gauss had refined and modified some of the results in 1829, but by using the principle (6.11.1) of virtual work and the calculus of variations.

10.17 *Hamilton's variational optics and mechanics*

In places above we have noted some contributions from outside France and/or after the 1830s; they form the main theme of the rest of the chapter. They came to the fore especially in optics, electricity and magnetism, where the offerings of Fresnel, Ampère and Poisson were rather passed over by their colleagues.

One of the first important responses to French work came from the youngest contributor. In 1824 William Rowan Hamilton presented a long paper to the Royal Irish Academy, which they published four years later. I know of no comparable piece of mastery from a mathematician aged 18 years. The approach that he followed was to characterize his career, as with Lagrange's (§6.11): to algebraize a theory which had been created in a geometric style. His initial target was Fresnel's optics, which he not only reworked but extended in certain ways.

The most remarkable prediction came in one of the three sequel papers. Fresnel had represented properties of double refraction by a fourth-order equation defined by the intersection of a sphere with a spheroid. Hamilton noted that at four points it took an infinity of tangent planes, so that a ray emerging at any one of them should generate a *cone* of light. His colleague Humphrey Lloyd verified this prediction experimentally, and thereby introduced conical refraction into optics.

This success in optics led Hamilton to extend his methods into 'a general method in dynamics, by which the study of the motions of all free systems of attracting or repelling points is reduced to the search and differentiation of one central relation, or characteristic function'. He sent a paper with this title to the Royal Society in 1834, and its own sequel there in the following year. The full cycle of his papers is gathered together in {Hamilton 1931}.

Hamilton's admirably clear title shows his purpose: it records the origins of Hamiltonians (as we now say) in mechanics. He let n masses m_i move from points (a_i, b_i, c_i), to (x_i, y_i, z_i) in time t, with r_{ij} as the distance between m_i and m_j, under some central forces of attraction and repulsion for which the potential was $F(r_{ij})$, (for example, $1/r_{ij}$ in Newtonian mechanics). The characteristic (or 'principal') function S was then defined as

$$S := \int_0^t \left[\textstyle\sum_{i,j} (m_i m_j F(r_{ij})) + \frac{1}{2}\sum_i (m_i(x_i'^2 + y_i'^2 + z_i'^2)\right] \mathrm{d}t, \tag{10.17.1}$$

where I follow his use of primes to indicate 'differential coefficients' with respect to t. Variation relative (only) to space produced the n basic equations of dynamics:

$$\textstyle\sum_i [m_i (x_i'\delta x_i - a_i'\delta a_i + y_i'\delta y_i - b_i'\delta b_i + z_i'\delta z_i - c_i'\delta c_i)] = \delta S, \tag{10.17.2}$$

which split into $6n$ equations on setting to zero each coefficient of a variation as usual. He went on to discuss the possible numbers of equations and their integrability, and the formulation of S and similar functions in general rather than rectangular coordinates.

While (10.17.2) looks somewhat like Lagrange's basic equation (6.11.2), Hamilton was quite clear about the differences in approach. He described his researches to the British Association in 1834 thus: 'Lagrange's function *states*, Mr. Hamilton's function would *solve* the problem. The one serves to form the *differential* equations of motion, the other would give their *integrals*' {Hamilton 1931, p. 214}. In one form or another, this approach was to become a major method in both mechanics and mathematical physics.

Jacobi was an important follower, using results such as his addition (10.4.4) to the Lagrange/Poisson theory of brackets to find new conditions for solving (10.17.2) and related equations. He also contributed to the three-body problem in 1843 by reducing it to the study of the motions of two fictitious bodies {CE 8.9}. In other work, he showed the power of elliptic functions as, or in, solutions of differential equations in dynamics.

10.18 *Green's function and Dirichlet's principle*

{Whittaker 1951, Chap. 7; CE 3.17}

Hamilton's methods gave further prestige to the admission of potentials, but the principal areas of interest in potential theory lay in electricity and magnetism and their connections. A small selection of examples is noted in this and the next two sections.

The greatest practitioner of the time, and proposer of the word 'potential', was sadly also the most obscure. Grinding flour in his father's mill in the scientific steppes of Nottingham, England, George Green mused on the latest findings in mathematical physics, especially Poisson's work on magnetism. He greatly advanced their theoretical basis in

An Essay on the Application of Mathematical Analysis to Electricity and Magnetism, published locally in 1828, his 38th year. And thereby he secured his obscurity: although he obviously knew that he had found gold, he did not publicize the fact, not even by writing a summary paper for one of the British scientific journals. In his own later work (that is, nine papers on aspects of mathematical physics and mechanics largely written during the 1830s when he was a mature student and then Fellow at Gonville and Caius College, Cambridge) he cited his book only occasionally, and for details. So it gained virtually no readers until after his death in 1841 (§10.19).

Green's most famous glory was the theorem now known after him. There are now various versions, of which his reads:

$$\iiint_B U\delta V \, \mathrm{d}v + \iint_S UV_s \, \mathrm{d}s = \iiint_B V\delta U \, \mathrm{d}v + \iint_S VU_s \, \mathrm{d}s, \qquad (10.18.1)$$

where U and V were two 'continuous functions' over the body B, partially differentiated with respect to the inward normal ds on its surface S, and 'δ' was his notation for the Laplacian operator. Still more original, however, was his search for a 'potential function' $f(r)$ (hence the later name 'potential theory'), now also known after him. He formulated the conditions on f as follows: it should satisfy Laplace's equation within B; $f(\infty)=0$ (and assumed to decrease monotonically to zero as r increased); it should possess no infinite 'differential coefficients' at points exterior to B; and it should equal a known function on S.

This was one of the greatest suggestions of 19th-century mathematics, for it went beyond Poisson's simplification (10.14.1) to grasp that the central issue in potential theory was to *relate properties inside volumes to properties on their surfaces*. Furthermore, theorems such as (10.18.1) played for multiple integrals a role similar to that of the fundamental theorem (8.6.3) of the calculus (hence the importance of integration by parts). Finally, the solutions of partial differential equations could be matched up (or not) to the form of boundary conditions in a much more explicit and systematic way. When all these insights gradually became recognized, they helped to stimulate a vast quantity of work.

But the mysteries of potential theory were highlighted again in 1840, when Gauss published a paper claiming that given potentials in and on a surface were uniquely determined by the distribution of matter inside. (The context, in magnetism, is noted in §10.19.) This result ran counter to a recent claim of non-uniqueness made by Michel Chasles; so a dispute arose. In the end Chasles was largely vindicated, but neither man fully appreciated the subtleties involved {de la Vallée Poussin 1962}.

Another aspect of potential theory where mathematical finesse was wanting concerns the 'Dirichlet problem': find a function f that satisfies Laplace's equation inside a body B and equals a given continuous function on its surface. This problem underpinned many of the assumptions made by potential theorists in their work (for example, Green had effectively assumed that his f existed). Its solution involved finding the minimum value of

$$\iiint_B (f_x^2 + f_y^2 + f_z^2)\,dv, \qquad (10.18.2)$$

for some (continuous) function f appropriate to the situation; but the assumptions about the existence of the minimum and the manipulations used in the theory turned out to be unrigorous (§14.3). The principle received its name from lectures on potential theory which Dirichlet began to give around 1840; another pioneer was William Thomson, to whom we now turn.

10.19 *Thomson: a world of flows*

Among the many mathematicians and physicists of the 19th century who admired Fourier's work, an extreme case was William Thomson, better remembered now as Lord Kelvin. He took Fourier's book on heat with him on holiday in 1839, his 16th year, and came home to write his first papers on aspects of Fourier series. Fourier's geometrical style fitted in well with Thomson's upbringing in Scotland, where an important tradition of 'common-sense' philosophy stressed the place of sensory experience. One of its methodological principles was 'the method of analogy', as Thomson called it {Wise 1981}.

One of the chief ideas that Thomson took over from Fourier was that differential coefficients of temperature were related to the flow of heat; for example, kdy/dx in the x-direction in the surface diffusion equation (10.11.5). He brought the same understanding to electricity and magnetism, writing the strength of their 'forces' in such terms (and indeed, Poisson on magnetic force as gradient of potential (§10.14) was another source). His enthusiasm for the approach deepened in 1844, when he discovered Green's book, three years after its author's death; later he arranged for it to be reprinted in instalments in Crelle's journal, between 1850 and 1854.

An important and immediate profit was a 'brain-saver' (his lovely phrase), the 'method of images'. Green had considered the effect on the surface S of an electrical conductor of a unit charge U placed outside *or* inside it; in his geometrical style Thomson saw that the analysis would be simplified if S were replaced by one or more 'image' points inside *or* outside it; however, whether as series or as integrals the sums were still troublesome. Soon he found a deeper approach, drawing upon his interest in energy physics noted in §10.8; he interpreted Poisson's electric (or magnetic) potential on S as the 'mechanical effect', that is, the work expended in bringing the electrical elements to S from infinity (the 'place' with zero potential) {Smith and Wise 1989, Chaps. 7–8}.

During this time Thomson, based in Glasgow, was in intense correspondence with Stokes at Cambridge. He had the essence of another important theorem involving a vector function (X, Y, Z) of a point (x, y, z) on a surface S bounded by a contour s:

$$\iint_S [l(Z_y - Y_z) + m(X_z - Z_x) + n(Y_z - X_y)]\,\mathrm{d}S$$
$$= \int_s (Xx_s + Yy_s + Zz_s)\,\mathrm{d}s, \qquad (10.19.1)$$

where (l, m, n) were the outward direction cosines of dS. It is now known after Stokes, who first published it, in 1854 – not in a normal paper or book but in a Cambridge prize examination paper for mathematics, of all places! {Cross 1985}. Since, however, the young Maxwell was one of the candidates, the examiners probably received at least one proof.

Stokes contributed with great distinction to hydrodynamics and elasticity theory, in manners redolent of Cauchy and Navier. Indeed, his name is also attached to the 'Navier-Stokes equations' for viscous fluids, although his version (of 1845) is actually different from Navier's (§10.16) because he produced them from a model rather like Cauchy's oblique-resultant version of elasticity theory. Yet it seems that he had not read his predecessors at that time; so for once French influence may not be present. The joint name is due to the fact that each man's equations reduce to the same one for incompressible fluids when the first of the continuity conditions (6.10.2) is applied.

Various models of the aether were proposed from the 1830s onwards; often they assumed the existence of tiny particles, whose manner of vibration propagated waves. Most theories drew upon the elastic compression or extension of parts of the aether, but in 1848 the Irishman James MacCullagh proposed that its rotation be taken as the basic notion. This idea was to gain favour 40 years later, when electromagnetism was the prime concern of aether studies (§15.10).

10.20 *Currents of German thought*
{CE 9.10}

Fourier's methods also gained approval in Germany, initially with Georg Ohm's book *Die galvanische Kette, mathematisch bearbeitet* (1827). We remember this work for the law relating to electric current, resistance and voltage, and indeed well on in the text the 'electrical stream' is stated to be proportional to the 'sum of all the tensions of the circuit'. However, the stream itself was his main concern, conceived as a flow in a Fourieran sense; and he found that the 'electroscopic force' satisfied the diffusion equation (10.11.4). In the 1840s he also interpreted Fourier's series (10.11.6) in acoustics, where the terms corresponded to supertonics of the fundamental tone and the coefficients characterize the sound and so distinguish, say, a clarinet from a violin.

Ohm pushed the similarity of structure with Fourier's heat theory too far: for example, he seemed to think that the occurrence of equilibrium necessarily required all circuits to take the same current. Credit for rejecting this assumption belongs to the 25-year-old Gustav Kirchhoff, when he produced his first major work in 1848 on the equality of currents in and out of a node (thereby allowing discontinuities in the values of currents). We owe to him Ohm's law as we now know it; for he rethought Ohm's vague idea of tensions as the electrostatic potential (V, say) of the circuit. Aware of the allure of energy physics, he also analysed the distribution of currents by minimizing an integral of the form (10.18.2) with V as the function.

This kind of analysis makes Kirchhoff sound like his exact contemporary Thomson, and indeed their work interacted in the 1850s with the application of old differential equations to these new areas, especially the electric telegraph. In 1854 Thomson proposed that V satisfied Fourier's diffusion equation (10.11.4) in this context, so allowing him to draw upon known solutions. (This oversimplified view of the phenomenon was to be corrected by Oliver Heaviside: §15.14.) Three years later, Kirchhoff pondered the effect on V of the electrical disturbance, and came up with the wave equation (6.4.1) together with a term in V_t (t for time) which was treated as negligible. The constant in the equation gave the velocity of propagation of the disturbance, which was found experimentally to be that of the velocity of light.

Another line of French influence upon Germany is evident in the way that Ampère's analysis of the constant currents in his cases of electrodynamic equilibrium (§10.13) was extended to variable currents by Franz Neumann and Wilhelm Weber in the early 1840s. They also analysed electromagnetic induction, which had been discovered by Michael Faraday a decade earlier, in terms of the change of potential between the primary and secondary circuits. Naturally they also considered Ohm (Kirchhoff's clarifications of Ohm came soon afterwards when he became a student of Neumann in Königsberg), and tried to formulate some basic properties of currents in terms of 'energy' conservation. They and others frequently used special functions in their analysis.

Neumann and Weber were also influenced by their compatriot Theodor Fechner (1801–87), who started his scientific career in these areas before moving on to a more famous life in psychophysics. He suggested that the electric current was composed of two flows of charges in opposite directions, which required that extensions of Ampère's formulae included their velocities. A place for motion in electrodynamics was now granted; it was to be of importance in the founding of relativity theory (§16.25).

Since Neumann and Weber were dealing with variable currents, they studied the dimensions of the various parameters and sought standard units. This brought them close to Gauss, especially Weber, who was also in Göttingen for many years. Gauss spent much of the 1840s on magnetism and geomagnetism, with an international *Magnetische Verein* to publish and compare data obtained in various countries. This was planetary mathematical physics *par excellence*.

Gauss made an influential mathematical contribution in 1840 with a 'theorem of the arithmetic mean', which greatly simplified expressions in potential theory. Given a collection of mass-points m_j distant r_j from an arbitrary point P, the potential V at P is found by taking an arbitrary point C as the centre of a sphere S passing through P. For V can be expressed using integrals over the surface of S, where the integrands depend on the potential function applicable and the location of each m_j inside, on, or outside S, and for standard potentials the integrals reduce to sums. But its simplification went much further, because of properties such as

$$\sum_j (m_j/r_j) = (\sum_j m_j)/\mathrm{PC}, \tag{10.20.1}$$

where PC served as a kind of mean over the distances r_j.

10.21 *A spectrum without unity*

This chapter has summarized the broadening of mechanics out to mathematical physics via Laplace's molecular programme. One contrast is worth stressing. The Laplacians hoped to find a *unified* model of phenomena using their principles (§10.10), but they could not deliver the theories required. None of the successful alternatives followed Laplace's methods; but in addition they differed *from each other* in various ways. Fresnel, Ampère and Cauchy as optician appealed to the aether; Cauchy as elastician oscillated between molecular and continuous models; Fourier, Navier and the other engineer-mathematicians usually adopted positivistic interpretations. Mathematical physics had become a rich but varied mansion.

An important example of these differences concerned models of the aether, especially for optics. Cauchy, Green in his later papers, Neumann, Stokes, Thomson, MacCullagh and others proposed various models for the vibrations of the elastic aether, both longitudinal and transverse, but no definitive version emerged, and the extent of their application was unclear.

Two features common to most figures conclude this survey. Firstly, there was a widespread tendency to deploy analogies – with Thomson as an explicit method – by appealing to structure-similarity from one theory to another. Secondly, most theories were based upon *linear* modelling: the basic differential equations, such as

(10.11.4) for diffusion, were linear ones; and properties such as cosine decomposition of the intensity of light or the relationships between stress and strain in elasticity theory were also linear in form. Of course, it was appreciated that the world was not really a linear place, but there were strong hopes that these powerful and general theories would continue to be fruitful; and in many ways they were to be vindicated for the rest of the century.

11

International mathematics, but the rise of Germany

Like §7, this chapter introduces its successors; but this time there are four of them, and they cover the rest of the 19th century. The division of labour is similar to the previous trio, except that mechanics and mathematical physics are split into separate chapters (§14 and §15); furthermore, the new developments in analysis, algebra and geometry brought novel connections between them, so that the divisions between §12 on analysis and §13 on algebras and geometries with analysis are more porous than between §8 and §9. From now on, the increasing quantity of work and (in many cases) levels of technical difficulty necessitate that the historical accounts are still briefer than those furnished hitherto in this book.

11.1 *The second sage of Göttingen*

{Laugwitz 1995}

Bernhard Riemann (1826–66) is a wonder. His collected works, consisting of his publications and a good selection of his manuscripts, fill just one volume, albeit a substantial one; yet the quality and depth of his insights is matched by nobody.[1]

The genius of Riemann shows itself most clearly in the simplicity of the matters to which he brought fundamental insights which had evaded many distinguished predecessors: that complex numbers and variables have a simple basis; that geometry determines space, not the other way round; and that Fourier series and integration theory needed tidying up. Behind these last two themes lay a belief that in mathematics one should handle collections of things with efficient mathematical tools.

Riemann prepared texts on these three subjects for his doctoral dissertations, in the above order: he read and published the first in 1851 for his *Dissertation* {Riemann 1851}, and three years later he prepared material for the other two for the higher-level *Habilitation*. But for the associated public lecture Gauss chose to hear about geometry rather than analysis, this being closer to his own concerns, and he was duly impressed. However, Riemann set both manuscripts aside and got on with other things.

In 1855 Riemann was appointed at Göttingen, but only as a lowly *Privat-Dozent*. Dirichlet succeeded the deceased Gauss the next year but lived only until 1859; Riemann then gained the chair, which he occupied for the rest of his life. However, that was only to be seven years: lung infections were developing, and even extended periods in Italy were not to save him. Indeed, he died at the Lago Maggiore, in his 40th year.

Riemann publicized certain of his ideas in lecture courses, and also published some profound papers on Abelian functions, sound, heat diffusion (with a non-linear model),

[1] As with Gauss, the manuscripts of Riemann are kept in Göttingen University Library; some would repay publication. The edition of his *Werke* appeared first in 1876, then in 1902, enlarged with a supplement a decade later, and again in 1990 in an extended version from Springer-Verlag with scholarly documentation by W. Purkert and E. Neuenschwander. Some editions of his lecture notes were published by K. Hattendorff and H. Weber; they became very influential, with repeated editions into the 1920s.

analytical number theory, and some other topics. After his death his close friend Richard Dedekind organized in 1867 the reprint of the 1851 *Dissertation*, and had the manuscripts for the *Habilitation* accepted as papers with the Göttingen Academy, where they appeared seemingly in the following year {Riemann 1868a, 1868b}. A decade later they appeared again when Dedekind and Heinrich Weber produced the first edition of Riemann's works.

The first thesis had already attracted a little attention; but the republication of the three doctoral papers in 1867–8 had a remarkable impact, for within a twelvemonth papers inspired by them were appearing, and this inspiration did not die down for decades. For the rest of the century, much analysis, and some aspects of geometry, algebra and other branches, were concerned with Riemannian questions. Several of them were answered by means of rigorous techniques; a chief source for them was the other main pole in German mathematics.

11.2 *Weierstrass versus Klein*

The career of Karl Weierstrass (1815–97) shows one massive discontinuity. In the mid-1850s he was lifted from obscurity as a schoolteacher to become professor at Berlin, and by the 1860s he was recognized as one of the finest mathematicians of his time. His research work covered all aspects of analysis, in particular foundations, differential equations, elliptic and Abelian functions, and the calculus of variations; he also contributed notably to parts of mechanics and to matrix theory.

The work on analysis (§12.3) made Weierstrass the successor to Cauchy; but in contrast to the teaching disasters at the Ecole Polytechnique in the 1820s (§8.6), he enjoyed a glittering educational career until his retirement in 1889. Somewhat unusually, he prosecuted his views almost entirely through lectures rather than in publications, which came in quantity only when he began to prepare his own edition of his works in the mid-1890s, shortly before his death. Adopting a rather naive view of pedagogy, he even discouraged students from taking notes during the lectures; so soon afterwards they would compose their own versions for personal use.

Just a few copies of these versions were known, mostly in libraries, until 1969 when I discovered 55 more volumes, which had been collected by Gösta Mittag-Leffler (1846–1927) and preserved in his house in Stockholm {Grattan-Guinness 1971}.[2] He was much responsible for the prominence of Weierstrassian analysis as Scandinavia rose from obscurity to significance in mathematics {CE 11.5}. Thanks to his wife's money, he was able in 1882 to found *Acta mathematica*, from its beginnings to the present a journal of major importance to mathematics. Much of the best work in Weierstrassian analysis was to appear in its pages. He also published several biographical articles on Weierstrass in *Acta mathematica*, especially in Volume 39 (1923). His antagonism to his compatriot Alfred Nobel has been asserted since the 1910s to be the reason why mathematics does not have a Nobel prize; however, Nobel offered them for inventions, so that mathematics was not an obvious candidate subject anyway.

Weierstrass's inheritance of Cauchy's style meant not only improving rigour but also avoiding geometry. In addition, he also advocated the study of exceptional cases of a situation, instead of letting them slide out of sight in sloppy theorems held to be true 'in general'. By contrast, Riemann was quite free in his geometrical/topological thought, and also quite ready to say things like 'in general'. This freer style of mathematics, as well as content, also had a great positive influence, largely among mathematicians outside Weierstrass's followers but with major figures of its

[2] Since the 1920s, Mittag-Leffler's house has been a research centre known as the Institut Mittag-Leffler. In it I also found massses of unknown materials concerning Weierstrass and Cantor (Mittag-Leffler's heroes), Kovalevskaya (whose career he helped by securing her a post at Stockholm University in 1883), and collections for some other figures. It would be nice if some day the Institut would appoint an archivist to set these magnificent materials in the order that they deserve – like Mittag-Leffler's own files, which were well maintained by his secretaries.

own. They included the algebraic geometer Alfred Clebsch, whose style was continued by his student Felix Klein (1849–1925), a central figure in this period.

In 1872, when he was only 23 years old, Klein launched at Erlangen a programme of classifying geometrical configurations by their invariance under different kinds of transformation (§13.10). His work was developed in close collaboration with the Norwegian Sophus Lie, whose later thoughts were to lead to the groups now known after him and the increasing use of algebra in this line of geometrical analysis. In the late 1880s Klein competed with a fabulous genius, Henri Poincaré, on finding geometrical and topological properties of curves and surfaces and of non-Euclidean geometry. He lost the contest, and thereafter concentrated rather more on teaching, with a distinguished record of lecture courses and graduate supervision (especially after his move to Göttingen in 1886) and the organization of projects within the mathematical profession (§16.3).[3]

11.3 *A school and a group*

Table 11.3.1 lists a good proportion of the major figures in mathematics in the late 19th century, and the areas of analysis with which they were concerned. The left-hand column concerns real and complex mathematical analysis in the Weierstrassian tradition, including the inauguration and use of Cantorian set theory (§12.4). The right-hand column covers analysis in the geometrical spirit as inspired by Riemann and personified in Klein. Much of the pure mathematics, and some of its applications, including other aspects of geometry in general, are covered by the collective work of this cohort – among whom Germans (especially Prussians) are very prominent.

[3] Klein's manuscripts are at Göttingen, like those of Gauss and Riemann (and also Cantor, Dedekind and Hilbert, among others). It is a massive collection, which at last has begun to receive the historical study that it deserves.

TABLE 11.3.1 ***German and foreign mathematicians in analysis and geometry, 1860s–1900.***

Dates and main initials are given, and nationalities of non-Germans.

abe	Abelian functions or integrals	ell	elliptic functions
		fnd	foundations of rea
alg	algebraic geometry and functions	fou	Fourier analysis
		geo	foundations of geometry
aut	automorphic functions	rea	real-variable analysis (functions, integrals, series)
com	complex-variable analysis	rie	Riemann surfaces
deq	differential equations	set	set theory
dgr	deq and groups		
AU	Austro-Hungary	NO	Norway
FR	France	RU	Russia
IT	Italy	SW	Sweden

Major German Analysts	**Major German Geometers**
Weierstrass, K. (1815–1897) abe com ell fnd rea	Klein, F. (1849–1925) alg aut ell rie
Cantor, G. (1845–1918) fou set fnd	Brill, A. von (1842–1935) alg
du Bois Reymond, P. (1831–1889) deq fou rea	Dyck, W. (von) (1856–1934) dgr rie
Heine, E. (1821–1881) fnd rea	Gordan, P. (1837–1912) abe
Schwarz, K.H.A. (1843–1921) com deq	Killing, W. (1847–1923) dgr
	Noether, M. (1844–1921) alg rie

Minor German Analysts	**Minor German Geometers**
Burkhardt, H. (1861–1914) fou rea	Christoffel, E.B. (1829–1900) abe com
Harnack, A. (1851–1888) alg rea	Fricke, R. (1861–1930) aut ell
Pringsheim, A. (1850–1941) rea fnd fou	Lindemann, F. (1852–1939) alg

TABLE 11.3.1 (*continued*)

Minor German Analysts (continued)	**Minor German Geometers (continued)**
Schlesinger, L. (1864–1933) aut deq	Meyer, F.W.F. (1856–1934) alg
Schönflies, A. (1853–1928) set	Pasch, M. (1843–1930) alg geo
Schottky, F. (1851–1935) abe com ell	

Foreign Analysts	**Foreign Geometers**
Casorati, F. (1835–1890) IT com	Betti, E. (1823–1892) IT alg ell
Dini, U. (1845–1918) IT fou rea	Bianchi, L. (1856–1928) IT dgr
Hermite, C. (1822–1901) FR dgr ell rea	Brioschi, F. (1824–1897) IT ell
Jordan, C. (1838–1922) FR rea	Castelnuovo, G. (1865–1952) IT alg
Kovalevskaya, S. (1850–1891) RU/SW deq abe	Cayley, A. (1821–1895) EN alg ell
Mittag-Leffler, G. (1846–1927) SW com ell	Cremona, L. (1830–1903) IT alg geo
Peano, G. (1858–1932) IT deq fnd geo set	Halphen, G. (1844–1889) FR alg rea
Picard, E. (1856–1941) FR deq com	Lie, S. (1842–1899) NO deq dgr
Stolz, O. (1842–1905) AU com fnd	Segre, C. (1863–1924) IT alg rie

German Straddlers	**Foreign Straddlers**
Fuchs, L. (1833–1902) abe com deq	Darboux, G. (1842–1917) FR deq rea
Hurwitz, A. (1859–1919) com ell rea rie	Painlevé, P. (1863–1933) FR alg deq
Lipschitz, R. (1832–1908) deq rea fou	Poincaré, H. (1854–1912) FR aut com deq set

Each column begins with its leading figure, although a different status applies to each. For the analysts, Weierstrass was the leader; his followers were *proud* to be known as Weierstrassians, and they constitute a school in the strict sense. By contrast, while Klein was the single most active geometer, he did not have Weierstrass's status as a leader among his group colleagues.

The intellectual differences between the two cohorts in the first part of the table hinge partly on the emphasis placed on rigour by Weierstrass and the encouragement of intuition by Klein. But they also have to do with the role of applications of mathematics, which were largely disdained in Berlin but were greatly encouraged by Klein, especially in his Göttingen period. These divisions are quite sharp, and show well the influence of education on research careers, especially those of the analysts. Sometimes figures in one cohort did contribute to the concerns of the other; however, only Wilhelm Killing completely transferred, pursuing Kleinian concerns after his training at Berlin. The division of interest is also evident with some mathematicians in other countries, shown in the second part of the Table. A few mathematicians, both German and foreign, straddled both areas of work to a notable extent; they are listed in the lower part of the Table. Finally, several figures in the Table worked in other branches of mathematics, sometimes as major figures.

This division is significant enough for the contributions of each cohort to be handled in different chapters. The Weierstrassians dominate the next one; but the geometric style of Klein and his associates, together with their use of algebraic structures (especially groups), places their work more comfortably in §13, although much of it involved solutions to differential equations.

11.4 *Some other figures*

Berlin could claim figures of note in other branches of mathematics apart from Weierstrass {Biermann 1988}. An important line of algebraists continued Ernst Kummer's line of thought, especially Leopold Kronecker (1823–91) and then Georg Frobenius (1847–1917). (In a faculty discussion about Kronecker's successor in 1892, Weierstrass described Klein as a 'hoodwinker'.) Of equal stature was Richard Dedekind (1831–1916), who chose to pass his career at the Technical High School in Braunschweig. He enjoyed elementary teaching, and indeed was inspired by it to some profound ideas on the foundations of analysis (§12.3); but his main contributions were to number theory, and to abstract algebras (§13 *passim*).

France was no longer the leading mathematical country, and knew it well; its defeat in the Franco-Prussian War of 1870–1 led to some estrangement between the two countries. The French wondered about the usefulness of the officers produced by the Ecole Polytechnique. Indeed, from the 1840s the Ecole Normale had gradually raised its status as a scientific centre, and in 1861 the 19-year-old Gaston Darboux recognized this by choosing it rather than the Ecole Polytechnique for his undergraduate training. A decade later he helped launch the *Bulletin des Sciences Mathématiques et Astronomiques*, which published both articles and extensive reviews of the literature. (The astronomers detached to form their own journal in 1884.) He later taught in the Paris Faculty of Sciences, assisting Liouville, and then at his Alma Mater, which founded its important *Annales Scientifiques* in 1875. A leader of French mathematics, in 1900 he became a perpetual secretary of the French Academy.

Of course, the Ecole Polytechnique still produced some major figures, of whom Poincaré (1854–1912), enrolled in 1875, was the most brilliant product. He then went to the Ecole des Mines, and fulfilled some duties for its professional Corps. His massive research activity included many parts of applied mathematics and mathematical physics; but he thereby somewhat isolated himself from his *confrères*, for whom the pure sides were predominant. A previous graduate (1859) was Emile Mathieu, whose fine contributions to mathematical physics (§14.3) did not even lead to posts in Paris. The same is true for Pierre Duhem, enrolled in the Ecole Normale in 1884, now better remembered for his remarkable contributions to the history of science.

As the Table hints, Italy continued to hold a very respectable place, especially after the unification of the country as a kingdom in 1860 {CE 11.8}. The importance of its analysts and algebraic geometers has one striking root: Riemann's sojourns in their warm climate, when he met Italian colleagues and drew them into his world.

In Britain, pure mathematics was largely led by Arthur Cayley and J.J. Sylvester, with their major contributions to matrix theory and invariants. But other branches, especially analysis, were seriously neglected, and fluxion-like doldrums descended on the island once again; like last time, it woke up only in the new century (§16.2). But in a puzzling contrast, its record in applied mathematics was far better; the best remembered movement was James Clerk Maxwell's work in electromagnetism, but other figures included William Thomson and Joseph Larmor, and later Horace Lamb and J.J. Thomson. Lord Rayleigh was a major figure in several other subjects.

11.5 *A new profession: journals and societies* {CE 11.12}

The increase in quantity of work carried out in all sciences and technologies was both great and rapid throughout Europe and in some colonial countries, especially from the 1870s. The growth in engineering generated more mathematics, and not only in mechanics and mathematical physics. The consequent expansion of towns and cities led many to found or enlarge institutions of higher education, and thenceforth to seek teachers and professors. We consider here the rapid increase in publications.

A few major journals have been mentioned already; some others need to be recorded. One of Alfred Clebsch's initiatives had been to found with Carl Neumann in 1869 the *Mathematische Annalen*, which immediately became a major organ for mathematics in German. Many papers on algebraic geometry by Klein and others appeared there, but also material of a Weierstrassian hue (for example, Georg Cantor's set theory). When Clebsch died in 1872 Klein took over, and thereby began a link with its Leipzig publisher, Teubner; it was to involve all the great projects which dominated most of his career and was to constitute perhaps the greatest alliance to date between a mathematician and a publishing house. An abrupt termination by Teubner in 1920 caused a switch to Springer, which is still important in mathematics.

A notable novelty had started in Berlin a few years earlier, in 1868: the *Jahrbuch über die Fortschritte der Mathematik*, the first general reviewing journal since Férussac's journal of the 1820s (§7.6). It covered a large number of journals and research books, and provides much valuable commentary, too little used by historians; however, its contemporary use was somewhat limited, for its reviews appeared about three years after the original texts.

Some journals were the publishing organ of a mathematical society. These were launched in London (1865), with a *Proceedings*; a year later in Moscow, with its *Sbornik*; and in 1870 the French Society, with a *Bulletin* {Gispert 1991}. Surprisingly, the German Mathematicians' Association and its remarkable *Jahresbericht* date only from 1890, although German mathematicians had long run an active section in the country's Society for Scientists and Doctors. Cantor was a principal stimulator: inspired by the hostile reception (as he saw it) of his set theory, he conceived of the society as a forum of freedom away from the dominant centres of Berlin and Göttingen, which were not used as venues for its annual September meetings.

Two interconnecting tendencies are evident in this period in all countries. Firstly, the inauguration of these societies and journals specifically for mathematics shows that the mathematician was becoming a professional in the full sense of the word; indeed, his chances of employment as such were increased by the expansion or creation of universities and like institutions, especially in Germany {Lorey 1916}. Secondly, a snobbish preference for pure mathematics over applications became more marked (although the boundary line between the two sides was not fixed, with mechanics seen by some as straddling both), so that the profession tended to separate into two components. In Berlin, and elsewhere in Germany, purism became an explicit creed for the professional, so precipitating a reaction from the other direction, with a strong movement against mathematics among German engineers (§14.1).

11.6 *Flourishing fringes: probability and statistics, and the history of mathematics*

We now note a further complication in the community; for at last probability and statistics began to gain a measure of notice appropriate to their importance. Two countries are notable (§12.11): Russia, where concern with limit theorems and their consequences was led by Pafnuty Chebyshev; and Britain, where the stimulus was initiated by Francis Galton and continued by Karl Pearson. Even then the absorption into mathematics was gradual, with several applications still gaining little recognition (for example, economics and production engineering). Most of the principal figures were not major mathematicians, and in some cases not even minor ones, and their areas of interest were marginal to the mathematics of their times – for example, eugenics, or variations in the weather. Thus, while the work used (and sometimes contributed to the development of) normal mathematical theories, it was often conducted outside the orthodox centres of activity.

In addition, a considerable amount of work in the history of mathematics was carried out during this period. The activity was launched by the rich Prince Baldassarre Boncompagni, who produced in Rome at his own expense a *Bullettino di bibliografia e di storia delle scienze mathematiche*

(20 volumes, 1868–87). It contained a mass of invaluable research, especially for the Renaissance period, his own major interest.

Britain made its contribution mainly through encyclopedias, in particular the *Encyclopaedia Britannica*. The editions from the 8th (1875–89) to the 13th (1926) are graced by many fine survey articles on mathematical subjects, especially on applications; some of the authors are major figures in §14 and §15.

But German was the principal language for historians, and Germany the main host country {CE 12.13}. The leading figure was Moritz Cantor (no relation to Georg); his *Vorlesungen über die Geschichte der Mathematik* up to 1799 (four volumes, 1880–1908, put out by Teubner) was the starting-point for many historical colleagues. He produced the first three volumes in two or three editions, but only edited the last one (covering 1759–99), which appeared in his 80th year. Between 1877 and 1913 he also edited a series of *Abhandlungen* on history, which contained several important papers and monographs in its 30 volumes.

The next major journal after Boncompagni was the *Bibliotheca mathematica*, which began in 1884 under the editorship of the Swede Gösta Eneström. It started as an appendix to Mittag-Leffler's *Acta mathematica*, but soon continued as an independent publication. Many valuable articles appeared, and interesting historical questions were posed.

Another important German enterprise, launched in 1889, was *Ostwald's Klassiker der exakten Wissenschaften*; it still continues. Classic works from all sciences (with mathematics prominent) were presented in German with commentary and notes.

A main focus of historical study was Greek mathematics, where the Danes H.G. Zeuthen and J.L. Heiberg were notable writers in German, with T.L. Heath in English and Paul Tannery in French. The Middle Ages and Renaissance also received much attention. The current century was served especially by editions of works of major mathematicians. The principal German ones were of Gauss (for which Klein was an editor), Jacobi and Steiner (in both of which Weierstrass was involved). Editions were produced in other countries; several of the French ones are awful, and the British ones included some done or at least started by the mathematician himself (for example, Cayley and Kelvin).

At the end of the century there were various other initiatives in history and in mathematics itself: more editions of works, a new generation of mathematicians around Jacques Hadamard and Emile Borel in Paris and David Hilbert in Göttingen, the main (and rapid) rise of mathematics in the United States of America, the development of mathematics in British colonies and in the Far East, and the International Congresses of Mathematicians. We pick these up in §16.

Among other enterprises led by Klein, the *Encyklopädie der mathematischen Wissenschaften* from the 1890s onwards (§16.3) must be mentioned as the principal source of historical information for the next four chapters. A major figure in this one, let him prepare us for them with this beautiful comparison of the other two figures {C.F. Klein 1926, p. 246}:

> Riemann is the man with the shining intuition. Through his all-embracing genius he surpasses all his contemporaries. When his interest is awakened, he starts afresh, without being led astray by intuition and without acknowledging the coercive pressure of systematization.
>
> Weierstrass is in the first place a logician; he advances slowly, systematically, step by step. Where he works, he strikes for the definitive form.

12

The rise of set theory: mathematical analysis, 1860–1900

12.1 *Riemann's draft ideas on trigonometric series, 1854/68*

When the manuscript of Riemann's thesis on real-variable analysis finally appeared as {Riemann 1868a}, readers faced a very cryptic text, but one full of extraordinary ideas. He had studied the convergence of Fourier series, as currently understood in Dirichlet's Theorem 8.7.2 with the represented function $f(x)$ restricted to a finite number of turning-points, discontinuities and infinities. He had discussed the matter with Dirichlet, and acknowledged this heritage by beginning his essay with a fairly good history of trigonometric series from the mid-18th century onwards. Since the Fourier coefficients were specified by integrals, he also improved upon Cauchy's definition (8.6.2) of the integral (§12.6). Then he advanced beyond Dirichlet in three ways.

Firstly, Riemann showed that the terms of a conditionally convergent series $\sum_r u_r$ of real numbers (that is, one for which $\sum_r |u_r|$ is divergent) could be rearranged to converge to *any* sum desired; Cauchy had mentioned the point in the 1820s, maybe because Dirichlet had found a particular example. Secondly, Riemann found expressions which defined functions containing an infinity of turning-points or discontinuities; he also checked whether or not they were integrable in his sense. Finally, sensing that seeking sufficient conditions more general than Dirichlet's was hard work, he sought necessary conditions in a most original way: he formed the second term-by-term integral $F(x)$ of the series (8.7.1) for $f(x)$,

$$F(x) := A + Bx + \tfrac{1}{2}a_0x^2 - \sum_{r=1}^{\infty} \frac{1}{r^2}(a_r \cos rx + b_r \sin rx),$$

$$-\pi \leq x \leq \pi \qquad (12.1.1)$$

and studied $f(x)$ by examining $F''(x)$. Two most remarkable consequences followed. Firstly, convergence at a given value of x depended only upon the behaviour of $f(x)$ in its neighbourhood, a result which became known as his 'localization theorem'. Secondly, no assumptions were made about the coefficients a_r and b_r (in particular, they need not be defined by the usual integral formulae (8.7.2)), so that there might not be any corresponding $f(x)$; thus trigonometric series were not necessarily Fourier series.

Since Riemann proved nothing rigorously, his suggestions needed rather refined tools to render them precise.

These were being forged for mathematical analysis in general by Weierstrass, to whom we now turn.

12.2 *Weierstrass, professor*

{Dugac 1973, CE 3.3}

Given the differences in the mathematics of the two men, and also in their ages (Riemann was ten years junior to Weierstrass), it is surprising to record that the mathematical public began to learn of both men in the same mathematical context and journal, and at around the same time: in connection with Abelian functions, in Crelle's journal, Weierstrass in 1854 and Riemann three years later (§13.13).

Some of Weierstrass's major ideas had come to him in the early 1840s, when he was in his mid-twenties; but he was employed only as a schoolteacher and published nothing in a major journal. However, his papers on Abelian functions brought him rapidly to international recognition; so did one on analytic functions (namely, those functions $f(z)$ of a complex variable that take a convergent power-series expansion in $z-a$ for each point z in a circle with centre a), which included a substantial and novel Cauchy-style discussion of the convergence of infinite products. Soon he had posts in the University and the Academy at Berlin, and began his glorious 30-year sequence of unpublished but heavily recorded lecture courses mentioned in §11.2, spiced with an occasional paper. He included several papers, and some manuscripts, in the second volume {Weierstrass 1895} of his collected works. An excellent recent edition of his 1886 course is provided in {Weierstrass 1988}; it covered both real and complex analysis.

Most of Weierstrass's students and followers were fellow-Germans; but his ideas were exported by foreign listeners such as Gösta Mittag-Leffler to Scandinavia and Salvatore Pincherle to Italy, or taken up by senior mathematicians abroad (for example, Ulisse Dini and Giuseppe Peano in Italy, and Camille Jordan in France). A further case is the

Russian-born Sonya Kovalevskaya, the only notable woman mathematician of this period. Berlin University would not allow a mere female to attend Weierstrass's lectures, never mind take their higher degree; so she studied with him privately, receiving her doctorate from Göttingen University in 1874. Her three theses dealt with Abelian integrals (§12.8), partial differential equations (§12.9), and the stability of the rings of Saturn (§14.15) {Cooke 1984}.

In the next section a summary is given of the principal ways in which Weierstrass and his followers enriched the foundations of the tradition in real-variable analysis established by Cauchy, of which the main strengths + and limitations – have been summarized in a numbered list in §8.19. Weierstrass continued the +'s with enthusiasm (necessary and/or sufficient conditions, prime place to limits, broad definitions), and dealt with some of the –'s.

12.3 *Weierstrass's refinements of Cauchy's programme for real-variable analysis*

{Bottazzini 1986; EMW, IA3 and IIA1–3; CE 3.3}

This summary is organized in seven parts.

1. Concerning 2+, the *theory of limits* was enhanced by distinguishing the least upper bound of a collection of values from its upper limit, and its greatest lower bound from its lower limit (the notations '$\underline{\lim}$' and 'lim' for the latter in each pair of notions date from this time).

2. Concerning 2–, *multiple limits* were handled with the care that Cauchy had brought to single ones; in particular, they were distinguished from repeated ones. Thus, for a series of functions $\sum_r u_r(x)$,

$$\lim_{x\to a, r\to\infty} \sum_r u_r(x), \quad \lim_{x\to a} \lim_{r\to\infty} \sum_r u_r(x), \quad \lim_{r\to\infty} \lim_{x\to a} \sum_r u_r(x) \tag{12.3.1}$$

could take different values, or some might be defined while others were not. The manners of convergence of such series were classified into uniform, non-uniform and quasi-uniform modes {Hardy 1918}; for example, uniformity over an interval $[a, b]$ of x required of the remainder function $r_n(x)$ after n terms that for any $\varepsilon > 0$ there was an integer N such that

$$|r_n(x)| < \varepsilon, \text{ for all } n > N \text{ and for } a \le x \le b. \quad (12.3.2)$$

Uniformity was important because it was needed to justify many of the main properties of analysis, such as differentiating under the integral sign or integrating a series term by term; but it was quite a severe restriction in, for example, potential theory (§14.3).

An example is provided by Cauchy's Theorem 8.5.1 on the continuity of the sum of a convergent series of continuous functions $f(x)$ over a finite range of values of x; for under the Weierstrassians' (ε, δ) way of defining continuity, the convergence had to be uniform. A companion refinement of continuity itself was effected by defining (non)-uniform continuity of $f(x)$ over the range in the Weierstrassian way: for any $\varepsilon > 0$, there is (or is not) a $\delta > 0$ such that

$$\text{if } |h| < \delta, \text{ then } \quad |f(x+h) - f(x)| < \varepsilon \quad \text{for } a \le x \le b. \quad (12.3.3)$$

3. Concerning 4–, distinctions similar to (12.3.1) were made between multiple and repeated integrals, and also when interchanging sums and integrals. Fourier series and integrals received much attention, especially from Paul du Bois-Reymond (a close associate of Weierstrass) and foreign followers such as Dini. One source for them was a study of 1864 by Rudolf Lipschitz, who took up Dirichlet's Theorem 8.7.2 on the convergence of the series and extended his conditions to cover an infinity of damped oscillations of $f(x)$ (where the magnitudes decrease to zero as x approaches the point of accumulation) obeying the following condition, which is named after Lipschitz:

$$|f(x+h) - f(x)| < Ah^k, \quad \text{with } A > 0,\ h > 0,\ k > 0. \quad (12.3.4)$$

4. *Tests of convergence* (§8.5) were reworked to check for these modes, and some new ones introduced. The most notable one was Weierstrass's own '*M*-test' (as it became known), published in 1880 {Weierstrass 1895, p. 202}, which stated that

$$\text{if } |u_r(x)| < M_r \text{ when } a \leq x \leq b, \text{ and } \textstyle\sum_r M_r \text{ is convergent,} \tag{12.3.5}$$

then $\sum_r u_r(x)$ is uniformly convergent over this range of values of x.

5. Weierstrass made two other important contributions to *the theory of functions*. Firstly, noting the appearance of functions with infinities of discontinuities and/or turning-points, he presented in 1872 the analytic expression of one which was continuous over a finite interval but not differentiable at any point on it {Weierstrass 1895, pp. 71–4}; finding other examples became quite a sport. A notable related case was Peano's expression (1884) for Dirichlet's characteristic function of the rationals (§8.7):

$$f(x) := \lim_{n\to\infty} [\phi(\sin n!\pi x)], \quad \text{where } \phi(x) := \lim_{t\to 0} \left(\frac{x^2}{x^2+t^2}\right), \tag{12.3.6}$$

takes the value 0 when x is rational and 1 when x is irrational.

Secondly, in 1885 Weierstrass published (for once) his 'approximation theorem', which stated that any continuous function can be approximated uniformly over a finite range of values by a sequence of polynomials of powers, or of sines and cosines. This result provided an important theoretical link between the major category of general functions and two of the principal ways of handling functions numerically (§12.10).

An important question for the generality of functions was this: can a relation $f(x, y) = 0$ between x and y be rewritten with y as a *function* of x – that is, taking only one value for each value of x? In 1878 Dini came up with a positive answer: it was possible if F were itself single-valued and continuous on x and y around a point, and took continuous partial derivatives there. Extensions and generalizations of this result became known collectively as the 'implicit function theorem'.

6. One part of analysis to benefit from Weierstrass's attention was *the foundations of the calculus of variations.* He lectured on it from 1865, although he published little. His main contributions were to restate in parametric form the basic integral (5.24.3),

$$\text{from } \delta\int_a^b f\left(x, y, \frac{dy}{dx}\right) dx = 0 \text{ to } \delta\int_{t_a}^{t_b} F\left(x, y, \frac{dx}{dt}, \frac{dy}{dt}\right) dt = 0, \tag{12.3.7}$$

and to distinguish the class of continuous functions F as admissible from that where the derivatives of F could be discontinuous. (In a strange hangover from Euler's use (§5.24), and in contradiction to item 5, 'discontinuous' here still referred to a function defined by more than one expression.) At the end of the century his follower Adolf Kneser introduced the adjectives 'weak' and 'strong' to distinguish these two cases, and to emphasize the importance of so doing {EMW IIA8a}.

Weierstrass also studied in detail conditions required for the existence of the second variation. Several mathematicians, notably Schwarz, extended this theory to functions of several variables and to sets of functions, and applied them to classical problems such as minimal surfaces and geodesics {Goldstine 1980, Chaps. 5–6}.

7. Concerning 3–, in order to prove theorems on the existence of limits, it was realized that chat about *rational and irrational numbers* had to be refined into formal definitions of irrationals. Weierstrass himself produced a rather clumsy one, but the version that was to become the most popular was furnished by Richard Dedekind in a pamphlet of 1872 on 'continuity and irrational numbers'. Given the rational numbers, associate each one with a point on the

continuous straight line by natural order, and cut the line at any point; then the numbers are divided into two classes, to left (L) and right (R). If L has no greatest member and R no smallest, then the cut corresponds to an irrational number, and thereby defines it.[1] The concern with irrational numbers led Felix Klein in 1895 to characterize Weierstrass's whole approach as 'the arithmeticization of mathematics'.

12.4 *Cantor's way in to set theory, 1872*

Another theory of irrational numbers came in 1872, from Georg Cantor, then in his 28th year. His senior colleague at Halle University, Eduard Heine, had drawn him away from number theory to the following Riemannian question in Weierstrassian analysis. A further consequence of Riemann's distinction between trigonometric and Fourier series was the *uniqueness* of the latter: if two functions differed in value for, say, just one value of x, then their series would be indistinguishable, since the integrals determining the coefficients would be equal. In 1870 Heine had proved this result in Weierstrassian style for any finite number of points – but what about an infinity of them? Cantor unintentionally opened up his whole life's work with his answer, given in a paper of 1872 {Cantor 1932, pp. 92–102}:

> THEOREM 12.4.1 *The set* P *of points at which a function may take a value different from that used in the determination of its coefficients can be infinite as long as it is of 'the first category'; that is, if its* n*th 'derived set'* $P^{(n)}$ *is empty for some finite integer* n.

[1] In lectures of 1861–2, Dedekind had already given a nice but little-known proof that √2 is irrational, different from the usual ones (Theorem 2.12.1) though also by proof by contradiction. Let b be the smallest positive integer for which a larger integer a exists such that

$$2b^2 = a^2, \text{ and put } c := a - b > 0. \text{ Then } (b-c)^2 = 2b^2;$$

but this contradicts the definition of b {Dedekind 1985, pp. 24–5}.

The language of this theorem is more Cantorian than is indicated just by the terms in quotation marks. For example, the phrase 'set of points' is his (*Punktmenge*), referring to a collection to which points (or values) may belong as individuals; so is the use of a letter to represent the set; and also the notion of derivation, adapting a word and notation (6.5.3) used by Lagrange in the calculus. P', the first derived set of P, was the set of its limit-points; if not empty, it took its own derived set, P'', and so on.

But those last three little words are the most crucial of all for Cantor's later work. Weierstrass's imperative to avoid saying 'in general' (§11.2) drove Cantor to consider what happened when derivation of P *never* led to an empty set. He perceived the answer to be some sort of infinitieth derived set, which, moreover, might not itself be empty. This led him to formulate, at this time in quite a naive way, a truly *infinite* sequence of derived sets, thus:

$$\begin{array}{c} P, P', \ldots P^{(\infty)}, P^{(\infty+1)}, \ldots P^{(2\infty)}, \ldots P^{(3\infty)}, \ldots \\ P^{(\infty^2)}, \ldots P^{(\infty^3)}, \ldots P^{(\infty^\infty)}, \ldots \end{array} \quad (12.4.1)$$

where I have deliberately not put commas after the ellipsis dots or a period, at the end. He did not publish this sequence in his paper, and probably had little idea of its meaning; for some time he referred to the indices as 'infinite symbols'. Clarification and extension of the sequence was to dominate his mathematical career.

In 1872 Cantor met Dedekind by accident while on holiday; and at times over the next 13 years they had a vigorous correspondence {Ferrierós 1993}; indeed, a few key suggestions made by Dedekind turned up without acknowledgement in Cantor's papers. An important six-part series appeared from 1879 to 1884 in *Mathematische Annalen*, with Klein as friendly editor {Cantor 1932, pp. 139–247}. For 1883–5, Mittag-Leffler's newly founded *Acta mathematica* was the venue, including a French translation of many of his former papers, which gave him a new public.

But Cantor's work then came to a sudden slow-down,

for he suffered some sort of mid-life crisis. He was also advised by Mittag-Leffler to withdraw a paper on order-types destined for *Acta mathematica* because it contained no proofs of major results {Grattan-Guinness 1970c} – well-meaning but unfortunate advice, which rapidly took him off Cantor's mailing list. Soon afterwards, Cantor stopped publishing in mathematical journals for several years (§12.15).

12.5 *The principal features of Cantor's set theory, 1872–85*

{Dauben 1979; CE 3.6}

Cantor called his set theory *Mannigfaltigkeitslehre* ('theory of manifolds') – rather unfortunately, for we shall see in §13.7 that the word came from Riemann, and referred to a rather different concept. His main findings of this period may be summarized, as with Weierstrass, in seven parts. I shall use some of his technical terms, many of which have been standard fare ever since; but also some modern notation (largely due to Peano: §12.13), since he was curiously uninterested in symbolizing the operations of his theory.

1. To his great surprise, Cantor showed in 1878 that the points (x, y) in a unit square could be put into one–one correspondence with those (z) on its unit side by intercalating their decimal expansions {Cantor 1932, pp. 119–33}:

$$x = 0 \cdot x_1 x_2 x_3 \ldots \quad \text{and} \quad y = 0 \cdot y_1 y_2 y_3 \ldots \\ \Rightarrow \quad z = 0 \cdot x_1 y_1 x_2 y_2 x_3 y_3 \ldots . \tag{12.5.1}$$

(Some homework is needed to distinguish, say, $0.299\dot{9}$ from $0.300\dot{0}$.) This correct but profoundly counter-intuitive finding messed up the normal understanding of different *dimensions*, including Cantor's own: 'I see it, but I do not believe it!', he wrote at the time to Dedekind. The mystery deepened in 1890 when Peano defined analytically a continuous curve which passed through every point in the unit square. Satisfactory new definitions of dimension were not to come for 50 years (§16.17).

2. From the sequence (12.4.1), Cantor developed *the topology of sets of points* by defining properties of a set P such as this central trio:

$$P \text{ is closed } or \text{ perfect } or \text{ dense in itself if } P \supseteq or = or \subseteq P'. \quad (12.5.2)$$

From there he proceeded to decomposition theorems, which stated that a set of a given type could be expressed as the disjoint union, finite or even infinite, of sets of other types. He also defined $P^{(\infty)}$ of (12.4.1) as the intersection of its predecessors.

The fundamental role given to limit-points soon attracted some colleagues, such as Weierstrass, who had already proved that a bounded infinite set contained at least one such point. This theorem, which Weierstrass published in 1876 {Weierstrass 1895, pp. 77–124}, was named after him and Bernhard Bolzano (whose work began to gain attention at last) by Cantor and by his school-friend and fellow Weierstrassian K.H.A. Schwarz, who worked on these topics in the early 1870s but later became strongly opposed to Cantor's theory.

3. One target for Schwarz's displeasure was Cantor's innovative study of *inequalities* between 'transfinite' (Cantor's word) numbers, inspired by the sequence (12.4.1) of 'infinite symbols'. The criterion was that two (ordered) sets were unequal in size if no one–one correspondence could be established between their members. To provide a theory for the thinking behind his infinite symbols, he proposed in 1883 a method of generating ordinals by starting with 1 and placing after each ordinal n a successor $n+1$, and then asserting the existence of a 'limit ordinal' after a denumerable sequence '...' of ordinals, allowing successorship to work again, asserting a limit ordinal, and so on in tandem. This led him to the 'transfinite ordinal numbers'

$$1, 2, \ldots \omega, \omega+1, \ldots 2\omega, \ldots 3\omega, \ldots \omega^2, \ldots \omega^3, \ldots \omega^\omega, \ldots \quad (12.5.3)$$

where the nervous '∞' was replaced by the new symbol 'ω'; he also studied in much more detail the structure of the numbers beyond ω^ω.

Another critic of Cantor was Leopold Kronecker. He did not make his reservations explicit, but as a constructivist (§13.3) he cannot have liked the way in which Cantor magically let ω appear in (12.5.3) after the ellipsis dots without drawing on a construction. He probably did not object to the invocation of the actual infinite as such, though for him God made the finite integers.

4. Cantor realized that sets may come in a variety of '*order-types*'; for example, that of the rationals differs from that of the 'well order' W of the ordinal numbers. He announced as a conjecture his 'well-ordering principle', that the members of *any* set could be reordered into W.

5. Cantor also examined *transfinite cardinals*, also available in different orders of infinity – thereby revolutionizing the traditional understanding of the infinite as 'of one size'. In particular, he showed that the 'cardinality' of the set *Ra* of rational numbers was denumerably infinite (that is, of the same order of infinity as the positive finite integers) but that of the set *Re* of real numbers was of a higher order. (His proof was not the 'diagonal argument'; its later arrival is noted at §12.15.) He also divided up (12.5.3) into 'number-classes': the first was the finite integers up to but excluding ω; the second contained ω and those successors α such that the ordinals from 1 to α could be reordered as a denumerable series to ω; and so on. He showed that the cardinality of each class was successively larger, and he now *defined* the transfinite cardinals as these numbers.

6. *Continuity* fell under Cantor's examination. In an important innovation he showed that the usual intuitive definition of a continuous set in terms of the inevitable existence of members intermediate between any two given ones was necessary but not sufficient; for it was satisfied by *Ra* – which, however, contained gaps in which irrational numbers were to be found. So he added extra properties to define continuous sets. He hoped to find a decomposition

theorem stating that *Re* could be expressed as the union of some disjoint sets, and then to add up their cardinalities to find that the cardinality of *Re* equalled that of the second number-class; he never obtained a proof, but this 'continuum hypothesis' (as it became known) gradually became recognized to be a major conjecture in mathematics (§16.8). His neat version of it is given as (12.15.6).

In connection with continuity, Cantor defined in 1883 an exotic object which became known as the 'ternary set': the set T of points z defined by the expansion

$$z = \sum_{r=1}^{\infty} \frac{c_r}{3^r}, \quad \text{where } c_r = 0 \text{ or } 2 \qquad (12.5.4)$$

for all combinations of 0's and 2's. Among its properties, T is discontinuous, denumerable, perfect and not everywhere dense in any interval.

7. Most of Cantor's theory dealt with sets of points, but from the early 1880s he was thinking of *more general sets*. In 1883 he envisioned a way of defining numbers from sets: from a given set S, abstract away the nature of the elements to reveal its order-type, and invoke the well-ordering principle to specify its ordinal; then abstract the order-type to obtain its cardinal. He restated this procedure in more detail in the mid 1890s (§12.15).

12.6 *The integral, and differential geometry*

{CE 3.7}

One further aspect of Cantor's theory related directly to Riemann. In his thesis {Riemann 1868a}, he served as Archimedes (FIG. 2.21.2) to Cauchy's definition (8.6.2) of the integral of $f(x)$ over $[a, b]$ as the limiting value of a sequence of partition-sums S_n by sandwiching the function between its maximum and minimum values in each interval of a given partition (FIG. 12.6.1, which he did not furnish); S_n lay between the corresponding lower and upper sums m_n and M_n over $[a, b]$. The existence of the integral of $f(x)$ over $[a, b]$ was guaranteed necessarily and sufficiently by the

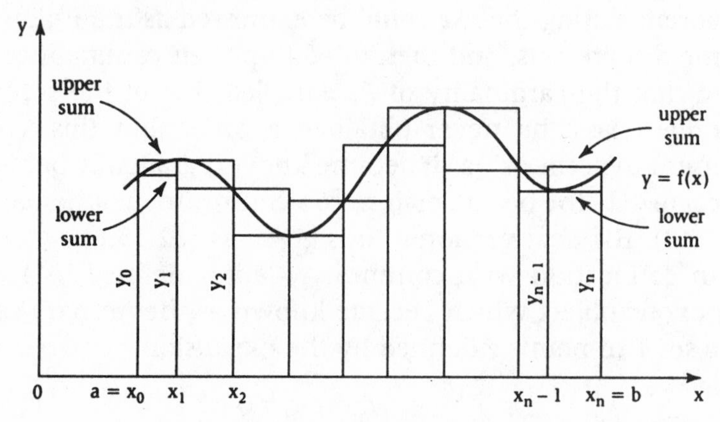

FIG. 12.6.1 *The Riemann integral of f(x) for $a \leq x \leq b$, 1854.*

convergence of these two sums to the same limiting value S, which of course was also the value of $\int_a^b f(x)\,dx$.

This dealt with part of limitation 4– of Cauchy's programme (§8.19). However, I stated it in the form given by Gaston Darboux in 1875; it became the normal treatment when the Weierstrassians handled the integral calculus. In his thesis, Riemann had spoken cryptically about 'the total size of the intervals' in which $f(x)$ varied in value by more than any assigned magnitude being 'made arbitrarily small'. Set theory and measure theory were being born here, the former to analyse their complexity and the latter their size.

Cantor and others developed these ideas further in extending the notion of integral to the 'content' of a set P of points in the plane (or in n dimensions in general) and using the new topological techniques to clarify the formulation. Unfortunately, the usual success was not experienced. One reason was that the definitions offered gave the same content to P and its closure set $(P \cup P')$; but if P were, say, the rational numbers within some interval $[a, b]$, then its content was zero while its closure was the interval itself, with content $b-a$. Subtle stuff, set theory, even to its chief founder; victory was to be achieved only from the 1890s, when Darboux's form using upper and lower sums was imitated (§16.9).

Darboux was also very interested in differential geometry, following in the tradition of Monge (§8.3). Between 1887 and 1896, Gauthier-Villars published four thick volumes of his *Leçons* given at the Paris Faculty of Sciences. He covered traditional topics such as principal curvatures, orthogonal and minimal surfaces, and geodetic lines. In his deployment of partial differential equations and the calculus of variations he made a few innovations, and he used some aspects of Weierstrassian rigour. He also noted the theory of coordinate systems developed by his senior colleague Gabriel Lamé. We noted in §10.16 that during the 1830s Lamé had extended Fourier's work on heat diffusion by using non-rectangular coordinate systems; by the 1860s he was a major figure in deploying systems of orthogonal curvilinear coordinates (u, v, w), where the constant values of u, v and w define surfaces in space which cut each other orthogonally. All this work was quite respectable, and indeed was quite a popular area of study {EMW IIIAB7 and D9}, but deeper insights on differential geometry were to come from a different tradition inspired by Riemann, which we shall record in §13.13 and continue in §15.8.

12.7 *Weierstrass on complex-variable analysis* {CE 3.12}

We saw in Theorem 8.11.1 an important contribution made in 1843 to Cauchy's theory of complex analysis by his compatriot P.A. Laurent: the expansion of a continuous function $f(z)$ in a series of positive *and negative* powers of z. At the same time, the young schoolteacher Weierstrass found essentially the same result, and over the following years he built up from it a general theory of complex-variable analysis that offered an alternative to Cauchy's reliance on basic differential equations and residues of functions (limitation 5– of §8.19). He taught his approach at Berlin in various forms from 1868.

In this summary I follow our modern convention of writing the complex variable as 'z'; at that time 'x' was frequently used for both real and complex variables. For Weierstrass, a function $f(z)$ was 'analytic' around $z=a$ if it took a series

$$f(z)=\sum_{r=-\infty}^{\infty} k_r(z-a)^r, \quad k_r \text{ constants, with } z\neq a; \qquad (12.7.1)$$

if only positive powers obtained, then $f(z)$ was analytic at $z=a$ also. This was the fundamental property of his theory, basing $f(z)$ on the coefficients k_r rather than on its derivatives. It made him the third mathematician after Newton (§5.5) and Lagrange (§6.5) to grant power series a prime place in analysis, although in their cases real-variable calculus was the concern.

An important bonus was that when it was convergent, the series for $f(z)$ was absolutely and uniformly so; hence the processes of analysis could be applied without loss of rigour (compare item 2 of §12.3). Since the convergence around $z=a$ was restricted to some circle C_a with a as centre, Weierstrass's approach was local; so in order to give it the required generality he introduced the idea of 'analytic continuation' (his name). Pick a point $z=b$ inside C_a; then by (12.7.1) $f(z)$ should take a similar circle C_b of convergence, only partly overlapping with C_a; to go further, pick c within C_b, and .... Thus a network of circles could be constructed, like the symbol used for the Olympic Games, say: it could cover the complex plane, or as much of it as was needed in a given situation.

Naturally, to Weierstrass this geometric way of describing the procedure is only heuristic. The proper version was expressed solely in analytical terms, with the complex numbers carefully defined in a manner based on his theory of real numbers (item 7 of §12.3), and the limits of convergence expressed in terms of inequalities. He also sought conditions under which each continuation, or at least one of them, would lead to single-valued expansions.

A major matter for Weierstrass was types of singularity of $f(z)$, as, for example, in his paper of 1876, containing also the Bolzano–Weierstrass theorem (item 2 of §12.3). When there were only a finite number of negative powers in its expansion (12.7.1), $z=a$ was called an 'inessential singularity' of f (basically the same as 'poles' in Cauchy's theory), and he showed that a function could be defined with poles at given points $z=c_r$ as a rational function with denominator of the form $\prod_r (z-c_r)$. However, he saw the possibility that $(z-a)^n f(z)$ was not analytic for *any* positive integral value of n, so that the series of negative powers was infinite; in this case $z=a$ was 'essential', and he showed in this paper that f was so badly behaved that $f(z)$ could come arbitrarily close to any specified value for $z=a$. (This result is also named after the Italian Felice Casorati, who had published it in 1868.) These investigations led to further work on finding an expression for $f(z)$ with given singularities of either kind. Cantorian point-set topology provided useful techniques.

Weierstrass's main followers included Mittag-Leffler. In France, the land of Cauchy, Emile Picard translated the 1876 paper into French in 1879, while Jules Tannery rendered a later one in 1881. Indeed, until the end of the century Weierstrass's approach to complex analysis held a leading place in many countries; then things began to change (§16.20).

12.8 *Elliptic and other functions*

{Houzel 1978}

The principal influence on the young Weierstrass was his teacher at Münster University, Christoph Gudermann (1798–1852). While not a major figure, Gudermann had studied elliptic functions extensively, introducing a version of uniform convergence {Manning 1975}. He also proposed the durable notations 'sn', 'cn' and 'dn' for Jacobi's principal functions; they became of principal concern to Weierstrass, often featuring in his own lecture courses. Mittag-Leffler wrote an excellent outline of the methods and their prehistory for his doctoral thesis at Helsinki University in 1876; it is available in the English translation {Mittag-Leffler 1923}. Schwarz edited in 1885 a booklet containing many of the *Formeln und Lehrsätze* involved; they were widely consulted, including in the French translation quickly published in *Acta mathematica*.

We saw in §8.12 that with Abel and Jacobi elliptic functions $x = f(s)$ were defined from integrals of the form

$$s = \int_{x_0}^{x} \frac{1}{\sqrt{R(v)}} \mathrm{d}v, \quad \text{with } R(v) \text{ a quartic in } v \qquad (12.8.1)$$

and took a pair of periods; and that complex variables became central to the development of the theory. Transformations from one variable to another also grew in significance, especially in Weierstrass's hands. He took v to be a bilinear function of a new variable q,

$$v := \frac{aq+b}{q+d}, \ a, b \text{ and } d \text{ constants, and put} \quad S(q) \equiv R(v) \qquad (12.8.2)$$

after transformation. He applied a string of conditions to the identity between R and S and several succeeding identities to obtain the following general formulation, which I state largely in his notations.

Given two real variables s and u, the elliptic function s of u is defined by the conditions

$$0 = \wp(\infty), \quad \text{and} \quad s := \wp(u), \quad \text{where } u = \int_s^{\infty} \frac{1}{\sqrt{S(s)}} \mathrm{d}s \qquad (12.8.3)$$

$$\text{and } S(s) := 4s^3 - g_2 s - g_3, \quad \text{with } g_2^3 > 27 g_3^2. \qquad (12.8.4)$$

The quantities g_2 and g_3 were known real-valued functions of the coefficients of R; the inequality between them was required to ensure that the zeros e_1, e_2 and e_3 of $S(s)$ were all real and distinct, so that u did not degenerate into a trigonometric or an exponential function. The periods ω_1 and $i\omega_2$ of $\wp(u)$, which required u to take real and imaginary values respectively, were given by the relations

$$\wp(\omega_1) = e_1 \quad \text{and} \ \wp(\omega_2) = -e_3, \text{ where } e_1 > e_2 > e_3. \tag{12.8.5}$$

Among many other properties, this one was especially elegant and fruitful:

$$\frac{\mathrm{d}s}{\mathrm{d}u} = \sqrt{(s-e_1)(s-e_2)(s-e_3)}. \tag{12.8.6}$$

On this basis Weierstrass rebuilt and extended the known theory of elliptic functions, and various related theories {EMW IIB3}. A major innovation was his 'sigma function' $\sigma(u)$, defined as

$$\wp(u) = -\frac{\mathrm{d}^2(\ln \sigma(u))}{\mathrm{d}u^2}; \tag{12.8.7}$$

for he hoped to develop from it the theory of elliptic functions (including elliptic functions as infinite products), as an alternative to the theta functions that Jacobi had introduced (§8.12). Jacobi's functions were not eclipsed, but Weierstrass's became a staple part of the normal presentation of elliptic functions.

Underlying Weierstrass's approach was his faith in the power-series expansion (12.7.1) of a complex (possibly real) function $f(z)$ in positive, and quite possibly also negative, powers of z. It also guided his version of Abelian functions, which we saw arising as integrals (8.13.4) of rational functions $F(x, y)$ of two real variables x and y. He defined them parametrically as infinite convergent power series of a new variable, and so came to formulate F as a quotient of them; this led him, for example, to a new definition of Abel's notion of genus implicit in Theorem 8.13.1 {EMW IIB2}. In 1874 he suggested to his private doctoral student

Sonya Kovalevskaya that she look for conditions under which Abelian integrals degenerated into elliptic or elementary ones.

I have summarized here Weierstrass's mature theory, as presented in his lecture courses. But his approach was nascent in his initial publication on the topic in 1854, when the mathematical community first learned of the arrival of a major new force.

12.9 *Differential equations*

{Forsyth 1890–1906; EMW IIA4–5 and 7; CE 3.14–3.15}

The theories just surveyed made much use of differential equations, which maintained their place as a staple part of mathematical analysis. Indeed, their status rose still further as complex analysis grew rapidly in importance; for example, the theory of the special functions such as the Bessel and the Legendre (§8.14) was often advanced by study of the pertinent differential equations {EMW IIA10}. Systems of equations received more attention, especially once determinants became available to provide new techniques.

Many mathematicians worked on these topics, especially in France, Germany, Italy and Scandinavia; elsewhere, some British mathematicians continued to use the algebraic operator methods which Boole had encouraged (§9.13). The overall picture is both vast in scale and complicated in detail and cross-connections. Certain methods inspired by Riemann's work on complex analysis are noted in §13.14 and §13.15, while various methods of solution appear in §14 and §15 in connection with applications; here are noted a few traces which show how Cauchy's style was enhanced by Weierstrass.

For ordinary equations, Cauchy's work on the existence of solutions to

$$\frac{dy}{dx} = f(x, y) \qquad (12.9.1)$$

exercised considerable influence after his death in 1857. His first method (8.15.2), based on analysing the limit of the partition sum $\sum_r f(x_r, y_r)\Delta x_r$ for continuous f, was developed by Lipschitz in 1868, who also used his own condition (12.3.4) on f.

Cauchy had also introduced a 'method of bounds' in 1831, which established a solution around some value $x = x_0$ by expressing y in a power series in $x - x_0$ and, in effect, seeking upper limits on the coefficients for it to converge. Taking f to be analytic in his sense (12.7.1), Weierstrass extended this method by replacing f by a multiple power-series expansion of a function in n variables and establishing the uniform convergence of such a series solution to a system of n first-order equations.

A third method, maybe also known to Cauchy but established by Joseph Liouville in 1838, built up the sequence

$$y_n(x) = y_0 + \int_{x_0}^{x} f(t, y_{n-1}(t))\, dt, \quad n \geq 1. \qquad (12.9.2)$$

It was extended by Picard during the 1880s, and gained popularity as the 'method of successive approximations'. It has some kinship with Laplace's method (6.7.2) of solving a system of ordinary equations, which carried the same name although it fulfils a different purpose.

Singular solutions were also studied in more detail. Their original representation (§6.3) as the envelope of the family of a complete solution was desimplified by further work on cases involving cusps and nodes and on the bearing of the initial conditions; Paul Painlevé (1863–1933) was significant here. These and other parts of the theory were extended into differential equations in a complex variable; for example, singularities at $x = X$ became branch points at $z = Z$, to be studied maybe by Weierstrassian methods of analytic continuation or by Riemann's methods, to be described in §13.13 {EMW IIB5}. But in 1883 Darboux showed that, in general, first-order equations did not have any such solutions.

For partial equations, Weierstrass's faith in power series suggested to him that his proof of the existence of solutions of ordinary equations could naturally be extended to partial ones, since an analytic function $f(x, y, u, \ldots)$ of several variables could be expressed as a multiple power series in *xyu*.... Kovalevskaya also took up this question for her thesis, and in 1874 confirmed his suspicion for a system of first-order partial equations where all the coefficients of the derivatives were analytic functions.

In 1889 another close associate, du Bois-Reymond, proposed a new classification of linear second-order equations with constant coefficients which imitated in algebra Apollonios's classification of the conic sections (§2.23). Let

$$az_{xx} + bz_{xy} + cz_{yy} + gz_x + hz_y + kz = 0, \quad \text{with } D := 4ac - b^2 \tag{12.9.3}$$

and the coefficients functions of x and y. Then the equation was 'elliptic' *or* 'parabolic' *or* 'hyperbolic' according to whether D was positive *or* zero *or* negative for all values of x and y. Each type could be transformed into special characteristic forms, thereby aiding its solution. This distinction had been intuited by Legendre a century earlier for equations with constant coefficients, and reductions had been found to some standard such equations, such as the elliptic Laplace's equation (6.9.1), the parabolic diffusion equation (10.11.4) and the hyperbolic wave equation (6.4.1); but this new classification, and later extensions of it, provided a powerful means of appraising second-order equations. Other forms of classification were proposed, and methods developed to reduce equations, both linear and non-linear, to special forms, and thereby aid the determination of general or at least particular solutions.

12.10 *Numerical and graphical methods*

{C.F. Klein 1928}

Even with this repertoire of solutions, many differential equations could not be solved exactly. So numerical and approximate solutions were often sought, and the equations related to solutions of difference equations; with luck, equations defining known functions might arise. A powerful figure in this area was Carl Runge (1856–1927), a student of Weierstrass's, who however turned to these interests later, in Klein's circle {Richenhagen 1985}. In 1915 he co-authored a fine survey {EMW IIC2} of the achievements.

These methods often involved interpolation, especially in the calculation of tables of values of functions. This became quite an industry during this period, not only for further tables of logarithmic and trigonometric functions but also for special functions {EMW 1D3}. A popular type of interpolation was by finite trigonometric series {EMW IIA9a} – a practical activity complementary to the concern of the analysts with Fourier series. Various new means of approximately evaluating series and integrals were devised, though often the rates of convergence were frustratingly slow, especially for mathematical astronomers analysing the motions of planets over a long period of time. In this context Henri Poincaré developed from the mid-1880s a good method for finding 'asymptotic solutions' in t^{-1} of differential equations which was effective when the time t was large {Schlissel 1977}.

Powerful methods of computation using geometry were developed, especially in France, where they became known as 'nomography' {CE 4.12}. An early example due to the engineer Louis Lalanne (1811–92) in 1843 concerned the general cubic equation

$$x^3+px+q=0, \text{ with discriminant } D:=p^3/27+q^2/4. \tag{12.10.1}$$

The normal treatment of the equation was to find the roots of x in terms of the constants p and q (§4.6); but Lalanne turned the tables in FIG. 12.10.1, which shows p and q as the axes and straight lines drawn for various values of x (which are indicated by each line) over a square between values from -1 to $+1$ on the horizontal axis p and vertical axis q.

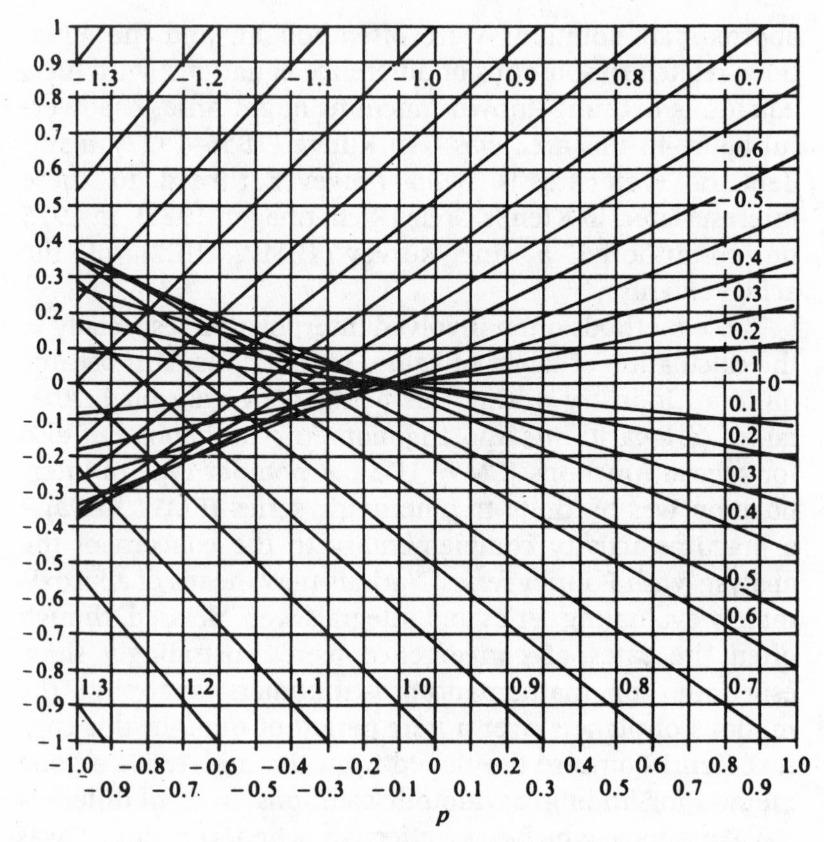

FIG. 12.10.1 *Lalanne's nomogram of a cubic equation, 1843.*

The density of lines is greatest to the left of centre; in fact three can intersect at any point there, giving three real roots of the cubic, with values read off from the slopes. Elsewhere lines do not intersect, and each slope is the single real root of the cubic with corresponding values of p and q. The boundary between the two regions is given by the curve $D = 0$.

From this and various other uses, nomography gained publicity, especially from the French mathematician Maurice d'Ocagne (1862–1938), in his textbook *Nomographie* (1891) and in other writings. Perhaps the most remarkable one is his article {d'Ocagne 1911} for the French version of the *Encyklopädie der mathematischen Wissenschaften*, in which he effected an extraordinary 250-page rewrite of {EMW IF1} on numerical mathematics in general.

Many of these activities in numerical mathematics were brought together at an event early in the life of the German Mathematicians' Association. In 1892 the 36-year-old Walther Dyck organized a massive exhibition of 'mathematical models, apparatus and instruments' at the Technical High School in Munich, where he was Professor; he also edited a marvellous catalogue {Dyck 1892, 1893}, including articles by several mathematicians and physicists. Runge and d'Ocagne were among the many exhibitors, who displayed a wide range of artefacts: calculating machines, plaster-and-string models of surfaces (§13.16), mechanical linkages (§14.7), planimeters to measure areas, gadgets to find roots of equations approximately, and so on. In his preface Dyck stressed that one of the motives was to bring pure and applied together, at a time when the separation was tending to increase – not least for the following reason.

12.11 *Mathematical statistics begins to arrive*

{CE 10.15}

One centre for the development of numerical methods was St Petersburg, in a group centred around Pafnuty Chebyshev (1821–94). But his main activities lay in probability theory and mathematical statistics, where he used techniques from mathematical analysis to obtain improved proofs of various principal theorems. As early as the mid-1840s, when he was in his mid-twenties, he had established Poisson's weak law of large numbers (§8.10) in a form more clearly oriented towards the notion of sampling than Poisson himself had shown: given a large number X of successes out of n trials (such as obtaining 'heads' when tossing a coin), and an average probability p_n of success over those trials, then the probability that $\frac{X}{n}$ was close in value to p_n tended towards 1 as $n \to \infty$. Among other contributions, he found in 1867 a more general proof of Laplace's central limit theorem (§8.8), although he tacitly assumed that the trials were independent. He also encouraged a fine school around him at St Petersburg {CE 10.6}, among whom Andrei Markov was to make remarkable contributions (§16.21).

Another important country in mathematical statistics during this period was England, in a continuation of its interest in 'political arithmetic' recorded in §5.21. An outstanding figure is Florence Nightingale (1820–1910), who used statistical thinking to show that British soldiers were far more likely to die from disease than from wounds, and therefore hygiene had to be improved in hospitals. FIG. 12.11.1 shows one of her 'coxcombs' (the modern name is 'polar-area diagram'), which she assembled from data for 1854 and 1855 collected during the Crimean War and published in 1858 in her book *Notes on Matters Affecting the Health, Efficiency and Hospital Administration of the British Army* {Cohen 1984}. Each sector represents a month, and the areas within it are proportional to the numbers of soldiers who died from (reading outwards from the centre) wounds, other causes, and – by far the majority – diseases.

Nightingale had to follow up her brilliant innovation with a battle against the male 'experts' who wished to avoid thinking of this kind in designing hospitals. Her contributions belonged to a gradual introduction of statistical methods of various kinds into medicine at this time {CE 10.12}. Twice she tried to have a chair of statistics founded at Oxford University; she failed, of course, but statistical theory began to enter many areas of human activity in Britain, due to the advocacy of men such as her almost exact contemporary, Francis Galton (1822–1911). Passionate for the collection of data and the appraisal of its variability, he sent a questionnaire to his colleague Fellows of the Royal Society, requesting from them details of their education, physiognomy, health, personality and scientific interests. In 1874 he distilled their replies into his *English Men of Science: Their Nature and Nurture* {Hilts 1975}. The

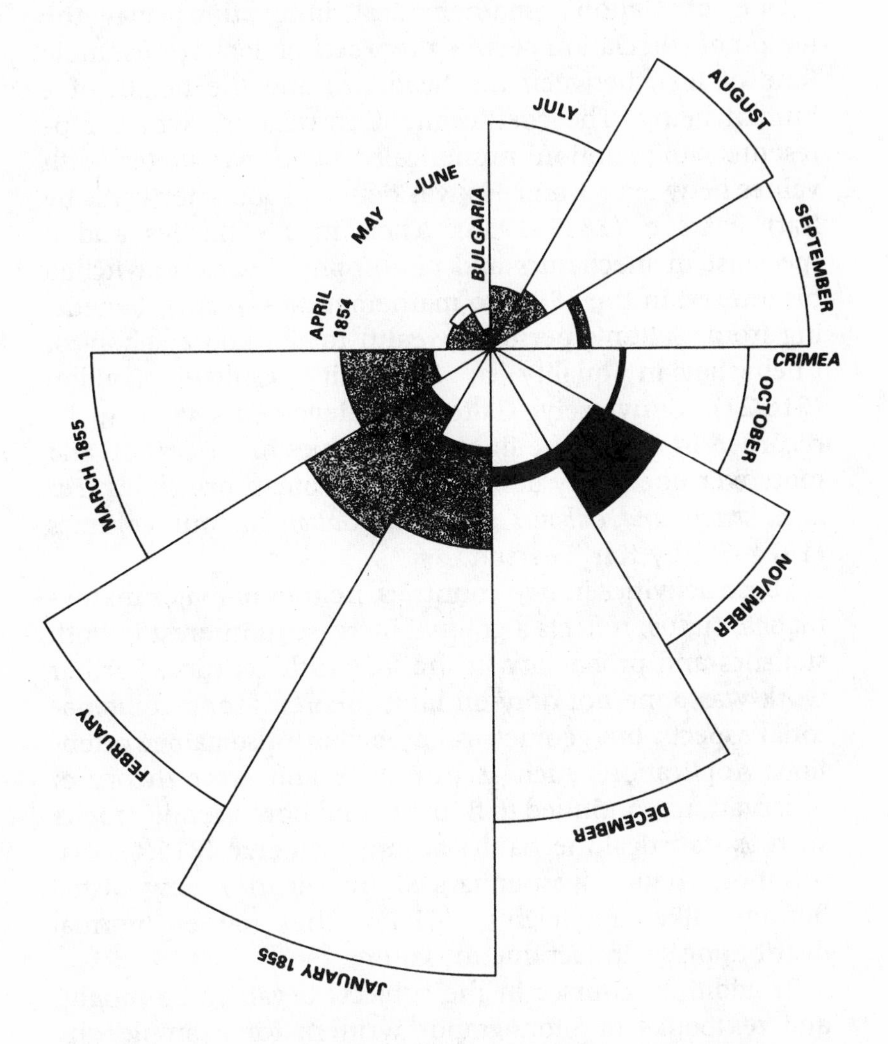

FIG. 12.11.1 *Nightingale's 'coxcomb' of the deaths of soldiers, 1858.*

subtitle of his book states one of the principal concerns, in the era of Charles Darwin (his cousin, and a respondent to the questionnaire); indeed, he had already given the title *Hereditary Genius* to a book of 1870, and he was to coin the word 'eugenics' in 1883.

One of Galton's mathematical innovations was the notion of correlation between two sets of data: for example (one of his), between the head size and the height of a human being. The coefficient of correlation, which represented this notion numerically by a parameter with values between −1 and +1, was defined soon afterwards by Karl Pearson (1857–1936). Then in his thirties and a specialist in mechanics and philosophy, Pearson switched his interest in the 1890s to mathematical statistics, benefiting from Galton's personal wealth to develop a school of Chebyshevian quality at University College, London (§16.21). Conversely, Galton's achievements were to be recorded in one of the finest biographies of a scientist, and moreover one partly designed on Galtonian principles: *The Life, Letters and Labours of Francis Galton* in four volumes (1914–30) by Karl Pearson.

These activities in two countries, neither of major mathematical status, reflects a gradual increase in interest in both statistics and probability in the late 19th century. Further work was done not only on limit theorems, and combinatorial aspects, but geometrical probability also gained attention. Applications such as insurance and error theory of observations continued to flourish, and new scientific topics such as statistical mechanics began to emerge (§15.4). Distribution theory was enhanced by (nearly) new distributions like Rayleigh's (§1.1); the name 'normal distribution' is in fact due to Galton.

In addition, courses in the subjects began to be taught, and textbooks or monographs written: for example, the *Calcul des probabilités* (1889) by Joseph Bertrand. The book-length report for the German Mathematicians' Association by the Austrian Emanuel Czuber (1851–1920) rounded off the century with a rich survey of methods and applications of probability theory {Czuber 1899}. Soon afterwards he gave a summary account for the *Encyklopädie der mathematischen Wissenschaften* {EMW ID1}, which was backed up by a two-part article {EMW ID4} by others on statistics and insurance. But the total of these pieces is only 140 pages, in a project containing thousands. Probability and statistics still had a long way to go professionally; the next steps are noted at §16.21.

12.12 *Analytic number theory: properties of the zeta function*

We turn from a new branch to an old part of mathematics, though one still with a modest roster of active personnel. Somewhat more attention was paid to number theory during this period than previously, especially on the algebraic side where it both benefited from and assisted in the development of abstract algebras of various kinds (§13.3, §13.20). Contributions on the analytical side were more patchy, although elliptic functions continued to be used {EMW IC3}.

Some very significant results were found about the distribution of prime numbers, which had been understood since Euler and Gauss as having to do with properties of the function $P(N)$ of the number of primes less than the large integer N. It appeared to be of the order of $(N/\ln N)$, and the guess was proposed that

$$\lim_{N\to\infty} \frac{P(N)}{N/\ln N} = 1. \tag{12.12.1}$$

In 1848, Chebyshev had shown that *if* there was a limit to the expression then it would be 1, and he also found bounds within which it lay.

Chebyshev made use of Euler's zeta function (5.18.5); a decade later, in his typical manner, Riemann ushered in a new era of study by restating it as a function of a *complex* variable:

$$\zeta(z) := \sum_{r=1}^{\infty} r^{z} = \prod_{p} \frac{1}{1-p^{-s}}, \quad p \text{ prime}, \ z := x + iy. \tag{12.12.2}$$

By a deft use of Fourier integrals and a casual invention of the Mellin transform (the second form of eqn (16.13.3)), he went further and guessed that

$$\text{if } \zeta(z)=0 \text{ and } x>0, \quad \text{then } x=\tfrac{1}{2}. \qquad (12.12.3)$$

Here Riemann made one of his most marvellous suggestions, for it remains as one of the great unsolved mysteries of mathematics; partial solutions to it have enriched both number theory and analysis. One of the great unsolved mysteries of history is the means by which he came to think of it in the first place: manuscripts show that he had even begun to calculate the value of y for the first three zeros of $\zeta(z)$.

While this conjecture remains open, the guess (12.12.1) was proved around 1896 by Jacques Hadamard and Charles de la Vallée Poussin, when it became 'the prime number theorem'. They also showed it to be equivalent to the conjecture that

$$\zeta(z)\neq 0 \text{ when } z=1+\mathrm{i}y, \text{ for any } y\neq 0. \qquad (12.12.4)$$

This and other work exemplifies the massive growth in importance of complex-variable analysis during this period. We have seen Weierstrass's style in this chapter; Riemann's is presented in the next one (§13.12–§13.14).

12.13 *Peano: mathematics in logical clothing*

{Borga and others 1985; CE 5.2}

The rest of this chapter is concerned with work which was driven largely by Cantor's set theory. An important propagandist was Peano, for he brought a new degree of symbolization and axiomatization to it and to its use in mathematical analysis. He was particularly impressed by the need for well-formed definitions, and by the need to make explicit the functional relationships between the many variables that frequently appeared in proofs in the subject. So every step had to be written down in detail.

For 20 years from the late 1880s, when he entered his thirties, Peano carried out his programme with the help of a large and powerful school of followers in Turin. The word 'school' is used in its strict sense: he called it a 'society of mathematicians', he was the boss, and many followers were his students at some time. They became known affectionately as the 'Peanists'. The chief lieutenants were Cesare Burali-Forti (1861–1931) (who wrote the first textbook on the programme in 1894), Mario Pieri (1860–1913) and Alessandro Padoa (1868–1937). They all used two special publications which he created: the journal *Rivista di Matematica* (8 volumes, 1891–1906), and a compendium of presentations of mathematical theories entitled *Formulario Mathematico* (5 editions, 1895–1908).

Peano's first logical essay, of 1888 (his 31st year), drew on Hermann Grassmann and some aspects of George Boole (§9.13); but at once he converted to Weierstrass's passion for rigour (although he never took a course at Berlin), and took little further notice of the subsequent progress of algebraic logic (§13.4). In particular, he used only Cantorian set theory to handle collections of things, in contrast to the employment of traditional part–whole theory of collections by algebraic logicians (following Boole); indeed, he criticized them for not distinguishing membership of a set from proper and improper inclusion of subsets. He also stressed the difference between an object and its unit set, and between sets and sets of sets in general.

In their exegeses the Peanists, especially the master himself, created many of the elements of 'mathematical logic' (his phrase) as we have it today,[2] thereby dealing with limitation 1– of Cauchy's programme (§8.19). They outlined the logic required for a calculus of propositions and predicates, including universal and existential quantification ('for all' or 'there is'); 'primitive propositions' covered both axioms and rules of inference. The algebra of variables that Fermat and Descartes had introduced (§4.18–§4.19) was

[2] The phrase 'mathematical logic' is due originally to De Morgan, in 1858 (see §9.13 for the context); but he had in mind a symbolized version of logic in general, in contrast to purely prosodic 'philosophical logic'. This latter was practised widely in Italy in the 19th century, but Peano's programme for logic has largely obliterated it from historical record.

now refined into one of 'apparent' and 'real' variables – that is, respectively bound or not bound by a quantifier.

The Peanists also provided notations for the basic logical connectives: for example, '⊃' for 'if ... then' (and also 'implies'), and '.' for 'and'. This latter symbol was cleverly built into a convention of dots used to bracket off parts of an expression into its components; the more dots, the 'bigger' the brackets. Properties were expressed as 'p_x', denoting a proposition p in which x is free – *not* necessarily a propositional function, therefore. Quantification over individuals x was often universal, and symbolized by the subscript '$_x$'; existential quantification could be defined from 'not (for all) not'. They used some symbols in dual pairs when they represented converse notions; for example, in 1889 Peano denoted the abstraction of sets ('the sets of entities such that ...') by '∍', which served as dual to set membership '∈' ('is a member of ...'). They also brought in symbols for set theory; some were the same as in logic, such as '⊃' for improper inclusion. Many of these symbols are still used, often with the same or similar referents as theirs.

Peano was very sharp on definitions, laying down rules for them to be properly formed in a mathematical theory. He distinguished various kinds, pointing out, for example, that nominally defining the upper limit of a sequence of values (for which kind Burali-Forti introduced the notation '$=_{\mathrm{Def}}$') differed from defining the equality of two such limits. The latter was an instance of the important 'definition by abstraction' (his phrase): some hypothesis is made about the nature of a given set of objects, and a property is defined which is reflexive, symmetric and transitive among them (like that of the upper limit). We now call this method 'definition by equivalence classes'.

The Peanist exegeses often looked like wallpaper, with prose kept to a minimum. A long paper of 1890 by Peano in *Mathematische Annalen*, dealing symbolically with Cauchy's first method of establishing the existence of a solution to a differential equation (§12.9), apparently astonished some readers by its appearance.

In their treatments of mathematics, the Peanists laid out the basic notions and sometimes axioms of set theory, arithmetic (including some transfinite arithmetic), certain aspects of mathematical analysis, Euclidean geometry (§13.18) and vector algebra; a few other theories were sometimes handled. They distinguished mathematics from logic by listing the basic notions in separate columns. However, the distinction between the two categories was not always clear, with set theory sloshing around between them; soon Bertrand Russell was to imagine that it did not exist (§16.6).

12.14 *Some philosophies of arithmetic*

{CE 5.2}

Peano's programme related to other studies of the time on the foundations of arithmetic. One was a logicist plan of the German Gottlob Frege (1849–1925), dating back to 1879, to show that arithmetic and parts of real-variable analysis could be formulated entirely within logic. This was conceived as a propositional and predicate calculus together with quantification of arbitrarily high order (over individuals, predicates, . . .). I use Cantor's word 'set', for Frege seems to have thought out parts of Cantorian set theory for himself. In these terms he defined integers thus: 0 as the set of the empty set, 1 as the set of unit sets, and so on. His work met a poor reception, for some reason,[3] but Peano reviewed him in 1895 and they had an exchange in the

[3] Frege is now a hero to philosophers, but mainly for his interesting insights into language in general. His neglect during his lifetime is normally explained as due to his unusual notation, which used a two-dimensional presentation of symbols in order to reduce substantially the need for brackets; but I learnt it in 20 minutes, and I cannot see why his contemporaries would have needed longer. I suspect that his unorthodox use of technical terms such as 'function' was a much more serious obstacle; in addition, maybe he aimed to construct foundations too deep even for mathematicians of like mind. Furthermore, advocating in 1903 that Euclidean geometry was the only kind possible (for context, see §13.8 and §13.18) was not designed to gain fans.

Rivista. Frege objected to definitions under hypothesis (though hardly with justice, since we need universes of discourse); he also lamented Peano's failure to distinguish the sense of a property or proposition from its reference, a matter central to his own philosophy. Curiously, the most obvious difference did not arise: Frege tried to define a mathematical theory out of logic, whereas Peano tried to keep logic and mathematics separate.

One of Peano's early axiomatizations (1889) was of arithmetic, taking '1' and 'successor of' as primitives and incorporating mathematical induction over sets of numbers as one axiom. (On this occasion he muddled the logic by including in it the defining properties of equivalence classes.) The year before, Richard Dedekind had published a short book on the same matter answering the question '*Was sind und was sollen die Zahlen*?' ('What Are the Numbers, and What Are They Good for?'). Working with sets ('systems'), he developed a theory of transformations of a set onto itself, and defined the integers from 1 onwards in terms of a particular kind called a 'chain'. The similarity with Peano showed up over mathematical induction; but here Dedekind was superior, for not only did he validate proof by induction but also found a way of justifying inductive *definitions* in the first place. This deep subtlety was not appreciated for some time.

Another difference shows Peano in better light: while Dedekind asserted that arithmetic was a part of logic, he never described that logic, and so left his position unclear. He cannot safely be bracketed with logicist contemporaries such as Frege or successors like Russell.

12.15 *Cantor's* Mengenlehre *in the 1890s: the diagonal argument and transfinite arithmetic*

After some rather barren years, in 1891 Cantor answered an 'elementary question' in set theory at the inaugural meeting of the German Mathematicians' Association, of which he had been a principal founder {Cantor 1932, pp. 278–81}. The result is among the best remembered of his innovations: the 'diagonal argument', which took a set and constructed from it one of larger cardinality. He considered the set M of all points E which could be expressed by the coordinates of a denumerably infinite coordinate space defined over a pair of 'characters' m and w (presumably *Mann* and *Weib*, 'man' and 'wife'); typically,

$$E = (m, m, w, m, w, w, m, w, m, w, m, \ldots). \quad (12.15.1)$$

To show that M was more than denumerably infinite, he took a denumerable sequence S of its members, and laid out their coordinates in a matrix-like array $a_{i,j}$. He then defined another member b_i by diagonalizing:

$$\text{if } a_{i,i} = m \text{ or } w, \quad \text{then } b_i = w \text{ or } m, \quad i \geq 1; \quad (12.15.2)$$

but this property guaranteed that b_i was not a member of S, as required. He also extended the proof by showing in the same way that the cardinality of the interval [0, 1] was less than that of the set of all its subsets.

Contrary to the common belief, Cantor's proof was direct, not by contradiction; he did not assume that S comprised the whole of M, although the result obviously held for M. The normal version of this argument today proves – directly – that the real numbers *Re* are more than denumerably infinite. Assume that *Re* within [0, 1] *are* denumerable, and so present each one, r_n, in a decimal expansion:

$$r_n = 0 \cdot a_{1n} a_{2n} a_{3n} \ldots a_{nn} \ldots . \quad (12.15.3)$$

(As at (12.5.1) on dimensions, some homework is needed to distinguish, say, 0.2999 from 0.3000.) But we can define another real number, s, within [0, 1] which differs from each r_n by doing so in (at least) the nth decimal place:

$$s = 0 \cdot s_1 s_2 s_3 \ldots s_n \ldots, \quad \text{where } s_n \neq a_{nn}; \quad (12.15.4)($$

hence (12.15.3) cannot furnish the complete list assumed.

The diagonal argument played an important role in Cantor's last major paper on set theory (*Mengenlehre* was now his name). It appeared in two parts in *Mathematische*

Annalen in 1895 and 1897, when he was in his early fifties {Cantor 1932, pp. 282–356}. Largely setting aside point-set topology, he concentrated on the structure of transfinite cardinal and ordinal arithmetic.

Cantor started with a general definition of a set which was to become notoriously unacceptable to most of his followers (§16.5–§16.6) for its reliance on mental acts: 'By a "set" I understand any collection M of definitely distinguished objects m of our perception or our thought (which shall be called the "elements" of M) gathered into a whole'. (This procedure is not to be confused with Peano's definition by abstraction in §12.13.) He then set the mind to work again on M, abstracting away the nature of the members m to yield its ordinal number '$\bar{M}$', and then the order itself to reveal its cardinal '$\bar{\bar{M}}$' {Article 1}. This elaboration of his outline of 1883 (§12.4) led him to envision set theory as a foundation for arithmetic, and by implication for (much) mathematics; it was to influence the philosophy of mathematics substantially. However, in contrast to Frege, he could not define zero this way, either cardinal or ordinal; presumably he could not imagine obtaining those numbers by abstracting from an empty set, a notion he used and even notated sometimes (as 'O') but never handled confidently. In addition, and like Frege, he ignored negative integers.

Unlike most of Cantor's previous papers, this one was replete with notations, including his Hebrew alephs for the transfinite cardinals {Article 6}:

$$\text{'}\aleph_0, \aleph_1, \aleph_2, \ldots \aleph_\nu, \ldots\text{'}; \qquad (12.15.5)$$

he even mooted a successor cardinal, '$\aleph_\omega$', without, however, giving a definition. The chief novelty was a method of 'covering' a set M to produce the set of all its subsets, whose cardinality C was shown by the diagonal argument to be greater than $\bar{\bar{M}}$ {art. 4}; indeed, that argument also suggested the natural notation '$2^{\bar{\bar{M}}}$' for C. He also gave an improved definition of a continuous set, requiring it now to be perfect and to contain an ordered subset like that of the rational numbers in between any given members {Article 11}. These considerations led him to state the continuum hypothesis, about the cardinality of the set X of real numbers, in its final and definitive form:

$$\text{theorem: } \bar{\bar{X}} = 2^{\aleph_0}; \quad \text{hypothesis: } 2^{\aleph_0} = \aleph_1? \quad (12.15.6)$$

Underlying this guess was another one: the well-ordering principle, upon which the generality of the concept of ordinal number depended, and that of set theory itself. Cantor still had no proof of it, but in the second part he discussed in detail various of the orderings in which a set may be laid, and the structure of the second number-class of ordinals. This line of thought was to lead to paradoxes, as we shall see in §16.5–§16.6.

12.16 *The selling and splitting of Cantor's* Mengenlehre

During the 1890s set theory grew rapidly in popularity. Cantor's two-part paper aided the interest; it was soon translated into French, and the first part also into Italian (in Peano's *Rivista*). However, it was his last published contribution; in 1899 he suffered a severe mental collapse, and spent parts of his remaining 20 years in clinics and sanitoria.

An important stimulus to the subject was provided in 1893 by Camille Jordan, when he started to publish the second edition of his *Cours d'analyse*. In the first edition he had placed the details of Cantor's point-set topology in an appendix to the third volume (1887); but now it shot to the opening chapters of the first one {Gispert 1983}. He also improved the understanding of the post-Riemann integral, as we shall see in §16.9 when recording its use by his younger compatriots.

Another source of popularity was the report written on the subject for the German Mathematicians' Association by Arthur Schönflies (1853–1928), Cantor's most ardent supporter {Schönflies 1900}; he had published a short summary {EMW IA5} the previous year. He also gave most space to the topological side, including applications to the theory of functions and the integral. He interpreted the diagonal argument as a proof by contradiction. Overall his treatment is not satisfying, but he recruited several newcomers.

The publicity given to set theory at the time looks in some ways like a drive. In particular, the International Congresses of Mathematicians were launched in 1897 at Zürich (partly because of Cantor's advocacy), and several lecturers made a point of using set theory. (In a talk 'on certain possible applications', Hadamard defined a well-ordered set, but made a mess of it!) In all Congresses up to Cambridge in 1912, some aspect or other of set theory was deployed by speakers in a seemingly concerted promotion.

Point-set topology and companion theories such as integration attracted the attention not only of pure mathematicians but also in certain areas of applied mathematics. In addition, several philosophers and mathematicians with philosophical interests took up the more general aspects, such as the structure of Cantor's transfinite arithmetic and his definitions of integers from sets. A notable member of this coterie was the Frenchman Louis Couturat (1868–1914), with a substantial survey *De l'infini mathématique* (1896), which gained plenty of readers.

In this way set theory tended to split into two parts. But the division was *false* to Cantor's own conception of *Mengenlehre*; for him, the topology and the transfinite were always indissolubly linked in the infinitieth derived set $P^{(\infty)}$ of a set P and its successor sets in the sequence (12.4.1) upon which he had stumbled in 1872 when pondering the consequences of Riemann's finding that trigonometric series were not always Fourier series.

13

Algebras and geometries: their relations and axioms, 1860–1900

13.1 *Algebras and geometries, separate and together*

The continued rise of (fairly) abstract algebras of various kinds is surveyed in §13.1–§13.6; then Riemann reigns over §13.7–§13.17, as he did in parts of the last chapter, through draft thesis texts, papers and lecture courses. But this time the subject matter is the foundations of geometry and complex-variable analysis, with some assistance from both abstract and linear algebras. Profound consequences were to follow: the distinction between (an) algebra and (a) geometry was to become more porous; and topology began to rise as a new branch of mathematics, somewhat between algebra and geometry, like trigonometry before it (§4.1).

The rest of the chapter deals with the 1890s, and belongs above all to David Hilbert. His theorems on invariants (1888), his report on algebraic number theory (1897) and his axiomatization of Euclidean geometry (1899) were culminations of much previous work of the previous 40 years. In §13.20 is noted the prominent place of linear combinations, which run as a thread through most of the algebras and some aspects of the geometries. The final section shows evidence of hope for *general* algebraic theories; let us start with one of the most important special ones.

13.2 *Group theory, generally speaking*

{Wussing 1984, Part 3; CE 6.4}

The advocacy of a group as a structure independent of any interpretation, which is the normal approach to group theory today, seems to have its origins just before the period covered here; for pioneers included Arthur Cayley in a paper of 1854 on the roots of unity, and the young Richard Dedekind in lectures at Göttingen two years later (unpublished, and dealing with Galois's work (§9.4) on equations). For Cayley, a group was a collection of objects 1, α, β, ... and a means of combining them to form a 'product' $\alpha\beta$, with the associative law holding and 1 serving as an identity (the existence of the inverse was assumed) {Cayley 1854}. However, neither author was then influential here, and the approach did not gain strong currency until well into the 1870s.

Two important early stimuli came from Paris. In 1866, his 47th year, Joseph Serret published the third edition of his *Cours d'algèbre supérieure,* in which he included a part on Galois theory. The textbook was very influential, with translations and later editions; however, his treatment was rather conservative and did not reflect the most recent advances {Kiernan 1971}. Some of those were due to Camille Jordan, who published his *Traité des substitutions et des équations algébriques* in 1870, when he was 33 years old. As his title states, he concentrated upon one manifestation of groups, that studied by Cauchy 30 years earlier (§9.4). However, he discussed several structural features of groups, especially normal subgroups N of a group G in which, for any member n of N, $n^{-1}gn$ was also a member of N for all g in G (note Charles Babbage's conjugate form (§9.12) again). He also described a wide variety of applications of the properties examined in detail. He emphasized mappings (one–one and many–one) from one group onto another one or onto itself, a procedure that would soon prove very fruitful for Klein and Lie (§13.9).

Among other early contributors was Lie's Norwegian compatriot Ludwig Sylow (1832–1918). From studying Galois's own work, he produced in 1875 theorems which pioneered analysis of the interior structure of finite groups: typically, if the order of a group were divisible by p^n (with p a prime), then it contained at least one subgroup of order p^n.

Thereafter, German mathematicians dominated the development of both group and Galois theory. The emphasis on general groups was publicized by Klein's young graduate student at Munich, Walther Dyck, who published important papers in Klein's *Mathematische Annalen* in 1882 and 1883. He studied infinite groups as well as finite ones in some detail. Other major contributions included the factor group of a group, later known as the 'quotient group' and defined as the group of co-sets of any of its normal subgroups; an important innovator here was Otto Hölder (1859–1937), a contemporary of Dyck whose training in Weierstrassian analysis had recently been supplemented by exposure to Klein at Leipzig {Nicholson 1993}.

13.3 *Dedekind versus Kronecker on algebraic numbers*

{Ellison and Ellison 1978}

Of the various algebraic structures which we now recognize, group theory far outstripped the others in development during this period; but substantial progress was also made in the study of algebraic numbers (that is, numbers that can be expressed as zeros of a polynomial with integer coefficients). We noted at (9.5.2) that, in order to prove Fermat's Last Theorem, Ernst Kummer had studied in 1847 means of uniquely factorizing cyclotomic integers for a given prime number λ (that is, polynomials with integer coefficients in a complex λth root α of unity). Thirty-five years later his student and successor as professor at Berlin, Leopold Kronecker, presented an extended theory in a huge paper of 1882, on the 'Foundations of an Arithmetic Theory of Algebraic Quantities', written to celebrate the 50th anniversary of Kummer's first doctorate (it was then customary to celebrate such occasions for major scientists). Kummer was then 72 years old, Kronecker a decade younger.

Treating cyclotomic integers as ordinary polynomials subject only to the condition $\alpha^\lambda = 1$, Kronecker extended them to the status of 'divisors' of 'algebraic quantities' (that is, multi-nomials in several variables $x, y, \ldots$) in a 'domain of rationality' (in modern parlance, fields). He focused upon this problem: take two polynomials $f(x)$ and $g(x)$ and a finite collection of polynomials $h_r(x)$, and determine numbers a_r and b_r from a given field D such that

$$f(x) + \sum_r a_r h_r(x) = g(x) + \sum_r b_r h_r(x) \quad \text{for all } x \text{ in } D; \tag{13.3.1}$$

if they can be found, then f and g belong to the same equivalence class of functions.

A different approach to such studies was taken by Dedekind, again consonant with his general philosophy of developing a mathematical theory in a very general, sometimes structural manner. He was inspired by both Kummer and Dirichlet's study of the unique factorization of polynomials. In collaboration with Heinrich Weber, his co-editor of the works of Riemann, he placed in Crelle's journal, also in 1882, a long paper on 'algebraic functions'. Among other things, they extended Kummer's number to that of an algebraic structure, called 'ideal' after Kummer's 'ideal complex number' (§9.5); it was composed of a set c_r of cyclotomic integers (they actually wrote of a similar concept called 'whole functions') of any power which was closed under addition and subtraction, and also under multiplication by any such integer c; that is, if c_r and c_s were members of an ideal, then so were $c_r \pm c_s$ and cc_r.

The contrast between the two approaches is marked. Kronecker's constructivist philosophy of mathematics (§12.5) led him to see eqn (13.3.1) as posing the *computational* task of finding the values of the a_r and the b_r. However, he restricted himself to a finite number of further extensions; thus he excluded the irrational numbers, which for him could not be constructed. Dedekind and Weber preferred the more structural approach in which the basic properties of ideals and extensions to algebraic functions were studied in terms of *forms* inherent in eqn (13.3.1) and elsewhere; Dedekind's later definitions of an ideal were still more structural, and set-theoretic. Gradually this kind of view came to prevail in an atmosphere in which laws and axioms were the fashion in algebras, and set theory was gaining popularity in mathematics.

But lacking even from their treatment was any conception of field as general as that of group; thus I have avoided speaking of 'algebraic number fields' or describing their treatment in terms of field extensions. Similarly, the word 'ring' has been off the vocabulary. They belong, respectively, to the end of the century (§13.19) and the beginning of the next (§16.19).

13.4 *The Peirces on algebras for logic*

{Roberts and others 1997}

The growing library of algebras obeying the associative law but differing in other respects drew Benjamin Peirce in the USA to compile a catalogue of the known and new kinds. Hamilton's quaternions (9.8.2) and their various extensions were an important stimulus to his research; he was instrumental in popularizing them in his country. The USA has hardly featured in this book up to now; and the economic difficulties for mathematics there are indicated by the fact that this Professor of Mathematics at Harvard University and Director of the Geodetic Survey of his country had to publish this work as a lithograph, written out by a lady with no mathematical training but a beautiful hand, from which 100 copies were printed at 12 pages a time on the lithographic stone.

Peirce formulated algebras of objects which combine under two operations, with multiplication associative over addition. An additive identity 0 was assumed to exist, but not necessarily a 1. Each algebra used between two and six units, and powers or products of any of them led to a linear combination of some or all of them; hence his title, 'linear associative algebra'. He laid out the basic laws of each algebra in an array; for example, one of his 'triple algebras' reads:

	i	j	k
i	i	j	k
j	j	k	0
k	k	0	0

(13.4.1)

Peirce defined two properties which have become standard; 'nilpotence' when $x^n = 0$ with $n \geq 2$ (k in (13.4.1) is an example), and 'idempotence' when $x^n = x$ with $n > 1$ (i there). This latter property obviously smacked of Boole's index law (9.13.4), and was of especial interest to Peirce's first serious reader. This was his son Charles, then in his early thirties and commencing a troubled career which would cover an extraordinary range of achievements in mathematics, science (especially for his father's Geodetic Survey), philosophy and, above all, logic {Brent 1993}.

For 15 years from the mid-1860s, Charles Peirce (1839–1914) published regularly on logic in American journals. His chief stimuli were his father's survey of algebras; Boole's algebra of logic, though with '$x + y$' understood to cover *all* classes x and y (Jevons' reading (§9.13), which he adopted independently); and his own recognition of the importance of 'relatives' in logic, at first independent of De Morgan's pioneering treatment of two-place relations (§9.13), though reinforced by it later. This last aspect, which he extended to many-place relatives, became his passion. A paper contemporary with his father's lithograph gave a wild but remarkable range of uses of relatives; later on he used his father's arrays to multiply (in the sense of compounding) relations. As with Boole, logical equivalence was important, because he was formulating and handling equations ('=' as 'if and only if'); but gradually 'if . . . then' took over prime role, denoted by the symbol '$-\!\!<$'. It was used also for classes m and n, say, '$m -\!\!< n$' indicating that m was part of n; although its properties were structurally similar to improper inclusion ('$\subseteq$' of (12.5.2) in Cantorian set theory), the underlying conception of a collection is quite different.

Peirce became deeply interested in aspects of Cantor's work (for example, continuity), but he did not *embrace* it in his logic. This is a major point of difference with Russell; they had little sympathy for each other's work. Another point of difference is quantification: whereas Russell was to follow the Peanists in thinking of 'there is' and 'for all' as operators binding terms on sets (§12.13), Peirce at first saw them as extended algebraic addition and multiplication, with notations to match: say,

$$\text{`}\Pi_s l_{rs}\text{' for `all } s\text{'s love } r\text{'; that is,} \quad \text{`"}s_1 \text{ loves } r\text{" and "}s_2 \text{ loves } r\text{" and} \ldots\text{'.} \tag{13.4.2}$$

Later, however, he moved towards the other position, probably because this one needed infinitely long formulae, and therefore an infinite language.

Perhaps the best professional period in Peirce's chequered career was between 1879 and 1884, when he held a post at Johns Hopkins University in Baltimore. He experienced difficulties in his relations there with his colleague J.J. Sylvester, who had founded in 1878 the *American Journal of Mathematics*, the first major mathematical journal in the country (§16.2). Peirce put some of his logic papers in it; and in 1881, a year after his father's death, he also gave publicity to the 1870 lithograph by reprinting it there, with his own notes. As we saw, its title did not relate to linear algebra as the phrase is now used; but that subject was beginning to flower at this time thanks to Sylvester and others, as we shall now see.

13.5 *Determinants and invariants, and then matrices* {CE 6.6–6.8}

By the 1860s, determinants were beginning to gain currency in mathematics as a valuable tool in various branches (§9.10), thanks especially to the influence of Jacobi. One result which was proved and reproved in various forms {May 1968} gave the formulae for the derivatives of a determinant when its elements were functions of a real variable. The topic was taught fairly widely; Kronecker introduced in his lectures the symbol 'δ_{ij}' now known after him, as the number defined as 1 when $i=j$ and as 0 otherwise.

A major use of determinants was in connection with invariants and matrices. The links may be exemplified by recalling Lagrange's transformation (6.16.4): take

$$F := ax^2 + 2bxy + cy^2, \text{ with discriminant } D := ac - b^2, \tag{13.5.1}$$

and subject x and y to linear transformations:

$$x = pu + qv \text{ and } y = ru + sv; \text{ if } F = au^2 + 2buv + cv^2, \tag{13.5.2}$$

$$\text{then } AC - B^2 = (ac - b^2)(pr - qs)^2, \tag{13.5.3}$$

a multiple of the old D. In terms of matrices, F uses the matrix of the three constants, with D as its determinant; D preserves its form, and so is an invariant under the transformation.

The extension of these properties to forms of higher degrees and more variables had become a major industry for Sylvester and Cayley from the early 1840s – the 'invariant twins', as they were known. How many 'invariants' are there for a given 'quantic' (Cayley's terms) such as F in (13.5.2), and what are they? What are their 'covariants' (also his term), the unvarying forms involving the variables of the quantic as well as its constants? As the degrees of the quantics became ever higher, the questions became ever harder to answer, sometimes involving thousands of terms; but the mathematics became more and more who-cares-ish, and the number of followers, even in Britain, was modest. Indeed, the most mathematically significant contributions were made by Germans, who thought about invariants rather than calculated them. In particular, Paul Gordan showed in 1868 that the invariants and covariants of a quantic of degree 2 in any number of variables reduced to a finite number of basic ones; a powerful successor theorem was to come from Hilbert 20 years later (§13.19).

The word 'matrix' was introduced in {Sylvester 1850}; and he chose it because *only* 'an oblong arrangement of terms consisting, suppose, of m lines and n columns' was to be 'a Matrix out of which we may form various systems of determinants'. When matrix *theory* came, quite soon, Cayley took over notations from determinant theory, framing the array to left and right by vertical lines.[1] In 1858, as an accompaniment to a suite of papers on quantics, he published an important paper in which he laid out basic laws for both square and rectangular matrices and established conditions for the existence of an inverse. He formulated functions of matrices **A**, including even $\sqrt{\mathbf{A}}$.[2] For square matrices he considered the equation of the latent roots λ, to use the term introduced in {Sylvester 1883} with the charming explanation, 'Latent roots of a matrix – latent in a somewhat similar sense as a vapour may be said to be latent in water or smoke in a tobacco leaf.'

For matrices of order 2 and 3, Cayley proved in 1858 that **A** annihilated its own equation: in modern notation,

$$\text{if } f(\lambda) := |\mathbf{A} - \lambda\mathbf{I}| = 0, \text{ then } f(\mathbf{A}) = \mathbf{0}. \qquad (13.5.4)$$

This theorem is now known after him and W.R. Hamilton, who had proved it for quaternions (9.8.2) corresponding to matrices of order 4.

13.6 *Weierstrass on the spectral theory of matrices*

{Hawkins 1977}

A different line of development of matrices took place on the Continent. The insights which Cauchy and Sturm had had in 1829 with (9.10.3) concerning the reduction of quadratic forms to sums of squares, but then largely ignored, came at last to flower in what is now called the 'spectral theory' of matrices (compare §16.13); at the time no name was given it.

[1] Cayley used '| |' to frame the array of elements of determinants and also of matrices; he also deployed '( )', '|| ||' and '{ }' for matrices, attaching '$^{-1}$' at the right to denote the inverse matrix. The generic notation of the form (a_{ij}) is due to E.H. Moore in 1896.

[2] Cayley's matrix function $\sqrt{\mathbf{A}}$ makes a striking analogy (probably not influence) with Leibniz, who had soon pondered about $\sqrt{(dx)}$ when forming his differential calculus in the 1670s (§5.5).

One of the new generation of pioneers was Weierstrass – Analyst Number One, indeed, but in fact he had a sharp eye for linear algebra: for example, he used determinants and invariants in his theory (12.8.3) of elliptic functions. In 1858, when in his early forties, he studied the known difficulties in quadratic forms having repeated latent roots. He took up in effect Sturm's generalized equation (9.10.3) to prove a result which in matrix notation reads:

THEOREM 13.6.1 *For two quadratic forms*

$$A := \sum_{i,j} a_{ij} x_i x_j \text{ and } B := \sum_{i,j} b_{ij} x_i x_j, \qquad (13.6.1)$$

$$\text{consider } f(s) := |s\mathbf{A} - \mathbf{B}|,$$

the associated characteristic equation, with roots s_i; **A** *is assumed to be positive definite (that is, always to take positive values). Then there exists a linear transformation with associated matrix* **P** *which converts each form into a sum of squares:*

$$A = \sum_i y_i^2 \text{ and } B = \sum_i s_i y_i^2, \quad \text{where } \mathbf{y} = \mathbf{Px}. \qquad (13.6.2)$$

His proof was inspired by work of Jacobi on solving systems of ordinary differential equations ($\mathbf{A\dot{x}} = \mathbf{Bx}$), which itself built upon the efforts of Cauchy and Sturm. Weierstrass himself extended it in 1868 into the theory which became known as 'elementary divisors'; if, say, s_7 occurs three times in the characteristic equation $f(s)$ in (13.6.1), then $(s - s_7)^3$ is one of these divisors, and the polynomial is then factorized into a *linear* product of them. They became important in the development of matrix theory, especially in the hands of Georg Frobenius from the late 1870s onwards. One major result was the theorem that two matrices **E** and **F** were similar (that is, a matrix **G** with a non-zero determinant exists such that **EG**=**GF**) if and only if they possessed the same elementary divisors; various versions of it were found in the 1870s.

Of the trio of theories reviewed in this trio of sections, determinants fared the best in both teaching and research.

Invariants *à la* invariant twins has been the least durable, although the general notion of invariance itself remained important. Matrices made the most fitful progress, despite the important contributions made in mid-century which have been noted in this and the previous sections. For example, in contrast to determinants, no textbooks on matrices were written before 1900, and seemingly little teaching was conducted. Matrix theory is one of the great late developers in mathematics (§16.4).

13.7 *Riemann on 'manifolds', 1854/68*

We switch emphasis now to geometry, with the topic for his *Habilitation* that Riemann was asked by Gauss to present in 1854 (§11.1) and which appeared posthumously under Dedekind's supervision as {Riemann 1868b}. We saw that consideration of space in his study of the 1820s on differential geometry had led Gauss to express ds^2, the square of the infinitesimal element of length s on a surface, as a quadratic form (8.16.1) *only* of quantities intrinsic to it. The young Riemann imperiously reversed this connection by starting out from a very general notion of an 'n-fold extended manifold' (*Mannigfaltigkeit*) over a finite but maybe infinite number n of dimensions, continuous or discrete, finite or infinite in extent, and treating it as a space in itself.

Many of the details were worked out for a continuous manifold M of finite number n dimensions carrying some coordinate system x_r of variables. Riemann defined ds upon it 'as the square root of an everywhere positive homogeneous function [F] of the second degree in the quantities dx, in which the coefficients are continuous functions of the quantities x'; the restriction of F to positive values avoided the possibility of an imaginary distance function. In order to follow Gauss in rendering ds independent of the space within which the x_r lay, Riemann specified them as running along the geodesic lines (those of shortest length) on M at a given point P on it. Then he expanded ds^2 in a

Taylor series in these x_r, so that F became a quadratic form in their differentials, which depended only upon the tangent plane to M at P. The 'curvature' C of M at P was defined in terms of the $\frac{1}{2}n(n-1)$ coefficients of F, in a manner which generalized Gauss's (8.16.1) to innovate differential geometry in n dimensions. Continuity, and even differentiability, were central assumptions to this approach, though they limited its generality.

With F in this form, Riemann could define any geometry solely in terms of the properties of C, ds and M, without need of a space in which M was located; for example, the sum of squares $\sum_r x_r^2$ characterized the 'flat' Euclidean geometry (as he later called it). He highlighted the non-Euclidean geometries alternative to those of Bolyai and Lobachevsky (§8.17), in which the sum of the angles of a triangle fell below two right angles, and where lines could be unbounded but not necessarily infinite in length because they could return to themselves, like a great circle on a sphere traversed over and over. If no metric was applied, topological aspects of the manifold such as connectivity could be considered (drawing on some features of his first thesis, as we shall see in §13.12). The publication of his thesis in 1868 was very opportune; for it initiated but also coincided with a rapid rise in interest in those neglected geometries.

13.8 *Non-Euclidean geometry becomes popular*

{Torretti 1978, Chap. 2}

Much of the initial publicity was given by the provincial French mathematician Jules Houël (1823–86), then in his forties. He wrote biographical articles on his fellow-outsiders Bolyai and Lobachevsky, and translated into French not only some of their writings but also Riemann's thesis on manifolds and three other new pieces of 1868.

Two of these were a pair of essays by Eugenio Beltrami (1835–1900) on interpreting non-Euclidean geometry'. Like Riemann, but independently, he started out from

Gauss's work on surfaces and created a concave surface of constant curvature by rotating Johann Bernoulli's tractrix (5.6.2) about its asymptote. This curve is similar in shape to Lobachevksy's curve m in FIG. 8.17.1, and Beltrami read the theories of such pioneers as unwitting forays into geodesics; but their surprising results (for example, the sum of the angles of a 'triangle' being less than two right angles) applied to more general contexts. He did not intend to prove that these geometries were consistent or that the parallel postulate was independent of the others in Euclidean geometry; but such questions were of concern to Houël, who explored them by seeking interpretations of the non-logical terms such as 'line' {Scanlan 1988}. His approach was to be refined at the end of the century by Hilbert (§13.18).

The publication of Riemann's thesis affected the completion of the other translated text, an essay 'On the Facts upon which Geometry Rests', also published by the Göttingen Academy. It was written by Hermann von Helmholtz, in his 48th year in 1868 and world-famous as a scientist. His interests in optics and acoustics gave his thinking a strongly empirical cast, apparent in this piece as a search for criteria to determine which geometry was correct for the space in which we live. The features of greatest relevance here are that he regarded non-Euclidean geometries as quite legitimate mathematics, and that he adopted without change Riemann's emphasis on the continuity and differentiability of manifolds. His own contribution was to start from the assumption that bodies could be rigidly moved in space, thereby giving prominence to the congruence of planar figures, and thence to argue that our experience of such motions, and of congruence, showed that the world was a Euclidean space. But three dimensions were not necessary: he imagined planar Euclidean spaces such as pond-skaters almost inhabit. This example was to be articulated by the English schoolmaster Edwin Abbott in his insightful and lovely guide to *Flatland* (1884), which can still delight and enrich our geometrical experience.

13.9 *Klein and Lie: from line geometries to continuous transformations*

{EMW IIA6 and IIIAB4b; CE 5.6 and 7.7}

Another aspect of geometry gaining interest at this time was the line geometry of Julius Plücker {Rowe 1989}. From the mid-1840s he had considered the equation of a line in the plane,

$$ax + by + c = 0, \tag{13.9.1}$$

and treated the quantities a, b and c as the variables. The theory addressed questions such as which lines touched a given circle at some point. The geometrical answer was the envelope of its tangents; 'line-complex' was the word often used (unfortunately, since complex numbers were not necessarily involved).

The algebraic task was to find equations to relate the line coordinates. Various methods are possible: Plücker found most suitable a procedure drawn from Grassmann's *Ausdehnungslehre* (§9.9). Fix two points in three-dimensional space by four homogeneous coordinates (x_1, x_2, x_3, x_4) and (y_1, y_2, y_3, y_4); then specify the line through them in a symmetric way by six line coordinates, all in the form of 2×2 determinants such as

$$p_{12} := x_1 y_2 - x_2 y_2. \text{ Now } p_{12}p_{34} + p_{13}p_{42} + p_{14}p_{23} = 0 \tag{13.9.2}$$

(since it is the expansion of a 4×4 determinant with each set of coordinates twice laid out as a row), so that only five of these quantities were independent.

Line geometry was becoming a topic of note, for families of lines were important in geometrical optics and the design of lenses {Atzema 1993}, in mechanics, and elsewhere. Plücker was writing up a comprehensive survey when he died suddenly in 1868 in his 68th year, and his manuscript became the responsibility of a student, who duly issued it as the two-volume book *Neue Geometrie des Raumes* (1868–9). The student was Klein, then just 20 years old. Plücker's use of algebraic forms to represent line-complexes of the first and second degree led Klein to consider transformations of the corresponding bilinear and, especially, quadratic forms $\sum k_r p_r^2 = 0$, in which the condition in (13.9.2) would be preserved. By a stroke of luck he sent his findings to his former supervisor Rudolf Lipschitz, who as an editor for Crelle's journal had just received the proofs of Weierstrass's article on Theorem 13.6.1 on the simultaneous transformation of such forms. This gave Klein a fine fillip, and focused his work on such transformations and the cases of invariance which they might exhibit.

In 1869 Klein had met Sophus Lie, a few years his senior, in Berlin; but their later work pursued lines of research far removed from the analysis that Weierstrass was beginning to make important there. Lie had been thinking about the 'fundamental tetrahedron', defined by the sextet of lines $p_r = 0$ as its edges $(1 \leq r \leq 6)$. Strongly influenced by some work on differential equations by Jacobi, he regarded the possibility of solving them as grounded in their invariance under certain infinitesimal transformations T. To take a simple case of n functions each in one variable x, the transform T of the function f_r to $f_r + \mathrm{d}f_r$ by means of a function g_r may be linked to ordinary differential equations thus:

$$\text{connect } f_r(x) \rightarrow (f_r(x) + \mathrm{d}f_r(x)) \text{ with } \mathrm{d}f_r(x) = g_r(x),\ 1 \leq r \leq n. \tag{13.9.3}$$

Among various extensions, Lie sought differential invariants (not his name), functions of several variables and their partial derivatives which remained invariant under appropriate T's; Riemann's quadratic form for $(\mathrm{d}s)^2$ (§13.7) is an example. He also studied 'contact transformations' (his name), which took not only points on a given curve to points but also tangent planes to tangent planes. Aware that the T's formed groups under composition (T_1 on T_2, and so on), he examined their algebra. He wrote up his theory in three fat volumes on the *Theorie der Transformationsgruppen* (1888–93), published by Teubner.

Lie's followers included Wilhelm Killing (1847–1923) from the late 1880s, and later the Frenchman Elie Cartan

(1869–1951). As a 'Lie algebra' (to use Hermann Weyl's later name), the functions involved in these transformations obey the same kinds of law as do the Lagrange and Poisson brackets in analytical mechanics, and so the square-bracket notation of (10.4.3)–(10.4.4) was imported into this theory. The methods furthered the cause of determinants and matrices, and also enriched the theory of partial differential equations in various ways {EMW IIA5}; the role of invariants assisted in the rise of vector calculus (§15.9) and of invariant theory itself.

13.10 *Klein's algebraic programme for geometries*

{C.F. Klein 1925b, Chap. 3}

One of Lie's findings was that lines of curvature on a surface could be mapped onto its asymptotes, and vice versa. This was a profound surprise to Klein, for it jerked him away from the view, developed during his Plücker days, that projective properties could ground all geometries. In order to find a different basis he turned to the group-theoretic aspects of transformations, for which he outlined a general theory in 1872. The work in question is known as his 'Erlangen programme' (in German, the *Erlanger Programm*), so-called because this 24-year-old had just been appointed Ordinary Professor at the university there, and by tradition he was required to circulate a written programme of work plans.[3]

In this work and related papers of the time, Klein advocated the idea of *defining* a geometry in terms of a given space S and the group G of transformations of figures in it. Even the dimension of S was a secondary factor, since studying its G was the same as embedding S in a larger space T and examining the groups of transformations on S that leave G invariant (tangents to curves became tangents

[3] This text by Klein is frequently misinterpreted as the inaugural lecture given upon this appointment; in fact, he spoke there on mathematical education, rehearsing ideas which in much riper form would be so prominent in his later life (§16.3).

to different curves, and so on). Different geometries were related by their spaces and/or transformations: for example, the group of affine transformations in the plane (that is, those that map straight lines into straight lines as well as points to points) could be enlarged by adding a line at infinity, to produce projective geometry in the plane. Euclidean and non-Euclidean geometries were specified in terms of the projective coordinates (a, b, c, d) of a plane P, that is,

$$ax + by + cz + d = 0. \text{ Let } S := a^2 + b^2 + c^2 - hd^2; \qquad (13.10.1)$$

h is a parameter whose possible real values led to this classification:

Value of h	Geometry of	Klein's name for it
Positive	Riemann	'Elliptic'
Zero	Euclid	'Parabolic'
Negative	Bolyai and Lobachevsky	'Hyperbolic'

The space S took over many of the duties of Riemann's function F in §13.8; for it helped to characterize the principal properties of each geometry such as length and angle, sometimes as an invariant.

The Erlangen programme gave projective geometry a new lease of life, but not immediately; the text did not gain wide attention until Klein reprinted it in his *Mathematische Annalen* in 1893. Geometry had declined in popularity in Germany, perhaps because of Weierstrass's and Kronecker's influence from Berlin; but French and Italian mathematicians had taken up Klein's ideas with greater alacrity – together with another line of thought from Riemann.

13.11 *The geometry, algebra and topology of complex-variable analysis*

The concerns of Klein and Lie also included the subject matter of this and the next six sections: a vast and remarkable network of studies involving Riemann's theory of complex variables, differential equations, the hypergeometric series, elliptic and Abelian functions, group theory, matrices, invariant theory, and Euclidean and non-Euclidean geometry. The disciplines of geometry, algebra and analysis gained profound new relations – ones that changed the face of mathematics. The technical details are formidable; and so are the historical ones, with a tight chronology and even elements of competition. With Riemann as the deceased father-figure but with Weierstrass's approach to analysis on site from time to time, the principal participants were Fuchs, Klein, Lie, Dedekind, Picard and Poincaré. As so often, Gauss had talked to himself on some of these problems; and a few of the results played a role in the story, for the edition of his collected works began to appear in 1863.

Klein's prominent role was probably a main reason why many of the relevant papers appeared in his *Mathematische Annalen*; it also assured ample and excellent discussion in the *Encyklopädie*, especially in sections IIB and IIIC–D. Later he gave a fine survey in his overview of 19th-century mathematics {C.F. Klein 1926, Chaps. 6–8}. Among other works of that time, {Brill and Noether 1894} is a masterly survey of much pertinent complex analysis and algebraic geometry; modern studies include a fine short summary {Kline 1972, Chaps. 35–9}, and a detailed look at the the geometry and algebra related to the differential equations {J.J. Gray 1986}.

13.12 *Riemann on complex-variable analysis, 1851/67* {CE 3.12}

At the age of 25, Riemann presented to Göttingen University a study of complex analysis for his inaugural *Dissertation* (first doctorate). It was printed in the usual way {Riemann 1851}, but probably did not circulate much before the reprint in 1867. However, it was to affect the later lines of thought of its author in various ways, some of which he mentioned in publications. He also diffused ideas through lecture courses given at Göttingen from the late 1850s.

Although Riemann included no diagrams in his thesis, geometrical thought permeated its considerations. For example, he adopted the intuitive idea of the continuity of a function $w(z)$ of a complex variable, whose derivative at the point z had to be independent of the path along which a neighbouring point $z + \mathrm{d}z$ slid to z. If

$$w = u + \mathrm{i}v \text{ and } z = x + \mathrm{i}y, \text{ then}$$

$$\mathrm{d}w = (u_x + \mathrm{i}v_x)\,\mathrm{d}x + (v_y - \mathrm{i}u_y)\,\mathrm{id}y; \tag{13.12.1}$$

and the value of the derivative $\mathrm{d}w/\mathrm{d}z$ was independent of x and y if and only if

$$u_x = v_y \text{ and } v_x = -u_y. \quad \text{Further, } u_{xx} + u_{yy} = 0 \text{ and } v_{xx} + v_{yy} = 0. \tag{13.12.2}$$

In the first two equations of (13.12.2), Riemann first laid down *fundamental* conditions on u and v for w to satisfy, now named after him and also Cauchy (who had given them some emphasis: compare (8.11.3)). Indeed, it became customary to define $w(z)$ as 'analytic' if they were satisfied – a striking contrast to Weierstrass's (12.7.1) in terms of power-series expansions of w. Riemann also briefly noted that the angles between two curves in the z-plane were (normally) preserved in the corresponding curves in the w-plane {Article 3}. The phrase 'conformal mapping' is attached to this property because it had already acquired an importance in chart-making (compare FIG. 4.8.2, on Mercator); it gained considerable further lustre from now on.

In the third and fourth equations of (13.12.2), Riemann showed that u and v satisfied Laplace's equation (6.9.1), thus allying complex analysis closely to potential theory; constant values for u defined equipotential curves, while constant values for v gave stream-lines. This also provided a justification for factorizing the Laplacian operator which was alternative to the algebraists' practice of 'separating the symbols' (§9.11):

$$(D^2_{xx} + D^2_{yy})u = (D_x + iD_y)(D_x - iD_y)u. \qquad (13.12.3)$$

The most remarkable innovation came in a difficult passage in which Riemann spread surfaces T over the complex plane and then classified them in terms of the number of 'cross-cuts', 'which cut simply – with no multiple points – from one boundary point through the interior to another boundary point' {Articles 5–6}. The conditional clause here shows that he had in mind means of dealing with multiple-valued functions; the method used was to determine the numbers of cross-cuts and of simply connected pieces into which T is split. One of its main consequences was that a function of a complex variable could be *defined* from a minimum of data – namely, the locations and types of its singularity; for example, an algebraic function possessed only isolated poles.

Riemann used this classification near the end of the thesis when he came close to proving the 'Riemann mapping theorem' (as it has become known), that there existed a means by which two 'simply connected surfaces' could be mapped conformally and one–one onto each other {Article 21}. His proof drew upon a major component of potential theory at the time: the principle (10.18.2) due to his mentor Dirichlet, which he used in this form. Let the function $\lambda(x, y)$ possess at most isolated discontinuities over the surface, and take the value 0 on its boundary; then the functions which minimized the integral

$$\int_T (\lambda^2_x + \lambda^2_y)\, dT \qquad (13.12.4)$$

also satisfied Laplace's equation {Articles 16–19}. However, a major flaw in known proofs was to be revealed in 1870 by Weierstrass, as we shall see in §14.3 in the context of applied mathematics, where the principle found its greatest use.

Riemann's geometric vision stood in marked contrast to Cauchy's analytical theory based upon residues (§8.11) –

and also to Weierstrass's different analytical approach based upon power-series expansions, which had been envisioned by the 1850s but had hardly been publicized. It is remarkable that Riemann came to such an original conception as his surfaces; but some found it congenial, especially Italian friends such as Enrico Betti, and also Felice Casorati.

13.13 *Riemann on the Riemann surface, 1857*

Riemann's next major contributions to geometric complex analysis came in 1857 (his 32nd year), in a group of papers {Riemann 1857a} on Abelian functions published together in Crelle's journal. We saw that Abel's formulation (8.13.5) determined an index M such that the sum of any number of indefinite integrals of rational functions $R(x, y)$ could be reduced to M of the same kind, where the M limits were algebraic functions of the original ones. But a major difficulty arose from the fact that y was not necessarily a single-valued function of x, or vice versa. Jacobi's method of inversion (§8.12) had become a standard means of handling it, though at the cost of using single-valued functions of several variables.

Riemann extended this notion by revising his 1851 formulation of a Riemann surface. He now thought it

> advantageous, that the manner of splitting a many-valued function be developed geometrically in the following manner. One imagines in the xy-plane another one spread on its pertaining surface (or of another thin body on the plane), which extends so far as the function is given and no further. By the continuation of this function will this surface likewise be further extended,

by the device of analytic continuation; this he also introduced, by means of power-series expansions and perhaps independently of Weierstrass (§12.7) {Neuenschwander 1980}. Then each of various values of a function at a point could be located as a branch on one of an equal number of surfaces around that point {Riemann 1857a, p. 90}. These were the 'Riemann surfaces' (as they have long been known), now in a much more extended form than that in which they were presented six years earlier in his thesis. Note that he wrote of another (continuous) *surface* over the plane at a branching point, not necessarily of further planes there; and indeed, this is the general theory that he intended and the one his successors would sometimes deploy.

In this paper Riemann gave some simple examples of how the complex surface might have to accommodate branches, and by ready use of Dirichlet's principle (13.12.4) he showed some of the different ways in terms of topological properties (he used the traditional name *analysis situs*: §8.18). FIG. 13.13.1 shows his four examples; they are

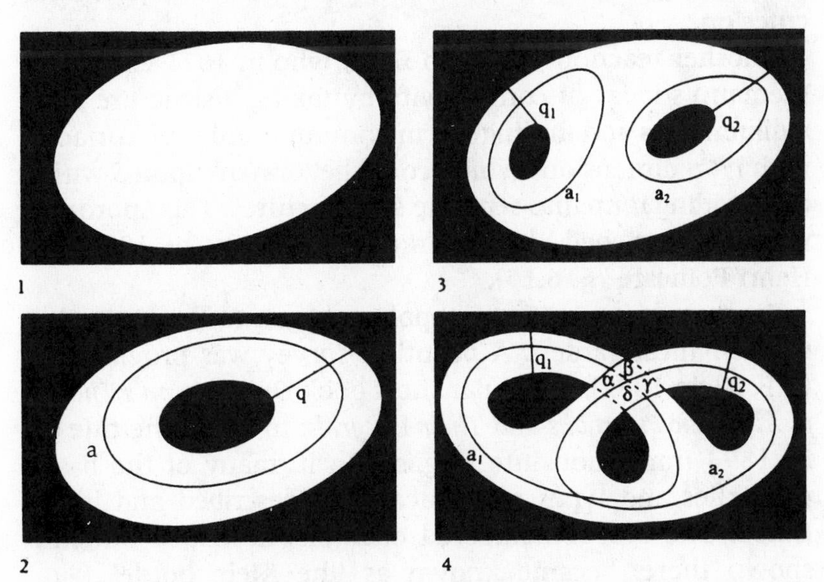

FIG. 13.13.1 *Riemann's original illustrations of his surfaces.* {Riemann 1857a, pp. 95–6}

distinguished by the number of simple closed curves that can be drawn upon them without dividing them up into separate pieces; in the third case the 'threefold connecting surface' had to be doubled to accommodate curves a_1 and

a_2. Most remarkably, he showed that this concept encompassed Abel's index number; the phrase 'genus of a surface' became attached to it. Euler's formula (8.18.1) for the numbers of vertices (V), faces (F) and edges (E) of a convex polyhedron was to be generalized to define the genus g of a rectilinear figure of any kind:

$$g := \tfrac{1}{2}(2 - V + E - F). \qquad (13.13.1)$$

The first major reaction to this paper was from Clebsch, who in 1863 translated the theory into geometric terms. He showed, for example, that the converse of Abel's theorem stated that the sum of integrals taken between the points of intersection of a fixed and a variable curve was constant. In this way Abel's analytic theorem, which had been inspired by Cauchy, was transformed into a topological criterion.

Another reaction was from Betti, who in 1871 extended Riemann's ideas of connectivity by taking a structure S in n dimensions and finding the maximum number of surfaces with $r < n$ dimensions which could be 'drawn' upon S without dividing it up into separate substructures. This approach was to be enriched in a remarkable way from the 1890s by Henri Poincaré (§16.15).

From such innovations, topology began to flower in the mathematical garden. A beautiful survey was provided in {C.F. Klein 1882}, a popular short book *On Riemann's Theory of Algebraic Functions and Their Integrals*, to quote the title of its 1893 translation into English. In it, many of the basic properties and types of surface were described and illustrated. The best-remembered example, described but not shown there, became known as 'the Klein bottle' (FIG. 13.3.2). The neck enters the bottle without intersecting it in a topological sense; more exotic even than Möbius's strip (§8.18), it has only one side and no edge. Klein also gave a rich survey of the roles played by Riemann surfaces in function theory; and he emphasized another connection emphasized by Riemann in referring throughout to potentials.

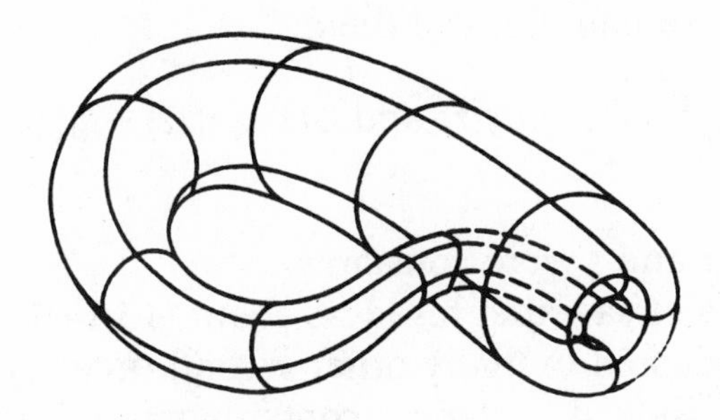

FIG. 13.13.2 *The Klein bottle, 1882.*

13.14 *Riemann on the hypergeometric equation, 1857*

A second major contribution came from Riemann in 1857: this time a paper on the hypergeometric function (8.14.3), put out by the Göttingen Academy {Riemann 1857b}. He started out from the differential equation which it satisfied, which I write as

$$z(1-z)w''(z) + [C-(A+B+1)z]w'(z) - ABw(z) = 0, \tag{13.14.1}$$

where z is a complex variable and A, B and C complex constants. The equation was second-order, so that there would be two independent particular solutions, and he clarified the relationships between Kummer's 24 solutions. He also used analytic continuation to explore the behaviour of w at the singular values 0, 1 and ∞ of z, where it would be multi-valued.

Since (13.14.1) was second-order, then any solution $h(z)$ could be expressed as a linear combination of two independent particular solutions, say

$$h(z) = af(z) + bg(z), \text{ with } a \text{ and } b \text{ complex constants.} \tag{13.4.2}$$

Now, as z passed around one of its singular points, f and g would change to, say, F and G which, as solutions also, were linear combinations of them:

$$F(z) = pf(z) + qg(z) \text{ and } G(z) = rf(z) + sg(z), \tag{13.14.3}$$

where p, q, r and s were constants.

Riemann also saw that his ideas about (13.14.1) could be applied or adapted to many other second-order differential equations; indeed, some contemporaries wondered whether every special function in mathematical analysis known at the time could be obtained as the solution of some special case of (13.14.1). Riemann now invoked some linear algebra, defining a '*P*-function' (his name) in a matrix-like notation to lay out relevant parameters:

$$\text{`}P\begin{Bmatrix} a & b & c & \\ \alpha & \beta & \gamma & x \\ \alpha' & \beta' & \gamma' & \end{Bmatrix}\text{'} \tag{13.14.4}$$

(including his 'x' for the complex variable, which I shall not retain). The first row gave the three values at which the function branched: for the hypergeometric function they were 0, ∞ and 1, and that row could be omitted from the array. The other two rows, to be read as three columns of pairs, gave the orders of singularity of two branches at the corresponding singular point (for example, for a singularity of order α, $(z-a)^{\alpha}Q(z)$ is single-valued at $z=a$).

Riemann also realized that more than two branches could occur at such points; but he stipulated that any three functions (P', P'' and P''') stood in a linear combination, and then showed that the function was a solution of (13.14.1) for appropriate values of the coefficients. He also wrote 2×2 'systems of coefficients' p, q, r and s for linear combinations of any variables t and u in the manner

$$\text{`}(S)(t, u)\text{'} \text{ for } pt+qu \text{ and } rt+su, \tag{13.14.5}$$

and recognized that the multiplication of the matrices '(A)', '(B)' and '(C)' corresponding to linear combinations of each of two of P', P'' and P''' satisfied the relation

$$'(C)(B)(A) = \begin{pmatrix} 1, & 0 \\ 0, & 1 \end{pmatrix}'; \qquad (13.14.6)$$

that is, the compound of the three transformations ended back at the start, as represented by the identity transformation {Riemann 1857b, pp. 69–72}.

These remarkable considerations brought Riemann the geometer not only to linear algebra but also (less explicitly) to group theory, for the transformations themselves formed a group. In France, the algebraist Jordan (§13.2) was quick to spot this; he called the matrices 'monodromic' (using an adjective proposed by Charles Hermite in 1851 to describe a function whose analytic continuation always gave only single values), and initiated a programme of study of their group-theoretic properties. This approach was well received; Helmholtz, for example, in his essay on geometry (§13.8), associated monodromy with congruence.

13.15 *Differential equations and automorphic functions*

{J.J. Gray 1984; CE 3.16}

We saw at (8.13.6) that, in extending Abel's results on Abelian functions, Joseph Liouville had found a finite expression for integrals of algebraic functions, including logarithmic terms. From the mid-1860s Lazarus Fuchs, then in his early thirties, extended this approach. Aware that some conditions for an ordinary differential equation of any degree in complex variables allowed it to have many solutions containing essential singularities, he sought conditions on their coefficients that would deliver an algebraic solution. Using methods of Jacobi inversion, he extended Riemann's use of matrices from 2×2 to $n \times n$, and by studying whether or not they could be converted into diagonal form he showed that logarithmic terms would or would not arise. His results were not exhaustive, but they helped to stimulate the study of the types of singular point that the solution of a linear ordinary differential equation might admit, and also solutions of such equations in complex variables {EMW IIB5}.

In 1880 Fuch's results attracted Henri Poincaré, then a 26-year-old early in his career {Hadamard 1922}. Noting limitations in Fuchs's treatment, he looked for the global form of solutions to differential equations (§14.17). He deployed a certain kind of function which he later called 'Fuchsian', but which had been only a special case for Fuchs himself. He greatly extended the use of group theory (to which he referred explicitly), and analysed Riemann surfaces carefully, including cuts between surface points. In this last context he enjoyed a wonderful 'aha' moment of discovery: as he boarded a bus in Caen, where he was working as a mining engineer, he realized that certain transformations of Fuchsian functions which he had used were identical to those deployed by Beltrami in 1868 when visualizing models of non-Euclidean geometry (§13.8).

Thus was forged another link with other new mathematics of the time. This one was especially innovative, in its application of a non-Euclidean geometry. Poincaré's inspiration exemplifies a striking contrast with the approach of Klein, who worked almost in competition in this topic at the time.

One point of contact between them concerned a process known since Riemann as 'uniformization' (not to be confused with uniformity, as studied by the Weierstrassians in analysis[4]). Means were sought to render multi-valued functions in terms of single-valued functions, and to make them resemble (newly) familiar geometrical objects such as a non-Euclidean geometry within a small space. This problem had long been solved in real-variable analysis by, for example, using parametric variables: the circle $x^2 + y^2 = 1$, which is double-valued in x and y, has the single-valued parametric coordinates ($\sin \theta$, $\cos \theta$) in terms of the angle

[4] Such a confusion does not arise in German anyway, since uniformization in the Riemannian sense was *Uniformisierung*, while Weierstrass's modes of uniformity were *gleichmässig*.

θ. However, in complex variables, problems were much harder to solve. Both Klein and Poincaré announced general uniformization theorems in the early 1880s, Klein by using Fuchsian functions; but in fact another 25 years were to pass before the matter was sorted out, after the topological ideas that Riemann had initiated had been extensively developed (§16.20).

13.16 *Klein and groups of permutations on the Riemann surface*

{CE 3.13}

In 1871 Fuchs's young Berlin compatriot Hermann Amandus Schwarz examined a related issue, seeking conditions under which the solutions to the hypergeometric equation (13.14.1) would be algebraic functions. His method led him to form a related differential equation, and to consider the effect on its solutions of applying bilinear (and thus normally conformal) transformations M of its independent complex variable z. One of these, from 1869, is still named after him and E.B. Christoffel, who anticipated him by a year. It came to take forms such as this:

THEOREM 13.16.1 *Take n points a_r in order of magnitude along the real axis of the z-plane, and n angles α_r measured in radians positively or negatively in the clockwise or anticlockwise direction, and let*

$$w := \int k \prod_{r=1}^{n} (a_r - z)^{-\alpha_r/\pi} \mathrm{d}z + b,$$

$$\text{with } a_r < a_{r+1}, \sum_{r=1}^{n} \alpha_r = 2\pi, \qquad (13.16.1)$$

and k and b complex constants. As z moves along the real axis, avoiding the points a_r by traversing small semicircles centred upon them and oriented towards the upper half-plane P, w tracks clockwise round a closed polygon G with angle of turn α_r from the rth to the $(r+1)$th sides at the vertex corresponding to a_r. P maps conformally into the *interior of G, which is treated as a Riemann surface if the edges of G cross.*

This transformation was significant as an early *specified* example of Riemann's existential mapping theorem (§13.12); various extensions of it were soon introduced. It was of great use in applications, for it transformed awkward configurations into the friendly upper half of the complex plane and thereby aided solution.

Other transformations by Schwarz led to pretty pictures; for example, he found that M mapped the upper half of the plane into regions of itself, such as triangles with circular arcs as sides, forming attractive tessellations of the plane (that is, divisions of it after the manner of tilings). Klein picked up this geometric aspect in the mid-1870s, and used his work on invariants and group theory (§13.10) to seek conditions under which a differential equation would give an algebraic solution. Then he was influenced by concurrent work on this problem in the theory of elliptic functions, which goes back to Jacobi's interest in their transformations (§8.12).

An elliptic function $E(z)$, now defined in terms of complex variables, has two periods, q_1 and q_2. What happens to its moduli,

$$k := E(q_1/q_2) \text{ and } k' := \sqrt{(1-k^2)}, \qquad (13.16.2)$$

if the quotient is increased by a prime factor p to the value pq_1/q_2? Both Jacobi and Abel had found that they were related by a 'modular equation' (to use a later name) involving a polynomial E of degree $p+1$. Galois's forays into group theory at that time (§9.4) had disclosed that if $p = 5$, 7 or 11, then the E could be transformed into one of degree p whose roots exhibited interesting properties in group theory (this is one origin for the term 'Galois group').

In the 1850s this subject became very popular, especially for the insights it cast on the properties of roots of the quintic equation {CE 4.11}, which Abel had shown not to be solvable by radicals (§9.4). But in addition, Hermite and Kronecker studied the modular equation for $p=7$ and found a seventh-degree function of its roots which was invariant under a particular group of 168 permutations. This algebraic hue was deepened in 1870 by Jordan, in his *Traité* (§13.2), who found various interesting subgroups. Then in 1878 Dedekind showed that many properties of elliptic functions, such as their periods, could be studied not by working directly on the functions themselves but by using this function $J(z)$ which was invariant under a bilinear transformation T:

$$J(z)=J\left(\frac{\alpha z+\beta}{\gamma z+\delta}\right), \quad \text{with } |\alpha\delta-\beta\gamma|=1, \qquad (13.16.3)$$

where α, β, γ and δ were integers determined from the coefficients of the hypergeometric equation (13.14.1). He showed that there was a region of the complex plane upon which $J(z)$ took each value once and only once; this simplified the study of the moduli k and k'.

Klein's contribution was to temper this algebraic emphasis with what was a geometric but also group-theoretic analysis of bilinear transformations of the upper half of the complex plane onto itself {CE 3.13}. In particular, he extended (13.16.3), and also Riemann's (13.14.6), by considering transformations satisfying this condition for a group of matrices:

$$\begin{pmatrix}\alpha & \beta\\ \gamma & \delta\end{pmatrix}=\begin{pmatrix}1 & 0\\ 0 & 1\end{pmatrix} (\text{mod } p), \quad p \text{ prime, with } |\alpha\delta-\beta\gamma|=p. \qquad (13.16.4)$$

Klein calculated the number $s(p)$ of permutations of the roots of the polynomial of degree p, and used Dedekind's insight to show how the permutations could be represented by specific regions in the complex plane or (when made necessary by multiple values) in the Riemann surface. The case $p=5$ (1876) gave a geometric form to Abel's and Gauss's findings about the roots of the quintic equation in terms of properties of the icosahedron (for example, $s(5) =$ 60 rotations about its centre corresponding to the permutations): Klein was to discourse with enthusiasm on this in a book of 1884 on 'the icosahedron and the solution of quintic equations'. The most spectacular case was $n = 7$ (1879), for to Hermite's and Kronecker's group of $s(7) =$ 168 rotations there corresponded a beautiful shape of genus 3 into which the Riemann surface was tessellated (FIG. 13.16.1).

This line of thought shows Klein's preferences in contrast

FIG. 13.16.1 *Klein's tessellation of the Riemann surface, 1879.*
{Klein 1926, p. 370}

to Poincaré's. Functions of the type $J(z)$ (his notation) became a principal concern; in 1890 he called them 'automorphic' to stress their invariant character. In general a function $h(z)$ was defined as automorphic for a collection $V_r(z)$ of analytic functions if

$$h(V_r(z)) = h(z) \text{ for each } r; \qquad (13.16.5)$$

further, they formed a group under composition ($V_r V_s$, and so on). The bilinear case (13.16.3) gained greatest attention, and is sometimes offered as the definition of automorphism. Many beautiful tessellations of the complex plane were found, and important cases of invariance disclosed, especially for elliptic functions (see {EMW IIB4}, by Robert Fricke, Klein's lieutenant in this research).

However, Klein did not follow Poincaré in focusing on the differential equations that the various functions satisfied. Nor did he emphasize the non-Euclidean aspects of some tessellations; for example, in his first account he did not mention the fact that the angles of the triangles in FIG. 13.16.1 add up to $\frac{41}{42}\pi$ radians. This omission is quite striking, for he had participated earlier in the decade in the discussion of these geometries, drawing upon his Erlangen programme (§13.10) to propose that *all* geometries be defined as projective ones under appropriate properties of invariance, instead of contrasting non-Euclidean with Euclidean geometry.

13.17 *Topology between algebra and geometry*

During this period, connections between algebra and geometry became much richer, with topology rising in importance between the two (similar to the way that trigonometry advanced in the 16th century: §4.1). One of the best known examples comes from John Venn's book *Symbolic Logic* (1881, 1894). He largely followed algebraic procedures, especially Boole's own (§9.13); but he also provided diagrams of logical relationships. The name 'Venn diagram' is well known, indeed, but it is a

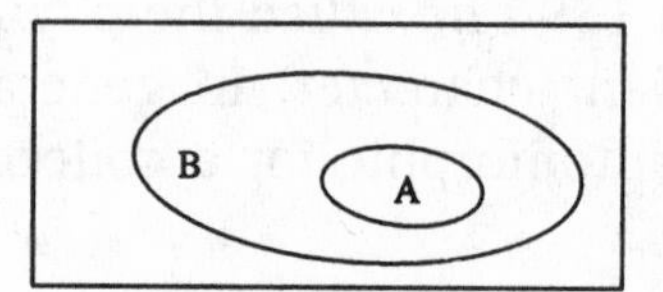

Eulerian diagram for two classes
given $A \subset B$, then
$A \cup . B = B$, $A + B = B - A$, etc.

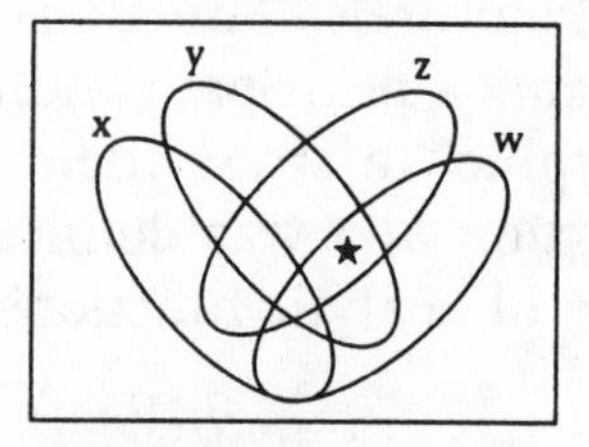

Venn's diagram for four classes,
with $w\bar{x}yz$ marked empty;
universal class inserted

Edwards's diagram for seven classes.

FIG. 13.17.1 *Three types of logic diagram.*

misnomer – it is applied to diagrams that Euler had introduced in 1768, in which some relationship between classes was pictured by the appropriate configuration of convex figures (for example, partial intersection by partial overlap) and conclusions were drawn about component classes (FIG. 13.17.1). In contrast, Venn showed *all* possible intersections of the classes, and then marked with an asterisk those intersections which were empty in a given case. When more than four classes were involved, convex figures were too simple to effect this representation; and such an apparently simple difficulty was too much for the topology of the time; an iterative representation for any number of classes was found only recently by A.W.F. Edwards – like Venn, a Fellow of Gonville and Caius

College, Cambridge {Edwards 1989}.

Another subject bridging algebra and geometry was combinatorics {CE 9.16}. In its youth in the 1870s, it was encouraged by the invariant twins thinking about chemistry. Cayley determined the number of ways in which atoms could be represented in a molecule as the coefficients of a certain polynomial, which became known as 'enumerative'. Then Sylvester, and also W.K. Clifford, sought connections between invariant theory and chemical valence: the word 'tree' in this context is due to Sylvester, on interpreting chemical notation as a graph.

The main source of links between geometry and algebra grew out of Riemann surfaces and their applications. This body of work is notable also for its extensions and generalizations: from interiors of circles to simply connected domains, from elliptic and Abelian functions to those of higher degree defined on general Riemann surfaces, and from the icosahedron and the quintic to the surfaces of 168 rotations and its seventh-order equation. Invariants of various degrees played essential roles, from complex numbers to corresponding polynomials in a real variable x. By these techniques, aspects of complex analysis and number theory were converted into algebraic geometry, especially for curves and surfaces in real-variable space, and for curves on Riemann surfaces.

Much of the initial interest was shown in linear and quadratric curves, and a rich vein of study was launched, for example by Brill and Noether. As early as the 1860s, Luigi Cremona was pioneering comparable studies of cubics and quartics, which became quite an Italian speciality {EMW IIIC4}. It is appropriate to end these comments with a pair of contrasting examples of the ways in which curves were treated.

One of them relates to Klein's tessellation of the Riemann surface (FIG. 13.16.1). Some of the subgroups of rotation exhibited properties of the associated quartic curve in the real plane: for example, a subgroup of order 28 corresponded to its bitangents (that is, the lines that touched it at two points in the plane). Now, in 1869, Steiner's grand-nephew Carl Geiser showed that these bitangents were closely related to a discovery made by Cayley in 1849 which had excited much interest, in Steiner and Cremona among others: that it is possible to draw 27 straight lines on a general cubic surface. Such results were not only mathematically profound but also visually exciting, and models began to be made in quantity of this and many other surfaces {CE 7.1}. Klein and his colleague Alexander Brill at the Munich Technische Hochschule set up a laboratory to produce them; they were marketed by Brill's brother, who owned a publishing firm.

By contrast, the algebraic approaches to curves ushered in by Kronecker and Dedekind in the early 1880s, while different from each other (§13.3), subsumed the geometry of algebraic geometry within their own algebra. This is particularly true of Dedekind's structural approach. It was deployed also by David Hilbert. Klein described him in 1892 as 'the rising man', and three years later recruited him to Göttingen from his native Königsberg. Then in his early thirties, Hilbert stayed there for the rest of his career. He is the dominant figure in the rest of this chapter.

13.18 *The full axiomatization of Euclidean geometry*

Humble Euclidean geometry and trigonometry of the plane were not neglected during the century, and many nice new results were found {Simon 1906; EMW IIIAB9}. FIG. 13.18.1 shows a lovely Euclidean theorem, which seems to date from the 1840s; it states that the line bisecting the right angle BAC also bisects the square on the hypotenuse BC.[5]

[5] This theorem should not be confused with Pythagoras's; neither is required to prove the other. However, a proof of the latter was developed from FIG. 13.18.1 by placing reverse images of ΔABC on and below DE. The innovator seems to have been the German military officer G.F. von Tempelhof in his *Anfangsgründe der Geometrie* (1769); it was repeated in some later books on Euclidean geometry, one of which may well be the origin of this theorem.

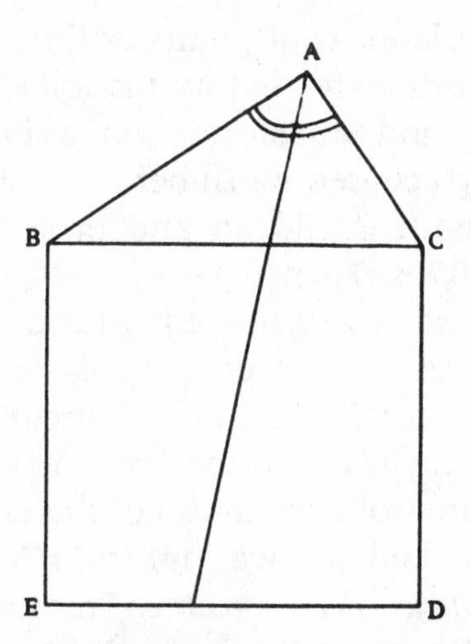

FIG. 13.18.1 *A lovely theorem in Euclidean geometry.*

Like linear algebra, Euclidean geometry was another English preoccupation: endless editions of the *Elements* appeared, with a measure of unoriginality reminiscent of the industry devoted to the Newtonian fluxional calculus during the previous century (§6.1). This infatuation led to a reaction from the schoolteachers: a severe dispute developed in their Association for Geometrical Teaching, not over the policy of teaching (only) Euclidean geometry, but over the issue of whether or not the proofs should be restricted to those given by Euclid or in his style {Price 1994, Chaps. 2–3}.

Meanwhile, as with invariant theory, the profound ideas came from Germany, especially concerning the insight that Euclid had not provided sufficient axioms for his purpose. In 1882 Moritz Pasch presented some extra ones, concerning properties of the order of points on a line and also their topology (for example, axioms that a given line AB contains at least one intermediate point C, and any points of AC are also points of AB). Peano took the matter further in 1888, early on in his crusade for axiomatization (§12.13), in a short book entitled 'On the Principles of Geometry Logically Expounded', with derivations laid out in detail, much attention given to forms of definition, and planes *or* lines defined as Cantorian sets of points *or* lines. A decade later this style was nicely extended by his follower Mario Pieri, using only 'point' and 'transformation' as basic notions; but he was unluckily eclipsed by Hilbert, who studied the axiomatization of both Euclidean and projective geometries from the early 1890s {Toepell 1986}.

Hilbert presented his system for Euclidean geometry in 1899, as 92 pages of a special book produced for the unveiling of a statue of Gauss and Wilhelm Weber in Göttingen. The publicity given to the occasion compensated for the unusual manner of publication of the essay, and within four years a second edition was out from Teubner as a normal book, *Grundlagen der Geometrie*. Translations into various languages also appeared; in English as {Hilbert 1902}, in the USA.

Hilbert found it necessary to assume 20 axioms {Chap. 1}, in five 'groups' (that is, collections): seven axioms on 'connection', such as two points determining a line; five on order (with acknowledgement to Pasch); Euclid's parallel postulate; six on congruence; and Archimedes' axiom (§2.20) to guarantee the continuity of the geometrical region. He proved that the axioms were consistent and independent {Chap. 2}, usually by the method deployed by Peano and others, of constructing some configuration in which one axiom(-group) was true and the one under study was false. He examined the roles of axioms: for example, the proof of Desargues's theorem (§4.20) needed the axioms of congruence {Chap. 5}. To handle proportions he laid out some of the chief properties of numbers, and in a paper published later in 1899 he axiomatized arithmetic in similar detail, with a 17-axiom system.

Hilbert's exegesis of geometry was very detailed – in fact, as Euclidean as Euclid had been reputed to be. But it also had a formal character in a way which Hilbert himself characterized in an oft-quoted remark: as far as the derivations were concerned, his three basic notions of points, lines and planes could just as well be 'tables, knives and beer-mugs'. Two aspects are worth noting, for the consequences they held for the continuation of this work on the foundations of mathematical theories in the 1900s (§16.4). Firstly, they show that his interest lay largely in the mathematical parts of his axiomatization; for logical connectives, such as 'not' or 'and', are *not* open to such multiple interpretations. Secondly, soon after both systems were published he added another axiom to them: 'completeness', which stated that 'the elements of geometry [or of arithmetic] form a system which is not capable of extension'. This idea does not correspond to 'completeness' in the sense of the term as used in his later foundational studies – it amounts to categoricity, that any two models of an axiom system were isomorphic (that is, their objects could be set in one–one correspondence).

This use of 'categoricity' entails only a minute time-slip; for it was soon to be introduced, in American mathematics. It was a component of the rise of model theory (§16.4) following the increased tendency to axiomatize mathematical theories {Cavaillès 1938}. Another branch of mathematics closely involved in this approach was abstract algebras, to which we now return for current news.

13.19 *Abstract groups in the 1890s*

Once again we find Hilbert at work. In 1888 he re-proved Gordan's finite basis theorem on the existence of a finite basis for invariants and covariants (§13.5) by methods drawing on abstract algebras. Gordan, who had calculated his way to victory, was not amused, but Hilbert saw his achievement as ushering in the 'critical' phase of the theory, to succeed the 'naive' phase of the British and the 'formal' one of his senior compatriot {Parshall 1989}.

An important stage in the establishment of abstract algebras was the publication in 1895 and 1896 of the two-volume *Lehrbuch der Algebra* by Heinrich Weber; a second edition soon followed. Here, both groups and fields were conceived as general structures, and the range of their properties and applications well conveyed. It was soon followed by a prodigious survey of algebraic number theory: {Hilbert 1897}, a 400-page report written for the German Mathematicians' Association. Gathering together the heritages from Galois, Kummer, Kronecker and Dirichlet, he reorganized the known results in new ways which were to influence the topic, and related ones such as algebraic function theory, for many years. One particularly durable notion, and term, was 'ring', implicit for some time in the work of Kronecker and others (§13.3), and now put forward as an algebraic structure from which the ideal could be defined {Chap. 9}.

Hilbert's report was one in a most distinguished series: soon afterwards there appeared {Schönflies 1900} on set theory, which was cited in §12.16. During the 1890s Schönflies played a prominent role in group theory of a particular kind: crystallography. Around 1850 the Frenchman Auguste Bravais (1811–63) had studied symmetries and related properties of crystals, classifying them by group-theoretic properties. His work had been one motive for Jordan to study groups 20 years later (§13.2), and this in turn led Klein in 1890 to encourage Schönflies, then his junior colleague at Göttingen, to take up this topic. During the 1890s Schönflies and the Russian mineralogist Eugraf Fedorov elaborated a full list of 230 crystallographic groups, and gave group theory one of its most attractive manifestations {EMW V7, with Schönflies as a co-author; CE 9.17}.

Kronecker's work on links between algebraic numbers and algebraic functions (§13.3) led to an important consequence when his leading student (and editor of his works), Kurt Hensel (1861–1941), found a 'new foundation for the theory of algebraic numbers' in 1899. He introduced a formal definition of a type of number which in the following form later became known as 'p-adic':

$$\sum_{r=h}^{\infty} A_r p^r, \quad \text{with } p \text{ prime, } h>0 \text{ and } 0 \leq A_r \leq (p-1). \tag{13.19.1}$$

(His original definition of the A_r was more complicated,

involving polynomials.) The finite number h of terms in inverse powers led later to important theorems in algebraic number theory about rational solutions of polynomial equations; and more abstract properties of ideals in a field of algebraic numbers were found, enriching the study of algebraic structures (§16.19).

13.20 *Linear combinations everywhere*

A frequent feature of the algebras described in this chapter is the appearance of linear combinations, finite or infinite, as a form or an equation:

$$\text{either } F := ax + by + \cdots + cz(+\cdots)$$

$$\text{or } ax + by + \cdots + cz(+\cdots) = d \tag{13.20.1}$$

in some interpretation or other of its letters and operations (see, for example, (13.3.2), (13.5.2), (13.9.1)–(13.9.2) and (13.14.2)–(13.14.3)). We recognize here the elements of vector space theory: this concept was not made explicit until the 1920s, but the ubiquity of the form F was noticed during the present period. For example, the Italian Salvatore Pincherle used F in his revival of interest in functional equations, which had flagged somewhat after the 1830s (§9.12–§9.13); he stressed distributive functions, and related their algebra to that of operators in general {EMW IIA11, written by him}. Boole, in his expansion theorems for eliciting the consequences of premises in his logic (§9.13), used F; Charles Peirce deployed them in his logic, and noted in 1882 that F occurred also in Cayley's definition of the product of two matrices, and also in some work by Sylvester on hypercomplex numbers.

Those numbers generalized quaternions and Grassmann's algebras (§9.8–§9.9) to any finite collection of basic units $x, y, \ldots, z$, one of which serves as the identity, closed under composition. They took the form F, with their coefficients $a, b, \ldots, c$ drawn from members of a field such as the real numbers. Their laws included

$$x^2 = \cdots = z^2 = -1,\ x(yz) = (xy)z,\ (by + cz)x = byx + czx. \tag{13.20.2}$$

They are satisfied also by, for example, Lie's groups of transformations (§13.9).

Around 1897 the Latvian mathematician Theodor Molien (1861–1941), and also Cartan, thought of applying hypercomplex numbers to group theory {Hawkins 1972}. Let the n members of a finite group G be $x_1, x_2, \ldots x_n$, and express the law of composition for a given member x_k,

$$x_i x_k = x_j, \text{ in the form } F \text{ as } \quad x_i x_k = 0x_1 + \cdots + 1x_j + \cdots + 0x_n. \tag{13.20.3}$$

Then the $n \times n$ matrix $R(x_k)$ whose ith row was given by the sequence of 0's and 1 corresponding to (13.20.3) for each i, and thus possessed a non-zero determinant, satisfied the property

$$R(x_k)R(x_l) = R(x_k x_l) \text{ where } k \neq l. \tag{13.20.4}$$

Properties of matrices, such as latent roots and the characteristic equation, could be used to explore the structure of the numbers. The collection of matrices $R(x_k)$ is now called a 'regular representation' of G over the field of coefficients involved.

Another contributor to this topic, Frobenius, was also the principal figure in another relationship put forward at this time: 'group characters' between finite groups and determinants (and also matrices) {Hawkins 1971}. It had been customary since Cayley's pioneering work (§13.2) to write out the products of two members of a group as an array: Benjamin Peirce's (13.4.1) shows the style, although that algebra was not a group. Frobenius's idea was to replace the array of products $ab, \ldots$ by another one in which ab^{-1} was given; to associate with each element b a variable x_b defined over some field of numbers; and finally to set up the determinant of the corresponding product variables, multiply it out by the usual rule and examine its factors. For example, the cyclic group C of three elements a, a^2 ($=b$, say), and a^3 ($=e$, the identity) could go to a determinant $D(x)$ defined over the field of complex numbers as follows:

C	e	b	a
e	e	b	a
a	a	e	b
b	b	a	e

$$\text{to } D(x) := \begin{vmatrix} x_e & x_b & x_a \\ x_a & x_e & x_b \\ x_b & x_a & x_e \end{vmatrix}. \quad (13.20.4)$$

When multiplied out, $D(x)$ gave a cubic which factorized into three linear terms, the coefficients of which were among the 'characters' of the group. For example, one term was $(x_e + \omega x_a + \omega^2 x_b)$, where ω is a complex cube root of unity; so 1, ω and ω^2 were the pertaining characters (in this case, all of them for that group).

13.21 *Universal algebras?*

Many algebras – was there a general one? In the late 1890s, in his mid-sixties, Dedekind extended his study of algebraic numbers A when he studied properties of the largest common divisors and greatest common multiples of collections of them. He saw that A displayed a structure which is now called a 'partial ordering'; the '$\leq$' over real numbers is a homely example. It satisfies these three properties

$$a \leq a; \text{ if } a > b, \text{ then } b \not> a; \text{ and if } a \leq b \text{ and } b \leq c, \text{ then } a \leq c; \quad (13.21.1)$$

in addition, there exist least and greatest elements 0 and 1 under the ordering. He called his structure a 'dual-group'; the modern term is 'lattice'. He found examples from a wide range of algebras, and saw the concept as very general {Mehrtens 1979}.

One of the cases noted by Dedekind was drawn from the *Vorlesungen über die Algebra der Logik* by Ernst Schröder (1841–1902), which appeared between 1890 and 1905 (part posthumously). In its three hefty volumes, Schröder brought to a pinnacle algebraic logic, including the predicate and relational calculi and quantification theory, with an extensive synthesis of Boole, De Morgan, Peirce and Robert Grassmann (§9.13, §13.4). His own considerable contributions included a systematic presentation of theorems in dual pairs on the page, like the projective geometers of yore (§8.4); a result involving 'there is' and 'or', say, generated a dual about 'for all' and 'and'.

The feature of interest to Dedekind concerned the symbol '⋹'. Schröder used it to represent both 'if . . . then' between propositions and part–whole inclusion between predicates and relations and their related classes; then the structure involved was a dual-group, with contradiction and tautology as the two extreme elements for propositions. He realized that this algebra was more general than Boolean ones, where equivalence of propositions (or equality of classes) was the chief connective; in addition, it did not necessarily satisfy the distributive law.

In 1897 Schröder announced that mathematics was reducible to his kind of logic. The method (or hope, anyway) was to lay out all combinations of relation, connective and proposition, and associated laws, in an 'absolute algebra'; and then to develop all the theories of mathematics one by one, like a cataloguer. This approach is typical of an algebraist: recall, for example, the taxonomy (13.4.1) of associative algebras by Benjamin Peirce.

It also characterizes our last figure, a newcomer: the English mathematician A.N. Whitehead, in his first book, *A Treatise on Universal Algebra with Applications* of 1898, his 38th year. His title was rather unhappy, as he did not follow the aspirations of the lattice theorists; Grassmann-style algebra was for him the strongest inspiration, helped along by Riemann's theory of manifolds (§13.7). He gave a good impression of the growing repertoire of algebras (with the form *F* often in place) and geometries, and the variety of relationships between them, including post-Boolean algebra, Grassmann's theory, some group and invariant theory, and non-Euclidean geometries (with emphasis on their metrics) and differential geometry. But overall the book is rather a mishmash, and soon he was to follow Russell into the quite different world of mathematical logic, where the relationship with mathematics was handled in a rather clearer way (§16.6).

At the end of his book Whitehead apologized for his slight treatment of vector analysis, due to the lack of availability of the few texts published. Its origins belong chiefly to the applied mathematics of the period, which we now examine.

14

An era of stability: mechanics, 1860–1900

14.1 *The constellation of mechanics*

During this period, mechanics continued to develop in all of its five branches (§6.1), from celestial through planetary, corporeal and engineering to molecular, and in its three traditions (§6.6) of Newtonian, energy and variational principles. Some important innovations were made, most notably by Henri Poincaré; but in general the contributions built upon theories established earlier in the century or in the previous one. However, the work does not thereby lack interest or quality: often the known theories were desimplified to treat small effects or to enter into more detail, tasks which required considerable mathematical prowess. This feature is especially marked in the engineering branch, which was very active (thanks to military funding and industrial development), with many textbooks and treatises, and even some special journals.

An inevitable consequence of this search for detail was an increase in the quantity of notional studies; for example (my invention, I hope), to analyse the motion of a smooth flat coin rolling inside the rough surface of a hollow ellipsoid balanced on the back of a hemispherical tortoise ambling at constant speed straight up a hill of uniform gradient on Saturn .... Material of this kind crept into, or was even produced for, textbooks.

An influential author, and not only in his own country, was E.J. Routh (1831–1907). He expanded upon his teaching at Cambridge in a two-part *Treatise on the Dynamics of Rigid Bodies* in editions from 1860, complementing it with a two-volume *Analytical Statics* in 1891–2 and one on *The Dynamics of the Particle* in 1898. All books contained many useful remarks on recent history, and also 'numerous examples', including this which I noticed after constructing the one above: 'A fly alights perpendicularly on a sheet of paper lying on a smooth plane ...' Routh exemplifies the puzzling contrast in British mathematics of this period (§11.4), especially at Cambridge, between its fine record in applications and its dreary teaching methods and poor reputation on the pure side. He made some important contributions to mechanics, as we shall see in §14.4 and §14.7; but he was also one of the princes of the coaching system which made mathematics so unacceptable to most students and held back the mathematical status of the country, and he opposed the reforms made in the 1900s (§16.2).

French-speaking engineers benefited in the 1870s from posthumous editions of the lectures of Poncelet (§10.8).

The German-speaking engineering fraternity was well served by Julius Weisbach (1806–71), who had built upon his teaching at the Bergakademie in Freiberg, Saxony to produce a splendid *Lehrbuch zur Ingenieur- und Maschinen-Mechanik* (editions from 1846 to 1901). Nevertheless, a strongly anti-mathematical movement built up among German engineers in the 1890s {Hensel and others 1989}. Klein's later initiatives at Göttingen were partly inspired by the need to encourage a reconciliation.

Another project of Klein, the *Encyklopädie der mathematischen Wissenschaften,* again comes into its own with a wonderful coverage. Its fourth Part treats everything imaginable in the corporeal and engineering branches. The sixth Part covers many aspects of the planetary and celestial branches, in two sections, respectively on geophysics and geodesy, and on astronomy, with articles numbered 'VI1' and 'VI2'.

As an example, take {EMW IV8}, a beautiful 65-page survey of physiological mechanics completed in 1904 by the Leipzig schoolteacher Otto Fischer (1861–1916). He listed and discussed a wide range of writing on the statics and dynamics of muscles in humans and animals, with consequences for ergonomics (as we now call it) and physiology (including that of plants): when mathematics was brought in, energy conservation and the calculation of moments seem to have been preferred. But the history of this topic has been little studied since, and Fischer has been forgotten.

This point can be generalized. As was mentioned in §11.5, the mathematical community grew greatly from 1860 in many countries, but with pure mathematics separating more from applied mathematics, and often snobbishly regarded as superior in status. Although Klein's role ensured that the *Encyklopädie* was not to suffer this way, many authors in its fourth and sixth Parts are too little known. Other historical writings of that time have suffered the same fate; for example, the splendid survey {Rühlmann 1881, 1885} of engineering mechanics, or {Wolf 1889, 1891}, still the finest text on the pendulum. The best-known history of mechanics written during this period, {Mach 1883}, stands upon the renown of its author and his philosophical advocacy of positivism, which was influential at the time {Blackmore 1972}. However, as a historical source it is unreliable on facts and frequently unsympathetic to – or even unaware of – the actual positions of his predecessors. An example of each defect appears in §14.4.

14.2 *A division of work*

One point of difference between the mechanics of this period and of earlier ones is the greater presence of physics in its concerns, making its distinction from mathematical physics more porous than was the case earlier (like that between geometries and algebras in §13). The two areas have been divided into separate chapters in order to prevent this one from becoming too jumbo; the main criterion is that theories treated in this one do not normally involve considerations of light, heat, electricity or magnetism. Thus elasticity theory is treated in this chapter, while studies of gravitation belong to the next one (indeed, serving in §15.16 as a conclusion to both).

Table 14.2.1 lists the principal figures in both areas, and their principal concerns across the divide. Several Continental mathematicians from the companion Table 11.3.1 for algebra and geometry were also active here, though only Henri Poincaré seems sufficiently important to merit a second entry. Among other figures, the Frenchman Joseph Boussinesq (1842–1929) wrote on hydrodynamics and heat diffusion. Riemann is again a significant person from the recent past {Bottazzini and Tazzioli 1995}.

The main difference between the two Tables is the prominence in this one of British figures (and one American), in contrast to their previous absence. Indeed, some other Britons are worth mentioning. In Scotland, P.G. Tait (1831–1901) worked closely with William Thomson (who became Lord Kelvin in 1891, more for his service to the Unionist cause in Scotland than for science, of course). Stokes was

TABLE 14.2.1 ***Continental and Anglophone mathematicians and mathematical physicists, 1860s–1900.***

Dates and main initials are given. Countries indicated are those of main activity, not necessarily also of birth. Only those interests of physicists are shown where mathematics played a significant role.

aco	acoustics	hyd	hydrodynamics/ hydraulics
cel	celestial mechanics	mec	mechanics, principles of
ele	electricity	mol	molecule/electron theory
els	elasticity/aether theory	opt	optics
emg	(electro)magnetism/ electrodynamics	pla	planetary mechanics/ physics
ene	energy physics	pot	potential theory
eng	engineering mechanics	the	thermodynamics
gas	gas dynamics		
hea	heat diffusion/radiation		

AU	Austro-Hungary	NE	Netherlands
EN	England	SC	Scotland
FR	France	SW	Switzerland
GE	Germany	US	United States of America
IR	Ireland		

Continentals
Boltzmann, L. (1844–1906) AU/GE gas mec the
Clausius, R. (1822–1888) SW/GE gas hea the pot
Duhem, P. (1861–1916) FR hyd the mec
Helmholtz, H. von (1821–1894) GE aco els ene opt
Hertz, H. (1857–1892) GE els emg mec
Kirchhoff, G.R. (1824–1887) GE ele els opt the
Lorentz, H.A. (1853–1928) NE ele emg mol pot
Mathieu, E. (1835–1890) FR cel els mec pot
Neumann, C. (1832–1925) GE ele emg hyd pot
Neumann, F. (1798–1895) GE ele opt pot
Poincaré, J.H. (1854–1912) FR cel ele els emg eng hea hyd mol opt pla pot

still at Cambridge, though his main achievements in research were behind him; more active was Routh, senior wrangler at Cambridge in 1854 with Maxwell second, and later the son-in-law of Airy. Others to pass through Cambridge included the Irishman Osborne Reynolds (1842–1912) and Horace Lamb (1849–1934); from the mid-1880s they held professorships for decades at the new Owen's College in Manchester, marking the beginnings of the distinguished record in applied mathematics at Manchester University. Slightly later came A.E.H. Love (1863–1940), who moved to Oxford University in 1899. An important goal at Cambridge was the Adams Prize, established in 1848 to commemorate the recent prediction of the existence of the planet Neptune by John Couch Adams (§10.5).

Among figures in the Table, there stands out Larmor, senior wrangler at Cambridge in 1880 (J.J. Thomson was second that year). He maintained an Irish tradition, drawing on Hamiltonians for mathematical methods (§10.17) and McCullagh for physics (§10.19), and working under the influence of his compatriot G.F. Fitzgerald. He also gave history a prominent place in his writings. During the 1900s he edited the works of Fitzgerald, and after their deaths he completed those started by his Cambridge senior colleague Stokes and by Kelvin; then in 1929 he published his own.

TABLE 14.2.1 *(continued) Anglophones*

Airy, G.B. (1801–1892) EN cel eng mec
Darwin, G.H. (1845–1912) EN cel hyd pla
Fitzgerald, G.F. (1851–1901) IR els emg mec
Gibbs, J.W. (1839–1903) US opt the
Larmor, J. (1857–1942) EE emg mag mec mol opt
Maxwell, J.C. (1831–1879) SC/EN cel emg eng gas hea opt
Rayleigh, 3rd Lord (1842–1919) EN aco hea hyd
Thomson, J.J. (1856–1940) EN ele els mol the
Thomson, W. (1824–1907) SC ene eng mag mec pot the

Three more points bearing on both chapters need noting here; another one is treated in the next section. Firstly, mathematical language was enriched by the regular use of determinants, but rarely of matrices; however, a halfway stage to matrix theory is evident when the contexts involved the latent roots of a matrix and the determinant arose in the associated characteristic equation (13.5.4). Vector algebra and analysis came in during this period, largely prompted by problems in mathematical physics (§15.9). Secondly, the general principle of conservation of energy advocated in mid-century (§10.8) bore principally upon physics, but its strong presence in mechanics is recorded here. Thirdly, while a few groups or circles existed (especially in some parts of the engineering branch of mechanics), there were none of the scale of the group influenced by Riemann and in touch with Klein, never mind a school like Weierstrass's; thus the presentations in these two chapters are less connected than those in the two preceding ones.

14.3 *The collapse of the foundations of potential theory*

The fourth point shows that pure mathematics could still affect applications. By the late 1860s, potential theory had crystallized into a stable body of theory, drawing upon both real- and complex-variable analysis, and applied to the shape of the Earth, gravitational forces in space, and electricity and/or magnetism. The existence of a function satisfying Laplace's equation, a function determined by its values on the boundary of a region, a function whose discontinuities are singularities (poles); all these could be specified. In such ways the achievements of Green, Gauss, Dirichlet and Riemann since the late 1820s were utilized and extended by a host of writers in a vast literature. The heritage was appreciated, too: for example, Green's ground-breaking book of 1828 (§10.18) was republished three times between 1870 and 1903, and also translated into German in 1895. Even a history was written, and a very fine one: {Bacharach 1883}, his thesis for the first doctorate at Würzburg University.

Various noteworthy innovations were made. One of these was the 'second' potential, U, proposed by Emile Mathieu in 1869 to solve the fourth-order equation for the vibrations of plates. He gave it this name to distinguish it from the normal 'first' potential P (often called 'Newtonian' because of its manner (6.9.1) of expressing inverse-square attraction): given a body with volume V and typical point A with coordinates (x, y, z), then these potentials at an external point at distance r from A are given by

$$P := \int_V \frac{1}{r} f(x, y, z)\, \mathrm{d}V \quad \text{and} \quad U := \int_V r f(x, y, z)\, \mathrm{d}V, \tag{14.3.1}$$

where f specified any relevant properties of V (for example, charge for electrostatic potential).

One figure prominent in potential theory (and also in the extension of mechanical principles into physics) was Carl Neumann. From the 1860s, when he was in his late twenties, he popularized the 'logarithmic potential' (his name) W of a planar region S and typical point B with coordinates (x, y) relative to an external point at distance R from B:

$$W := \int_S \ln\left(\frac{1}{R}\right) F(x, y)\, \mathrm{d}S, \quad \text{with} \quad W_{xx} + W_{yy} = 0. \tag{14.3.2}$$

Here, F played a role corresponding to that of f in (14.3.1), and W complemented P, which solved Laplace's equation in three dimensions; indeed, Neumann often presented theorems about P and W in dual pairs. Captivated by the Dirichlet principle (13.12.4), he used it in 1865 as the basis of his vision of functions of complex variables, including those of Abel, Jacobi and Riemann. Then, in the first issue (1869) of the *Mathematische Annalen*, the 27-year-old Heinrich Weber used it to develop certain types of special function.

However, serious difficulties were arising over foundations. Weierstrass's follower Hermann Amandus Schwarz noticed that the alleged proof of the Dirichlet principle, guaranteeing the existence of functions of a complex variable with certain singularities through a variational procedure applied to minimize an integral, was not valid. In 1868 he quickly replaced the principle by a proof based on Cauchy's theorem for a function of a complex variable defined by an integral, and eventually published his methods in 1870; unfortunately this theorem as established by Cauchy also has its flaws.

It was Riemann's student Emil Prym (1841–1915) who published the first detailed proof of the simplest case of the principle, also in 1870, for a function with discontinuities in its given values on the boundary of a circle. But he went on to show that there were cases where the energy integral (13.12.4) was infinite for certain families of functions; thus the process used by Dirichlet and Riemann could not even begin to minimize it!

But worse was to follow that same year, when Weierstrass took rigour a step beyond even Dirichlet's level in an address to the Berlin Academy. He quoted Dirichlet's last series of lectures in Göttingen as transcribed by Richard Dedekind; but then he demolished the method of proof used in the principle by this simple counter-example. Consider a function $\phi(x)$ which is continuous, along with its derivative, over $[-1, 1]$, and takes *differing* values a and b at $x = -1$ and 1 respectively. By Dirichlet's principle, such a function should minimize

$$J := \int_{-1}^{1} (x\phi'(x))^2 \, dx \quad \text{as} \lim_{\varepsilon \to 0} J = 0; \tag{14.3.3}$$

however, for the case

$$\phi(x) := \frac{a+b}{2} + \left(\frac{b-a}{2}\right)\left(\frac{\tan^{-1} x/\varepsilon}{\tan^{-1} 1/\varepsilon}\right), \text{ where } \varepsilon > 0,\ a \neq b, \tag{14.3.4}$$

he showed that the limit in (14.3.3) required that $\phi(x)$ be a constant function, which infringed the specification $a \neq b$. He spoke of the 'lower limit' (*untere Grenze*) of the integral rather than of its greatest lower bound (compare item 1 of §12.3); but in any case the torpedo had struck home.

Weierstrass did not publish his result until preparing his edition of his works {Weierstrass 1895, pp. 49–54}; but it gradually circulated from his lecture courses. Faced with the collapse of this component of standard methods of proof, Neumann immediately constructed a proof in 1870 for regions in the plane, bounded by curves or sections of curves not necessarily circles but restricted so that the region was in a sense convex and the curves intersected at non-zero angles. This was a smoothing process called 'the method of the arithmetic mean', based upon Gauss's theorem (10.20.1) of 1840 and developed later by Weierstrass.

Schwarz simultaneously expanded the regions to which his alleged proof applied. Being interested primarily in geometry, he discovered a reflection principle in which a region with a function that is real on a straight-line boundary on the real axis and lies to one side of it could be reflected in this line by taking the complex conjugate of the function. He also used his solution to the circle (or anything that could be mapped to a circle by a complex function – squares, triangles, polygons, etc.) and alternatively he used the boundary values from one of two intersecting regions to solve the problem in the other, switching back and forth between the regions until in the limit the functions coincided on the common region. In addition, he deployed his transformation (13.16.1) of the complex plane to determine sufficient conditions under certain circumstances for the existence of Green's function. Christoffel's motivation to (13.16.1) was similar, in that he studied the occurrence of steady temperatures, which satisfied Laplace's equation in the plane.

In a book of 1877 on 'logarithmic and Newtonian potentials', Neumann summed up the problems facing potential and complex-function theories as follows. The basic theorem on transforming integrals from integrals over a region to integrals over the boundary of a region as proposed by Green's theorem (10.18.1) could be established only for regions bounded by simple curves and surfaces; but it was in doubt for general surfaces, even those defined by algebraic equations of the third degree or higher. Secondly, the methods of proof used by Gauss involved surfaces whose nature was unknown, and so the range of applicability of Green's results (or similar ones of his own) was unclear. Thirdly, since Weierstrass had undermined the generality of many methods due to Green, Gauss, Dirichlet and Riemann, the whole subject became extraordinarily difficult, requiring lots of majorant arguments using upper bounding values in inequalities to circumvent the restrictions imposed by uniformity (item 2 of §12.3). This was a necessary contrast to the splendid but pre-critical early simplicity and vision, and many years of effort had yet to establish a new Dirichlet principle; so Neumann had to replace the one page on the principle in his book *Das Dirichletsche Prinzip* (1865) with 120 pages of intricate detail in the second edition in 1884.

Thus a major technique for mechanics and (especially) mathematical physics was shown to have no secure basis for its procedures. Even new approaches, such as Poincaré's 'method of sweeping out' masses within a region onto a surface within it, and thereby determining potentials there (1890), did not resolve or even avoid the difficulties {EMW IV24}. Moreover, Weierstrass's typical failure to publish in 1870 meant that some users of the principle were unaware of the difficulty – and sometimes even of its existence. The techniques of set theory and measure theory, as developed early in the 20th century (§16.9–§16.12), were to provide more powerful tools, though they helped also to detach potential theory from its applications.

14.4 *Principles for mechanics*

{EMW IV1 and 6 and 11}

Of the three traditions in mechanics, the one most heavily dependent on potential theory was that based on variational methods. Nevertheless, it maintained a distinguished place, principally in formulations following Hamilton (§10.17). Some methods for expressing the motion of masses under constraints were developed to cope with 'non-holonomic' systems of equations (that is, systems in which the expressions for the coordinates of each mass involve first-order derivatives with respect to time); but normally they were Bad News for solvers, and holonomic systems were preferred. Now, in 1876 Routh found a way to transform only *some* of the variables in a Lagrangian formulation (6.11.2) of a mechanical situation into a Hamiltonian one. This greatly increased flexibility in the choice of variables: for example, with a bit of luck non-holonomy could be avoided. His procedure appeared in a work on the *Stability of Motion* (1876), an Adams Prize essay at Cambridge University in his 47th year. In this work he also improved the Lagrange/Laplace analysis (6.8.2) of a century earlier by examining the most difficult cases, when the associated matrix possessed repeated latent roots.

Energy mechanics was bolstered by the great popularity of the general principle (10.8.2) of conservation of energy from the 1850s. A widely influential example is the two-part *Treatise on Natural Philosophy* by W. Thomson and P.G. Tait, which came out in 1867 and became affectionately known as 'T and T''. (Helmholtz, no less, was the co-translator into German in 1871–4.) They produced a summary *Elements* in 1873, and then a revised two-volume edition of the parent work, {Thomson and Tait 1879, 1883}, which was reprinted for decades; articles from it are cited regularly here, in their bizarre numbering system.

While they discussed all three traditions in detail, it is clear that energy occupied centre-stage in their overall conception (compare §15.3). Indeed, the term 'kinetic energy' is theirs (1862, introduced in the popular Christian journal *Good Words*), with 'statical' as companion adjective ('potential energy' had already been proposed in 1853 by their Scottish compatriot, Rankine, with 'actual' as companion). Their coverage, which began with dynamics and then treated statics, included elements of potential theory itself as a ground in statics, including derivations of Green's theorem, Laplace's and Poisson's equations, and presentations of Legendre functions (then still 'Laplace's coefficients') {Articles 482–550}. Many engineering applications in dynamics were also included, for example the motion of gyrostats and flywheels {Articles $345^{\text{x–xxviii}}$}.

Their treatment of Newton's laws, which followed that of energy principles, actually quoted the great Man in his Latin {Articles 250–64}; but they read into him an advocacy of energy principles which he had never uttered. In line with Scottish 'common-sense' philosophy, they adopted a more straightforwardly empirical cast than Newton himself had followed; the 'Division' of the book containing this discussion was called 'kinematics'. During this period positivist or empiricist interpretations of mechanics were widely accepted; in particular, Mach also read Newton this way but with far less sympathy. In his history of mechanics {Mach 1883} he regarded Newton's appeal to absolute space and time as insupportable by observation. He also took Newton's definition of (inertial) mass in the *Principia* {Book 1, Definition 1} as (volume × density) to be circular – a typical example of his lack of historical sympathy mentioned in §14.1, in supposing Newton to be so crass – and replaced it by treating the mass of a body as specified by the context of the universe in which it was situated. This led him and others to think in terms of frames of reference for mechanical and physical systems in 'Mach's principle', to quote the name introduced later by his follower Albert Einstein (§16.26). During this period, moving frames of reference were frequently used in dynamics, continuing the practice encouraged by Laplace and Coriolis (§10.7).

Helmholtz also stressed empiricism, but within the context of neo-Kantian philosophy, which allowed certain non-tautological propositions to be true *a priori*. From the early 1880s, in a search for a general foundation for mechanics in Newton's laws based *only* upon mass, space and time, he tried to elevate the principle (6.6.4) of least action to this status as a proposition. While not successful, his foray influenced Heinrich Hertz, his student at Munich, to marry the laws and the principle to Mach's philosophy and articulate an empiricist approach. Wishing to eliminate force and action at a distance, he worked only with space, time and energy (preferably kinetic) and the notion of mechanical constraints. He assumed that bodies of mass m traversed space in as straight a manner as possible and with uniform velocity. The variational element in this theory centred on the idea that $\int dm/R^2$ was to be minimized along the path traversed, where R was the radius of curvature at a point of the path of m. (It was in this context that he introduced the term 'non-holonomic' referred to above.) Unfortunately, he had to spoil the philosophy by positing the existence of 'hidden bodies' moving under certain constraints in order to make the theory work.

Hertz's theory was presented mainly in his *Die Prinzipien der Mechanik*, published posthumously in 1894, two years after his early death in his 37th year. In using optimization principles, it was in tune with two theories of some ancestry which gained a measure of attention from the 1880s {EMW IV1, Part 4}. One was based upon a formulation of the principle of virtual work by Fourier (§10.2) which treated it as an inequality rather than an equation, and so required the expression for work to take 0 as its maximum value when the system was in equilibrium. The other, due to Gauss in 1829, claimed that if the constraints on a system of masses m displaced each one a distance D in an infinitesimal interval of time, then in equilibrium the 'constraint' (*Zwang*), given by $\sum_m (mD^2)$, was minimized. These ideas provide fine examples of the use of inequalities in mechanics.

14.5 *The mechanics of economic equilibrium*

{Ingrao and Israel 1990; CE 10.18}

One student of Fourier's inequality around 1830 was the young Augustin Cournot (1801–77). A few years later he brought his background to bear on mathematical economics in his *Recherches sur les principes mathématiques de la théorie des richesses* (1838). Seeking general principles for economics rather than local details of the market-place, he drew on analogies from mechanics (especially the engineering branch) to characterize various situations: for example, friction in the negotiation of exchange, unavoidable loss, and limits of production and activity.

Cournot was following a general tradition, from at least as far back as the 18th century, of expressing economic thought in mechanical terms. His most important borrowing was equilibrium, which he applied to the buying and selling of a good in an isolated market. The numbers of a good demanded or supplied were conceived as functions ($D(p)$ and $S(p)$, say) of the unit price p; they were assumed to be continuous, even differentiable, and independent of time. Demand increased p while supply made it fall; hence $D'(p) > 0$ and $S'(p) < 0$, so that the curves corresponding to the functions, convex or concave curves (and definable from inequalities), met at a point E of equilibrium, which gave the optimal value of p required. Sometimes Cournot used the method of hit-and-miss (*tâtonnement*) to find E, but often he deployed the calculus to construct some function $f(x)$ of an appropriate variable x and apply the usual first-order condition $f'(x) = 0$. For example, to maximize the gain function $G(p)$ of a good for its seller, he defined it as

$$G(p) := pD(p) \quad \text{and solved } G'(p) = pD'(p) + D(p) = 0. \tag{14.5.1}$$

He also examined conditions under which $G(p)$ would be favourable *or* unfavourable to the seller, when $G'(p) > or < 0$.

Although Cournot is an important pioneer in mathematical economics, with durable ideas such as these, his contributions met with little response at the time; for though economics was being widely studied, it was usually with a resolute avoidance of mathematics. The next significant stages did not emerge until the 1870s, in Switzerland and Britain.

The most influential pioneer was Cournot's countryman Léon Walras (1834–1910). While studying at the Ecole des Mines in Paris, he read Poinsot's textbook on statics (§10.2) – one of the later editions, including also papers on mechanical equilibrium – and he took it as his bible for the development of mathematical economics. Unable to find employment in France, he passed his career at the University of Lausanne in Switzerland. Following Poinsot's rather formal style and adopting Cournot's innovations, he sought general conditions under which equilibrium could be found, though he did not fully examine the existence or uniqueness of an equilibrate situation. When the calculus could be used, he would form some function and optimize it. He introduced the notion of 'scarcity' (*rareté*) $R(g)$ of a good g to a consumer to express the consumer's desire for it, and he interpreted $R'(g)$ as the differential increment in pleasure obtained by consuming an extra unit of g. He also assumed a law of diminishing returns in the form that $R'(g) > 0$ but $R''(g) \leq 0$.

Upon his retirement in 1892, Walras's chair was taken by the Paris-born Italian Vilfredo Pareto (1848–1923). After training in engineering in Turin, Pareto had passed a career in the railway industry before taking up economics; and within a short time he had one of its leading chairs. His background helped him to think about cycles of processes in economics analogous to ones in technology: consumption, exchange, production, capitalization. More concerned with sociological factors than was Walras, he initiated a second phase of inspiration from mechanics by following variational methods. *Homo œconomicus* was like a mass-point acting under the action of economic and social forces, and his equilibrium was to be expressed via the principle (6.11.1) of virtual work, as follows. If the scarcity of good g to him after consuming amount x_g was $f(x_g)$, and x_g increased by an infinitesimal amount $\mathrm{d}x_g$, then his state of equilibrium over a range of goods $x_a, x_b, \ldots$ was given by

$$\mathrm{d}\Phi = \sum_a f(x_a)\,\mathrm{d}x_a \tag{14.5.1}$$

for some assumed potential function Φ, where y runs through the values $a, b, \ldots$. This move to potentials was also inspired by energy physics, which was passing on to economics both its own analogies and its habit of borrowing from mechanics {Mirowski 1989}.

Alive to the new discipline, the *Encyklopädie der mathematischen Wissenschaften* commissioned a short piece from Pareto, which duly appeared in 1902 as {EMW 1G2}. The extended version of this article prepared for the French *Encyclopédie*, {Pareto 1911}, is an early classic of mathematical economics; unfortunately it is incomplete.[1]

14.6 *British mathematical economics: Edgeworth and the Marshall school*

An important pioneer in Britain was Stanley Jevons, last seen as a logician (§9.13). He also published a string of works on macroeconomic questions, especially the *Principles of Political Economy* of 1871 (his 37th year), contemporary with Walras's first writings. Indeed, he adopted a similar notion to scarcity, which he called 'utility', and this has became the more durable term; he also used the derivatives of his various functions (utility, demand, and so on), attaching the adjective 'marginal' to the corresponding concepts. Among his other analogies with mechanics, he

[1] The circumstances behind this little-known fact are as follows. Pareto's article was one victim of the collapse of the *Encyclopédie* during the Great War (§16.3). Like its German predecessor, each 'tome' of that work was published in sections made up of multiples of 32-page signatures when ready; and his article as printed finishes, by coincidence with a full sentence, on the bottom line of page 640 (= 32 × 20). Had it been the proper ending, a centred bar line would have been added. Apparently the rest of the article is lost.

likened marginal utility to the force of gravitation (but why?), and compared an economic transaction agreed between two parties with the balance of the lever (10.2.1). Much more concerned than Walras with economic data, one of his principal studies dealt with the rate of decline of the supply of coal.

In the years prior to his early death in 1882, Jevons encouraged the Irishman Francis Edgeworth (1845–1926), whose main contribution was a book of 1881 with the title *Mathematical Psychics* – presumably interpreting the subject as the personal and social analogue to mathematical physics. Prior to Pareto, he applied variational methods with equations like (14.5.2). He also gave prominence to the economic indifference curves, defined after the manner of an equipotential surface: the corresponding equation related the quantities of two goods from which the consumer gained the *same* value of utility.

The peak of Edgeworth's career was his appointment in 1891 to a chair at Oxford University, where economic theory surpassed even mathematics in dullness. One of his supporters was his Cambridge counterpart Alfred Marshall (1842–1924); however, their interests did not fully coincide, and relations were often uneasy. While Marshall had earlier adopted marginalist notions, he preferred analogies from biology to those from mechanics or even physics; for example, he studied economic cycles as if they were organisms. But one innovation was inspired by stability questions in celestial mechanics (§14.17): the notion of 'partial equilibrium', which could obtain in an economic situation only over a period of time.

Like Cournot, Marshall also studied microeconomic problems. He made Cournot's equations (14.5.1) more specific in order to define 'elasticity of demand' (his mechanics-inspired phrase) $D(p)$ with respect to price p:

$$(0<)L := -\frac{pD'(p)}{D(p)}, \quad \text{or } L := -\left[\frac{\mathrm{d}D(p)}{D(p)}\right] \Big/ \left[\frac{\mathrm{d}p}{p}\right] \qquad (14.6.1)$$

in differentials. Cournot's case was given by $L = 1$, and his thinking about inequalities was expressed by describing demand as elastic *or* inelastic if $L >$ *or* < 1.

Marshall led an influential school at Cambridge; Edgeworth was rather a loner. Another feature which distinguishes Edgeworth from all other figures in economics was that he was influenced by psychology (for example, he held that belief has a bearing upon economic decisions). Well aware of the uncertainty of psychological data, he came to consider probabilistic and statistical aspects of economics after his book was published. He followed broadly the ideas of Pearson (§12.11), making use of laws of errors and the normal distribution; yet even then many of his applications were drawn from outside economics.

This separation of economics and statistics is strange to the modern mind, for now they are so intimately linked; yet the economists Cournot and Jevons did not even work with the probabilists Cournot and Jevons! Furthermore, the possibility of connections was fiercely attacked (though with poor understanding of the contexts) by figures such as Joseph Bertrand, author of an influential textbook on probability (§12.11). For these and other reasons, mathematical economics did not achieve the level of intellectual eminence of the other parts of the expanding world of mechanics which its practitioners were trying so hard – perhaps too hard – to imitate.

14.7 *Frameworks and mechanisms*

Equilibrium is closely allied to stability, which is the central feature of the quintet of contexts described here. They were significant to the engineering part of mechanics in particular, and most of them drew inspiration from the development of railways.

1. The need to build large covered railway stations focused a fresh degree of attention on the *stability of frameworks* {EMW IV5 and 29; CE 8.2}. For example (an important one), the Swiss engineer Carl Culmann (1821–81)

produced in 1866 his *Graphische Statik* (as the topic had become known), outlining a range of geometrical methods of analysing and simplifying a proposed structure. He emphasized dual diagrams, which also attracted Rankine

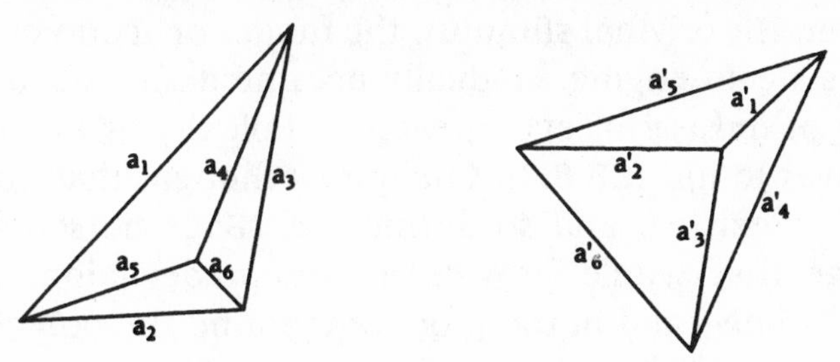

FIG. 14.7.1 *Maxwell's 'reciprocal figures', 1864.*

and Maxwell around the same time. FIG. 14.7.1 shows one of Maxwell's simple cases: F and F′ are 'reciprocal figures' because a one–one correspondence can be set up between their edges with the same angle between each intersecting pair (in this Figure each pair of edges is parallel) and between the vertices of one figure and the component polygons of the other (in this case, between each exterior vertex and a component triangle in the other figure, and between the interior vertex and the whole figure). Engineers were not always impressed, however; the anti-mathematical movement in Germany (§14.1) used it as evidence to support their case!

2. The increased need to build tunnels and embankments for railways (and also for roads and canals) similarly rekindled interest in *earth pressure stability* {EMW IV28}. As in the era of Coulomb a century earlier (§6.11), the methods were often trigonometrical, and were used to analyse components of force and seek conditions for equilibrium; Culmann's book included an account. Rankine desimplified the procedures by dividing the body of earth into sections and handling each one separately. But some more sophisticated mathematics was used: for example, in his report on the subject to the German Mathematicians' Association, Fritz Kötter deployed elements of variational

mechanics, including force potentials {F.W.F. Kötter 1893}.

3. One of Coulomb's immediate followers had been Gaspard de Prony, who had analysed, as a piece of difference mathematics, the conversion of circular into rectilinear motion. His original stimulus, the modes of action of James Watt's steam engine, gradually became desimplified into a study of *linkages* of ever greater complexity as no conversion was found {CE 8.3}. Chebyshev thought that no exact solution existed, and so during the 1850s he sought the linkage that would provide minimum deviation from a straight line – and in the process he found the polynomials named after him.

However, Chebyshev's belief was refuted in the early 1870s by his student L. Lipkin, who found an exact linkage in a seven-part piece (FIG. 14.7.2); set on two fixed pivots P

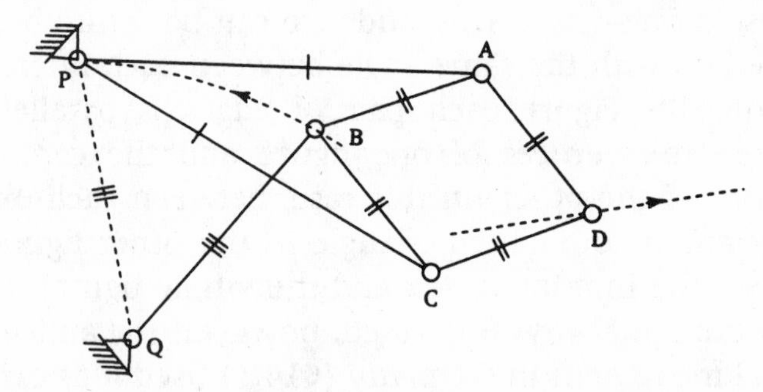

FIG. 14.7.2 *Peaucellier and Lipkin's exact linkage.*

and Q, it makes D move in the straight line marked. In fact it had already been published a decade earlier by the French army office J.-N. Peaucellier, but for some reason he had been ignored. Their achievements did not close off the subject; on the contrary, interest extended into the 'kinematics of mechanisms', when the task was to calculate the velocity and acceleration of any point of a mechanism at any moment in time and analyse special kinds of movement such as screw motion. The German engineers Franz Reuleaux and L.E.H. Burmester published influential textbooks; yet their work became a further target of their anti-mathematical colleagues!

4. Another contribution of Watt via his steam engine was the *governor*, a pair of mounted balls rotating horizontally on the machine about a common pivot which would cut off power to the engine when they moved too quickly. A very important early example of a feedback (or servo-) mechanism on a machine, its analysis attracted some British mathematicians, so building upon G.B. Airy's study in 1840 (§10.9). In the 1860s Maxwell studied stability in this and other contexts {Fuller 1982–1986}, such as the pendulum clock and Saturn's rings (§14.15). Regulators of various kinds became widely used in machines {EMW IV10, Articles 14–18}.

The underlying general considerations were examined by Routh in his Adams Prize essay of 1876, mentioned in §14.4. They involved roots of polynomials (later, when the analysis was formulated in terms of matrices, their latent roots). A sufficient condition is that the real part of each root must be negative; Adolf Hurwitz found a lovely formulation and proof using determinants in 1895.

5. Of concern to Rankine, Routh's schoolmaster son George, and some others was the *stability of bicycles*, which grew rapidly in importance from the 1880s as an affordable means of personal transport, even to the labouring classes {EMW IV9}. Mathematically the problem is very difficult, with several variables needed to describe the configuration and tricky friction factors to be accommodated. Analyses were largely confined to versions of Newton's second law and to expressions of energy consumption. The reactions of the manufacturers to this sometimes notional research is not recorded; but opposition, or at least uninterest, is conceivable.

14.8 *Elasticity theory*

{EMW IV23–5 and 27}

Many of the case studies just described involved the elastic properties of materials, and so drew upon the statics and dynamics of elastic and flexible solids. This topic continued to hold the important place in mathematics established for it during the 1820s (§10.16). In 1864, Barré de Saint-Venant published an edition of Navier's textbook in which he drowned out his hero with his own notes, one of them a *book-length footnote* on torsion theory. In 1883 he was more considerate when translating the 1862 textbook of Alfred Clebsch. Important German lecture courses were published, by G.R. Kirchhoff (1876) and Helmholtz (posthumously, 1902). Among the British, T and T′ included a very extensive survey of statical properties in their *Treatise* {Articles 573–740}, and Thomson wrote a magisterial survey article in 1878 for the *Encyclopaedia Britannica*. Love produced a textbook in 1892–3 which was translated into German in 1907; and the first general history was written by Isaac Todhunter, the first volume appearing in 1886 and the second after his death in 1893, completed by Karl Pearson.

Many of the earlier principal concerns were continued: Cauchy's stress–strain model (10.16.1) (these names are due to Rankine, in 1851 and 1856 respectively); molecular and truly continuous representation of the solids involved; and the handling of quadratic forms and conversion to sums of squares to find the principal stresses and strains, with determinants now regularly used. Applications to physics increased (for example, in cystallography), and also to engineering (for example, bending strength and torsion). Linear modelling still predominated, although non-linear theories began to be produced and helped to develop differential geometry and, ultimately, tensor calculus (§15.9). From the late 1850s Thomson, Rankine and others began to bring considerations of thermodynamics (§15.3) to bear.

Reductions to known equations and forms were effected where possible. Kirchhoff presented a very nice reduction in Crelle's journal in 1859, showing that the equations defining the equilibrate state of a thin elastic rod which has been bent and twisted at its ends could be transformed into Euler's equations for the rotation of a solid body about a point (§6.6), with length along the rod replacing time as the independent variable.

An interesting example of the competition between geometry and algebra is evident in the analysis of stress. The calculation of its components as second derivatives of certain functions was proposed in 1862 by Airy and extended by Maxwell and others; graphical methods were also offered, broadly following the one shown in FIG. 14.7.1 and others, but they were of limited use.

Some changes and innovations in solutions of the pertaining (differential) equations were made, especially ones drawing upon potential theory. Green functions and the Dirichlet principle were widely used – with cautions from Weber and Neumann, as was noted in §14.3 – to solve forms of the wave equation. An important theorem, involving a form of duality, was published in 1872 by Enrico Betti, one of several Italians active in this field:

> THEOREM 14.8.1 *Imagine two collections of forces* F_1 *and* F_2 *acting on an elastic solid, and causing collections of displacements* s_1 *and* s_2 *of its points (or molecules). Then the work done by* F_1 *over* s_2 *equals the work done by* F_2 *over* s_1.

This was proved again independently a year later by Lord Rayleigh, in terms of reciprocal families of vibration in a system, as part of his preoccupation with acoustics (§15.7). Similar theorems were sought and proved for other types of physical phenomenon – another fine example of the intertwining of mechanics and mathematical physics during this period.

14.9 *Hydrodynamics: from stream-lines to shock waves*

{EMW IV15–16 (by Love); CE 8.5}

The similarities between elastic and fluid objects continued to be a source of inspiration for both; most of the figures noted in the last section also appear in this one. The analogies extended to optics and acoustics (§15.7), and later to aerodynamics (§16.23); again, thermodynamical aspects began to be considered. Surprisingly few textbooks were produced, so Lamb's *Treatise on the Mathematical Theory of the Motion of Fluids* was welcomed on its appearance in 1879 (his 31st year), to the extent of a German translation in 1884 and further editions, under the title *Hydrodynamics*, from 1895 onwards. Its extensive references make it also a valuable historical source.

Linearity remained predominant; Laplace's and the wave equations were principal starting-points, with a widening repertoire of initial and boundary conditions. Properties of wave propagation, and of surface and standing waves, continued to be studied in detail. The current popularity for energy mechanics was manifest in the study of its conservation in ideal fluids and dissipation in real ones. Reynolds made a valuable contribution in 1877 when he increased the importance of group velocity (§10.3) by pointing out that it was proportional to the rate of transmission of energy in a deep fluid body.

Solutions were often provided by Fourier analysis and, in particular, the special functions, the latter including complex-variable properties. Potential theory provided powerful techniques; ready use was found for Green's theorem, and Thomson imitated the structure of his method of images in electrostatics (§10.19) in a method of 'sources and sinks' for fluid flow. In addition, the complex functional solution to Laplace's equation (14.3.2) over the plane to indicate wave propagation,

$$W = f(x + \mathrm{i}y) + g(x - \mathrm{i}y), \tag{14.9.1}$$

attracted especial attention both for the important place of equipotential curves and stream-lines in Riemann's conception of functions of a complex variable (§13.12), and also for its interpretation as fluid flow without rotation when W was the velocity potential. Helmholtz encouraged this line of thought in 1859, and followed up with the notion of 'free stream-lines', where curves and surfaces of constant pressure were considered to be independent of any bounding surfaces. This led him, and others including Kirchhoff and Rayleigh, into detailed studies of jets.

Another topic to benefit was the analysis of fluid flow around obstacles. Riemann was the main innovator, with a paper published in 1860 by the Göttingen Academy: he treated the one-dimensional flow of gas in a pipe in which there was a sudden change of width at one point, by forming a pair of non-linear first-order partial differential equations. Christoffel and others continued this line of thought with work on 'discontinuity waves' (now called 'shock waves') which constituted a most important foray into non-linear modelling during this period {EMW IV19}. He extended the treatment in a paper of 1876 to three-dimensional fluids, and also to certain kinds of elastic solid.

The emphasis upon discontinuity led Christoffel also to record in this paper a profound idea which he had been communicating in his lecture courses and which deserves more publicity than it has ever had. He proposed that a differential equation should be regarded as an *integral* equation, so that the possibility of it taking discontinuous and/or singular solutions could be appraised in terms of the equation itself rather than relative to the properties of any proposed solution form.

This largely German line of work in hydrodynamics was picked up by the French, especially Henri Hugoniot (1851–87) and Pierre Duhem. It also led to a set of *Leçons sur la propagation des ondes et les équations d'hydrodynamique* from Jacques Hadamard in 1903.

14.10 *Hydraulics: analysing the flows*

{EMW IV20–21; Unwin 1911}

Alongside hydrodynamics, the engineers continued their studies of large fluid bodies (§10.9) and the action and efficiency of machines such as turbines driven by water (on which James Thomson was an authority); they also took up newer topics, such as the impact of jets. Boussinesq was a notable contributor, with an essay 'on the theory of flowing waters' which won a prize at the French Academy in 1877, his 36th year; a decade later he was elected a full member. Like his predecessors he relied on empirical formulae to some extent; but he adjoined them to normal principles such as the parallelogram of forces or the energy equation to extend or even innovate the analysis of fluid friction, and the effects of overflow and blockages.

In later work Boussinesq studied forms of turbulent flow in water; Reynolds extended it to include the effect of pulses. His name is still attached to a dimensionless 'number' involving the density, viscosity and mean velocity of a fluid which he introduced in 1883 to characterize the flow of a fluid. His idea, and excellent associated experimental work, individuated the notion of flow, which was becoming a major concern in hydraulics, especially with the increased use of viscous fluids (for example, of oil as a lubricant in machines).

Another principal area of concern for hydraulics was the motion of ships {EMW IV22 and VI1, 5}. In the 1850s, Isambard Brunel's enormous passenger ship *The Great Eastern* had led the civil engineer William Froude (1810–79) to study the effects of water resistance and the properties of pitching and rolling. In an important programme of both mathematical and experimental research, which was continued by Froude's son Robert, he formulated various hypotheses about relationships between the design of the hull, the types of action of propeller, and the kinds of motion. Fourier series were sometimes deployed in this last context; elsewhere the Froudes used differential equations or series expansions. For them and their colleagues solutions could be only approximate; geometrical presentations akin to nomography (§12.10) were used quite frequently {Watts 1911}. The study involved not only hydraulics but also descriptive geometry, especially in preparing designs of ships; Rankine produced *Shipbuilding Theoretical and Practical* in 1866, an important contribution to a rapidly growing literature which now even included specialist journals.

One class of phenomena involving ships where hydrodynamics met up with hydraulics was waves and tides. Laplace's theory (§6.10) was developed and extended by

Airy, Thomson, George Darwin and Rayleigh in various ways, with particular emphasis on waves of long period {EMW VI1,6, Part B, partly by Darwin}. Thomson took his enthusiasm for Fourier analysis (§10.19) to the extent of designing in the 1870s a machine, which he called a 'harmonic analyser'; it represented mechanically the early sine and cosine terms by a system of interconnected gears and pulleys and thereby moved a weight (or alternatively an ink-loaded pen) in the manner which these terms and their coefficients described. He included an account of it in 1879 in an appendix to the revised edition of T and T′ {Appendix B′, Section VII}; at the exhibition of mathematical instruments in 1892 (§12.10) several versions of the machine were on display, from Thomson and other designers, showing a range of periodic effects. Soon afterwards the finest example of the genre was designed, by the Americans A.A. Michelson and S.W. Stratton; it could handle up to 80 terms of a Fourier series and draw out their sum-function, and also work in reverse.

In this appendix {Sections V–VI} Thomson also described his vision of numerically integrating ordinary differential equations, starting out from this second-order case for $u(x)$:

$$\mathrm{d}\left(\frac{\mathrm{d}u(x)/\mathrm{d}x}{P(x)}\right) = u(x)\,\mathrm{d}x, \quad \text{with } P(x) \text{ known.} \tag{14.10.1}$$

He offered a two-stage process using intermediate functions $g_1(x)$ and $g_2(x)$ given by

$$g_1(x)\,\mathrm{d}x = \mathrm{d}g_2(x) \quad \text{and} \quad g_2(x)P(x)\,\mathrm{d}x = \mathrm{d}g_1(x); \tag{14.10.2}$$

elimination shows that $g_1(x)$ is a solution. So two 'integrators' (a type of machine which his elder brother James was good at designing) were set up to effect the integrations of (14.10.2), the starting positions representing the initial conditions. The function produced by the second integrator was put into the first one, and then the complete solution of (14.10.2) was generated. Furthermore, the second integrand was the product of two functions, like the formulae

(8.7.2) for the coefficients of the Fourier series in the harmonic analyser. The procedure could be generalized to nth-order equations by a linked train of integrators.

The analyser and integrator are remarkable cases of structure-similarity, taken beyond the mathematics and its physical interpretation to a kinematic imitation *in technology* which could 'substitute brass for brain', as Thomson put it. But, at least in its early days, the integrator often could not deliver, even for $n=2$; for too little torque was available to transfer $g_2(x)$ from one machine to the other. Mechanical integration had found physical limitations.

14.11 *Guessing in ballistics*

{EMW IV18, Part 1; CE 8.11}

Some kinds of phenomenon remained extremely difficult to analyse. One was the passage of projectiles, which technology and the needs of war found means of sending ever further. (This was 'exterior' ballistics; the 'interior' aspects, concerning the means of launching, will be noted in §15.3.) A vast amount of teaching, research and preparation of tables was financed during this period. Among many factors, allowance had to be made for the rotation of the Earth, with the role of the Coriolis force (§10.7) now well understood. However, other aspects remained more intractable. An example was 'drift' – the tendency, maybe gyroscopic in nature, for a bullet *not* to travel after launch in the vertical plane through the rifle shaft.

Grant x, y, v and g their usual reference in this context, and symbolize differentiation with respect to time by overdots; then the angle i of inclination of the path to the horizontal and the basic equations of motion may be written as

$$\tan i := \mathrm{d}y/\mathrm{d}x, \quad \ddot{x} = -f(v)\cos i \quad \text{and} \quad \ddot{y} = -f(v)\sin i - g, \tag{14.11.1}$$

where $f(v)$ was the law of air resistance as a function of velocity v. It could be eliminated, but the resulting equation was messy. So what sort of function was it, how could one discover it experimentally, and could it be expressed mathematically without complicating the differential equations too much? Was the normally accepted form, $av + bv^2$ with a and b constants, argued by analogy with fluid resistance (§6.10), too simple?

In 1880 the Italian mathematician and army officer Francesco Sciacci (1839–1907) came up with a different and influential simplification of the basic equations. Starting from a suggestion by Euler that $\cos i$ did not vary too much over the path of the projectile and so could be replaced by a suitable average value I of i, he defined the 'pseudo-velocity' $\dot{x} \sec I$ and used it in the original equations (14.11.1) as a new variable to obtain good approximations to the path, thereby making it possible to calculate reasonably reliable range tables. However, even this technique turned out to be too inaccurate over long ranges: for a high angle of fire, $i > 15°$, the assumptions behind pseudo-velocity were too coarse, and in any case (14.11.1) did not take into account the rotation of the Earth. So the enemy in the line of fire still had some chance of survival.

14.12 *The pendulum and the planet*

One of the analogies used by William Froude in studying the motion of ships was with the motion of the pendulum. We saw in §10.6 that it had submitted to highly refined analyses of its tiniest movements, and this kind of study was now continued by many figures {EMW IV7}. In particular, Charles Peirce wrote many pages for the United States Coast Survey; one of his special concerns was the effect of the flexure of the framework holding the pendulum.

This Survey was one of many national institutions devoted to the development of maps and charts. Again building upon past experience (§10.6), a wide range of projections was used, with many choices of rectified parallels or meridians (that is, lines of latitude or longitude whose lengths are to be preserved) {EMW VI1,4}. Usually they were conformal, satisfying the Cauchy–Riemann equations (13.12.2); Peirce published a remarkable example in 1879 with his 'quincuncial' projection, which he named after the arrangement of five points in the form :·: and its repetition in oblique rows. It depicted the whole Earth twice over (FIG. 14.12.1): the top quincunx has the north pole at its centre and the south pole at each of its four corners, while the bottom one reverses the poles.

Inspired by Schwarz's 1869 transformation (13.16.1) of

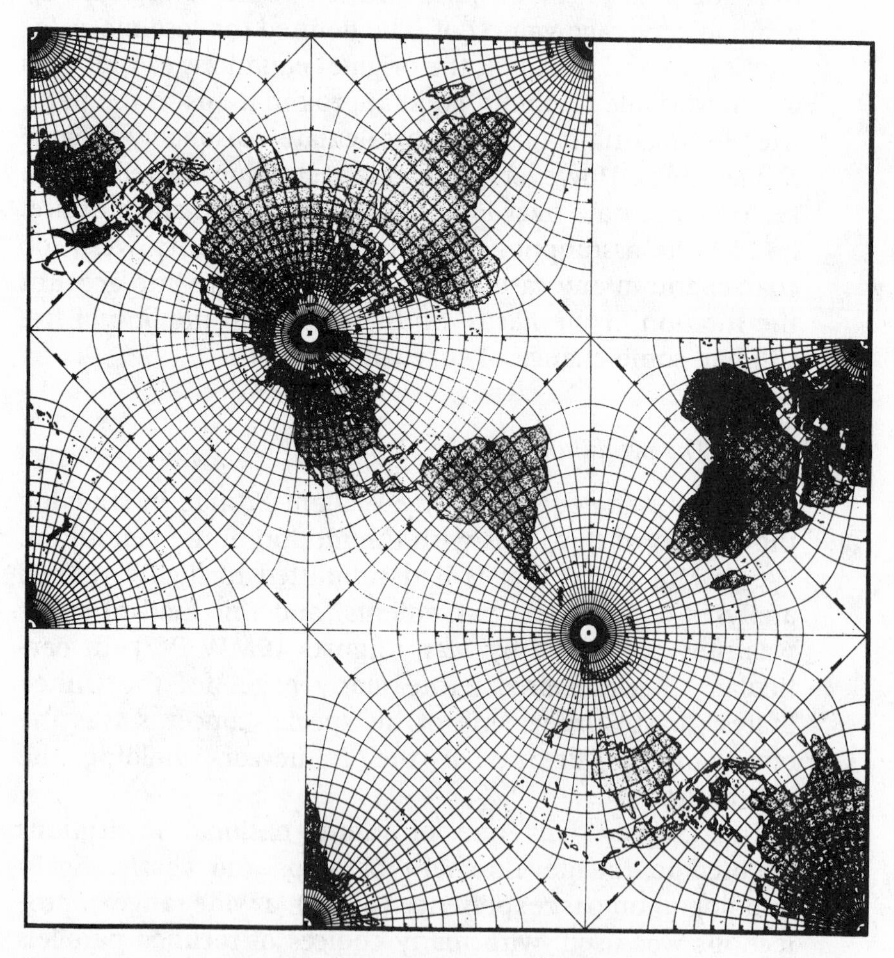

FIG. 14.12.1 *Peirce's 'quincuncial' projection, 1879.*

a circle onto a polygon of n sides, Peirce's projection took the interior of a circle onto the interior of a square. A point P on the Earth, a distance p from the north pole with longitude θ and latitude λ, was mapped onto a point (p, θ) of the plane through the equator construed as a complex w-plane, and then onto the point (x, y) in the z-plane by an elliptic function of the first kind {Pierpont 1896}. In the notation that Gudermann had introduced (§12.8), the relationships are

$$\tan\left(\frac{p}{2}\right) e^{i\theta} = \mathrm{cn}\left(z, \frac{1}{\sqrt{2}}\right), \quad \text{where } w = p\, e^{i\theta} \text{ and } z = x + iy. \tag{14.12.1}$$

Large areas were distorted less than by competitor projections such as the Mercator. It is fitting that this first-rate mathematical problem was solved in this way by a first-rate mind; his map deserves to be widely known.

The heritage in topography and navigation was enriched in similar ways, especially with the improving quality of instruments {EMW VI2,4–5}. The standard formulae of spherical trigonometry (§10.6) were adapted and desimplified to allow for new techniques, such as tachymetry, where location and altitude were measured simultaneously. Much more attention was paid to the irregularities in the shape of the Earth's surface, usually by placing appropriate local surfaces above or below the 'reference ellipsoid' (its basic assumed shape) and modifying both trigonometry and the surface and volume integrals of potential theory when calculating exterior potentials. In addition, it was well understood that data could still be coarse; so least-squares techniques were deployed regularly {CE 10.13}. Aspects of geophysics also appeared in theories: for example, the Earth's constitution as solid and/or liquid (§15.6), and its magnetic field.

The differential geometry of surfaces, as practised by Darboux and others (§12.6), was also used. In particular, Christoffel graced the pages of Crelle's journal in 1864 with a method of determining a curved surface given the principal radii of curvature at each point; it gave a theoretical grounding to some of the procedures followed by the topographers.

14.13 *Celestial mechanics: a variety of approaches*

The previous two sections related to specific topics in planetary mechanics; the rest of the chapter is devoted to its general theory, and to the celestial branch. As well as the *Encyklopädie der mathematischen Wissenschaften*, a splendid guide of the 1890s is provided by the 2,000 quarto pages of *Traité de mécanique céleste* by Félix Tisserand, published in Paris by Gauthier-Villars between 1889 and 1896, the year of his death in his 52nd year. Formally based on his 'theoretical mechanics' lecture course at the Paris Faculty of Sciences, he drew also on his experience as Director of the Paris Observatory. His choice of Laplace's title (§10.4) was deliberate, for he too wished to present the state of the art; and, while he did not have the reputation or originality of his predecessor, he provided a survey of comparable range, with a much stronger historical sense and many references, and in a text which is far easier to read! His book is our regular guide from now on.

Pride of place was still given to the equations of motion based on Newton's second law of motion and the expansion (6.7.1) in trigonometric series of the distance function between two planets; but a wider choice of independent variables and coordinate systems was now available, including moving ones {EMW VI2,9 and 12–13}. For methods, French solutions with very long series tended to lose out to the German compact methods (§10.5). The Swede Peter Hansen (1795–1874) had been prominent in his advocacy of moving frames and compact methods. His successors included Hugo Gyldén (1841–96), Director of the Stockholm Observatory (and unusual as a contributor of applied mathematics to *Acta mathematica*). Gyldén also tried always to keep the time variable t inside the sine and cosine terms, for when 'released' into terms of the form $t \sin(at+b)$ it could slow the convergence of series, or even lead to divergence, when it look large values in analyses of motion over long periods of time {EMW VI2,15}.

14.14 *The stability of rotating bodies* {EMW VI2,21}

Tisserand's second volume (1892) was devoted to the general theory of the shapes of the heavenly bodies. Here potential theory was conspicuous: not only Legendre functions to solve Laplace's equation, but also the theorems of Gauss and Green. He recorded recent work (by Darwin, among others) on questions such as the variation in rotation of the Earth's axis. This matter related to a problem which he also discussed; the shapes that rotating masses and fluids could adopt when in equilibrium.

Inspired by the finding by Jacobi and development by Liouville (§10.4) that a mass of fluid could rotate in equilibrium in the shape of an ellipsoid with three unequal axes, T and T′ wondered, in their revised edition of 1879, whether other shapes were possible; however, they came to no definitive answers {Articles 770–8″}. But in a huge paper published in *Acta mathematica* in 1885, his 32nd year, Poincaré presented a theorem on 'the exchange of stabilities' in which he defined a parameter that characterized the stability of a body, and considered the effects of its variation. For example, let the axes of an ellipsoid be $a \leq b \leq c$; form the two variables $x = a/c$ and $y = b/c$, and plot their values for each ellipsoid as a point on the plane. Maclaurin's ellipsoids of revolution (§5.25) lie on the line $x = y$, while Jacobi's fall on a continuous curve C which crosses it at a point P; but Poincaré found new solids, for example pear-shaped ones, which were stable when identified with points on C to one side of P but unstable on the other side. His work excited Darwin to wonder if the planetary bodies in our system might have evolved through such shapes before finally settling in their Maclaurin-like forms.

Meanwhile, at St. Petersburg in 1882, Chebyshev invited his student A.M. Lyapunov to see if non-ellipsoidal shapes were possible. The response led to an essay of 1884, independent of Poincaré's study but similar to it in certain ways. Then in 1892, when he was 36 years old, Lyapunov presented to Kharkhov University a wonderful doctoral dissertation 'On the General Problem of the Stability of Motion'; it was to become much better known in a French translation of 1907 from the original Russian. Like Liouville, and especially after reading T and T', Lyapunov started out from the energy equation; specifically, he sought conditions under which a shape could contain minimal energy. But his full panoply of results went far beyond the original problem about shapes. He analysed different types of stability: the 'asymptotic' form, for example, in which bounded oscillations not only occur around a position of equilibrium but also damp down to nothing (a matter of concern to engineers analysing governors). He also pioneered certain methods of solving non-linear differential equations by sequences of linear approximations. For these and other reasons, the recent centenary English edition {Lyapunov 1992} of his thesis is not only a historical record but also a source of research ideas that is still fruitful.

14.15 *The stability of Saturn's rings*

{Cooke 1984, Chap. 4}

Tisserand also discussed this problem in his third volume. Assuming that the rings collectively took a uniform elliptical cross-section, Laplace had used potential theory to show that stability required the ellipse to be broader than it was high, and the mean density of Saturn to be low for a planet. The young Maxwell took this idea further in an essay for the Adams Prize in 1857, arguing that much of the rings' mass had to be concentrated in some sector (after the manner of an orbiting satellite) and they could not be composed of continuous material. Sonya Kovalevskaya chose this as one of the topics for her doctorate (§12.9) around 1874, and soon found herself with a theorem; but then she lost interest, and published only a decade later after receiving encouragement from Gyldén. Starting out from the energy equation for the rings, where the potential was determined (in terms of elliptic integrals) for the centre of Saturn, she made deft use of Fourier series and the divergence theorem (10.15.1) to produce equations which could provide details about the shape of the cross-section, which she took to be an oval. However, the method landed her in infinite matrices, which were still to come (§16.13).

As this highly discontinuous chronology shows, the problem aroused only occasional interest, presumably because the observational equipment available was still too coarse to permit any hypothesis to be tested effectively. But it is an instructive example of progress and difficulties in study over the century; indeed, it completes a cycle, for Laplace had published an account of his ideas in his *Mécanique céleste* in 1799, while Tisserand slightly extended Kovalevskaya's in a paper of 1880 and in his *Mécanique céleste* in 1892.

14.16 *Theories of the Moon*
{EMW VI2,14}

Tisserand's third volume (1894) was given over entirely to this troublesome astronomical object; his historical sensitivity is especially evident, for he presented (in modern symbolism) all the main theories that had been proposed from Euler to his own time. Libration was normally handled using the Euler equations for rotation of a rigid body (§6.6) {EMW VI2,20a}; but all other motions continued to be described by a large variety of variable and coordinate systems, and also by several selections of principal parameters, mentioned in §14.13.

We saw in §10.5 that Charles Delaunay had found that tidal friction between the seas of the Earth and the seabeds had resolved a discrepancy between prediction and observation of its secular acceleration; Darwin analysed its properties in more detail. But Delaunay's French predilection for very long trigonometric expansions indeed met with much less enthusiasm, and Germanic approximations once again came to the fore. Hansen was a prominent participant in this movement, especially for calculating tables of the Moon's positions. The American G.W. Hill (1838–1914) went further: in the spirit of to-hell-with-it, he decided that the Sun was infinitely massive and infinitely far away while the Moon was infinitely small and nearby, and he produced some relatively simple differential equations which were very effective for many purposes. Darwin liked them, and drew them to the attention of his junior colleague E.W. Brown (1866–1938), who developed a fine theory of this kind when he moved to the USA in the 1890s. In 1914 he wrote an excellent survey of lunar theory for the *Encyklopädie der mathematischen Wissenschaften*, which is cited at the head of this section.

14.17 *Poincaré's 'qualitative theory' of differential equations* {Hadamard 1922}

Brown's simplification of lunar theory marked a return to special configurations of the three-body problem. After some decades of approximate solutions of special cases {EMW VI2,19}, this problem received a major new assault in the late 1880s from Poincaré, as Tisserand reported in his fourth volume (1896). Some mathematical background needs to be sketched.

The interaction between celestial mechanics and differential equations was a continual preoccupation throughout Poincaré's career {J.J. Gray 1993}. One major link goes back to 1880, when he began to rethink the understanding of solutions to differential equations in general, especially non-linear ones. In his 'qualitative theory' (his name) he thought of solutions as curves and surfaces rather than as analytical expressions; and he looked at their global properties, rather than properties local to (for example) a range of convergence of a variable. He took the first-order ordinary differential equation

$$\frac{\mathrm{d}x}{X(x, y)} = \frac{\mathrm{d}y}{Y(x, y)}, \quad \text{with solutions } F(x, y) = z \qquad (14.17.1)$$

where z was the constant of integration, and considered the singular points P with coordinates (x, y) where $X = Y = 0$. Projecting the xy-plane onto a sphere S about its own centre so as to accommodate infinite values of x and/or y more easily, he envisioned the solutions as 'flows' upon S. Now he treated the solution as a surface in space with z as the third variable, and the equation as defining a plane for the particular forms of X and Y involved. This led him to discover previously unsuspected kinds of behaviour of singular solution through P, somewhat similar to properties of surfaces known to topographers (FIG. 14.17.1): 'saddle-point'

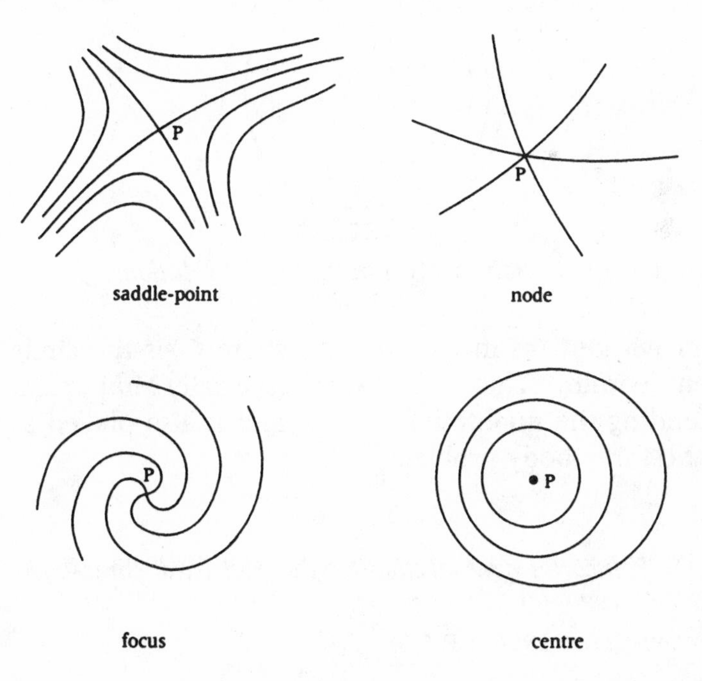

FIG. 14.17.1 *Examples of Poincaré's four kinds of singular curve.*

U (unstable motion swirling towards but then away from P), 'node' N (stable motion of various curves to P), 'focus' F (stable motion spiralling onto P), and 'centre' (closed curves surrounding P). Each kind served as a sort of qualitative invariant of its orbit, in that it obtained for *any* planar section. Furthermore, Euler's theorem (8.18.1) on convex polyhedra became

$$n(\mathrm{N}) + n(\mathrm{F}) - n(\mathrm{U}) = 2, \qquad (14.17.2)$$

where 'n' denotes the number of each kind of singular solution of the given equation.

Among properties of the solutions themselves, Poincaré stressed 'limiting cycles', those yielding closed curves C_0 which are solutions to the equation and which other solution curves may approach asymptotically, spiralling onto C_0

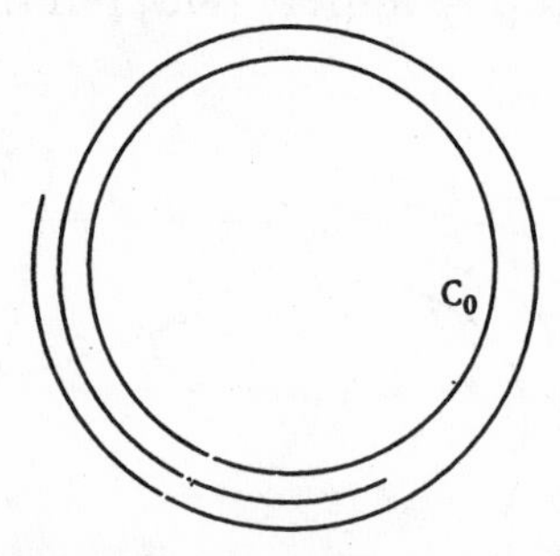

FIG. 14.17.2 *Poincaré's 'limiting cycle' solutions.*

from without (as in FIG. 14.17.2, where C_0 is the circle) or from within. This kind was especially important in extending the qualitative theory, and it also played a role in the three-body problem.

14.18 *Poincaré's new solutions of the 'restricted' three-body problem, 1889–90*

{Barrow-Green 1996; CE 8.9}

The opportunity had been created for Poincaré by Mittag-Leffler in 1886, seemingly as a means of raising the international profile of his journal *Acta mathematica* and Swedish mathematics in general: a prize problem, to be awarded by King Oscar II of Sweden in 1889 on his 60th birthday. Weierstrass, Hermite and Mittag-Leffler were the judges, and four problems were set: three in and around the pure theory of differential equations (Fuchs's methods of §13.15, for example), and the three-body problem. Poincaré might have written on any of them; he chose the last, and duly won the prize in January of that year, his 36th.

However, to prepare it for printing Mittag-Leffler gave the manuscript to his young assistant Lars Phragmèn, who told Poincaré that he found some analyses difficult to follow. This reaction seems to have led Poincaré to review the paper – and in the process he found a major mistake. By then the paper had been printed, but Poincaré had to rewrite it substantially and request a complete reprinting, which duly appeared as 270 pages of Volume 13 (1890) of *Acta mathematica*. However, his prize money had to go to the printer for additional costs, and the most famous theorem in this version had not been present in its predecessor!

Poincaré's choice of problem must have been inspired not only by his work on the qualitative theory, but also by two applications of differential equations to the three-body problem which had recently been reprinted in *Acta mathematica*. In connection with lunar theory, Hill (1886) had found periodic orbits for the Moon by taking its mass to be tiny relative to that of the Earth, which in turn he took to be tiny with respect to that of the Sun; his work ushered in a new era in the study of periodic orbits. On the three-body problem itself, the German astronomer and former Weierstrass student Heinrich Bruns (1887) had shown that the only algebraic solutions of the differential equations were the 'classical' ones (motion of the centre of gravity, and conservation of angular momentum and of energy). In his prize essay, Poincaré assumed that two bodies were of substantial mass (F with m and S with $1-m$, say), and he studied possible orbits for the third one, T, whose own mass was too negligible to affect the orbits of F and S. He called the problem 'restricted', meaning that one planet was reduced to a light point-mass, so that Newton's laws of motion were no longer satisfied.

Every part of Poincaré's analysis was innovative. For example, he formulated the differential equations of motion in a new version of the variational approach using a more general potential function in a power series in m whose terms covered all bodies and perturbing forces. Since the motions were continuous, these equations defined a surface of genus 0 (like a sphere) upon which the solution orbits could flow in the sense of his qualitative theory of differential equations. Instead of considering the total orbit O though space, he examined just the curve C produced by points at which O successively punctured a (plane) surface E set in its path. For $m = 0$, T executed a periodic orbit P around S, when C degenerated into a single point on E; for $m > 0$, he showed that solutions existed which oscillated with Hill-like periodicity around P, producing a curve on E. This situation obtained even in his very general conception of periodicity, which was defined in terms of the *relative* configuration of the bodies; two arrangements differing only by an angle, such as the 90° between ∴ and ∵, corresponded to the same solution of the equations, and thereby were regarded as periodic with time. He also distinguished between three kinds of periodic solution: zero inclination of the orbits to the ecliptic and small eccentricity; this kind but with finite eccentricity; and non-zero inclination.

Even more striking features of the essay lay in Poincaré's revisions. They concerned asymptotic solution curves, which gradually approached or veered away from a periodic solution, thereby producing a situation involving limiting cycles. He showed that the series for such solutions, which might include terms of the form $t \sin(at + b)$ (despite the strictures of Gyldén noted in §14.13), could not be uniformly convergent; they therefore lost the safety that Weierstrassian analysis had provided (item 2 of §12.3). But more importantly, they could include curves which returned to the neighbourhood of a periodic solution and thereby generate curves C in the page like those of FIG. 14.17.2. While writing for Oscar he had thought that all C's were closed; but now he corrected himself to the extent of realizing that even an infinite number of crossings were possible. FIG. 14.18.1 shows a solution curve L, given by an

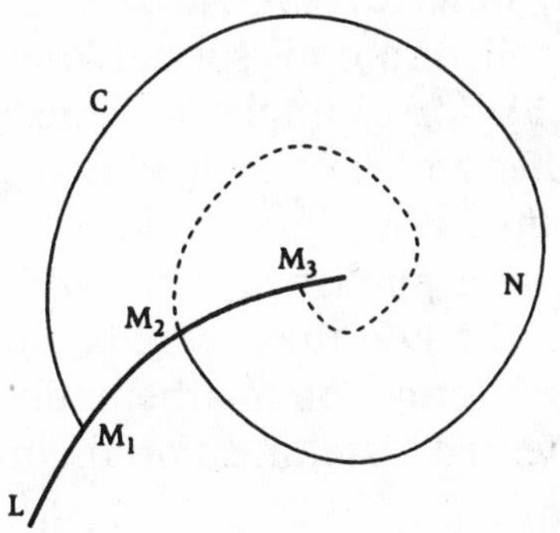

FIG. 14.18.1 *Possible curves in the restricted three-body problem, 1890.*

algebraic equation $F=0$, which is being crossed clockwise by another curve C; touching might happen (when the crossing might change sense), but only on a finite number of occasions since F is given by an algebraic function. So C is confined within the sector M_1NM_2; hence the points of crossing $M_1, M_2, \ldots$ must have a limit-point, M_0. (This is one place where Poincaré used Cantor's point-set topology: §12.5.) By definition, M_0 is its own successor crossing-point; hence its corresponding curve, C_0, is closed, and serves as a limiting cycle for C.

This finding led Poincaré to a result which became known as the 'Poincaré recurrence theorem', which brought in a new type of stability. The bounded curve C, and so also the orbit of one of the bodies, could return infinitely many times to any neighbourhood of a previous position. The proof was based on the following very informal idea, whose mathematical refinement (principally measure theory) was to cost his successors some effort. Take a string of points along C and centre (say) a sphere on each one; if they never intersect, then at some stage their total volume will exceed that of the sector bounding C (like M_1NM_2 above), which is impossible. Hence some spheres intersect, so that their centres are close together; therefore the orbit keeps crossing C, as required.

Consequences of Poincaré's reasoning soon began to be heard, especially in lectures in Paris. At the Ecole Normale, Paul Painlevé applied to it his special interest in differential equations (§12.3); for example, he considered initial positions of the bodies which could lead two of them to collide. Poincaré wrote up three volumes of *Les Méthodes nouvelles de mécanique céleste* and published them with Gauthier-Villars between 1892 and 1899; they were based upon his course at the Faculty of Sciences on 'mathematics and astronomy', which complemented Tisserand's on theoretical mechanics.

14.19 *Other bodies in the heavens*
{EMW VI2, 9–11, 16–19, 23}

In the rest of his fourth volume, Tisserand described the latest work on the orbits of Saturn and Jupiter, including important lengthy studies by Hill and his Canadian-born colleague Simon Newcomb. He also reviewed knowledge of the satellites and minor planets; improvements in equipment had led to much greater concern with the orbits of these objects. Other items on the astronomers' agenda included the real and apparent motions of stars, and the paths of meteors and comets. The trajectories of meteors and the orbits of comets were found to correlate sufficiently well to demonstrate that annual showers of meteors (then called 'shooting stars') were caused by cometary debris {EMW VI2,18a}. Often the methods of calculation were approximate, and where necessary tailored to the paucity of data; but sometimes more classical ones were usable. For example, double stars had risen in status from the speculation of their existence by John Michell in 1767 (§6.19) to a veritable catalogue of cases; and analysis of their orbits as a two-body problem, Kepler's equation (4.11.4) and all, was quite accurate enough for objects that far away. In many contexts trigonometry maintained its prominence; for example, in the preparation of star charts.

By the late 19th century the study of these and many other topics in astronomy (indeed, of mechanics in general) was being informed by considerations of light, heat, electricity and magnetism. The widening world of mechanics was intersecting with that of mathematical physics, to which we now turn.

15

An era of media: mathematical physics, 1860–1900

15.1 *The constellation of mathematical physics*

The order of sections in this chapter reflects a change in the relative importance of the various branches of physics during this period. Heat theory was making its greatest mark around 1860, especially with the advent of thermodynamics; but from the mid-1870s electricity and magnetism, and their interaction in electromagnetism, came more to the fore, including a new reading of optics.

Many of the theories, especially in electricity and magnetism, continued to assume the existence of the aether as the medium within which the phenomena took place. But *how* did such processes occur? A very good question, of course, and one that received interesting old and new answers, usually drawing upon elasticity and potential theories.

As regards principles, mechanics remained the parent discipline to such an extent that one might regard *mathematical physics as mechanics in fancy dress* – often the togs of energy mechanics, but sometimes variational and Newtonian. Potential theory was prominent; often line and surface integrals were used, together with variational methods. For example, J.J. Larmor advocated the principle of least action throughout his career, from dynamics through optics to electron theory (§15.13).

All the mathematical methods noted hitherto are present (or, in the case of matrices, normally absent); the most significant newcomers are the vector calculus (§15.9) and some new operator methods (§15.15). Differential equations were still usually linear, though sometimes getting more complicated. The rivalry between algebraic and geometric styles remained active, with the former somewhat less confined than hitherto to the reformulation of known theories.

Several of the principal figures of the previous chapters remain important here; in addition, Riemann is again significantly in the background, with a few papers and, in particular, when editions of his lectures on mathematical physics began to appear in 1869. Newcomers include Rudolf Clausius, whose contributions were focused upon physics.

For literature, the recommendations of §10.1 can be repeated for {Whittaker 1951, Chaps. 7–13}, largely for the physics; and especially for {Burkhardt 1908, but now Part 2} for his amazing coverage of the mathematics to around 1890 in this, the largest and perhaps the greatest of all reports written for the German Mathematicians' Association. {Klein 1927} covers much of the mathematics well; but the greatest single source is the fifth Part of Klein's

Encyklopädie der mathematischen Wissenschaften, on physics. Its importance to this chapter is comparable to that of its neighbouring Parts to the previous chapter. Its editor was Arnold Sommerfeld (1868–1951), invited to this post by Klein a few years after he had served as Klein's assistant at Göttingen in the mid-1890s. Sommerfeld was to edit a very comprehensive record of the physics of the period, containing over 3,250 pages on its completion in 1926; several eminent physicists contributed. This chapter cannot treat even all the branches in which mathematics was used, but it broadly follows his order of topics.

15.2 *Cycles of heat process*
{EMW V3}

We saw in §10.19 that during the 1840s the young William Thomson discovered and publicized the work of George Green. A few years later he performed a similar service for Sadi Carnot by recognizing the importance of his insights of 1824 (§10.8) on the 'motive power of machines'; they were to play a role in the birth of thermodynamics in the hands of Thomson and others.

One of Carnot's most influential innovations came from his insights into the way in which a body G of gas or hot air behaved under the action of heat. He showed that expansion (or contraction) of G under a constant quantity of heat (the word 'adiabatic' has become attached to this process) alternated with 'isothermal' expansion under constant temperature. In FIG. 15.2.1, G is trapped by a piston S inside a cylinder Y, which comes into contact alternately with bodies H and C at higher and lower temperatures. Starting at level i, it rises through h to r, and then falls to l before returning to i.

Carnot gave only a verbal description of this cycle; but mathematical commentators found a geometrical version helpful, like the 'indicator diagrams' relating the pressure and volume of G that James Watt had already been using

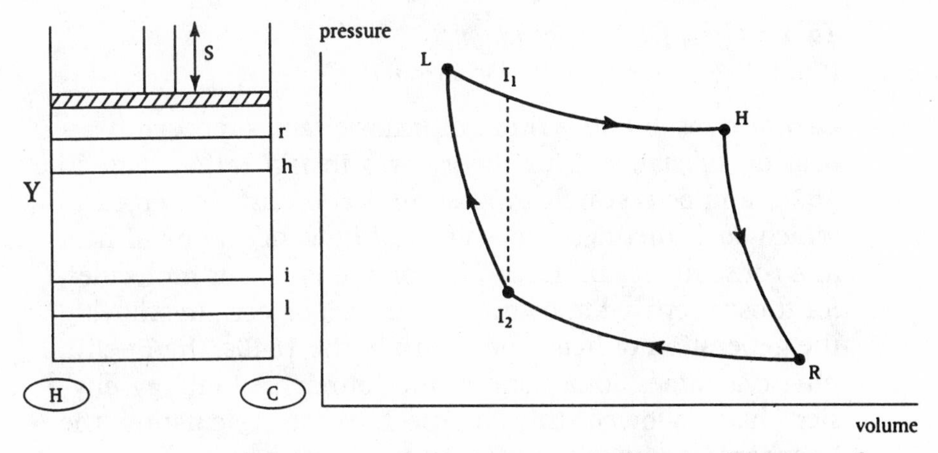

FIG. 15.2.1 *Action of piston S in cylinder Y following contact with hotter and colder bodies H and C.*

FIG. 15.2.2 *Indicator diagram of the behaviour of a gas body.*

The sequence proceeds as follows:

Action on Y	*Motion of piston S*	*Gas body G*	*Cycle*
H contacts Y	From i up to h	isothermal	I_1H
H removed	From h up to r	adiabatic	HR
C contacts Y	From r down to i	isothermal	RI_2
C removed	From i down to l	adiabatic	I_2L
H contacts Y	From l up to i	isothermal	LI_1

for analysing the action of steam engines. Benoît Clapeyron deployed them in 1832, in the first detailed study of Carnot; and they became very important when Carnot's work was rediscovered by Thomson. FIG. 15.2.2 shows one such diagram, in which the lettering parallels the lettering for the piston in FIG. 15.2.1. Various versions were given, arising from somewhat different theories. Carnot himself started at I_1, and did not assume that heat was conserved; neither did Clerk Maxwell, who started at H. But Thomson followed Clapeyron in starting at L, and had to assume conservation. The point is not trivial, since a basic notion in energy mechanics is involved, as we shall now see.

15.3 *Origins for thermodynamics*

{Brush 1976, Book 1; EMW V3; CE 9.5}

Carnot himself had written of 'caloric', the supposed heat-bearing substance. This theory was in fact rather *passé* in 1824, and as research progressed a consensus emerged in which such theories were replaced by conceptions of heat as a consequence of the motion of matter – of its molecules, for those prepared to countenance such objects. In studying the generation of heat from work in the 1840s, the English physicist James Joule, one of the founders of energy physics, had followed this approach when calculating the 'mechanical equivalent of heat' to measure it.

Once alerted to Carnot, Thomson immediately began to rethink his theory mechanically. In a purple paragraph in his first paper he wondered about a 'thermo-dynamic' engine, and announced that 'energy' could never be destroyed {Thomson 1849, p. 118}: the débuts of these terms in this scientific context (compare §14.4). This latter view soon became a dogma for him and for Rankine; they also thought much about irreversible processes, the one aspect of heat phenomena which seemed to differ[1] from those of mechanics.

In the empirical spirit of Scottish 'common-sense' philosophy, Thomson and Rankine did not enquire of the inner constitution of the matter they were analysing. The same is true of Rudolf Clausius, an early major figure on the Continent, who also began to consider the exchange between heat and work in 1850, his 29th year. Basic properties of heat exchange and recent experimental work convinced him that the amount W of work made available by the conversion of a quantity Q of heat at absolute temperature T

[1] The acceleration term in Newton's second law (§5.9) is often construed to suggest that mechanics admits reversibility: $d^2x/dt^2 = d^2x/d(-t)^2$, and so time can run backwards. However, this view fails to make allowance for the crucial place of the initial conditions for a mechanical process; they cannot thereby become final ones. The importance of this point for indeterminism was first recognized in {Popper 1950}.

could be expressed as the ratio

$$W = Q/T. \tag{15.3.1}$$

This form led him to introduce a parameter E (which he later called 'entropy', after the Greek word *tropi* for 'conversion') to help calculate the portion of energy not convertible to work, and to formulate two basic laws of thermodynamics in a paper of 1865. Mathematically, the laws take forms such as this: given a body B of (say) gas, with kinetic energy U and mechanical equivalence J of heat, then for an irreversible process

1. the energy of B is constant: $$\mathrm{d}Q = \mathrm{d}U + \frac{\mathrm{d}W}{J}, \tag{15.3.2}$$
2. the entropy of B tends to a maximum: $$\int \frac{\mathrm{d}Q}{T} < 0. \tag{15.3.3}$$

For a reversible process the inequality became an equation. Clausius and others found expressions for Q with separate terms involving the positions and the velocities of the molecules of B.

In order to give thermodynamics a general foundation, some of the basic principles of mechanics were deployed. For example, Hermann von Helmholtz and J.J. Thomson tried to use the principle of least action (6.6.4) and d'Alembert's principle (§5.25) in their formulations, though they faced some of the same difficulties as had their predecessors. So, in Crelle's journal in 1884, Helmholtz suggested dividing the independent variables of a system into 'cyclic' and 'non-cyclic' ones (respectively u and v, say): u was defined as cyclic if it were absent from the energy equation and only the time derivative $\dot{u}$ were present in the kinetic energy expression, and if $\ddot{u}$, $\dot{v}$ and $\ddot{v}$ remained small when the system was disturbed. These and related conditions, especially for a system containing only one u, led to special forms for the variational Lagrange equations (6.11.2) of motion and for the principle of least action.

As in his youthful work on the conservation of energy (§10.8), Helmholtz assumed that all differential forms were exact, and so could assume a potential. However, Ludwig Boltzmann soon played Carnot to Helmholtz's Lagrange (§6.12) by pointing to cases involving disequilibrium where the work differential dQ could not be so construed, so that the generality of (15.3.2) was compromised.

Among other figures, from 1873 J.W. Gibbs in the USA studied entropy intensively, and interpreted indicator diagrams in great detail. He laid more emphasis on conditions for equilibrium rather than on the action of heat. His work came to reinforce the influence of physics upon economics (§14.5) via his student Irving Fisher (1867–1947), who in a doctoral thesis of 1892 and later work deployed line integrals and exact differentials to interpret utility like energy, and also Gibbs's vector analysis (§15.9} to treat marginal utility as a force-like vector {Mirowski 1989, Chap. 5}.

Sadi Carnot's motivation from hot-air engines was not forgotten under all this theory. On the contrary, the engineering community made extensive use of his ideas in the design (especially the safety) of steam engines, and also introduced refrigerators {EMW V5}. For example, one of Boltzmann's cases of a system with one cyclic variable was a steam engine carrying a governor. Thermodynamics was also applied to 'interior ballistics', concerning the manner of explosion of gunpowder in the constant volume of the firing chamber {EMW IV18, Part 2}.

15.4 *The mean and probable behaviours of gases*

{Truesdell 1975; CE 9.14}

By 1857, Clausius's attention was being drawn to the behaviour of gases and vapours. Maybe mindful of Laplace's molecular physics (§10.10), he endowed the molecules of gas with many dynamical properties: rotation and vibration as well as straight- and curved-line motion. His initial ideas on gas dynamics assumed that the molecules travelled at great speed. However, it was pointed out to him by the Dutch meteorologist C.H.D. Buijs-Ballot that the paths could not be very straight, since various cases were known in chemistry where intermixed gases were slow to merge. This piece of statistical thinking led Clausius to a radical reformulation of his theory in terms of average values of quantities, as the best way of analysing such an unruly collection of physical objects as a gas body.

For Clausius, each molecule was now a sphere of diameter d, and N of them lay within a tiny layer L of the containing space around a particular molecule M. Hence the probability that M would hit another molecule was proportional to the cross-sectional area of L which the molecules occupied, namely Nd^2. Thus the 'mean free path' (his phrase) of straight travel for M before collision was inversely proportional to Nd^2. From here he formed integrals for the mean free path though a finite layer. He and others started to reformulate the entropy of molecules in a gas body, replacing a physical notion of energy with a quantitative statistical measure of their disorder.

Clausius assumed that all molecules moved with the same velocity. Clerk Maxwell went further in 1860, his 30th year, to assert that the repeated collisions changed their velocities, which therefore could be estimated only statistically. He took up a derivation of the normal distribution function (§8.8) $f(x) = \exp(-x^2)$ sketched by John Herschel in 1850 from some general assumptions about how errors x in data were distributed about a true value. Using those assumptions to form a functional equation for the variation of the component velocity u of a molecule in a given coordinate direction, Maxwell proved that f was the only function that satisfied certain conditions of containment of the gas.

Stimulated by correspondence with P.G. Tait, Maxwell continued this statistical thinking in his *Theory of Heat* (1871) by presenting his 'demon'. A gas is enclosed in a vessel which is divided into two parts by a screen containing an eyehole covered by a shutter. The demon is a clever devil, for he is endowed with the ability to observe all molecules simultaneously, and opens and closes the shutter such that swifter molecules go into one half of the vessel, and slower ones into the other half, thereby producing an uneven distribution of temperature in impudent contradiction to the second law. This thought experiment showed Maxwell that the second law (15.3.3) was a 'statistical certainty' – a clever phrase, sometimes misunderstood, since in referring to it *only* as a sure bet, he allowed the possibility that the demon would win without trying if the molecules spontaneously fluctuated in behaviour that way.

Boltzmann took this approach a stage further in 1872, his 28th year, with a result which, in his later notation and formulation, followed here, is known as his '*H*-theorem'. Take a function $F(u, v, w)$ of the components of velocity in all three coordinate directions, and form the integral

$$H(t) := \int_{-\infty}^{\infty}\int_{-\infty}^{\infty}\int_{-\infty}^{\infty} F(u, v, w) \ln F(u, v, w)\, du\, dv\, dw; \qquad (15.4.1)$$

all possible values of velocity were thus covered, but there was a marked dependence of the variation F of velocity on time t. Boltzmann argued that this function gave a good mathematical characterization of the entropy of a gas body: that when a gas was not in steady-state equilibrium, then $H'(t) < 0$, and that at equilibrium the usual first-order condition $H'(t) = 0$ led to the conclusion that F would reduce to Maxwell's normal distribution function in each variable.

A vast discussion ensued about the consequences for physics: reversibility or not of phenomena, the possible bearing of recurrence in equilibrium *à la* Poincaré (§14.18), and so on. In a book with the interesting title *Elementary Principles in Statistical Mechanics Developed with Special Reference to the Rational Foundation of Thermodynamics* (1902), Gibbs extended the concept to an 'ensemble' – a collection consisting of a given gas body and variants of it which possessed the same quantity of energy but different distributions of position and velocity of its molecules. He estimated the probability of one of these variants occurring at a given instant or in a given interval of time.

This line of statistical theory developed further into the

'ergodic hypothesis', first formulated by Maxwell, which stated that the molecules of a given gas body containing a known amount E of energy would eventually take all values of position and velocity consistent with E; among its various advantages, this greatly simplified the calculation of average values of physical properties. An important airing of this new theory was in the *Encyklopädie der mathematischen Wissenschaften*; for in an event rare in its history, Klein commissioned an extra article, from the Austrian Paul Ehrenfest and his Russian-born wife Tatiana. It appeared in 1911 as a 90-page appendix, not to the Part on physics, but to the one on mechanics {EMW IV32}, of which Klein was co-editor.

15.5 *Clausius on the 'potential' and the 'virial'*

Other lines of thought drew more traditionally upon mechanics. For example, Clausius was a keen exploiter of potential theory. In a well-read book *Die Potentialfunction und das Potential* (1859), he reformulated the basic notion of Green and Gauss (§10.18) as follows: the Newtonian potential P of attraction between a mass M and system of masses m_j,

$$P := \sum_j \frac{m_j M}{r_j} = MV \qquad (15.5.1)$$

was the 'potential', while V was the 'potential function'. A decade later he considered a real imperfect gas composed of molecules of constant mass m of non-negligible size moving at point (x, y, z) with finite velocity v within a bounded space under forces (X, Y, Z). He converted the energy equation into a form using the averages (marked by overbars) of the pertaining quantities over some interval of time:

$$-\frac{1}{2}\sum_m \overline{(xX + yY + zZ)} = \frac{1}{2}\sum_m (m\overline{v^2}). \qquad (15.5.2)$$

He called the first, work-like, term the 'virial', after the Latin *vis*; the summation covered forces both internal and external to the gas body, and could be expressed by separate virials if desired. Being a moment (force × distance), it was independent of any coordinate system.

An early application of virials was made by the Dutch physicist Johannes van der Waals (1837–1923) in his doctoral thesis of 1873 at Leiden. He found an 'equation of state' of a gas body G at pressure p, volume v and absolute temperature T. After further work by Lorentz and others to the mid-1890s, the equation took the form

$$pv = KNT\left(A + \frac{B}{v} + \frac{C}{v^2} + \cdots\right), \tag{15.5.3}$$

with K a universal constant, where N was the number of molecules in G, and the constants A, B, C, ... were to be determined empirically (as was done by van der Waals himself, Boltzmann and others). For a perfect gas $B = C = \cdots = 0$, and the equation simplifies to the classical Boyle–Mariotte law.

In addition, Maxwell and then Boltzmann emphasized three properties, which they associated with corresponding ones in mechanics by analogies: viscosity/momentum, conduction/energy, and diffusion/mass. The name 'transport' became attached to the correspondence, because the free path of the molecule was held to provide the mechanism in physics. While this approach remained rather analogical, it did offer insights into the mysteries of gas behaviour.

15.6 *From gases to matter*

{Brush 1976, Book 2}

As these studies of gases developed, they led to disciplines whose differences are explained as follows:

1. The *kinetic* (sometimes called 'dynamical') *theory of gases* was the study of the behaviour of gases (and vapours) in various physical conditions, such as trapped in a bottle by a cork and left in the sun for a while. When these concerns were applied to a wider range of substances, the name

'kinetic theory of matter' was sometimes used – for example, in the article {EMW V8}, which Boltzmann co-authored in 1905.

2. *Thermodynamics* was the analysis of mechanical and physical heat processes in matter which a system exhibits over some period of time (maybe an indefinitely long one), and included the study of categories such as energy which may (not) be conserved; non-thermal effects were also examined. Again, the matter might not be gaseous; indeed, some students starting out from energy conservation made no assumptions about its state.

3. In *statistical mechanics*, matter was assumed to be composed of a large number of molecules (or atoms), and statistical theory was deployed to deduce its macroscopic properties {CE 9.14}. While often thermal or radiative, these properties could be of a more general character, especially in quantum mechanics (§16.27–§16.28).

The rise of statistics within this trefoil of disciplines is nicely exemplified by the position of three main figures on gas dynamics. Clausius's 'mean free path' of a collection of molecules exactly captures his position, for it is a locational parameter, like the mean or median, but no more. Maxwell added statistical dispersion to this approach by considering also the distribution of paths. Boltzmann's H-theorem (15.4.1) took the approach still further, towards the asymptotic world of Laplace's central limit theorem (§8.8), for it asserted a property of a large number of such paths.

Thermodynamics had already been applied beyond gases and vapours, by Thomson in the 1850s, when he brought it to bear upon elasticity theory. Clausius acted similarly for chemistry in the 1860s when analysing chemical equilibrium in terms of potentials. But the decisive figure was Gibbs, in a massive two-part study of 'the equilibrium of heterogeneous substances' (1876–8) which started out by adding to Clausius's equation (15.3.2) terms of the form $P\,dm$, where P was the chemical potential of the element, now with variable mass m {CE 9.16}. His exposition was well laced with indicator diagrams. Unfortunately, his essay appeared in a new journal published in Connecticut, thus retarding a general acquaintance with his work; contributions from Helmholtz and others from the 1880s gained more attention. But Gibbs's work was translated into German and French around 1900.

By then, thermodynamics was present in hydrodynamics: when, for example, Pierre Duhem (an early commentator on Gibbs) and others {EMW IV19} considered heat effects in shock waves (§14.9). Similarly, the interior constitution of the planets became a major question: were they solid, liquid or gas, or some mixture thereof? This question inspired thermodynamical analyses in which a planet was construed as a (partly) gaseous body enclosed within a solid shell, and theory based on the virial (15.5.2) could profitably be developed {EMW VI2,24; CE 9.6}.

15.7 *Acoustics and optics*
{EMW IV26 (by Lamb); CE 9.8}

Acoustics gained the attention of heat theorists, especially concerning the possible adiabatic expansions and contractions of air as sound passed through it; but the traditional non-thermal properties were also studied. A mathematical innovation was made by Emile Mathieu in 1868: a new and important type of special function, now named after him, which arose when he formed a differential equation to represent the vibration of membranes of elliptical shape. Otherwise Fourier analysis and forms of the energy equation remained popular.

The period was graced by two magisterial works which both summed up current knowledge and set agenda for a long time to come. In 1863 Helmholtz published the first edition of his book *On the Sensations of Tone* (to quote the title of its English translation of 1875). He concentrated upon physiological aspects; but, for example, he also contrasted the vibration of a piano string when struck by an elastic hammer with that of a violin string when stroked by a bow.

Then, in 1877 and 1878, Lord Rayleigh put out the first edition of his two-volume *Theory of Sound*: the German translation followed three years later. He deployed the energy equation and variational methods, along with potential theory and its principal theorems, and rehearsed various solutions of the wave equation (6.4.1), especially Fourier series. He gave special attention to vibrations in all their forms; not only of strings, plates and rods, but also of fluids in tubes or in chambers, and resonators. He complemented the hydrodynamical study of the flow of jets (§14.9) by analysing the noise that they produced, and also examined the noise from telephone cables. Analogies from optics were considered, including mathematically: for example, the possible modes of reflection, refraction and diffraction of sound. A second edition came out in 1894 and 1896.

In optics itself Helmholtz was an important contributor to the physiological side, especially in the 1850s, when he was in his thirties. Most parts of physical optics continued along lines laid down by Fresnel and his successors, with the interference properties of spectral emissions quite popular {EMW V24}; extensions included the study of rainbows and haloes {EMW VII,12}. Geometrical optics benefited perhaps rather more from new advances in science. The design and testing of lenses required more exact mathematics as the pertaining physics and technology became more refined. In particular, the increased use of cameras gave a new lease of life to descriptive geometry in 'photogrammetry', the analysis of spatial objects as depicted (and maybe distorted) on photographs; {Finsterwalder 1899} reported on developments for the German Mathematicians' Association.

The mathematical aspects of physical optics included analyses by Sonya Kovalevskaya and others of double refraction by solving hyperbolic differential equations {Gårding 1989}. But the most important role for mathematics during this period was the subsuming of optics under electromagnetism, which we shall consider in §15.12.

15.8 *Heat theory and differential forms*
{EMW V4; CE 9.4, 3.4}

Heat diffusion was affected by the rise in thermodynamics, for example when in 1870 Rankine analysed the propagation of shock waves in a perfect gas. But, like optics, it normally followed the lines laid down early in the century by Fourier and then Thomson (§10.11, §10.19). The most notable change was a gradual introduction of potential theory, especially for appraising heat flows in bodies and fluids {Carslaw 1921}. One figure in this movement is Joseph Boussinesq, who published in 1901 and 1903 a two-volume treatise with the striking title *Théorie analytique de la chaleur, mise en harmonie avec la thermodynamique et avec la théorie mécanique de la lumière.*

The main mathematical innovation in this topic came from Riemann. We saw in §13.7 that in his 1854 thesis on geometry he expressed his ideas on manifolds in purely prosodic form. In a paper of 1861 on non-linear heat diffusion, first published in the 1876 edition of his works, he presented the equation involved in its general bilinear differential form, and considered its transformation from rectilinear coordinates x_i to general spatial variables s_i. That is, given

$$F := \sum_{i,j} \alpha_{i,j}\, \mathrm{d}x_i\, \mathrm{d}x_j = \sum_{i,j} \beta_{i,j}\, \mathrm{d}s_i\, \mathrm{d}s_j, \qquad (15.8.1)$$

what are the relationships between the coefficients and partial derivatives of the s_i with respect to the x_i? Answers not only gave some more publicity to matrices, but also helped articulate theories of cosmology and gravitation. In particular, it was picked up in 1869 and again in 1882 by Elwin Christoffel, who found invariants for F under various transformations. This led him to express principal invariant notions (for example, curvature) with the help of clever notations using both subscripts and superscripts which are still known after him. His theory was soon extended by the Italian mathematician Gregorio Ricci-Curbastro (1853–1925) into the 'absolute differential calculus', and applied to various areas of applied mathematics; later it was to become a staple mathematical component of both tensor calculus and relativity theory (§16.25–§16.26).

15.9 *Algebras and notations for vectors* {EMW IV14; CE 6.2}

Alongside Christoffel's notations, the geometry and analysis of vectors at last began to be established. The definitions of addition and subtraction for vector algebra are obvious: the difficulties lie in multiplication and division, for which the answer is two forms of the first and none of the second. It had been proposed by a few figures in mid-century such as Barré de Saint-Venant (Hermann Grassmann's response (§9.9) had been somewhat different), but none of them exploited their gold; this was left chiefly to Gibbs in the USA, and Maxwell and Oliver Heaviside in Britain, all three of whom started out from quaternions. Questions of balance between geometry and algebra are once again evident in their approaches.

Maxwell pioneered elements of vector analysis during the 1870s, as part of his fundamental contributions to electricity and magnetism (§15.11). Although he used quaternions, he proposed the names 'convergence' and 'curl' (later, 'rotation') for the scalar and vector parts (9.8.5) respectively; he also gave the name 'grad' to the vectorial differential operator in (9.8.3).

Gibbs publicized his ideas in an odd way. While he gave many lectures at Yale University, he privately printed his *Elements of Vector Analysis* (the origin of the name) as a three-part pamphlet between 1881 and 1884. However, he circulated it widely, as he did with offprints of some related papers, so the news did spread. Acknowledging a large debt to Grassmann, he presented his theory as a 'multiple algebra'. For two vectors α and β he defined both a 'direct product $\alpha.\beta$' and a 'skew product $\alpha \times \beta$' (today the adjectives 'scalar' and 'vector' are used). He went further in defining 'linear vector functions', a vector multiplying a vector in a way that in effect substitutes for multiplying a vector by a matrix, and obeying the property of linear combination which we saw at (13.20.1) to be common to many new algebras of this period. Since his motivation lay in the electromagnetic theory of light, he wedded his theory to the differential calculus, defining, for example,

$$\text{for } V := (X, Y, Z), \text{ '}\nabla.\mathrm{V}\text{'} := X_x + Y_y + Z_z, \tag{15.9.1}$$

where '∇.' was his notation for the grad operator.

Heaviside learned of Grassmann's work only when reading about it in Gibbs's booklet in 1888 (his 39th year). By then, he had been publishing his version of the algebra for five years, as an aid to his work on electromagnetic theory (§15.14). Definitions such as (15.9.1) were more important to him than the vector algebra itself. Among his most durable suggestions, for the operator in the second expression in (15.9.1) he popularized the word 'divergence' (the negative of Maxwell's 'convergence', in order to make it a positive quantity); he adopted the symbol '∇', with no period, for grad. He also introduced bold-face letters ('Clarendons') to denote vectors.

As was mentioned in §9.8, quaternions flourished until the 1920s; figures of the calibre of Thomson preferred to stay with them, and Tait was one of their most prominent advocates. To many commentators during this period, vector analysis was little more than a modification of the strict separation of the scalar and vector parts of quaternions; its significance did not emerge quickly. A large miscellany of technical terms and notations in the new theory, which endured for a long time and resisted attempts to establish an international convention, also hindered their acceptance. The first textbook on vector analysis appeared only in 1901, from Gibbs's student E.B. Wilson and based upon his lectures. But Gibbs had been invited by the Göttingen mathematician Max Abraham to write an article for the *Encyklopädie der mathematischen Wissenschaften*. He had declined, perhaps for reasons of age; but a piece did appear, also in 1901 in the Part on mechanics. Abraham himself wrote it, under the interesting title 'Geometrical Fundamental Concepts'; it is cited at the head of this section.

15.10 *Modelling the electromagnetic aether*

{EMW V12; CE 9.10}

From now on, electricity and magnetism run this chapter. We start with two German papers of 1858 which were of durable importance in different ways in developing electromagnetic theory.

Firstly, Riemann wrote a cryptic 'contribution', published posthumously in 1867. Assuming a finite velocity α of propagation within a medium, he adjoined to Poisson's equation (10.6.2) a term representing the acceleration of the electrostatic potential function $U(t)$ at a point distant r from the conductor, and found as solution with a differentiable function f

$$U(t) = f(t - (r/\alpha))/r. \tag{15.10.1}$$

Then the form $P := -\int_0^t \varepsilon\varepsilon' F(t - (\tau/\alpha), \tau)\, d\tau$

gave the potential between an electrical element E of mass ε at time t of emission and another one E′ of mass ε' at time $t' > t$ of reception of the effect from E, where F was the reciprocal of the distance between the locations of E at t and E′ at t'. For a collection of such elements, the sum of such integrals was taken. The name 'retarded potential' became attached to P.

Although Clausius soon found Riemann's theory unsatisfactory, it was seen as important for analysing phenomena involving finite propagation of some kind. In particular, it was related to two principal theories of propagation of electromagnetic force: Weber's, which saw it as a function of relative and absolute velocities of the carrying molecules (in (15.16.3) below); and Franz Neumann's, where it depended upon their positions and absolute velocities.

Secondly, an 1858 paper by Helmholtz on vortex motion in fluids published in Crelle's journal soon inspired Thomson to conjecture that the aether was a fluid capable of vorticial motions, and that within this plenum the ultimate particles of matter might be composed of 'vortex atoms' linking and knitting themselves together in various ways. He had already been nudged in this direction by a theory of 'molecular vortices' advocated from the 1840s by Rankine, as a model of heat in terms of a supposed 'atmosphere' around each molecule. Another inspiration was the 'rough geometrical method' that Michael Faraday had used from the 1830s to envision the action of 'lines of force' in electromagnetic induction.[2]

Faraday was also an important influence upon Maxwell's initial ideas about electromagnetism in the mid-1850s, for he had rejected molecular theories of action at a distance such as Weber's and Neumann's on the ground that Newton's third law, action = reaction, was not always satisfied. Maxwell developed his ideas at intervals in a string of papers, starting with 'On Faraday's Lines of Force' in 1855–6 (when he was in his mid-twenties) and culminating in his two-volume *A Treatise on Electricity and Magnetism* of 1873, written by the Professor of Experimental Physics at Cambridge University but published by the Clarendon Press at Oxford.

Many of Maxwell's concerns and models were driven by analogies with mechanics – explicitly so, in this Scottish-born enthusiast for the common-sense philosophy of his country. Also like Thomson (§10.19), he read Fourier's

[2] Faraday marvellously exhibits topological thinking at a time when theory existed only as scraps of mathematics (compare §13.17) – and in a physicist notoriously incapable of handling even simple mathematics in this or any other application.

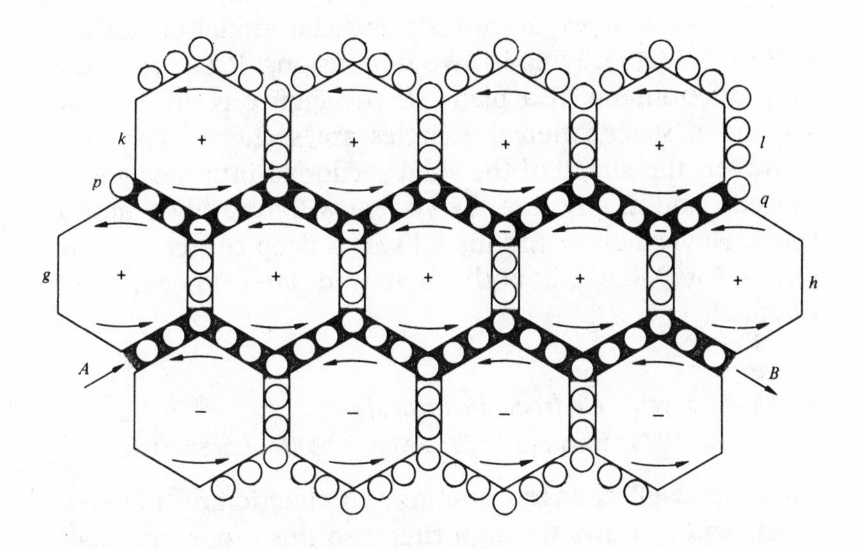

FIG. 15.10.1 *Maxwell's electro-mechanical model of electromagnetic induction, 1861.*

book on heat diffusion at a young age. An early example of his approach, dating from 1861, is shown in FIG. 15.10.1, which shows an electro-mechanical model of electromagnetic induction in an incompressible fluid. The flow of current was represented by the rotation without slipping of the tiny idle wheels along the zigzag from A to B. These motions generated rotations of the tiny hexagonal vortices (a shape suitable for representing Thomson-like molecular vortices), as shown by the arrows, in contrary directions and thereby producing opposite polarities. The electromotive force was represented by the forces exerted by the surfaces of the vortices on the idle wheels, which played a role akin to electric charges. Similarly, the magnetic force was represented by the centrifugal force arising from their rotation: they created Faraday-like lines of force, which in turn generated further motions (that is, currents) along companion zigzags.

This model aroused much admiration. However, while it was not, of course, to be taken literally, it inspires the same reservation as does Descartes's vorticial model of the universe (FIG. 5.8.1): lying in two dimensions, it can represent only phenomena in a plane. If the Figure is an arbitrary section of space, then the circles are sections of spheres; however, the shape of the solid sectioned into hexagons is unclear, and in any case the system will lock when motion starts. Nevertheless, the model was a deep source of inspiration for Maxwell, both then and later {Siegel 1991, Chap. 2}.

15.11 *Maxwell's electromagnetic field*

{Buchwald 1979; Harman 1987; EMW V13 (by Lorentz)}

Later Maxwell changed ontology to electric and magnetic fields, and to flows through them; so this model gradually lost some of its importance, and his later ideas about vortex motion were not tied to it. His next theory assumed an elastic aether, with properties drawn by analogy from elasticity theory and hydrodynamics as well as from vortices. Thus continuum took over from particle mechanics, a change which attracted the charge of 'mysticism' from Thomson. Passing over intermediate stages, where both theory and equations varied somewhat, I take the version presented in 1873 in the *Treatise*. In the following summary of some of the main features, I retain Maxwell's choice of mathematical letters, rendering them in Heaviside's Clarendons; but I replace his quaternions (which he described in some detail) by vector notation.

1. When the aether was disturbed from a state of equilibrium, quantities of 'flux' or 'displacement' occurred across a surface bounding two media, like the motion of a fluid (thus unlike the displacement of an object from one position to another), creating a stress-like effect. Such displacements induced 'intensity' or 'force' within the aether.

Discontinuities in electric flow created electric charge.

2. The amount of 'electric quantity' (or 'electricity') was constant in a given volume, like an incompressible fluid. It was characterized by 'dielectric inductive capacity' K, 'magnetic inductive capacity' μ, and electric conductivity C, all positive quantities.

3. The flux was continuous across a boundary as long as neither medium was dielectric; however, the forces were discontinuous if K or μ took different values in the two media but the flux remained continuous. The only exception was when the boundary separated a conductor C from a non-conductor such as a dielectric; then the 'true charge' of C was given by the surface integral $\int_C \mathbf{D} \cdot \mathbf{dS}$. 'Apparent charge' was produced similarly across two dielectrics of different capacities, or when that in a dielectric varied.

4. 'Electric displacement' **D** arose from a disturbance of the aether from equilibrium, like changes in potential energy; it created a 'current of conduction' K proportional to 'electromotive force' or 'intensity' **E**. 'Magnetic induction' **B** was caused in the aether by vorticial motion in the aether, akin to kinetic energy; it was proportional to the 'magnetic force' **H**, which was produced by the ensuing centrifugal forces. These five vectors, all functions of place and time, were related thus:

$$\mathbf{D} = \left(\frac{K}{4\pi}\right)\mathbf{E}, \quad \mathbf{K} = C\mathbf{E} \text{ and, } \quad \mathbf{B} = \mu\mathbf{H}; \qquad (15.11.1)$$

the middle equation was Ohm's law (§10.20).

5. When charge flowed into one plate of a capacitor (such as a condensor) and out of the other, and the gap G between the plates was filled with dielectric, the open circuit was completed by **D** across G because **E** displaced the distribution of the dielectric, like the polarization of a magnet by a magnetic field. The relationship between the 'total current' and **H** was given by

$$4\pi(\mathbf{K} + \mathbf{D}) = \text{curl } \mathbf{H}. \qquad (15.11.2)$$

6. Electromagnetic induction was caused by the change in the angular velocities of the rotating vortices. The 'electro-tonic state', which Faraday had imagined the molecules of the conducting material to enter during induction, was represented mathematically by the vector potential 'electromagnetic momentum' $\mathbf{A}(t)$ at a point moving with velocity $\mathbf{v}$ at time t under 'electric potential' Ψ. It was specified by the conditions

$$\text{curl } \mathbf{A}=\mathbf{B}, \quad \text{div } \mathbf{A} = 0 \text{ and } \mathbf{E}=\mathbf{v}\times\mathbf{B}-\frac{\partial \mathbf{A}}{\partial t}-\text{grad } \Psi. \tag{15.11.3}$$

Other equations related these and other quantities, and their rates of change, to electric current, magnetization, electromagnetic induction and electromotive and mechanical forces. Their formulation could change with the manner of defining the fundamental units, an important issue then in rather a chaotic state; Maxwell made proposals in his *Treatise* and elsewhere. For mathematical methods, he not only used quaternions but also devoted a short chapter of his book to variational principles in mechanics, especially Hamiltonians like (10.17.2). However, in a theory in which flux was such an important notion and current flowed along wires, he gave greater place to potential theory, especially integral theorems, which he not only stated but also proved. He used Stokes's theorem (10.19.1) to show, for example, that the integral of $\mathbf{A}$ around a closed contour bounding a surface S equalled the flux of magnetic induction through S.

Using these and other principles drawn from his and others' theoretical and experimental researches, Maxwell covered most known electric and/or magnetic phenomena, including geomagnetism, electric circuits, and the resistance of wires. Principal mathematical results included Ampère's equations for electrodynamic equilibrium (§10.13), and among techniques he developed Thomson's method of electrical images (§10.19) in terms of spherical harmonics.

Maxwell's account aroused both interest and perplexity in good measure among his readers. The relationship between aether and matter was perhaps the central one, but within his aether theory itself the association of electric conduction with heat was a mystery. The displacement current (as his notion is normally called) **D** was a major enigma, perhaps to Maxwell himself; his 1862 version, for example, had used FIG. 15.11.1 to suggest that the rotating vortices carried the idle-wheel particles making up the electric current **I**, and that **E** cancelled the effect of **H** within the gap **G**, in the manner of reversed polarization, by moving in the opposite direction to that of **H** due to the converse rotation of molecular vortices {Siegel 1991, Chaps. 3–4}. This conception was quite different from that outlined in item 4 above, and had led to a different equation.

15.12 *Electromagnetic optics?*
{EMW V22}

Some changes led to modifications of Maxwell's basic equations, including those relating to his theory of light. A major influence was his calculation in 1861 that the velocity with which electromagnetic waves were propagated (α in Riemann's (15.10.1)) was close to that of light; for he concluded that optics (and the spectrum of light) had to became *part* of electromagnetism, not just possess similar-looking theories. He followed the orthodoxy established by Fresnel (§10.12) in proposing that light was caused by transverse vibrations of the aether, and put forward equations for its propagation in uniform and in crystalline media. In his early stages molecular vortices were prominent {Siegel 1991, Chap. 5}. Further analyses included, for example, one of the rotation by a magnetic field of a plane of polarized light, which he launched from a form of Lagrange's equations (6.11.2).

Maxwell's optics aroused much attention on the Continent. Modifications had begun to come in even before the

Treatise appeared. For example, Maxwell gave no explanation of reflection and refraction; so Helmholtz proposed in Crelle's journal in 1870 that **D** and **B** be continuous along the normal to the bounding surface between two media and **E** and **H** along it, and the young Hendrik Lorentz amplified the idea mathematically five years later in his doctoral thesis. Soon afterwards, and independently, G.F. Fitzgerald introduced a different theory based upon the principle of least action to express the equilibrate state of the aether, and obtained surface ('superficial') integrals to express conditions of reflection and refraction.

Both approaches drew on James MacCullagh's theory of the aether (§10.19), which had started out from rotation rather than compression or extension (in the new vectorial language, from curl **e** rather than div **e**, where **e** was the disturbance of a point or molecule of the aether from its position of equilibrium). As we shall soon see, other workers were to adopt this idea, which had rather lain fallow during its first 40 years. In Paris, Henri Poincaré included optics in papers and Sorbonne lectures from the late 1880s, as did Pierre Duhem out at Lille.

Individual branches of optics benefited from the new theory. For example, in the mid-1890s Sommerfeld made two valuable contributions to diffraction: an integral for calculating the diffraction effects of a semi-infinite plane screen (*Habilitation* of 1895, his 28th year), and soon afterwards an analysis of the behaviour of X-rays around a wedge-shaped slit. But despite these advances, electromagnetism and optics remained apart in many ways. An important context was energy: Maxwell saw it in electromagnetism as being exchanged between parts of the aether or bodies within it, whereas in optics it was free to travel and suffer only effects such as rays splitting into two with double refraction. Such differences, together with the difficulties with electromagnetism itself, were to lead Maxwell's successors not only to modify his theory and equations but even to replace his aether model, at least partly, by theories involving particles – of various kinds.

15.13 *Particles rather than fields: Larmor and Lorentz in the 1890s*

{Schaffner 1972 (a source book); EMW V14 (by Lorentz) and 18}

One spectacular vindication of Maxwell's theory was the experimental corroboration by Hertz in 1888 of the existence of electromagnetic waves; his procedure was soon analysed mathematically by Fitzgerald in terms of retarded potentials (15.10.1). But negative reactions were also evident: Hertz himself modified the form of Maxwell's basic equations for bodies moving through the aether.

Many convolutions around and away from Maxwell's approach were effected. For example, in the mid-1880s Emile Mathieu found that Maxwell's theory of dielectrics lay outside the normal elasticity theories of the time, and Eugenio Beltrami could not derive some of his equations from the assumption of a homogeneous elastic medium. In the early 1890s, Sommerfeld preferred MacCullagh's model of the aether over Maxwell's, while Boltzmann lectured on redefining the electric and magnetic fields at a point in terms of the velocity and angular velocity of an element there.

Similar changes took place in Britain. J.J. Thomson, who succeeded Lord Rayleigh in Maxwell's Cambridge chair in 1884 and edited the third edition of the *Treatise* in 1892, moved towards theories using particles in his *Applications of Dynamics to Physics and Chemistry* (1888). In this book he followed Gibbs (§15.6) in using Hamiltonians, in Routh's modified form (§14.4); but in order to understand conduction better, and to describe the manner in which a chemical mixture might attain equilibrium, he worked with molecules and their constituent atoms.

The preference for particles was pursued by Larmor and Lorentz, both around the mid-1890s. Larmor came to disbelieve in Ampèrian infinitesimal elements of current, and preferred to understand the transport of electrical charge in terms of electrons rather than some supposed discontinuity in/of the aether; as usual for him, the principle of least action in Hamilton's form (10.17.1) determined the electrons' paths in an assumed equilibrate situation {Buchwald 1981}. He used striking mathematical analogies to describe an electron, like a simple pole of a complex-variable function (§8.11), and moving through the aether 'much in the way that a knot slips along a rope'.

Larmor was also concerned with the effect that the motion of bodies had on the aether: whether the aether remained immobile in its Newtonian absolute spacetime prison, or whether it was dragged along, either partially (as Fresnel had thought) or totally (Stokes's view). A major issue was the experimental work by A.A. Michelson and E.W. Morley in the late 1880s, which did not detect any drag effect. Larmor presented his approach in his book *Aether and Matter* {Larmor 1900}, which won the Adams Prize the previous year (his 43rd). His subtitle admirably expressed the main views and assumptions of many mathematical physicists of the time: 'A Development of the Dynamical Relations of the Aether to Material Systems on the Basis of the Atomic Constitution of Matter Including a Discussion of the Influence of the Earth's Motion on Optical Phenomena'. He chose to develop the model of the aether proposed by his compatriot MacCullagh. For the 'propagation of light through moving matter', for example, he found various equations which gave in {Chap. 11} this expression for the velocity U of a wave-train relative to a medium moving with velocity v in the same direction relative to its velocity V with respect to a stationary medium:

$$U = V - p^2 v - (p - p^3)\frac{v^2}{c} + \mathrm{O}\left(\left(\frac{v}{c}\right)^3\right), \quad \text{where } p := \frac{V}{c} \qquad (15.13.1)$$

and c was the velocity of light.

Lorentz's approach was to propose in 1895 that the aether be left standing still, but that tiny 'ions' travelled though it; he added appropriate terms to Maxwell's equations, which he took in Hertz's form. Then he adopted the 'contraction hypothesis', now named after Lorentz and

Fitzgerald, that a body moving with velocity u though this aether shrank in size in the direction of its travel and changed its time-keeping by certain functions of u, and obtained equations similar in form to Larmor's. In 1905 Albert Einstein was to rethink the space-time framework in which they had been set (§16.25).

15.14 *Heaviside and the telegraph equation*

{Yavetz 1995}

Another student of the equations required to analyse the motion of bodies in the aether was Oliver Heaviside (1850–1925). His dates match Klein's to within months; but in contrast to that great German organizer he was one of the most remarkable of even English eccentrics. Despite receiving no formal education, he taught himself a formidable amount of mathematics. After a short career spent with his elder brother Arthur in telegraph companies in the north-east of England, he was forced by deafness to retire in the mid-1870s, and he spent the rest of his life in poverty. But he continued his researches, which offended views held at the General Post Office in Britain; so he replied in a lingo replete with idiosyncratic analogies and sarcasms, such as 'even Cambridge mathematicians deserve justice'. But this was harsh on his supporters Rayleigh and William Thomson; they and others arranged for him to receive a civil pension. In his last years he lived in a little house in Devon, where boulders substituted for chairs around the dinner-table.

Heaviside published much of his research work as numerous short articles from the mid-1880s in the engineering journal *The Electrician*. He gathered many of them together with other papers in two volumes of *Electrical Papers* (1892) and three of *Electromagnetic Theory* (1893–1912), the latter dedicated to Fitzgerald. His accounts included extensive explanations of his vector notations (§15.9).

Heaviside was one of the most passionate adherents of Maxwell's theory – but also one of its most perceptive critics. A major contribution was his proof in 1884 that the flux of energy at a point in a medium was given by $c(\mathbf{E}\times\mathbf{H})/4\pi$, and therefore was propagated, with the velocity c of light, in the direction perpendicular to $\mathbf{E}$ and $\mathbf{H}$. (The result had also been found some months earlier by the English physicist J.H. Poynting, after whom it is usually named.) Heaviside came to construe this quantity as an 'energy current' travelling sideways as a slab through the aether and generating current within a wire in its presence: 'we reverse', as he put it, the Maxwellian orthodoxy that current generated the surrounding field, by means of the mysterious displacement current. He did not develop this idea fully, but he spoke of currents as one might consider railway lines: the lines go *as guides* from London to Edinburgh, but nothing *in* them makes the journey.[3]

The manner in which current did (or might) travel along wires was Heaviside's main professional concern, in connection with the development of telephone systems and especially the submarine telegraph. One of his major contributions, in 1876, was to formulate correctly the 'telegraph equation' to represent the process within a perfectly insulated uniform cable with capacitance C, self-inductance L and resistance R (the first two names his own): the voltage V at time t and at the point distant x along it from the source of propagation satisfied

$$V_{xx}/C = LV_{tt} + RV_t. \tag{15.14.1}$$

Among previous failures, William Thomson, an important pioneer in submarine telegraphy, had allowed his enthusiasm for Fourier (§10.19) to let him propose in the 1850s that the equation (10.11.4) for heat diffusion represented this phenomenon also, and so he had overlooked self-inductance.

Heaviside's solutions and manipulations of (15.14.1) and related equations and expressions had a great impact on

[3] I owe this point about Heaviside to discussions with I. Catt, for whom displacement current is an unnecessary and indeed indefensible concept.

the design and deployment of cables. He predicted, for example, that electromagnetic induction would increase the strength of the signal, and proposed that induction coils be inserted at intervals along the cable to act as boosters. He studied the form that propagated waves would take, and sought means to minimize distortion of the signal. But his methods of working were often as eccentric as their author, and led to formidable problems for his mathematician contemporaries.

15.15 *Another English operator: Heaviside's 'broad and bold' methods*

{Lützen 1979; CE 9.11}

Especially from the late 1880s, Heaviside revived the English liking for operator methods. We saw in §9.13 that it had become quite an industry in mid century with Boole and others, but by the 1870s it had fallen away. Like his predecessors, Heaviside took a differential equation to be an nth-degree differential operator polynomial Y upon the sought function $F(t)$, which he expressed formally as

$$F(t) = \frac{1}{Y(p)} f(t), \quad \text{with } p := \frac{\mathrm{d}}{\mathrm{d}t}, f \text{ given.} \qquad (15.15.1)$$

By partial-fraction expansions and the like he found many properties. His most general result, named after him, reads, in the notation of (15.15.1) and somewhat cleaned up, as follows:

THEOREM 15.15.1 (Heaviside's expansion theorem). *Let $t>0$, $f(t)=1$ and $F^{(r)}(0)=0$ for $1 \le r \le (n-1)$. If the zeros of $Y(p)$ are a_r, such that*

$$Y(p) = \prod_{r=1}^{n} (p - a_r), \qquad (15.15.2)$$

$$\textit{then } F(t) = \frac{1}{Y(0)} + \sum_{r=1}^{n} \frac{\exp(a_r t)}{a_r Y'(a_r)}. \qquad (15.15.3)$$

(This result was already known in its essentials to Cauchy, in the context explained in §9.11, and to some others.) This expansion assumed that the a_r were distinct and non-zero; he extended it and other theorems to cover repeated zeros, variable coefficients, other functions f, and systems of ordinary equations and partial equations such as (15.14.1). Cases where f was discontinuous were of especial concern, for his interest in energy slabs, and in the initial propagation of signals, led him to introduce step functions $S(t)$, which are zero for $t \leq T$ (fixed) and take a constant non-zero value thereafter. In addition, differential equations of second order and higher led him to fractional differentiation (that is, p^m when m is not an integer), for which he built upon work by Liouville and others to produce series expansions in p involving the gamma function (5.18.4) in the coefficients. He also formed operator versions of Fourier series and special functions: for example (one of his, and in his operator notation),

$$\frac{2}{\pi}\sum_{r=1}^{\infty}\frac{1}{r}\sin \pi r x = \left[\coth \Delta - \frac{1}{\Delta}\right]1, \quad \text{with } \Delta := d/dx. \tag{15.15.4}$$

Heaviside's proofs of this and other results used power-series expansions of basic functions, in forms grounded in Arbogast's operator form (9.11.1) of Taylor's series. So he often produced (power) series, frequently divergent; the 'bold and broad' world of 'physical mathematics', rather than the 'narrow' regime of rigour. Thus, while he checked his solutions by numerical calculation or re-derivation using more orthodox methods when possible, much mystery remained. The search for more rigorous theory was left to contemporaries; but the main answer, in terms of the Laplace transform, was not to emerge for decades (§16.13).

15.16 *Matter, motion, aether: properties of gravitation* {EMW V2}

The symbiotic relationship between mechanics and physics was brought to an especially close convolution by

Maxwell's theory and its alternatives. Flux and displacement in/of the aether avoid the mysterious action at a distance; but they lead to mysteries of their own, for which the motions of the supposed electrons or ions might provide a middle way for theorizing {van Lunteren 1991, Chap. 8}. Such issues aroused much concern, especially during the 1880s, among several of the main figures mentioned above and many others too.

This closing section of this chapter picks out examples related to properties of the gravitational constant G, the factor which linked force F and masses m and M distance r apart according to Newton's inverse-square law:

$$F=\frac{GmM}{r^2}. \tag{15.16.1}$$

The mathematics involved was usually based upon potential theory, in connection with solutions of Laplace's equation and/or the determination of equipotential surfaces; but various other techniques were used, including numerical approximations. The findings and speculations of greatest mathematical interest may be set out as follows.

1. A variety of attempts was made to *calculate the value of* G. Some arose in connection with determining the masses of planets and satellites {EMW VI2,22}; others were direct efforts, either static (for example, deviation from the vertical of a bob) or dynamic (swinging of a pendulum). An experiment by Poynting in 1891, using very large bodies, also exemplifies the rise of statistical thinking in physics (§15.4); for, impressed by the ways in which quantities always seemed to vary, he held that G was a fiction and regarded his work as leading to a calculation of the mean density of the Earth.

2. The *independence of G from properties of bodies* was a major question. Poynting's experiment bore upon the magnitudes of their masses; the interpretation of the law for infinite masses was pondered, and even negative masses entertained. Other pertinent properties of bodies included their chemical and crystalline structure, and temperature.

3. The *effect of absorption in bodies* upon the transmission of gravity had been considered by Laplace in 1825, in his *Traité de mécanique céleste*: he had argued for multiplying the right-hand side of (15.16.1) by e^{-ar}, where a was the positive absorption coefficient of the body in question. This suggestion was recalled in a range of considerations of the effect of other bodies intervening in the transmission.

4. Of the laws assuming *action at a distance but replacing the Newtonian* (15.16.1), the most popular were Weber's (§15.10) and a variant due to Riemann, both dating from the 1860s:

$$P = \left(\frac{GmM}{r^2}\right)\left(1 - \frac{F(r)}{c^2}\right) + O\left(\frac{1}{c^3}\right), \tag{15.16.2}$$

where $F(r) = \dot{r}^2$ (Weber) or $F(r) = \dot{x}^2 + \dot{y}^2 + \dot{z}^2$ (Riemann), (15.16.3)

and (x, y, z) marks the relative coordinates of m and M. Cases were argued for each of them, and linear combinations were also proposed.

5. The possibility of *drag* through the aether (or the medium, for the agnostic) was a related question, which united mechanics with electromagnetism. Various mechanical models of transmission were proposed: vibrations, pulsations, energy loss, and the separation of molecules and atoms in the aether. Lorentz's contraction hypothesis (§15.13) belongs to this context.

6. A *degree of freedom* in (15.16.1) was troublesome. Double m, M and r; then 4 cancels out top and bottom, so that F and G are left unaltered. How then are we to distinguish the original situation from any other specified by such common multiples?

It is fitting to leave the last word to Larmor, not only for his unusually strong historical interest but especially for his efforts in 1900 to sum up the state of the art. In his book *Aether and Matter* {Larmor 1900}, and also in an address to the British Association for the Advancement of Science on

'The Methods of Mathematical Physics', he advocated as usual the principle of least action and electron theory; but he also rehearsed various other ideas, especially different kinds of aether modelling, vortex atoms, energy principles, equations to represent the passage of matter through a medium, and consequences of Michelson and Morley's negative finding. On this last point we might view him with the hindsight of the young Einstein in 1905 (§16.25); but, without denigrating his importance, it would be an error of historical determinism to assume that his distinguished predecessors were waiting for Einstein to happen.

16

The new century, to the Great War and beyond

16.1 *Scope of the chapter*

The main narrative of this book ends with this chapter, which therefore has some special features. For the first time since §6, a chapter covers all areas of the mathematics developed in its period. However, we cannot stop brusquely at the Great War with all the accounts: for some, 1914 is a convenient stopping-point, but many run through the War period or even beyond. In addition, a few developments up to fairly modern times are mentioned, and supplemented by passages in the next chapter. However, as was stated in §1.6, no attempt is made to capture developments in mathematics after the War, since the story is far too vast and complicated (§17.1) even to be sketched in the manner followed in this book.

As with so many other areas of science (and also the arts), the range and importance of the achievements effected during this short period is quite astonishing. So this chapter concentrates largely on new mathematical topics and on older ones in which some *substantial* advance or change occurred. The most notable general features are the considerable impact made by set theory and by relativity theory, the rise of mathematical statistics, and a certain increase in interest in axiomatizing theories. Topics which continued to advance largely within established frameworks, such as many parts of analysis, geometries, mechanics and mathematical physics, are at most treated only briefly.

To atone for the omissions, some new topics are picked up from their background in the late 19th century – integral equations and functional analysis, for example. Among the available historical literature, the *Encyklopädie der mathematischen Wissenschaften* continues to play a sterling role for nearly (§16.6) all branches of mathematics;[1] indeed, its own publication is part of the story, as we shall soon see.

[1] Another valuable source is the second edition of Ernesto Pascal's *Reportorium der höheren Mathematik,* written by many authors and published by Teubner in several volumes between 1910 and 1929. The original

(*Continued overleaf*)

16.2 *Institutions and communities: the rise of the USA*

In the late 19th century, a spirit of international collaboration became evident in the mathematical community. However, it seems that the Franco-Prussian War of 1870–1 had prevented the leadership from falling on either France or Germany, which were the leading mathematical countries. So the mantle arrived, to some extent, on the island of Sicily. Italy was an important country in its own right; in addition, the Palermo geometer G.B. Guccia (1855–1914) created the Mathematical Circle of Palermo and made it the prime society in the world. When he died it had a membership of around a thousand, far more than any other mathematical society could boast, and its *Rendiconti* was one of the leading mathematical journals (Brigaglia and Masotto 1982).

Another important example of international collaboration was the series of International Congresses of Mathematicians. Beginning in Zürich in 1897, it was followed three years later in Paris (as one of many meetings for academic disciplines held there that year, in connection with the Paris Universal Exhibition), then continuing every four years at Heidelberg, Rome and Cambridge. Georg Cantor was one of the initiators; the War destroyed the dream of his friend Gösta Mittag-Leffler to hold the next one in Stockholm in 1916. The bitterness of the War led to German mathematicians being prevented from participating in 1920 and 1924; normality returned only at the Bologna Congress in 1928, but the numbering of Congresses was abandoned.

Table 16.2.1 lists principal new figures who emerged during the period of this chapter, and indicates the branches in which they initially worked. Since some of the new topics had started in the late 19th century, the inclusion here of a few figures such as Vito Volterra is a little tardy.

---

version (1898–1900), written by Pascal (an Italian analyst and algebraist), had soon been translated into German. Some additional information on the period of this chapter can be found in (Temple 1981) and (Pier 1994).

In Germany, cohorts of new figures emerged from Berlin and Göttingen (where David Hilbert directed dozens of doctorate students over the years), but also some with different backgrounds, for example Albert Einstein. The *Mathematische Annalen* was perhaps the most dynamic journal, with Hilbert and Felix Klein among its editorial board members. The list under France shows the new generation of mathematicians who appeared in Paris from the mid-1890s, centred on Jacques Hadamard and Emile Borel, and enamoured of set theory and its uses in mathematical analysis (§16.7–§16.14).

Some countries were new entrants to Western mathematics; for example, the *Tohuku Mathematical Journal*, started in 1905, was the most important initiative in Japan. But the most striking novelty was the rise of the USA {Parshall and Rowe 1994; CE 11.10}. The presence of Englishman J.J. Sylvester at Johns Hopkins University in Baltimore from 1876 to 1884 had provided some stimulus, including the founding of the *American Journal of Mathematics*; the *Annals of Mathematics* improved the picture on its launch in 1884. When Bryn Mawr College for women opened in 1885, a doctoral programme for mathematics was in place, and the Englishwoman Charlotte Angas Scott (1858–1931) went over there to teach. But few notable American mathematicians had emerged before the 1890s.

However, a change began in 1888, when the New York Mathematical Society was founded, and three years later launched a *Bulletin* presenting a 'Historical and critical review of the mathematical science'. In 1894 its national status was recognized when it converted to the American Mathematical Society; the *Bulletin* was suitably renamed, and a *Transactions* for longer papers began in 1900 and soon gained European as well as American authors. Although New York was a major city, Harvard, Chicago and Johns Hopkins Universities were perhaps the most significant centres, with E.H. Moore at Chicago a pivotal figure.

Although a late starter, the rise of the USA was astonishingly rapid. Probably it was a consequence of the definitive

TABLE 16.2.1 ***Principal new figures, 1900–1914, and areas of initial research interest.***

Each mathematician is listed under the country in which he (or, for G.C. Young, she) worked during this period. The country of birth is indicated when different.

In the headings below, 'ana' refers to real and complex variables in general (including the calculus of variations and potential theory); 'deq' includes integral as well as differential equations, 'har' covers Fourier analysis and harmonic analysis, 'geo' incorporates differential geometry, and 'set' treats both the topological and general sides of set theory.

alg	algebras	mec	mechanics
ana	analysis	num	number theory
deq	differential equations	phy	(mathematical) physics
geo	geometries	pro	probability and/or statistics
fun	functions and/or measure theory	set	set theory and/or logic
har	harmonic analysis	top	topology

Britain

Eddington, S. (1882–1944) mec phy
Hardy, G.H. (1877–1937) har num
Littlewood, J.E. (1885–1977) har num
Russell, B. (1872–1970) set

USA

Bateman, H. (1882–1946) Britain har
Birkhoff, G.D. (1844–1944) ana pro
Bliss, G.A. (1876–1951) ana geo
Bôcher, M. (1867–1918) alg ana
Dickson, L.E. (1874–1954) num
Huntington, E.V. (1874–1952) alg set
Moore, E.H. (1862–1932) geo set
Osgood, W.F. (1864–1943) ana fun
Veblen, O. (1880–1960) geo set
Wedderburn, J. (1882–1948) Britain alg

Italy

Burali-Forti, C. (1861–1931) geo set
Enriques, F. (1871–1946) geo set
Levi-Cività, T. (1873–1941) geo mec
Volterra, V. (1860–1940) ana deq fun mec

France

Baire, R. (1874–1932) fun har
Borel, E. (1871–1956) fun pro
Couturat, L. (1868–1914) set
Denjoy, A. (1884–1974) fun set
Fatou, P. (1878–1929) fun har
Fréchet, M. (1878–1973) har top
Hadamard, J. (1865–1963) deq mec set
Lebesgue, H. (1875–1941) fun
Montel, P. (1876–1975) fun
Valiron, G. (1884–1955) deq fun

Germany

Bernstein, F. (1878–1956) pro set
Blumenthal, O. (1876–1944) fun
Carathéodory, C. (1873–1950) ana deq fun
Dehn, M. (1878–1952) alg geo top
Hamel, G. (1877–1954) mec
Hausdorff, F. (1868–1942) fun set
Landau, E. (1877–1938) ana num
Planck, M. (1857–1947) mec phy
Prandtl, L. (1875–1953) mec phy
Schmidt, E. (1876–1959) fun har set
Schur, I. (1875–1941) alg fun num
Schwarzschild, K. (1873–1916) mec phy
Steinitz, E. (1871–1926) alg
Weyl, H. (1885–1955) phy set top
Zermelo, E. (1871–1953) phy set

TABLE 16.2.1 (*continued*)

Hungary

Féjer, L. (1880–1959) ana fun har
Haar, A. (1885–1933) alg ana har
Riesz, F. (1880–1956) deq fun har
Riesz, M. (1886–1969) deq har

Others

Brouwer, L. (1881–1966) Netherlands set top
Einstein, A. (1879–1955) Germany/Switzerland geo phy
Fischer, E. (1875–1954) Austria/Bohemia har
Fredholm, I. (1866–1927) Sweden deq har
Hahn, H. (1879–1934) Austria deq fun top
Plancherel, M. (1885–1967) Switzerland fun har
de la Vallée Poussin, C. (1866–1962) Belgium har ana
von Mises, R. von (1883–1953) (various) mec pro
Young, G.C. (1868–1944) Britain/Switzerland fun har
Young, W.H. (1863–1942) Britain/Switzerland fun har

formation of the country during this period (the last ten contiguous states joined the Union between 1889 and 1912) and the great expansion in engineering, the size of cities, and schooling provided by public funds. German mathematics was its single main incentive, and connections with Klein important; many Americans took doctorates at German universities.

In Britain the status of mathematics teaching at Cambridge was improved by a reform of around 1906 which the analysts A.R. Forsyth and E.W. Hobson encouraged (and E.J. Routh resisted). The new generation was to benefit, in the form of G.H. Hardy and J.E. Littlewood, whose collaborative work in analysis and number theory from 1912 to Hardy's death in 1947 is one of the greatest in the history of mathematics.

Another striking British partnership is that of Grace and William Young, the first husband-and-wife team in research mathematics {Grattan-Guinness 1972}. They worked in set theory and analysis, from 1900 to 1925, but there the similarities with Hardy and Littlewood end: unable to find proper recognition in Britain, they lived most of their life in Switzerland, living off William's previous earnings as a Cambridge coach and their investments. In addition to numerous papers and an important book (§16.9), they also produced six children, of whom L.C. Young is a major mathematician in his own right. Grace is of note also as the first woman to hold a doctorate awarded at Göttingen, in 1895, under a Prussian programme to encourage women into higher mathematics; 30 years later, her mathematician daughter Cecily could receive only the title of a doctorate from Cambridge University . . .

16.3 *Projects with Klein and others: the* Encyklopädie

Klein was in his prime at Göttingen as an organizer during this period, and the Leipzig publishing house of Teubner was kept extremely busy. His largest enterprise was the *Encyklopädie der mathematischen Wissenschaften mit Einschluss ihrer Anwendungen* – to quote its full title, with the explicit 'inclusion of its applications'. It was launched in 1894 in connection with the German Mathematicians' Association as a detailed report on all areas of mathematics at the time. An academic commission was formed, with representatives from the Association and from several major Academies in Germany (but not Berlin) and Austria; its first President was Klein's former student the algebraist Franz Meyer (1856–1934). The *Encyklopädie* was divided into six Parts, each with its own editor(s) and publishing histories:

1; on arithmetic and algebra (editor Meyer): around 1,200 pages, 1898–1904;

2; on analysis (co-editor Heinrich Burkhardt, author of the wonderful report on mathematical methods cited in §14 and §15): 3,050 pages, 1900–27;

3; on geometry (co-editor Meyer): 5,100 pages, 1903–34;

4; on mechanics (co-editor Klein): 3,200 pages,1901–35;

5; on physics (editor Arnold Sommerfeld: §15.1): 3,250 pages, 1903–26;

6; section 1, on geodesy and geophysics: 950 pages, 1905–22;

6; section 2, on astronomy (co-editor Karl Schwarzschild): 2,200 pages, 1905–34.

The total is around 19,000 pages, but it is not only the quantity that shines: many articles were the first of their kind on their topic, and several are still the last and/or best. Some of them also contained much information on the deeper historical background, although no author discussed general questions of historical method.

French mathematicians soon began to prepare their own translation and elaboration of the project, as the *Encyclopédie des sciences mathématiques,* put out by Gauthier-Villars with Teubner. All Parts were started, and several of the revisions were very remarkable: an example from numerical mathematics was mentioned in §12.10, while some of the articles on set theory and functions were so good that Arthur Rosenthal (1887–1959) translated them back into German as an extra article {EMW IIC9} in the second Part, with his own additions. But the death in 1914 of the general editor Jules Molk, and the general circumstances of the War, led to the collapse of the *Encyclopédie* around 1920, with some articles literally stopping in mid-sentence at the end of a signature of 32 pages (for example, Pareto's in §14.5). The published pages were reprinted during the 1990s by the French house of Jacques Gabay.

Although a few British and American authors contributed to the German original, no English edition was prepared. The Youngs started translating articles on analysis, but met with total apathy from their compatriots.

The Part of the *Encyklopädie* on mechanics was the last to be completed, in 1935, when Constantin Carathéodory was President. A second edition had recently started, with foundations of mathematics, algebra and number theory as the initial branches; but the effects of the Second World War finished it off soon afterwards with little published.

The original edition had been planned to take a seventh Part on history, philosophy and education; unfortunately not a line appeared. But Klein's other major activity lay in this area. At the Rome Congress in 1908, he helped launch an International Commission for Mathematics Education {Grattan-Guinness 1993}. Up to 1920, nearly 200 books and pamphlets were produced, and over 300 reports; they treated not only developments in many countries but also themes which are of topical interest even today, such as teaching mathematics to girls. The activities were reported on a regular basis in the Swiss journal *L'Enseignement Mathématique*, which was founded in 1899.

In connection with this Commission, from the early 1900s Klein discoursed with great insight on various branches of 'elementary mathematics from an advanced standpoint'. These discourses were made available as lithographs, and then in print in three volumes {Klein 1925a, 1925b, 1928}. During the Great War he also lectured on several aspects of 'the development of mathematics in the 19th century', which led to two more volumes {1926, 1927}. All these works are still in print (two of them in English translation), and deservedly so. In his last years he received various awards, including an honorary doctorate from Berlin University of all places (the Faculty for Philosophy and Law . . .).

Historical work continued unabated, especially in German-speaking countries. Gustav Eneström's journal *Bibliotheca mathematica* reached a zenith with its third series (1900–15) of 14 volumes, although he himself rather tainted the picture with excessive criticisms of Moritz Cantor's monumental *Vorlesungen* on mathematics, the fourth and last volume (§6.1) of which, covering 1759–99, appeared in 1908. Cantor himself continued to edit until 1913 his series of *Abhandlungen* on history; many valuable pieces appeared, some of book length. The Danish historian J.L. Heiberg made perhaps the most sensational discovery of the period: in connection with his new edition of Archimedes, he identified in 1906 an important unknown manuscript (*The Method,* discussed in item 3 in §2.21). Among other figures, Pierre Duhem worked intensively at the history of medieval mechanics and astronomy, much of it unfinished at his death in 1916, though published afterwards.

Various other editions of works were prepared during this period: those of Descartes and Fermat excited admiration but also controversy concerning historical methods. In 1911 there began to appear the edition of the works of Euler which Eneström had been planning for years; it is still in progress (§5.16).

16.4 *Hilbert's various problems*

To mark the new century (or, to the mathematically informed, to mark the end of the old one), Hilbert decided to present to the Paris Congress of 1900 a review of current progress and open problems. His lecture was given on the morning of 8 August, in the section on the history and bibliography of mathematics, with Moritz Cantor in the chair. It was the star event, and the text was soon published and also translated into French and English (in the USA). He spoke on 10 problems; the published version contained 23, covering set theory, some aspects of differential geometry and the calculus of variations, Lie groups, algebraic and analytic number theory, algebraic geometry and functional analysis. They provided a good focus of attention for several decades; solutions, or attempts at them, were to enrich other branches of mathematics as well as their own. Some of the problems which were soon successfully attacked are mentioned in this chapter; but major progress on many has occurred only since the Second World War, and a few are still open {Alexandrov 1971, which also contains Hilbert's lecture}.

However, even granted that he did not intend to offer a complete list of major problems, Hilbert's coverage of the mathematics *of his time* was rather disappointing. Some gaps reflect the state of the mathematical profession of the time: for example, no questions dealt specifically with probability and statistics, or with engineering mathematics. Even then, it is surprising that the enigma of Maxwell's equations (§15.11–§15.13) was not on his list; two related problems are too broad to focus on it (the 6th, the 'mathematical treatment of the axioms of physics'; and the 20th, on 'general boundary-value problems').

Some of Hilbert's own work of these years, when he was in his early forties, dealt with his 2nd problem, 'the consistency of arithmetical axioms' {CE 5.5}. After noting the 'geometrical way' of presenting arithmetic by showing number as points on a line, and the 'genetic way' in terms of basic laws, he turned to his preferred 'axiomatic way'. Building on his success in 1899 with the foundations of geometry and arithmetic (§13.18), he presented his axioms in a formalist way, in order to show that a contradiction could not arise in the resulting theory. At the Heidelberg Congress in 1904 he gave a version of the Peano axioms for the integers with the notions '1' and '=' as primitive and '*f*' as the successor operator, and then argued that the theory based on the sequence of signs 1, 11, 111, . . . and the relation '<', together with given methods of inference, could not lead to contradictory consequences.

But the presentation was rather underwhelming: the metatheory was not clearly isolated from the theory, and the logic was largely left to look after itself {Peckhaus 1990}. Hilbert abandoned the line of investigation until the late 1910s, when a more developed version of his 'metamathematics' was to clarify his position and enrich the philosophy of mathematics in general (§17.10). During the 1900s the most stimulating work of this kind was produced in the USA, where E.V. Huntington, Moore and Oswald Veblen extended their innovations in model theory, originally inspired by Hilbert's work on geometry (§13.18), to study a variety of axiom systems from algebra, logic and geometry {Scanlan 1991; CE 5.8}.

16.5 *Too general set theory: Zermelo on a paradox and axioms* {CE 5.4}

It is curious that Hilbert published rather unclear papers, for during the summer semester of 1905 he gave a marvellous lecture course on 'Logical principles of mathematical thought', in which he handled the logic much more carefully, and presented axiom systems for both arithmetic and geometry {Hilbert 1905}.[2] He also reported 'Zermelo's paradox', which may be stated as follows. Consider sets in general. Some may belong to themselves, such as the set of all sets; some may not, such as the set of all oranges. Now consider the set Z of all sets which do not belong to themselves; it easily follows that Z belongs to itself if and only if it does not do so.

This double contradiction really *was* a paradox. Ernst Zermelo seems to have found it in 1899 (his 29th year), but he did not publish it, and his achievement came to light only in the 1970s. A strong supporter of Hilbert's emphasis on axioms, he thought that the axiomatic method was the best way to avoid such paradoxes and to give set theory the broad foundation for mathematics that Cantor had envisioned, but without Cantor's reliance on mental acts. He presented his axiom system in a paper of 1908 in *Mathematische Annalen*. Some axioms justified elementary operations, such as the union of sets and the formation of subsets.

[2] An edition of this lecture course by Hilbert is in preparation. It survives in two transcripts by students; one of them was the 22-year-old Max Born, who later did rather well in physics (§16.28).

Among others, the important axiom of 'separation' was meant to avoid paradoxes by limiting the propositional functions *f* which were permitted to define sets; yet, in the rather casual Göttingen philosophy of logic, he required *f* only to be 'definite', without clarifying that it could be defined only by a finite number of logical connectives and quantifiers.

16.6 *Too general set theory: Russell on paradoxes and logic*

{Rodriguez-Consuegra 1991; CE 5.2–5.3}

This lacuna in Zermelo's system was soon filled, by Hermann Weyl in 1910; but it had already been spotted by Bertrand Russell. On the morning of 3 August 1900, at a session of the International Congress of Philosophy in Paris which had immediately preceded that for Mathematicians, the 28-year-old Russell had heard papers by Peano and his three main disciples. Their way of both axiomatizing theories and dressing them up in logic and set theory gave him just the tools he had been seeking for some years for a general foundation for mathematics. Upon learning their system, he noted with understandable surprise that they had failed to develop a comprehensive logic of relations; dismissing with grand disregard the substantial achievements in this area of the algebraic logicians (§9.13, §13.4), he developed broadly the same theory in Peano's and Cantor's terms. (These similarities were to be demonstrated in a doctoral thesis written in 1913 by an American youth, Norbert Wiener (1894–1964) {Grattan-Guinness 1975}.) Now Russell could formulate his ambition more precisely.

We saw in §12.13 that the 'Peanists' (as they were known) had kept mathematical and logical notions separate; however, the boundary-line was not clearly drawn, especially where set theory itself was concerned, and Russell imagined that none existed. His 'logicist' thesis (not his name) asserted that *all* mathematical theories could have their concepts defined within this enriched logic, and also that all their methods of proof and inference could be formulated there also. General set theory and point-set topology were essential components, although the former was shorn of Cantor's idealistic aspects.[3]

The mathematical basis of Russell's logicism lay in his definitions of cardinal numbers as sets of similar sets. They were the same as Gottlob Frege's (§12.14), which he had not yet studied: 0 as the set of the empty set, 1 of unit sets, 2 of duos, and so on, including (for him) Cantor's transfinite cardinals. Like Frege, he distinguished between zero, the empty set, and literally nothing, an important foundational matter which had perplexed predecessors, even Cantor (§12.15). He also defined finite and transfinite ordinal numbers in terms of sets of well-ordered sets. He founded set theory in his logic by defining all sets in terms of propositional functions and relations (the set of integers from 'is an integer', and so on), and quantification.

Russell then proffered a definition of irrational numbers, so that the continuum of real numbers was formed. Quite an amount of mathematics could now be developed from these bases, as he showed in 1903 in the book *The Principles of Mathematics*. After laying out his logic and set theory, he covered finite and transfinite arithmetic; the continuum and irrational numbers; projective, descriptive and Euclidean geometries; and some elements of dynamics, including Hertz's formulation (§14.4). Then, in 1905, a theory of definite descriptions furnished a clear means of defining single-valued mathematical functions ('*the* so-and-so') in terms of propositional functions.

However, bad news had come with the good. In 1901, Russell had found Zermelo's paradox for himself. Unlike Zermelo, he published it, in his book; since then it has been

[3] A voluminous literature of variable quality has developed over Russell's work. The principal source of information on Russell studies is the journal *Russell* of the Russell Archives in Hamilton, Canada. The most important source for understanding Russell's progress is Volumes 2–6 of the edition of his *Collected Papers*, currently in preparation and now published by Routledge of London. Most of the volumes on logic and philosophy are published.

known after him. Aware as a logician of the prime importance of paradoxes, he found others, creating some by re-reading results of predecessors. Two principal ones concern the supposedly greatest ordinal and greatest cardinal number N, for each of which

$$N = N, \text{ but also } N > N; \tag{16.6.1}$$

Russell named these paradoxes after Cesare Burali-Forti and Cantor, respectively, neither of whom had viewed his own result as paradoxical {Garciadiego 1992}.

Over the rest of the decade, and in collaboration with his former teacher A.N. Whitehead (§13.21), Russell tried various methods of avoiding – and, as a philosopher, Solving – all the paradoxes that he had collected, and yet preserving as much mathematics as possible. The final answer was provided in a vast work called *Principia mathematica*, published in three volumes between 1910 and 1913. The paradoxes were avoided by a theory of types, in which propositional functions and relations were divided up in such a way that the corresponding sets could be members *only* of a set of sets, . . ., and individuals only of 'simple' sets. The theory was not only very complicated; it severely compromised the universality of logic. So Russell's hope of a Solution was not really satisfied.

In addition, despite its 1,900 pages, the coverage of mathematics in *Principia mathematica* was much less than in *The Principles*. Most of the details were confined to a marvellous and detailed presentation of Cantorian set theory and transfinite arithmetic; but, for example, mathematical analysis was never reached. Whitehead had started a fourth volume on geometries, but later he abandoned it; however, only some parts would have been covered. Thus the book would still have been a long overture to an unwritten logicist opera. We have lamented in this book the large gap between logic and mathematics: in claiming that no gap existed, Russell went too much in the other direction.

Not surprisingly, the impact of logicism upon mathematicians was rather slight (§17.9). In particular, the

French mathematical community abhorred logic, with Henri Poincaré especially sarcastic; only Louis Couturat took a serious interest in it, and become very isolated as a result. There is no article on logic in their *Encyclopédie* – but then, there is none in the parent *Encyklopädie* . . .

16.7 *Controversies over choices*

{G.H. Moore 1982; CE 5.4}

Russell called one of his axioms 'multiplicative', for he came across it, in 1904, when defining the product of an infinitude of numbers his way in terms of sets of sets. It permitted him to take an infinite set of sets, choose one member of each set independently of the other choices, and treat the collection of chosen members as a set on a par with the original ones. The axiom gave him special difficulties, because it did not fit easily within logicism; for his logic always used only *finitely* long expressions, which seemed to be too short to state the infinity of independent choices that the axiom permitted {Grattan-Guinness 1977}.

As with Russell's paradox, this axiom was also proposed by Zermelo: he found a form of it independently, with his friend Erhard Schmidt (1876–1959), and published it in 1904 in *Mathematische Annalen*. Calling it later the 'axiom of choice', which became its standard name, he and Schmidt needed to prove another outstanding query of Cantor's, which Hilbert had attached to his first problem in 1900: the well-ordering principle, that the members of *any* set could be laid out in the same order as that of the positive ordinals (12.5.3).

The axiom aroused much controversy from its publication in 1904. The following questions attracted especial attention:

1. Is the infinity of independent choices a legitimate mathematical procedure? Is the choice set defined, or merely constructed? Answers: depends upon one's attitude to the nature of mathematical objects.

2. Different forms of the axiom were found; are they in fact logically equivalent, and are some philosophically more acceptable than others? Respective answers: yes, and yes for some figures concerning certain issues (for example, on whether the choices are made simultaneously or successively).

3. If acceptable, where in mathematics is the axiom needed? Answer: in many places in set theory and mathematical analysis, and also in other branches of mathematics.

4. Can theorems using the axiom be re-proved without it? Answer: sometimes yes, often no; some theorems, especially the well-ordering principle itself, turned out to be logically equivalent to the axiom.

5. Does it lead to paradoxes of its own? Answer: well, certain consequences were surprising (§16.9).

Temperature and tempers over the axiom gradually cooled, with a variety of attitudes being adopted; no general compromise was reached. For example, Borel followed a constructivist view of mathematics rather like that of Kronecker (§13.3): he worked only with sets put together by at most a denumerable number of unions and complementations ($A \cup B$ and $A - B$), so to him the axiom was unacceptable. By contrast, his colleague Henri Lebesgue shared Zermelo's content over its need, and even used it quite casually.

16.8 *The fruitfulness of Cantor's* Mengenlehre

Borel had been an early contributor to the diffusion of set theory in the mid-1890s (§12.16). He applied it to complex-variable functions in his doctoral thesis of 1894 (his 23rd year), where he also threw off a remark which was later recognized as an important pioneering example of a covering theorem: if every point on a line can be covered by a denumerable number of subintervals, then in fact a finite number could do the job. Arthur Schönflies unfortunately gave the name 'Heine–Borel theorem' to this result.[4] The adjective 'compact' became attached to sets or spaces of this kind (§16.12).

Borel's constructivistic view of sets was taken up a decade later in a more severe form by the Dutchman L.E.J. Brouwer. From around 1910 he reworked point-set topology without recourse to the law of excluded middle or acceptance of Cantor's transfinite ordinals beyond the second number-class. Soon afterwards, he began to apply this 'intuitionist' philosophy (his unfortunate name) to mathematics in general {CE 5.6}; luckily, though ironically, he did not apply it to his own work on topology (§16.17).

One of Brouwer's targets was Cantor's most ardent supporter, Schönflies. He, and other followers such as the Felixes Bernstein and Hausdorff, also paid much attention to the further reaches of Cantor's sequence (12.5.3) of transfinite ordinals, and to cardinal exponentials (that is, numbers of the form a^b defined by his method of covering (§12.15), where a and/or b are transfinite). In particular, various efforts were made to prove his continuum hypothesis (12.15.6), which Hilbert had put forward as his first problem. Yet that result was not to be proved during this period; its status as undecidable (that is, neither true nor false) relative to Zermelo's or others' axiom systems was not to emerge until the 1960s.

The publicity for all aspects of Cantor's *Mengenlehre* engendered at the successive Congresses was reinforced in two books published just before the War. Schönflies's *Entwickelung der Mengenlehre und ihrer Anwendungen* (1913), the second edition of his report, was the culmination, in his early sixties, of his long support for Cantor. The superb *Grundzüge der Mengenlehre* (1914) by the 46-year-old

[4] Schönflies's name for this theorem {Schönflies 1900, p. 119} is unhelpful. Heine had proved in 1872 a typical Weierstrassian theorem involving no set topology, that a function which is continuous over a finite interval is also uniformly continuous there; its covering technique is the trivial one of dividing the interval into a finite number of subintervals.

Hausdorff was dedicated to Cantor, then in his seventies and to die in 1918.

The next six sections of this chapter deal with the application of set topology to real-variable mathematical analysis {Kline 1972, Chaps. 44–7}. French mathematicians feature very prominently {EMW IIC9, the retranslation into German of their articles}. The chief innovative work was Borel's *Leçons sur la théorie des fonctions* given recently at the Ecole Normale and published in 1898 (his 28th year) by Gauthier-Villars. Its success and importance led the publisher to invite him to direct a 'collection of monographs on the theory of functions'. Complex-variable analysis is considered in §16.20.

16.9 *From integral to measure*

{Hawkins 1970; CE 3.7}

In his book, Borel sought to generalize the notion of integral by treating it as the 'measure' function of a set E, a non-negative real number $m(E)$ satisfying three properties:

1. that the measure of a line would be its usual length;
2. that two congruent sets had the same measure; and
3. that the measure of a finite or denumerable union of mutually disjoint sets was the sum of the component measures.

This still left open the definition of measure itself, which Lebesgue provided in his doctoral thesis of 1902 and elaborated two years later in the Borel collection, as his *Leçons sur l'intégration* delivered at the Collège de France. Equivalent definitions were soon found independently by the Italian mathematician Giuseppe Vitali (1875–1932) and by the Youngs, whose *Theory of the Sets of Points* (1906) was the first book on point-set topology and measure in English.

We recall from FIG. 12.6.1 that Riemann's definition of the integral of $f(x)$ over $[x_0, x_n]$ of x worked by sandwiching it between upper and lower sums, and that Gaston Darboux had expressed the procedure in more formal terms. In 1892, Camille Jordan had restated this procedure in terms of sets: basically, the 'outer content' of E was given by the smallest possible area of a finite number of rectangles that could contain E, the inner content was given by the largest area of those rectangles contained by it, and the integral existed if the two contents were equal. Lebesgue went further in FIG. 16.9.1, by taking the set P_r of points x such

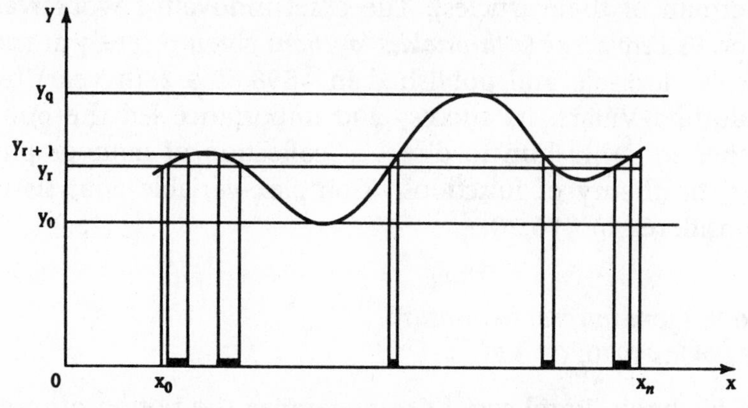

FIG. 16.9.1 *Diagram for the Lebesgue integral.*

that $y_r < f(x) \leq y_{r+1}$ and covering it with a finite *or denumerable* number of intervals; then he defined the 'outer measure' $m(P_r)$ of P_r, Jordan-style, and its 'inner measure' as $((b-a)-m(P_r))$, and defined it as 'measurable' if the two contents were defined and equal in value. Then $f(x)$ was said to be a 'measurable function' if every set P_r were measurable, and its measure-type integral was given by

$$\int_{x_0}^{x_n} f(x)\,\mathrm{d}x := \lim_{r\to\infty} \left(\sum_r y_r m(P_r)\right). \qquad (16.9.1)$$

Lebesgue's major innovation was that, instead of partitioning the interval $[x_0, x_n]$ of x, he divided up $[y_0, y_n]$ of $f(x)$, where the discontinuities and turning values of f were occurring. He did not provide any diagram; for the very simple function in FIG. 16.9.1, each set P_r is composed of intervals, and his procedure comes into its own for highly 'misbehaving' functions, with point-set topology rather than geometry providing the required tools. Some functions of this kind which had no Riemann integral were now integrable: for example, Dirichlet's characteristic function of the rationals (§8.7) over $[x_0, x_n]$ now took measure zero. Extensions to infinite intervals and values of $f(x)$, and to multiple integrals, were made.

In the 1910s, Arnaud Denjoy became especially concerned with generalizing Riemann's definition of the integral, and also with extending the differential calculus by means of point-set topology; Grace Young was one of his followers in the latter enterprise. The basic theorems of the differential and integral calculus were reworked using the new broader definitions. They included the property that the sum of the measures of a convergent infinite series of functions equalled the measure of their sum. It turned out that far weaker conditions were needed to justify the basic processes of analysis (for example, integrating an infinite series term by term) than the uniformity often needed in Weierstrass's theory.

Of the two extensions of Weierstrass's approach to analysis, the weakening of conditions was far more useful: one was much more likely to meet non-uniformity than exotic functions. So it is very surprising that the latter was the main motive for the creation of measure theory. Another feature which surprised contemporaries themselves was that the axiom of choice could be used to construct sets which did not have a measure under the above conditions. Vitali found one in 1905, and in 1914 Hausdorff went further when he presented in the *Mathematische Annalen* his so-called 'paradox': a hemisphere could be made congruent with a third of its parent sphere {Czyz 1994}. Critics of this axiom had good reason to fidget.

16.10 *Functions, series and convergence*

{Medvedev 1993; CE 3.3}

One matter considered by Borel in 1898 became the subject of the doctoral thesis written by René Baire in 1899; he developed it five years later as *Leçons sur les fonctions discontinues* in the Borel collection. Consider the convergence

$$F(x) := \lim_{r\to\infty} [f_r(x)] \quad \text{for all } x \text{ within } [a, b]; \qquad (16.10.1)$$

if the f_r were a sequence of Fourier series of r terms, say, then F might be discontinuous. Baire used this fact to classify functions by the limit process, taking continuous ones as the zeroth class. (Weierstrass's approximation theorem (item 5 of §12.5) provided the class below it, for a sequence of polynomials took a continuous function as limit.) Functions definable by (16.10.1) formed the first class, and might be discontinuous for a perfect set of values of x; those similarly definable by a sum of first-class functions were the second class; and so on, with a function in each class defined as the limit of functions from *any* lower class. He also imitated Cantor's specification (12.4.1) of derived sets up to $P^{(\infty)}$ by admitting an ωth class of functions defined by (16.10.1), where each f_r belonged to the rth class; then, as with Cantor's sequence (12.5.3) of transfinite ordinals, one could go still further.

Like Borel and Lebesgue in measure theory, Baire assumed no particular kind of expression for functions. However, examples such as Peano's (12.3.6) for the characteristic function of the rationals raised the question of the range of possible expressions for functions, and in 1905 Lebesgue exhibited both examples for each Baire class, and also one that lay outside the classification.

Another fashion of the time was finding further examples of continuous non-differentiable functions, to add to Weierstrass's (§12.3). A geometric method of generating one by an iterative procedure (FIG. 16.10.1) was provided in a paper by Mittag-Leffler's student Helge von Koch (1870–1924). Seen at the time as a nice curiosity, its use of self-repetition is recognized today as heralding fractal sets {CE 3.8}.

These discussions of functions often related to their representation by series such as (16.10.1). Dispensing with

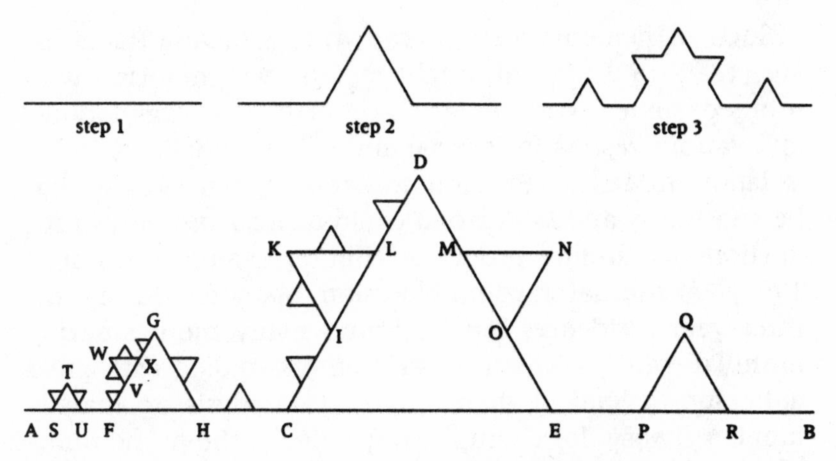

FIG. 16.10.1 *On the way to von Koch's continuous non-differentiable function.*
{von Koch 1904}

uniform convergence was not the only issue; in his *Leçons sur les séries divergentes* of 1901, Borel developed some remarks made in his parent book of 1898 by clarifying the relationship between the various methods of summing series. We recall from §8.5 that Cauchy had banned divergent series as illegitimate in mathematics, and they had remained in exile for most of the century. But upon reading Cauchy, Borel realized that the same series could have different sum*s* for different methods of summation {Tucciarone 1973}. To take a long-known simple example as re-interpreted shortly before Borel by the Italian mathematician Ernesto Cesaro (1859–1906), it *is* correct to claim that

$$1 - 1 + 1 - 1 + 1 - 1 + \cdots = \tfrac{1}{2} \qquad (16.10.2)$$

if the summation is done by arithmetic means; that is, if the 'sum' A of $\sum_r u_r$ is given by

$$A = \lim_{n\to\infty} t_n, \text{ where } nt_n = \sum_{j=1}^{n} s_j \text{ and } s_j := \sum_{r=1}^{j} u_r. \qquad (16.10.3)$$

The latter sum is the usual method of summation, for which (16.10.2) indeed takes no value.

Such refinements were unknown to Srinivasa Ramanujan (1887–1920), a self-taught Indian mathematician who intuited some extraordinary non-orthodox summations, and various results for special and elliptic functions, in his isolation in Madras. He came to Cambridge in 1914 so that he and Hardy and Littlewood could make sense of his productions by finding proofs or counter-examples {Kanigel 1991}. A mutual incomprehension obtained during his three-year residence, but he found many more amazing formulae which his patrons and others studied with great difficulty. Several of them related to analysis or analytic number theory; for example, respectively, the prime number theorem (12.12.3), and the partitioning problem (about the number of distinct ways in which an integer can be the sum of smaller integers).

At less exotic levels of investigation, other kinds of orthodox convergence were studied. They proved particularly useful in the development of function theory from a different point of view, to which we now turn.

16.11 *Harmonic analysis from Fourier series*

{Mackay 1978; CE 3.11}

Fourier now joins Cantor as a father-figure for the next trio of sections, where the new ideas on function, integral and series were applied to some old and new problems in Fourier series:

$$f(x) = \sum_{r=0}^{\infty} (a_r \cos rx + b_r \sin rx), \quad 0 \le x \le 2\pi. \qquad (16.11.1)$$

Lebesgue's *Leçons sur les séries trigonométriques* (1906) in the Borel collection surveyed all classical aspects of Fourier and trigonometric series: Dirichlet's sufficient conditions for convergence (Theorem 8.7.2), Riemann's double-integral function (12.1.1), Cantor's Theorem 12.4.1 on exceptional points of non-convergence, non-orthodox methods of summation (especially with Leopold Féjer), and so on. Now, as a result of Cantor, mathematicians came to see functions as having *sets* of discontinuities, turning-points, or whatever, rather than just *numbers* of them, as in Dirichlet's conditions; the study of Fourier series, and many related matters in analysis, was thus greatly enriched {Cooke 1993}.

Harmonic analysis can be seen as an extension of Fourier series. The name is due to William Thomson, in connection with his harmonic analyser described in T and T′ and elsewhere (§14.10), he was alluding to (16.11.1) as the sum of simple solutions of a differential equation. The phrase came now to refer to any such expansion, such as those of the special functions, with the relationship between the sum function and its components as a main concern.

The theory of function spaces took this approach further, independent of physical interpretations. Equation (16.11.1) was understood as an example, in this case infinite, of the linear combination

$$ax + by + \cdots + cz(+\cdots) = e \tag{16.11.2}$$

which we found at (13.20.1) to be so popular in recent algebras. Each trigonometric function was like a base vector, orthogonal to all the others by the properties

$$\int_0^{2\pi} {\sin \atop \cos} rx {\sin \atop \cos} sx \, \mathrm{d}x = \pi\delta_{rs} \text{ and } \int_0^{2\pi} \sin rx \cos sx = 0 \tag{16.11.3}$$

(where I use the Kronecker delta (§13.5) for convenience). But one part of the analogy did not hold: these were *infinite* expansions, so that straightforward ideas about a finite number of dimensions were too naive, and the 'space' which the trigonometric functions and their sum functions occupied would have to be analysed. Convergence was also considered, especially the convergence of series of functions; and the Fourier coefficients were defined by the integrals (8.7.2), so that integration played a role. Point-set topology and measure theory provided ideal tools for these and many related studies.

16.12 *Fréchet and the rise of the 'functional calculus'*, 1906 {Siegmund-Schultze 1982; CE 3.9}

These areas of research became very fashionable during the 1900s; many of the European figures listed in Table 16.2.1 participated, some as a main activity. A few main results are outlined here.

One advance was to generalize (16.11.1) to any set of functions $f_r(x)$ orthonormal over some interval $[a, b]$: given constants c_r and

$$s_n(x) := \sum_{r=0}^{n} c_r f_r(x) \text{ over } [a, b], \text{ with } \int_a^b f_r(x) f_s(x)\, dx = \delta_{rs}, \tag{16.12.1}$$

then entertain the possibility that there is a function $F(x)$ such that

$$F(x) = \sum_{r=0}^{\infty} c_r f_r(x)\, dx \text{ over } [a, b], \quad \text{with } c_r = \int_a^b f_r(x)\, F(x)\, dx, \tag{16.12.2}$$

and wonder what properties F may (not) have in common with its base functions f_r. Maurice Fréchet seized on these questions and more in his doctoral thesis in Paris, which appeared in the Palermo *Rendiconti* as {Fréchet 1906}, when he was 29 years old. The opening pages are striking for the very simple way in which he explained the scenario. Here is the start {p. 4}:

> Let us consider a set E formed from any elements whatever (numbers, points, functions, lines, surfaces, etc.) but such that one knows how to discern distinct elements. To each element A of this set let there correspond a determinate number $U(A)$; we thus define what we shall call a uniform *functional operation* in E.
>
> The study of these operations is the subject of the *Functional Calculus*.

This name came from Volterra's compatriot Salvatore Pincherle, who also had been working on function spaces. He had emphasized algebraic aspects, especially operations $U(A)$ which were distributive over the set of A's. His work, which he reported in the *Encyklopädie* in 1905 as {EMW IIA11}, helped revive interest in functional equations, which had slumbered since the English work decades earlier (§9.12).

Fréchet focused more on the topological aspects. In particular {pp. 6–7},

> a set is *compact* when it contains only a finite number of elements or when each infinitude of its elements gives rise to at least one limit element. When a set is at once compact and closed we shall call it an *extremal* set [. . .]

In general, he replaced Cantor's emphasis on limits, with its notion of the limit point of a set, by topological notions such as the neighbourhood of an element and compactness. These and other of his ideas were heavily exploited, and have remained standard; extremality is now usually known as 'sequentially compact' {Taylor 1983}. Here is an influential sequence of examples from that time.

Spaces of functions were defined by appropriate properties. Popular for its relevance to Fourier series was this one:

$$G(x) \text{ is } \textit{square-integrable over } [a, b] \text{ if } \int_a^b (G(x))^2\, dx \text{ is finite,} \tag{16.12.3}$$

and belongs to the L^2-*space* of functions. Properties of its subspaces, and relationships with other spaces (such as L^p for $p \neq 2$) became a major concern.

In this context, Parseval's formula (8.7.4), hitherto known as a nice summation for Fourier series, gained new significance in this form:

$$\sum_{r=1}^{\infty} c_r^2 = \int_a^b (F(x))^2\, dx; \tag{16.12.4}$$

for, if true, it showed that, like the $f_r(x)$, $F(x)$ belonged to L^2 and so made it 'complete'. An important and pioneering converse theorem was found independently by Ernst Fischer and Frigyes Riesz in 1907:

THEOREM 16.12.1 *Assign to each function* $\mathrm{f_r(x)}$ *a constant* $\mathrm{c_r}$. *Then* $\sum_r \mathrm{c_r^2}$ *converges if and only if there exists a square-integrable function* F(x) *satisfying* (16.12.4).

Two years later, Riesz showed that integrals were the only functionals that took continuous functions onto the real numbers and satisfied the property of linear combination; extending this 'representation theorem' (as it became known) excited much interest {J.D. Gray 1985}.

New forms of summation of series were defined; not only cases such as (16.10.2) by arithmetic means, but, in particular, ones involving measure. Further modes of convergence were defined. For example, Fischer introduced this one in proving the above theorem:

$$s_n(x) \to f(x) \text{ over } E \text{ in mean square} \quad (16.12.5)$$
$$\text{if } \int_E [s_n(x) - f(x)]^2 \, dx \to 0 \text{ as } n \to \infty.$$

In addition, much use was made of convergence 'almost everywhere'; that is, to all values of x over an interval or set E except for one of measure zero (under whatever definition of measure was used, and therefore whatever other properties such sets possessed). Among many theorems was one relating convergence almost everywhere of a sequence of measurable functions over E to its uniform convergence over a measurable subset of E. Proved by Borel and Lebesgue, it was found again in 1911 by the 42-year-old Russian mathematician Dmitri Egorov (1869–1931), and marked the international début of an important school of function theorists in Moscow. His follower Nikolai Lusin (1883–1950) soon appeared on the scene {Phillips 1978}.

By the early 1910s, enough material was available for book-length presentations. Volterra wrote up his version for the Borel collection as *Leçons sur les fonctions de lignes* (1913), based on lectures given at the Paris Faculty of

Sciences: 'line function', an interest since the 1880s, was his name for a functional whose values depended upon those of a collection of functions. Hausdorff's *Grundzüge der Mengenlehre* of 1914 (§16.8) included an account of 'point sets in general spaces' which used a definition of neighbourhood which is known after him. He also explored connections between compact sets and Borel's theorem (§16.8) on finite coverings.

16.13 *Integral equations and infinite matrices*

{EMW IIC13; CE 3.10}

As we saw in §8.7, an essential companion to Fourier's series was his integral theorem, in forms such as

$$\text{if } g(q) = \int_{-\infty}^{\infty} f(p) \sin pq \, dp,$$

$$\text{then } 2\pi f(x) = \int_{-\infty}^{\infty} g(q) \sin qx \, dq; \qquad (16.13.1)$$

it gained new life in the new century as an important example of an integral equation with a known solution. Once again, such equations had arisen from time to time in the 19th century, often in connection with Fourier analysis, special functions or potential theory with Dirichlet's problem, and sometimes with variable limits of integration. However, there was no general theory; solutions were often sought by differentiating back to get a differential equation.

Motivated by a problem in hydrodynamics, Volterra had taken up integral equations in the 1880s. He focused on two linear equations: the 'second kind' (as Hilbert was to call it), to determine f from

$$g(x) = f(x) + \int_a^b K(x, p) f(p) \, dp, \qquad (16.13.2)$$

with g and K known, a and b constants; the 'first kind' lacked the term in f alone, and so was an extension of (16.13.1). After some success with solutions by successive approximations, his great innovation came in 1896 when he made the classic Leibnizian move of treating the integral as an infinite sum, forming the corresponding sequence of equations and trying to solve them. This brought him to the theory of infinite matrices {Bernkopf 1968}. It was not completely new; indeed, yet again it goes back to Fourier: before finding the coefficients in (16.11.1) by the method of integration term by term, he had differentiated it successively to form an infinity of linear equations and found them by a wild routine. The theory had popped up occasionally thereafter; among others, with Sonya Kovalevskaya in the 1870s (§14.15) and Poincaré in the 1880s, soon followed by von Koch. Now it gained a new level of importance.

Mittag-Leffler's former student Ivar Fredholm lectured on these equations regularly at Stockholm, and wrote an important paper in 1903 for *Acta mathematica*. Hilbert heard about them too late to make up a 24th problem for his Paris lecture, but he devoted much of the decade to an intensive programme of research, placing six substantial papers with the Göttingen Academy between 1904 and 1910. He concentrated on solving (16.13.2) under various classical assumptions about K, such as it being a continuous function or symmetric in its variables; but from his fourth paper, of 1906, he also made much use of infinite matrices, applying known theory about quadratic forms (§13.5) to make reductions to sums of squares. He used ideas from functional analysis, such as convergence (16.12.5) in mean square; indeed although somewhat unfairly to others, the L^2 space in (16.12.4) is often named after him.

Hilbert's most active student of these equations was Schmidt. His doctoral thesis of 1905 led to an important theorem known after them both and heralding the 'spectral theory' (their name) of these equations; it gave an expansion of g in the equation of the first kind in terms of functions defined from K and the latent roots of the associated matrix. Then in 1908 he pioneered the idea of treating the coefficients as members of a space and using its properties to assess whether or not the system of equations could be solved. He also made progress on non-linear equations.

One issue concerned the generality of solutions, where measure theory and functional analysis came in useful. The Fourier integral theorem (16.13.1) itself needed a general proof; one was supplied by the young Swiss mathematician Michel Plancherel in 1910 in the *Rendiconti*, in the context of a general study of orthonormal functions and their expansion properties such as the Riesz–Fischer theorem.

Publicity for integral equations came aplenty from late in the decade onwards: a tract by Maxime Bôcher for Cambridge University Press (1909), and reports by Harry Bateman for the British Association {the superb Bateman 1911} and by Hans Hahn for the German Mathematicians' Association (1911). As with function spaces, Volterra gave a lecture course at the Paris Faculty of Sciences and wrote it up for the Borel collection in 1913, this time as *Leçons sur les équations intégrales*; in the same year Frigyes Riesz contributed a volume on infinite matrices. Another substantial book was a Teubner reprint in 1912 of Hilbert's sextet of Göttingen papers.

Also studying integral equations at this time, Hilbert thought in 1904 that he had defused the torpedo (14.3.3) that Weierstrass had launched against methods of solving the Dirichlet problem in potential theory, by treating the required minimal surface as an exercise in the calculus of variations (his 19th problem in 1900). However, both Hadamard and Lebesgue found restrictions to his form of solution.

Another context was Heaviside's broad and bold Theorem 15.15.1 for electromagnetism. It was finally replaced by a rigorous version in which the Laplace transform finally came to light after much gestation {Lützen 1979}. The solution of F in

$$f(p) = \int_0^\infty e^{-qp} F(q)\,\mathrm{d}q$$
$$\text{or } f(p) = \int_0^\infty q^{-p} F(q)\,\mathrm{d}q,\ p > 0, \tag{16.13.3}$$

was given by a complex-variable integral containing f, with its contour determined by the properties of f, especially its poles. There were some beneficial side-effects, such as new means of summing divergent series (§16.10). Bateman made some progress towards the solution, following work by another former Mittag-Leffler student, the Finn R.H. Mellin (1854–1933), after whom the second integral in (16.13.3) is known. However, a full general theory did not emerge until the mid-1920s, and it did not enter regular use until 30 years after that! {CE 4.8}.

16.14 *From function spaces to algebras*

The presence of algebras of various kinds increased during the 1910s and afterwards. Infinite matrices provide a fine example, for some properties differ from those of finite matrices (for example, conditions for the existence and uniqueness of an inverse matrix). Normal matrix theory was still in the doldrums, although it did appear in some textbooks on determinants. Bôcher included it with determinants and invariant theory in a nice *Introduction to Higher Algebra* (1907) based upon his teaching at Harvard University. But textbooks were infrequent until the 1930s, and the *general* teaching of the subject began only in the second half of the century.

Another American, Moore, was developing his own approach to functional analysis mixed in with touches of model theory. Starting out from analogies between integral equations and infinite matrices, he developed a 'general analysis' as an underlying theory. He drew heavily on 'these theories of Cantor, [which] are permeating Modern Mathematics' {E.H. Moore 1910, p. 2}, and also on Fréchet's topological formulation of functional analysis.

A later important book on function spaces, to succeed Volterra's, appeared in 1922 in the Borel collection. Written by Volterra's follower the Frenchman Paul Lévy (1886–1971), it carried the title *Leçons d'analyse fonctionnelle* – the origin of the name 'functional analysis'. This name was a natural extension of Pincherle's 'functional calculus', and continued the same concern with operators. The approach was adopted from the late 1920s by John von Neumann (1903–1957), with a variety of applications made to quantum mechanics. The 1928 International Congress at Bologna gave the subject a fine airing, with lectures delivered by Volterra, Hadamard, Fréchet and Hilbert; presumably Congress organizer Pincherle was responsible for this publicity.

16.15 *Poincaré on the algebra of topology*

{Bollinger 1972; EMW IIIAB3; CE 7.10}

The stress placed by Fréchet and others on topological aspects of analysis was part of the rise of topology in general. Hilbert's 16th problem dealt with the topology of algebraic curves and surfaces (§16.18), while the 18th posed the issue of tiling a space with congruent polyhedra.

The autonomy of topology from both geometry and algebra (§13.17) increased when Poincaré published in 1895 an extraordinary paper on 'Analysis situs' in the surprising location of the *Journal de l'Ecole Polytechnique*, followed up to 1904 by a quintet of 'complements' in various other journals. They are all reprinted in Volume 6 of his *Oeuvres* {Poincaré 1953}, which is cited by page number in the selected highlights below.

An important difference of tradition is involved in Poincaré's work. The topology described in the sections above grew out of the Weierstrassian tradition of rigour and the use of Cantor's set theory; by contrast, Poincaré used his more intuitive style, as in his study around 1890 of the three-body problem (§14.18), in tune with Riemann's and Klein's more intuitive way of working with functions and surfaces (§13.13–§13.16).

The first of Poincaré's innovations was to define a surface in n dimensions from p equations in n variables, and its interior in terms of q inequalities:

$$F_i(x_1, \ldots, x_n) = 0,\ 1 \leq i \leq p(<n), \tag{16.15.1}$$

$$\text{and} \quad \Phi_j(x_1, \ldots, x_n) > 0,\ 1 \leq j \leq q(<n).$$

All functions were assumed to be continuous and to have continuous derivatives, giving a smooth domain; Weierstrass's approximation theorem implicitly gave backing from analysis. The set of points so defined was called a 'variety' if the pth-order Jacobian determinants of the partial derivatives of the F_i were never zero for the same set of values of the x_k, for otherwise the variables would not be independent.

Upon this basis Poincaré built a theory of the bounded varieties V and their boundaries B, and their continuous deformation into 'slightly different' ones; two varieties that could be so related were 'homeomorphic' (meaning 'of the same shape') {p. 199}. He defined an 'opposite variety', produced if the order of two of the equations in (16.15.1), or two of the independent variables, were exchanged; its Jacobian would also change sign. His main idea concerned the decomposition of B into a collection v_r of continuous varieties of $n-1$ dimensions; he expressed this situation by the relation, in his notation {p. 206},

$$`k_1v_1 + k_2v_2 + \cdots + k_nv_n \sim 0' \qquad (16.15.2)$$

which he called a 'homology'. Each k_r was an integer, negative or positive according to whether v_r was opposite or not; its absolute value gave the number s of times that varieties which differed slightly (or not at all) from v_r could be found on B. If no relation existed between the v_r, they were 'linearly independent'.

It is striking that we see yet again the form (16.11.2) of linear combination. Poincaré also developed an algebra for (16.15.2), with methods for adding and multiplying the relations, and conditions for the existence of a homology and for the independence of a variety from others. In the latter context he defined the 'Betti numbers' (his name, in honour of Enrico Betti's work (§13.13) on the connectivity of surfaces of n dimensions) P_r of C_n, from the maximal number $P_r - 1$ of linearly independent varieties of dimension $r \leq n$ of C_n {p. 271}. Then he took Riemann's notion (13.13.1) of the genus of a surface, which had been defined from a re-reading of Euler's formula (8.18.1) relating the numbers of vertices, edges and faces of a polygon, and generalized it to this definition of the 'characteristic' $N(C_n)$ of C_n:

$$N(C_r) := P_0 - P_1 + \cdots + (-1)^n P_n.$$

$$\text{Further, } P_r = P_{n-r}, \ 0 \leq r \leq n \qquad (16.15.3)$$

when C_n was closed (that is, topologically similar to Euclidean space).

In the first complementary paper, placed in the Palermo *Rendiconti* in 1899, Poincaré began a new combinatorial approach to topology by assuming that a variety could be covered by a collection of triangles in two dimensions, and of polyhedra for higher dimensions {p. 294}. The simplest way of expressing his procedure starts out from dimension zero, where a simplex is a point: in one dimension it is a line; in two dimensions a triangle, with its boundary composed of its three simplicial sides in order (the reversed order gives the opposite boundary); and so on. He also considered continuous changes into curvilinear polygons and polyhedra. The relation (16.15.2) was now extended by replacing '0' with the complex 'C_n' of n dimensions, and being read as composed of a 'chain' of simplexes v_r, each one counted as before. Now '0' itself was a complex with no boundary: for example, the perimeter of a clockface divided into a chain of 12 simplexes by the markers of the hours taken clockwise (or into the opposite chain when taken anticlockwise).

In these papers Poincaré put out various conjectures. One which is still unsolved in general occurs in the fifth complement, of 1904 {p. 475}: that in three dimensions any variety of a certain kind is homeomorphic with a sphere.

16.16 *The rise of topology: some responses to Poincaré*

{Dieudonné 1989; EMW IIIAB13}

This theory was presented by Poincaré with his usual mixture of amazing insight and irritating vagueness: for example, the proof of the duality property $P_r = P_{n-r}$ (16.15.3) was incomplete. In reaction to its wondrous innovations, algebraic topology gradually but steadily gained attention. A number of Europeans writing in German picked it up; an important example is the article of 1907 on 'Analysis situs' in the *Encyklopädie der mathematischen Wissenschaften* {IIIAB3} by Max Dehn and his Danish friend Paul Heegard (1871–1948), cited at the head of §16.15. They took the algebraic aspects beyond laws to the abstract group structure, and also introduced the words 'homotopy' (for Poincaré's 'homeomorphy') and 'isotopy', for the properties that two complexes could be transformed into each other by a sequence of transformations of certain kinds.

The next year the Austrian mathematician Heinrich Tietze (1880–1964) wrote an important long paper on 'topological invariants', in which he strongly emphasized algebraic structures {Tietze 1908}. He introduced the adjective 'orientable' to replace Poincaré's overly geometric talk of 'one-' and 'two-sided' varieties. He also guessed, in contrast to Poincaré, that a homology could not be decomposed in a structurally unique way; later workers took it as a 'central conjecture' on the algebraic structure of homologies, and it led Dehn in 1912 to a guess about groups which is described in §16.19 below.

Elsewhere in his paper, Tietze treated knots from the topological point of view. This topic had appeared from time to time, including in two surprising contexts. One related to T and T′: Thomson's idea of vortex atoms (§15.10) had led Tait to classify many types of knot. The other source, pioneered in lectures by the Austrian analyst Wilhelm Wirtinger (1865–1945), was functions of *two* complex variables; Riemann's theory of surfaces, which could accommodate the many values of a function of one variable with sheets and helices above the base plane (§13.13), was extended to 'Riemann spaces', in which the two surfaces could literally tie themselves together. Knot theory flourished in the mid-1920s, with cases like FIG. 16.16.1, which shows the trefoil knot for such a function {Epple 1995}.

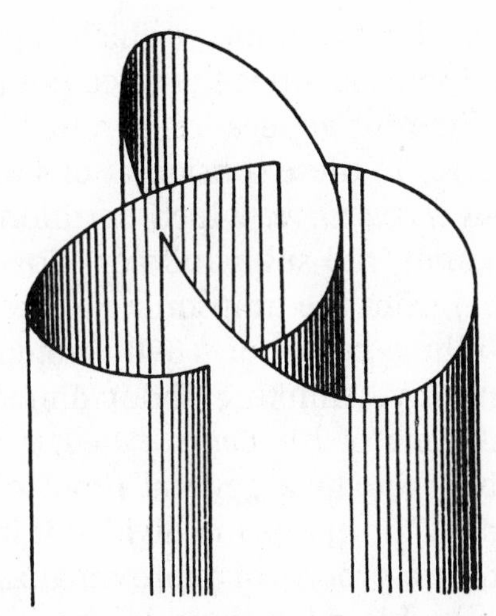

FIG. 16.16.1 *The trefoil knot for a function of two complex variables.*
{Artin 1925, Article 6}

16.17 *Brouwer on complexes and dimensions, 1908–12*
{Johnson 1979, 1981}

Brouwer was another major figure in the rise in topology during this period. When in his late twenties he published a suite of difficult though fundamental papers, mainly in *Mathematische Annalen*. Luckily, he ignored his own philosophical demands that mathematics should be confined to his intuitionistic limitations (§16.8). Here are three main features, with some immediate prehistory.

1. Taking up Klein's programme (§13.9) of studying invariants under groups of transformations, Brouwer studied continuous ones which mapped an n-dimensional complex one–one onto itself. This led him in 1911 to a *fixed-point theorem*: that in such a transformation at least one point is left unmoved. Extending such theorems to still more general domains such as function spaces became a major concern; Birkhoff was an early student. Among applications, one which had partly motivated Brouwer himself and received much attention afterwards was the existence of a solution to an ordinary differential equation (§12.9): treat the equation as a transformation of a continuous

function, then under conditions (which have to be established) the solution function is the fixed point in a function space. A quite different application is noted in §16.22.

2. A major factor in these considerations was the concept of the *continuous closed curve,* which partitioned a manifold (he used this name) into submanifolds. However, Cantor's mapping (12.5.1) from the unit square to the unit line, and Peano's space-filling curve of 1890 (noted at (12.5.1)), showed that intuitive thinking about dimension was fallible. In the editions of his *Cours d'analyse* (§12.16) from 1887, Jordan had sought a general proof of the theorem that such a curve C on the plane divided it into an interior and an exterior, by a process of approximating to C by polygons analogous to his definitions of the inner and outer measure of the integral (§16.9). Schönflies sought a more general proof of the 'Jordan curve theorem' (as it soon became known), using point-set topology.

However, {Brouwer 1910}, published in *Mathematische Annalen,* contained counter-examples such as FIG. 16.17.1, where the plane was divided into three regions by the maze-like closed curve to which they form the common boundary. They were distinguished by colour: white, black hatching, and the one marked with 'PP' and 'QQ' in red hatching. Brouwer's own proof of Jordan's theorem used his new topological ideas; in its most general form, the theorem now stated that a closed curve C of dimension n which divided a manifold M of $n+1$ dimensions into two submanifolds mapped continuously onto another such curve (thus the one-dimensional Betti number of $M-C$ was 2).

3. This work involved disagreements with Schönflies, and also Lebesgue, on another inheritance from Cantor: *the invariance of dimension under transformation* {CE 7.11}. Introducing in 1911 the terms 'simplex' and 'complex' used earlier, and defining a notion called 'mapping degree', he characterized invariance in terms of the non-existence of a continuous 'simplicial mapping' between simplexes and complexes of different definitions. A more general proof of

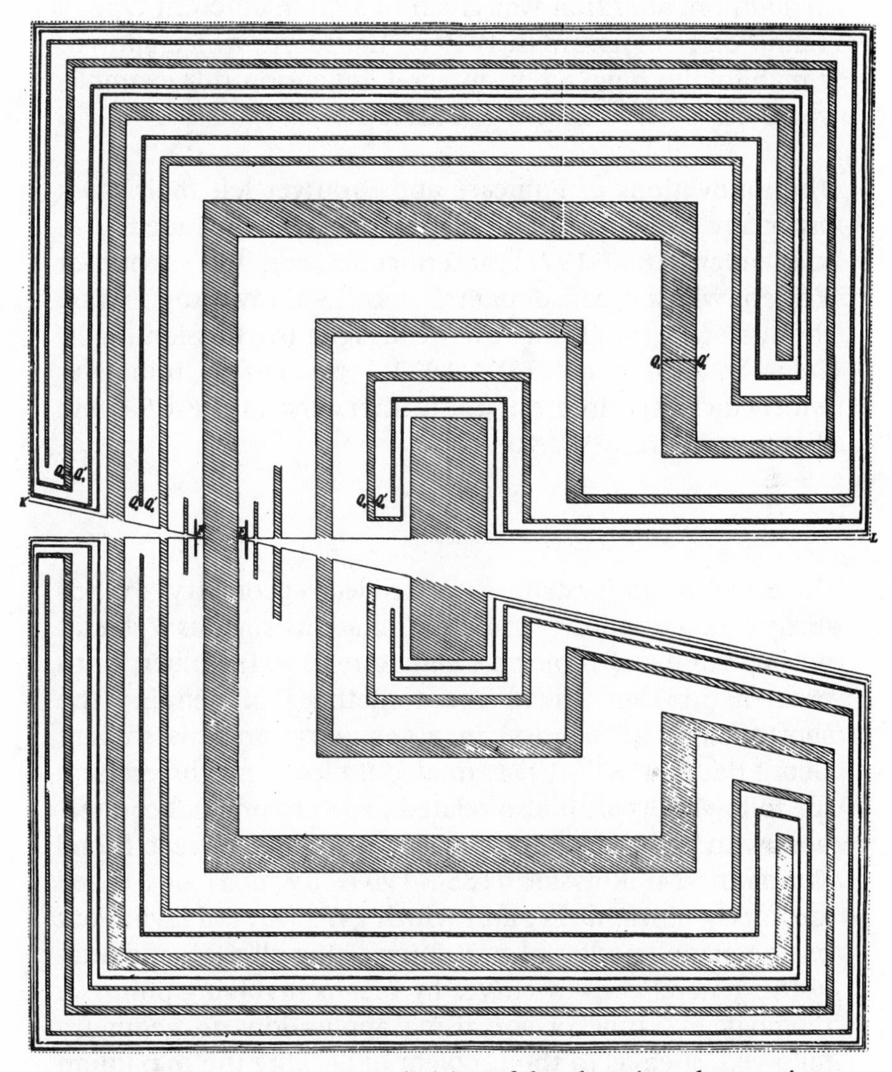

FIG. 16.17.1 *Brouwer's division of the plane into three regions.* {Brouwer 1910}

1913 worked by defining dimension recursively in terms of dividing up a manifold of a given dimension by closed curves into submanifolds of lesser dimensions. During the 1920s, further work by Brouwer and others both extended the treatment and led to definitions of dimension number.

In addition, attention was given to a quite different type of definition put forward in 1919 by Hausdorff, which allowed a manifold to have a non-integral dimension (for example, von Koch's curve shown in FIG. 16.10.1).

The innovations of Poincaré and Brouwer left their mark especially on the USA. Pioneers there included J.A. Alexander (1888–1971) at Princeton and his supervisor Veblen, whose treatise *Analysis situs* (1922) was the first on the subject. At that time their colleague the Russian *émigré* Solomon Lefschetz (1884–1972) was taking initiatives which increased interest still further, first in the USA and then elsewhere {CE 7.10}.

16.18 *The topology of geometries*

Hilbert's axiom system for Euclidean geometry excited further examinations, and also imitations such as Veblen's on the axioms of projective geometry (1904). Hilbert's 4th problem in 1900 asked about methods of defining the shortest lines (geodesics) in a geometry, and his student Georg Hamel (1877–1954) treated it nicely in a dissertation the following year. It also related to a very original connection with number theory made by Hilbert's great friend Hermann Minkowski (1864–1909), which he called 'geometry of numbers', and which gave ways of laying out on the plane families of non-intersecting ellipses, and analysing general convex curves by means involving point-set topology. A range of surprising applications to geometry followed, such as to the problem of packing the maximum number of circles into a plane region; in addition, Carathéodory led a line of thought in 1913 in functional analysis in which functions were regarded as points in a function space whose convexity was defined by suitable inequalities.

Two more of Hilbert's problems brought links between geometry and topology. In the 16th he wondered about the 'topology of algebraic curves and surfaces', citing German rather than Italian work (which he had trouble in reading) on the matter. The 15th concerned questions such as determining the numbers of curves or surfaces which touched or cut other curves or surfaces in various circumstances. They go back to Apollonios (§2.23) and had continued quietly thereafter, for example with Monge (§8.3). They were revived from the 1870s when the Hamburg schoolmaster Hermann Schubert (1848–1911) generalized and also algebraized the theory into 'enumerative geometry', a study of the number of objects (such as tangent points) common to sets of geometrical objects (such as circles) {EMW IIIC3}. He found various invariant numbers, but not a general theory about their existence – hence the 15th problem. The Italian Francesco Severi (1879–1961) came up with general formulations in 1912; but the topic thrived only from the 1930s, when a new phase of research in topology and abstract algebras subsumed it under algebraic geometry. To the earlier stages of this latter area we now turn.

16.19 *More abstract algebras*

We have seen how abstract algebras entered into other subjects, especially topology. The influence was two-way with Tietze's concern of 1908 about uniquely decomposing a homology, for it helped Dehn in 1912 to wonder in the *Mathematische Annalen* about the generators of a group (that is, those elements whose combinations produced all elements of the group) and relations between them: could each member of an infinite group be produced in a finite number of steps? This question has become known as the 'word problem', because it refers to the order of letters in the 'word' representing the operations which engender the product; it is a kind of decision problem in the modern algebra of logic and computability, important for its bearing upon topics such as encoding and decoding.

This was the first of Dehn's 'fundamental problems' about such groups; two others dealt with transformations of the form utu^{-1} which Charles Babbage had initiated

(§9.12), and with isomorphisms between groups. Among other developments, the structure of the crystallographic groups, mentioned in Hilbert's 18th problem, was not only enhanced but also applied in the early 1910s to studying the diffraction of X-rays {EMW V7; CE 9.17}. The theory of finite groups itself was extensively examined by the Englishman William Burnside (1852–1927); his book *Theory of Groups* (1897, and especially the 2nd edition of 1911) is still a classic text. The theory of group representations (§13.20) was greatly extended by Georg Frobenius's student Isaac Schur (1875–1941), a leading Jewish mathematician of the time.

The study of abstract algebras other than groups, which had grown during recent decades (§13.21), continued to increase in the new century. Those using the linear combination (16.11.2) drew the attention of Joseph Wedderburn (1882–1948). As a student at Edinburgh, he had learnt quaternions as enthused by Professor Tait (§15.9), and after studying in Germany he went to Chicago (where he came under the influence of Moore) and then passed most of his career at Princeton University. His main innovation, from 1907 (his 26th year), was to analyse *directly* the structure of a linear system S instead of following Elie Cartan and others in treating properties of matrices and determinants associated with S (§13.20). This approach led to the modern technical sense of the word 'algebra' mentioned in item 2 of §1.5 to refer to S; it comes from 'linear (associative) algebra' of Benjamin Peirce (§13.4), who was another important influence on Wedderburn. He defined a 'complex' (Poincaré's word after (16.15.3) – in a different context, note) as the set of quantities

$$k_1 x_1 + k_2 x_2 + \cdots + k_n x_n, \qquad (16.19.1)$$

where the x_r belonged to an S and the k_r to a field F, and looked at the properties of an S; he especially favoured an S which possessed Peircean idempotent elements x for which $x^n = x$. He also studied relationships between algebras; his best-remembered theorem states that if S is a division algebra (that is, each $x \neq 0$ has an inverse element relative to an identity element), then it is a field {Parshall 1985}.

However, in the 1900s field theory was still not well explored. The 39-year-old Ernst Steinitz attended to this matter in a huge paper published in Crelle's journal in 1910; one stimulus was Kurt Hensel's introduction (13.19.1) two decades earlier of the (field of) p-adic numbers. The defining properties of combination of elements $a, b, c, \ldots$ under two operations, + and × (I use '*' to represent either) were these: uniqueness and closure, and the laws

$$\text{commutative } a*b=b*a, \quad \text{associative } (a*b)*c=a*(b*c), \tag{16.19.2}$$

$$\text{distributive } a\times(b+c)=a\times b+a\times c,$$

$$\text{units } a*e=a, \text{ and inverses } a*a^{-1}=e. \tag{16.19.3}$$

He also studied various applications, especially this one: given a field of polynomials P with coefficients in F which cannot be factorized into such polynomials of lower degree, under what conditions does an extension of F exist containing all the zeros of P, so that P can be factorized in it down to linear terms? The matter had deceived Wedderburn for one, but it went back, for example, to Kummer's ideal numbers (§9.5). The Cantorian name 'perfect' has become attached to this property; Steinitz did not use it here, although he did examine 'well-ordered fields' at length.

Four years later, the German mathematician Adolf Fraenkel (1891–1965) began to pay the same kind of attention to rings. A ring is an algebraic structure which is a commutative group under '+' and also satisfies

$$a\times(b\times c)=(a\times b)\times c, \quad a\times(b+c)=a\times b+a\times c$$

$$\text{and} \quad (a+b)\times c=a\times c+b\times c. \tag{16.19.4}$$

However, he soon supplanted this concern with a preoccupation with set theory and transfinite arithmetic, topics on which he was to achieve world eminence – from the 1930s as the Jewish mathematician Abraham Fraenkel.

Wedderburn's structural approach was soon picked up by L.E. Dickson, and appeared in various versions later; rather obscurely in the Bourbaki circle of mathematicians from the 1940s, in their ambitious hope to capture 'mathematical structure' {Corry 1996}, and somewhat more clearly in recent times in the theory of categories. In some ways it imitates Lagrange's desire to make mathematics algebraic; as with him (§6.16), success has been only partial. By contrast, the focus on properties of algebras, of all kinds, soon came to dominate abstract algebra; key figures of the 1920s include Emmy Noether (1882–1935) and Emil Artin (1898–1962), whose ideas were codified in the textbook *Moderne Algebra* of 1931 by B.L. van der Waerden, which is still in print in its 9th edition.

Much of this work also nourished algebraic number theory, to which Hilbert devoted problems 9–12 in 1900. The last of these dealt with the decomposition of algebraic number fields, building on Kronecker and Dedekind (§13.3); the new results were given a Hilbert-type treatment (§13.19) in a double report to the German Mathematicians' Association in 1926–7 by Hensel's student Helmut Hasse, which was reprinted by Teubner as the book {Hasse 1930}. He also covered some material relating to another two of these problems: the 9th, on reciprocity laws, and the 11th, on properties of quadratic forms. The 10th problem dealt with solving Diophantine equations; the achievements filled the second of the extraordinary three-volume catalogue-like *History of the Theory of Numbers*, which Dickson published between 1919 and 1923.

16.20 *Complex-variable analysis and differential equations*

Only a few features of a vast area of work can be pointed out here, beyond the spectacular advances recorded in §16.12–§16.13. Some applications appear in §16.23–§16.28.

Weierstrass's approach to complex analysis based on the power series (12.7.1) remained influential: for example, Borel and Hadamard looked closely at properties of analytic continuation and the status of Taylor's series. But the other approaches still bore fruit; in particular, much work in analytic number theory centred upon Riemann's series (12.12.2) and his conjecture about the location of its zeros {EMW IIC8}. In addition, Hilbert's 22nd problem concerned automorphic functions. Klein and Poincaré had tried to elucidate analytic functions; instead of spreading their multiple values around a Riemann surface, even in ways as pretty as Klein's tessellation in FIG. 13.16.1, single-valued representations by parameters were sought. The possibility of so 'uniformizing' them in general (§13.15) was proved in 1907 by Poincaré and also by the German mathematician Paul Koebe (1882–1945) {J.J. Gray 1994; EMW IIC3–4}.

Poincaré's recurrence theorem on the orbits of planets (§14.18) led to a further development in 1915, when the French Academy proposed this prize problem: start with a fixed point on the complex plane with value z, take some function f, and find what kind of set of points is given by the values $z, f(z), f^2(z), \ldots$. The problem had been conceived as an extension of Poincaré's curves of recurrent points; but the papers that were submitted by Pierre Fatou and Gaston Julia (1893–1978) went much further, producing sets of points which are now recognized as important contributions to fractal theory and the study of dynamical systems {Alexander 1994}.

Certain mixtures of Weierstrass's and Riemann's theories of complex analysis, and even reconciliations between them, began to appear, with Cauchy's approach (§8.11) also in place. Fine evidence for this change is given in the article {EMW IIB1} of 1901 on the subject in the *Encyklopädie der mathematischen Wissenschaften*; for its author was W.F. Osgood, one of the many American mathematicians who had studied in Germany (in his case, Klein's Göttingen). His textbook *Functionenlehre* of 1906 (his 43rd year), which ran from set theory through both real and complex analysis (including uniformization) to aspects of potential theory, soon became a classic, and appeared in editions to the fifth of 1928. Several volumes in the Borel collection also kept readers up to date on topics such as analytic continuation and residues.

Around 1900, Osgood also studied the calculus of variations. He was partly inspired by Hilbert's current efforts to establish the existence of optimal values following the principles of Weierstrass (item 6 of §12.3). This was the last of the 23 problems; the 19th and 21st also related to the subject {EMW II8a (by Zermelo and Hahn); Goldstine 1980, Chap. 7}. Some of Hilbert's results bore upon his concurrent work on integral equations (§16.13); the rise of functional analysis also led to profitable interaction, as Hadamard explained in his *Leçons sur le calcul des variations* (1910), edited by Fréchet but not published in the Borel collection.

Partly through Osgood's initiatives, the subject grew in importance in the USA. G.A. Bliss soon came to lead an influential school at Chicago, and in the late 1920s Marston Morse (1892–1977) extended the theory to 'the large' by deploying algebraic topology in the variational analysis of functions in certain kinds of space.

Morse was led to this approach partly by the three-body problem, to which Poincaré made a final contribution in 1912 (the year of his death) in the Palermo *Rendiconti* with a guess that also bore upon variational methods. Hold a flat rubber ring in your hand, and distort it by twisting the inner rim one way in its plane and the outer rim the other way; then, claimed Poincaré, at least two points on the ring will remain fixed. His motivation for the theorem related to his belief in the existence of periodic solutions to the problem (§14.18); but it was of consequence also for the existence of closed shortest curves on convex surfaces, with applications to problems in dynamics. Birkhoff was especially active here; he proved the guess in 1912, and surveyed its uses in a marvellous monograph of 1927 on *Dynamical Systems*.

Paul Painlevé proposed other methods of solving the three-body problem {dell'Aglio 1993}. We recall from

§14.18 that he had considered collisions of the bodies; this possibility was studied in more detail by Levi-Cività, who considered singularities in the solution. His work connected up to another line of thought from Painlevé: that the equations could be formally solved by power series of a complex variable and Weierstrass's way of using analytic continuation. This possibility was realized by the Finnish astronomer Karl Sundman (1873–1949) in papers between 1909 and 1913. However, as Birkhoff and other critics pointed out, such solutions were useless for practical work, and the problem remained unsolved in the sense envisioned by Newton. It still is.

All these studies drew frequently on differential equations. The general theory was enriched in many ways. The distinction between elliptic, parabolic and hyperbolic types of linear equation of the second order was often deployed; Paul du Bois-Reymond's original version (12.9.3), formulated in terms of the sign of the discriminant of the second-order terms, had been extended as follows. By changing its n independent variables if necessary, transform the equation into a sum of squares; then, for example, the hyperbolic type (H) arises if n terms appear but do not take the same sign. Hadamard worked intensively on this type during the 1910s, and presented his findings as *Lectures on Cauchy's Problem in Linear Partial Differential Equations* (1923), delivered at Yale University and published by its Press. He focused upon H because the wave equation (6.4.1) is an example of it; the book followed his lectures of 1903 on hydrodynamics (§14.9). 'Cauchy's problem' had concerned the existence of solutions to *ordinary* differential equations (§8.15); Hadamard's was to specify the initial and boundary conditions to guarantee the existence of a solution of H of the desired type.

16.21 *Probability and mathematical statistics*

{Krüger and others 1987}

For the rest of this chapter, the balance shifts towards applied mathematics. Hilbert's 6th problem was rather vaguely addressed to the 'mathematical treatment of the axioms of physics'; probability and mechanics were placed 'in the first line'. It is strange to us to see probability theory so characterized, but so it was also by Borel. He wished to apply probability theory, especially to physics, and thereby make it an autonomous branch of mathematics; a four-volume treatise to this effect appeared from him between 1925 and 1939. In 1909 he had formulated the notion of probability as a measure with properties similar to those required of the Lebesgue integral; in 1933 his approach was to be converted to an axiomatization by the Soviet mathematician Andrei Kolmogorov (§17.11).

Similar hopes for autonomy were evident in mathematical statistics, especially in the (often fractious!) school that developed around Karl Pearson at University College, London, in the chair endowed by Francis Galton (§12.11). Many of the now standard parameters were introduced or at least popularized by Pearson and his colleagues around the turn of the century: for example, median, 'standard deviation' (his name), some distributions (including the Rayleigh distribution: §1.1), and the correlation coefficient r between two sets of data. Graphical presentations of data by 'histograms' (his name again) and frequency polygons were emphasized. He also proposed in 1900 a 'chi-squared' test for the goodness of fit between observed data and values predicted from a hypothesis.

Among Pearson's followers, G. Udny Yule (1871–1951) introduced in 1900 a parameter Q (for 'Quetelet'), to which his name is still attached, for associating two pairs of properties and their negations (for example: from the four, do smokers tend also to be non-drinkers?). Like Pearson's r, Q lies between -1 (answer 'no') and $+1$ (answer 'yes'), and no association is suggested when $Q = 0$. Yule published the first of many editions of his fine *Introduction to the Theory of Statistics* in 1910.

Other followers made durable contributions. In 1904, Charles Spearman (1863–1945) proposed a method for correlating pairs of sets of data by their respective rankings by order, in a pioneering application of statistics to psychology. In 1909, W.S. Gosset(t) (1876–1937), writing under the pseudonym 'Student' as he worked for the Guinness brewery in Dublin, estimated the mean of the population of data from a small sample of them. In the 1920s his idea was extended by R.A. Fisher (1890–1962) to the 'Student *t*-test' for testing hypotheses; Fisher also examined the role of of 'variance' (his name for the square of the standard deviation).

The school applied its theories to social and biological sciences, and to prosecute their views Pearson founded in 1901 the journal *Biometrika*. One motive was to combat opposition to his use of statistics from influential biologists. Fisher continued the approach, even carrying out some of his own breeding experiments; his textbook *Statistical Methods for Research Workers* (1925) was a classic in this and other areas of statistics, with many editions over 30 years. Nevertheless, the *general* use of statistical methods in biology, and also in psychology and medicine, dates largely from the 1940s {CE 10.9–10.12}.

The Russian school in statistics inspired by Chebyshev (§12.11) also made important contributions {CE 10.6}. One of them was due to his follower Andrei Markov (1856–1922) in 1906. A process S is evolving in successive discrete stages, but relating these stages is in general hopelessly complicated. Markov had the following insight: let the distinct states be represented from the past through the present to the future by the values of a sequence of discrete random variables . . . $x_{-1}, x_0, x_1, \ldots$. Then maybe S can be well represented under the assumption that the distribution of values of x_1 depends *only* upon x_0 (and of those of x_2 only upon x_1, and so on). This sequential arrangement became known as a 'Markov chain' in his honour, partly for the ways in which he proved major results such as the central limit theorem and laws of large numbers (§8.8) for such sequences of values. But his main achievement was to show the versatility of this seemingly special kind of sequence; only specific cases had been formulated by predecessors (for example, the Ehrenfests had modelled diffusion processes in connection with their ergodic hypothesis: §15.4).

The analogous version for continuous variables, now known as a 'Markov process', also gained much attention after the War, in particular from Wiener; and, as with probability itself, Kolmogorov came along in the 1930s with a general formulation. By then, probability and statistics was developing into a profession – and one rather separate from those for pure and applied mathematics (§17.11).

16.22 *Economic control: mechanical or statistical?*
{Mirowski 1994; CE 10.18}

The dominance of analogies from mechanics and physics (§14.5–§14.6, §15.3) continued as before; one of Léon Walras's last writings was the paper 'Economique et mécanique' (1909). The name 'neo-classical' became attached to this approach, with the group around Alfred Marshall at Cambridge exercising much influence.

However, sometimes there was little agreement about which analogies were the most fruitful. A good example is the economic production function; how should it be defined? For example, Marshall's colleague W.E. Johnson (1858–1931) presented it in 1913 in terms of gradient, a function cutting across equipotential surfaces; but other formulations were offered, of quite different characters, some mechanicalish but not all mathematical {Mirowski 1989, Chap. 6}. Apart from the forays of Francis Edgeworth at Oxford, connections with statistics were still avoided; not until the 1930s did this approach gain substantial favour.

A very striking area of non-contact between economics and statistics was production management and logistics. Incredible as it may seem, more than a century passed after the Industrial Revolution before statistical quality control began to be applied generally {CE 10.14}. An important pioneer was W.A. Shewhart (1891–1967) at the Bell Laboratories in the USA, with a wonderful book of 1931 under the ungainly title *Economic Quality Control of Manufactured Product*. Realizing that everything varies all the time, even if by only minute amounts, he presented tests for deciding whether or not a production system was delivering faulty goods with a frequency in excess of chance expectation, and thus whether repairs were called for.

One of Shewhart's followers was W. Edwards Deming (1900–93), a physicist turned demographer who was asked in the late 1940s to advise on the method of voting in the first Japanese election after the Second World War. While there, he also taught Shewhart's theory, married to his own humanitarian ideas on factory and business management; the consequences constitute a major theme in the economic history of our own time.

By contrast, early in this century the American economist F.W. Taylor had advocated methods which were impersonal in character, imposing hierarchical structures in the workplace; they also made little allowance for variation of any kind. They were widely used by the American armed forces during the Great War, and practised then and especially thereafter by Henry Ford in his automobile factories. In reaction, during the Second World War a much freer approach was adopted, to which the name 'operational research' was attached; it inspired (naive) expectations of the merit of planning and organization within a capitalist system, so that the social and economic disasters of the 1920s might be avoided {CE 6.12}.

The launch of operational research was partly linked to the concurrent upsurge of linear programming, in which the optimal value of a linear function of variables is sought subject to linear relationships which define a convex region of possible solutions (FIG. 16.22.1) {CE 6.11}. The needs of the Second World War provided much stimulus (minimizing the cost of transportation of matériel, and so on); so it would have served well during the Great War. Furthermore, Minkowski had then been highlighting convexity (§16.20), and just about enough linear algebra was available to provide effective techniques; a basic theorem on

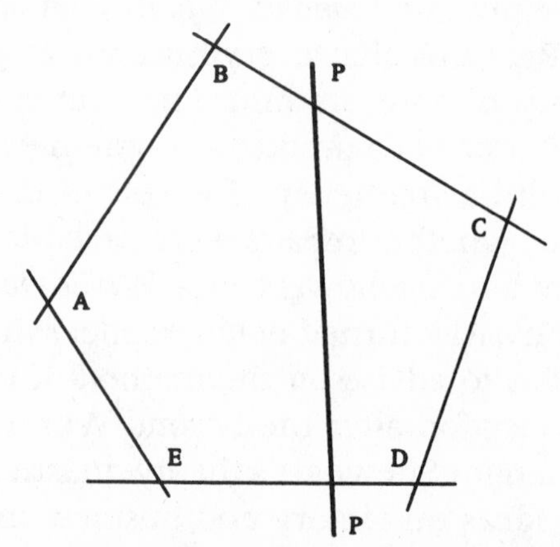

FIG. 16.22.1 *Basic linear programming configuration.* The function represented by the line PQ will take optimal values at the points A and C of the convex region ABCDE.

linear independence was even proved in 1906 by the Hungarian mathematician Gyula Farkas (1847–1930). Yet all these opportunities were missed {Grattan-Guinness 1994}. The delay is especially striking because the basic insight goes back to the 1820s and Fourier, who drew diagrams just like FIG. 16.22.1; he had started out from his formulation of the principle of virtual work in terms of inequalities (§10.2). This version of the principle, and Gauss's 'least constraint' principle of 1829 (§14.4), which also used inequalities, were examined around 1900; yet still the subject slept (§17.22).

16.23 *Mechanics and aerodynamics*

The main context for these missed opportunities just mentioned was mechanics. Its foundations became a popular topic around 1900, especially among German writers {EMW IV1}. In addition to inequalities, issues included formulations of least action and other principles in general curvilinear coordinate systems; by then a wide range of systems, including many special ones, were being deployed in mechanics and elsewhere {EMW IIIAB7}. However, a Hilbert-style axiomatization of mechanics had to wait until the 1920s, and Hamel.

All parts of mechanics flourished in the 1900s. The *Encyklopädie der mathematischen Wissenschaften* provides detailed accounts, and in one case it played an active role, with the influential 200-page article {EMW IV10} of 1911 on the 'dynamic problems of the theory of machines'. Its author, the 28-year-old Richard von Mises, classified machines and instruments according to whether the forces involved were generated by, say, water, steam, friction or electricity. He also noted the newer mathematical ways of analysing mechanical actions, such as by integral equations.

Von Mises included 'air machines' in his survey. The rapid growth in aerodynamics posed many exciting problems for him and other students of hydrodynamics and the stability of dynamical systems {EMW IV17; CE 8.12}; he gave a pioneer lecture course in the subject at the time of his article. Two kinds of problem attracted particular attention. One was the behaviour of air around a wing: the creation (or not) of lift, and the properties of vortices and dangers of turbulence in the wake of its passage. Nicolai Zhukovsky (1847–1921) built up a powerful group in Moscow, as did Ludwig Prandtl (1875–1953) in Göttingen. Wing theory grew out of these studies; occasional use was made of integral equations, while to Zhukovsky is due the conformal mapping of a circle into the shapes of wing sections.

The other problem, concerning the stability of flight, became something of a British speciality because of E.J. Routh's study of dynamical stability long before (§14.7). His student G.H. Bryan (1864–1928) wrote an important paper in 1904 (the year after the Wright brothers' pioneering flight in the USA) on the consequences of disturbance from level flight. He found equations whose roots should lead to damped oscillations of the aircraft, and thereby ensure stability. His compatriot Leonard Bairstow (1880–1963) at the National Physical Laboratory extended the methods by finding in 1913 new ways of approximating to the roots of such equations.

The next year saw the launch of the Great War, which encouraged development to a degree that no commercial initiative could have matched. In Britain, for example, the Air Battalion of the Royal Engineers possessed around a dozen usable aircraft in 1911; six years later the Royal Flying Corps was planning for 3,600 fighters and bombers, and the Royal Air Force was soon to be established by Jan Smuts. Bairstow had much to describe in his *Applied Aerodynamics* of 1920.

16.24 *Physics and thermodynamics*

Bryan and Prandtl's other research interests included thermodynamics: Bryan wrote the general article on it for the *Encyklopädie* in 1903, while Prandtl co-authored the one on technical aspects two years later {EMW V3 and 5 respectively}. This subject continued to penetrate ever more strongly into physics, and especially into statistical mechanics {CE 9.14}.

An important area was meteorology {EMVI1,6 (part by George Darwin) and 8; CE 9.7}. After various attempts from the 1880s by figures such as Hertz and Helmholtz to modify the basic equations (15.3.1)–(15.3.3), in 1902 Hertz's friend the Norwegian physicist Vilhelm Bjerknes (1862–1951) conjoined various notions from heat theory, such as specific heats of gases, to the basic equations (6.9.1)–(6.9.2) of hydrodynamics to find formulae for pressure and temperature.

However, finding solutions proved largely beyond the powers of the analysis of the time; so numerical methods were brought into play, although linear approximation to the equations required horrendously large systems. In Britain, Lewis Fry Richardson (1873–1950) envisioned thousands of human calculators working on the reduction of the equations. He also stressed the importance for theorizing of self-similar shapes, of which von Koch's FIGS. 16.10.1 is a simple example; this insight has made him a hero for the modern advocates of fractal theory. But meteorology has always remained an especially difficult challenge to mathematical physicists. As in a weather cycle, statistical methods have now regained the prominence they had, in their primitive forms, before the advent of Hertz {Sheynin 1984}.

16.25 *The invariants of Einstein's relativity theory*
{EMW V19 (by Pauli); CE 9.13}

Perhaps the most famous scientific development during this period was the introduction of special relativity in 1905 by the 26-year-old Patent Officer Albert Einstein (Berne), and of general relativity around 1916 by Professor Albert Einstein (Berlin). The topic has been studied very extensively from all technical and social points of view: a few relating to the mathematics are noted here.

1. In 1905, Einstein set out the special theory in a paper on 'the electrodynamics of moving bodies'. It addressed the question of 'matter and motion' (to quote the title of a popular book by Maxwell); but a more direct influence on him seems to have been the German physicist and engineer Auguste Föppl (1854–1924), whose *Einführung in die Maxwellsche Theorie der Electrizität* (1894) not only popularized the theory in Germany but also included a chapter on vector analysis and a section on 'the electrodynamics of moving conductors'. It could have led Einstein to notice the following asymmetry (a feature of physics of which he was fond): the electric field was absent from the Maxwellian analysis of the motion of an electric conductor past a stationary magnet, but it appeared in the analysis of the motion of a magnet past a stationary electric conductor.

In his 1905 paper, Einstein expressed the special theory in terms of the simultaneity of events in different locations. He asserted that the velocity of light, c, was the same for all observers, and that bodies did not drag themselves through the aether; instead, the means of measuring their position was contracted. He formulated his basic equations in terms of moving frames of reference. I see a point P with space-time coordinates (x, y, z, t); you, travelling at uniform velocity v along my x-axis, see it at (X, Y, Z, T). Then the coordinates are related as follows:

$$X = \frac{1}{p}(x - vt),\ Y = y,\ Z = z \text{ and } T = \frac{1}{p}\left(t - \frac{vx}{c^2}\right),$$

$$\text{where } p := \sqrt{\left(1 - \frac{v^2}{c^2}\right)}. \qquad (16.25.1)$$

Hence, among other consequences, the sum V of velocities v and w along the x-axis was given by

$$V = \frac{v + w}{1 + vw/c^2}; \qquad (16.25.2)$$

$$\text{hence } V \leq c, \quad \text{and } V = c \text{ only if } (v - c)(w - c) = 0. \qquad (16.25.3)$$

He also showed that a form of Maxwell's equations (15.11.1) for electromagnetism was invariant under this transformation.

Equations of this kind had already been published by the German physicist Voldemar Voigt in 1887 in a paper on the Doppler effect in a compressible elastic medium, by Hendrik Lorentz (after whom they are usually known: §15.13) in various forms between 1892 and 1904, and as (15.13.1) by J.J. Larmor in his *Aether and Matter* {Larmor 1900}. Einstein's finding may well have been independent, and in any case none of his predecessors had envisioned his kind of radical consequences: that the Newtonian manner of defining velocity from absolute space and time was reversed, with velocity as the invariant from which space and time were specified, and that talk of an aether was unnecessary.

2. The framework satisfying (16.25.1) became called 'Galilean', in allusion to Galileo's emphasis on conditions for uniform motion (§4.14). The 'principle of relativity' asserted that if two coordinate systems F_1 and F_2 moved in

this way relative to each other, then natural phenomena could be described in each one by the same laws.

3. Moves for extending the theory were soon afoot. Minkowski argued in 1907 that all physics should now be rethought in terms of space-time rather than space with time, with the metric for the infinitesimal distance ds given by

$$ds^2 := dx^2 + dy^2 + dz^2 - c^2dt^2. \qquad (16.25.4)$$

At that time Einstein himself proposed that energy E and mass m were related by the equation $E = mc^2$. He also announced a 'principle of equivalence', which asserted in a weak form that inertial and gravitational mass were equal; more generally it asserted that if object F_1 moved with uniform acceleration relative to F_2, then it could be treated as at rest if it was assumed that it produced a gravitational field {Norton 1985}.

4. In 1916 Einstein fulfilled the promise of this principle when he published 'general relativity' (his name), in which the two reference frames could be accelerating. (The adjective 'special' then became attached to the previous theory.) In a way he united Riemann's conception of geometry (§13.7) with Minkowski, in that he assumed that space-time was curved in a way that was characterized by exactly the bilinear form (15.8.1) for infinitesimal distance that Riemann had handled:

$$ds^2 = \sum_{r=1}^{4} \sum_{s=1}^{4} g_{rs}\, dx_r\, dx_s, \qquad (16.25.5)$$

where the x_r were the space-time variables and the g_r gravitational tensors. Einstein saw them as the components of the gravitational potential dependent upon the distribution of mass and energy, and regarded their determination as the chief aim of the theory.

5. To develop general relativity, Einstein drew upon, and even advanced, the tensor methods that Levi-Cività and others had been using (I am not deploying the special notations), and used them to express both the covariance

of the coefficients and the invariance of the expressions. Sophus Lie's theory of differential invariants (§13.9), in forms extended beyond (16.25.5) to sums of products of several differentials, was expressed in vectorial language {Reich 1993, 1994}. In addition, from this time onwards tensor *calculus*, and not just specific tensors, was used in relativity theory.

16.26 *(Mis)interpreting relativity theory*
{Hentschel 1991; Vizgin 1994}

Among the first contributors was the German astronomer Karl Schwarzschild (1873–1916), with solutions to the field equations already in 1916, produced on the Russian front where he caught a fatal illness. More generally, Levi-Cività himself, and also Weyl, re-read the theory in terms of curvature properties and so made it look like fancy differential geometry. This view became and remains popular, but it masks the *central* place of invariants. For both 'relativity' theories assert *absolutism* in the sense of invariance: a dynamical situation looks the same to any two different observers, under conditions of constant velocity for the special theory, and under acceleration for the general theory. In addition, the latter theory provides absolute measures of distance, in contrast to the degree of freedom in Newtonian mechanics (item 6 of §15.16).

Unfortunately, Einstein took a different philosophical line in developing his theory. In a paper of 1918 on the general theory he proposed 'Mach's principle' (his name), in which he followed Ernst Mach's opposition to Newton's assertion of absolute space and time by asserting that motions of bodies can be analysed only relative to the total mass and energy of the universe of which they form part. This is why the theories have the absurd name 'relativity'. It was chosen by Einstein himself, around 1912, when he was overly attracted to the fetish then prevailing in various disciplines for relativistic talk {Feuer 1971}. His choice was to generate several of the misinterpretations of the theory among the general public, and even by some scientists.

These remarks are not based upon historical hindsight – Einstein's choice of name was deplored soon after the event by commentators such as Sommerfeld. Furthermore, the invariant character of the theory had been stressed during the years of its development: for example, Klein pointed out that the Lorentz equations represented a certain group of transformations invariant in the sense of his Erlangen programme (§13.10) {Klein 1927, pp. 70–85, and already in a lecture of 1910}. A name like 'invariance' or 'covariance' theory would have been far more appropriate, and we would not then 'understand' the theory as asserting that Grand Central Station is pulling up at the next train.

16.27 *From electrons to quanta*

{Hermann 1971}

The invariant feature of Lorentz's equations had also been noted by Poincaré. He disliked them partly because he *did* take space and time to be relative and also admitted the aether, and partly because they invoked a hypothesis, which he wished to avoid in science; for similar reasons he rejected relativity theory. But he was deeply drawn to the problem area that it involved, and in 1906 he published his own analysis of the 'dynamic theory of the electron', which he published at length in a paper in the Palermo *Rendiconti* and in shorter forms in his philosophical articles {Miller 1973}. Embracing both the very large of relativity theory and the very small of the electron, he took an electron to be a hollow sphere which was deformed into an ellipsoid of revolution in the direction of its travel, modified the potential function to allow for this effect, and found that Lorentz's hypothesis was the only one that was compatible with the impossibility of detecting absolute motion (as the Michelson–Morley experiment had suggested) because of the group property (to which Klein was also to point).

At this time, molecular physics was growing to a new level of prominence in science, playing roles in a range of discoveries and theories which included cathode rays and X-rays, and the isolation of radium. Some of the ensuing developments drew on various branches of mathematics; a few examples are mentioned here.

The general notion of the quantum, the unit of *discrete* measure of energy, came mainly from Max Planck (1858–1947), impressed by the finding that so many substances emitted 'spectral' lines of radiation at specific and separate frequencies {EMW V26}. He considered activity inside a closed space S such as the interior of a heated furnace containing a gas whose molecules were treated as oscillating 'resonators'. S was 'black' (that is, all radiation was absorbed), so that the emitted levels U of radiation were functions *only* of the absolute temperature T and frequency ν. But which functions? Ludwig Boltzmann (§15.4) was an important influence in two assumptions lying behind Planck's answer. Firstly, since the molecule was a discrete, albeit tiny, object, then it took discrete levels of energy; hence the theories of classical mechanics were inapplicable, and numerical methods should complement the calculus as a main mathematical tool. Secondly, entropy must be understood probabilistically, as in §15.4, so that thermodynamics and statistical mechanics were united {CE 9.14}. So Planck assumed the existence of a tiny unit h of action (now known as 'Planck's constant') such that $NU/h\nu$ was always an integer, where N was the (large) number of resonators. He found an expression for U involving binomial terms for the numbers of resonators with the same energy levels, which he converted to an exponential form by using a cousin of Stirling's formula (5.7.1).

Einstein proposed in 1906 that the quantum of radiation, which he called the 'photon', be taken as the principal concept. His further attempt to link it with a quantum of electricity was opposed by the German physicist Willy Wien (1864–1928), in his article of 1909 for the *Encyklopädie der mathematischen Wissenschaften* {EMV V23} on radiation, on the grounds that a theory of atoms was needed. This point was taken up by the Viennese physicist Artur Haas (1884–1941) in 1910, who sought to link the energy of the electron with Planck's elementary quantity $h\nu$ of energy. His work is of historical interest for unusual reasons: it was done to complement, and thereby render acceptable for the award of a doctorate, an excellent history of the conservation of energy which he had already prepared {Haas 1909, cited in §10.8}. But his physics was soon to be eclipsed.

16.28 *From the wave equation to Schrödinger's equation: the quantum phase in physics*

{van der Waerden 1967 (a source book); CE 9.15}

Enter Niels Bohr (1885–1962), who from 1913 also adopted the notion of the quantum to study the structure of the atom. His view of the electron as a simple harmonic resonator led him to express its displacement r in terms of the wave equation (6.4.1) in the form

$$\frac{\mathrm{d}^2 r}{\mathrm{d}t^2} + n^2 r = \Phi(t), \qquad (16.28.1)$$

where n was the frequency (incorporating h) and $\Phi(t)$ the force impressed upon the electron; the special feature of energy levels enabled him to determine its elliptical orbits around the nucleus, and also to analyse the stability of the atom {Hoyer 1973}.

After some interruption caused by the Great War, quantum mechanics developed rapidly in the 1920s; neighbouring topics such as electromagnetism were incorporated. German-speaking scientists were very prominent. The rate of progress was sufficient to lead to two book-length articles in the *Encyklopädie*: Max Born (1882–1970) on the theory of atoms in 1922 {EMW V25}, and Adolf Smekal (1895–1959) on quantum statistics three years later {EMW V28}. Important mathematical methods included functional analysis and integral equations, especially the parts relating to latent roots, which corresponded to the spectral lines of radiation; it is curious that Hilbert had used the word 'spectrum' to describe a collection of roots in his work on integral equations (§16.13) *before* this physical interpretation emerged. His student Richard Courant (1888–1972) added Hilbert's name to his own in his book *Methoden der mathematischen Physik* (1924), which many physicists found very timely. But other branches of mathematics played roles also, as theories of various kinds began to develop.

Sommerfeld followed traditional mathematical physics in forming differential and integral equations; he accommodated the quantum chiefly by claiming that if a resonator with momentum p and location q took a period of vibration T, then $(\int_{t=0}^{t=T} p\,dq/h)$ was an integer. Erwin Schrödinger (1887–1961) went further in 1926 when he found that if the electron, of mass m, moved with kinetic energy E under a potential V, then its displacement $r(x, y, z)$ at time t was given by

$$r_{xx} + r_{yy} + r_{zz} + \frac{2m}{h^2}(E - V)r = 0. \qquad (16.28.2)$$

The means, and even the possibility, of solving 'Schrödinger's equation' have remained major questions. The hydrogen atom, with its single electron rotating around the proton, fell quite easily; but other elements, with their more complicated electronic structures, posed far harder challenges, of which some are still unmet.

Meanwhile, from 1925 Sommerfeld's student Werner Heisenberg (1901–76) took a different line, assuming that the electron admits n discrete energy levels, and makes transitions between the (say) nth and $(n - m)$th levels with a frequency expressed by the function $\omega(n, n - m), m > 0$. Contemporary understanding of its motion suggested that its position $x(t)$ was described by the wave equation (6.4.1) with additional terms in x, and took a Fourier series as solution:

$$x(t) = \sum_{r=-\infty}^{\infty} a_r(n) \exp[ir\omega(n, m-n)t]. \qquad (16.28.3)$$

His manipulation of this solution was rather mysterious: a key asumption was that the location of the electron depended upon the *order* of the transitions, so that the law of commutativity did not hold {MacKinnon 1977}.

Confronted by this enigmatic contribution, Born and his student Pascual Jordan (1902–80) realized that it could be re-expressed in terms of matrix theory, of which Heisenberg had apparently been unaware. The momentum of the electron was now given by square matrix $\mathbf{p}$ of order n, where the coefficient p_{jk} represented the capacity of the electron to change from level j to level k when $j \neq k$: it involved the expression $\exp(2\pi i \nu_{jk} t)$, where ν_{jk} was the associated frequency of resonance. The coefficient p_{jj} marked the component of position for level j, and was handled similarly; so was the location matrix $\mathbf{q}$ and its coefficients q_{jk}. Now, $\mathbf{p}$ and $\mathbf{q}$ obeyed the basic laws of variational mechanics of Lagrange and Hamilton (§6.6, §10.17), which meant among other things that they could be multiplied together; so Heisenberg's law of non-commutativity became the matrix relationship

$$\mathbf{pq} - \mathbf{qp} = \frac{h}{2\pi i}\mathbf{I_n}, \qquad (16.28.4)$$

where $\mathbf{I_n}$ is the identity matrix of order n. The idea of the quantum allowed a partial explanation of this equation: the right-hand side expressed an 'intrinsic' energy level for the electron that it could never lose. But the place of i was one of the mysteries of the theory, and was only partly resolved at that time by Schrödinger's demonstration that its results could be correlated with solutions of his own equation (16.28.2).

The algebraic line was continued by the initial contribution of the Englishman Paul Dirac (1902–84). In 1925 he took Heisenberg's approach a stage further by founding $\mathbf{p}$

and **q** in an algebra in which the properties (10.4.3)–(10.4.4) of Poisson and Lagrange brackets played major roles. Thus an idea of the 1800s, created to help analyse the continuous and determined orbits of the placid planets around the Sun in a vast and mysterious galaxy, reappeared in the next century to help analyse the quantized and probabilistic orbits of the ferocious electrons around the nucleus of the tiny and mysterious atom.

Other conundrums arose or remained in place. In some formulations the electron was better understood as a wave (in Schrödinger's (16.28.2), r was a 'wave function'), but in others as a particle. Again, the place of probability and statistics needed clarification: Heisenberg formulated (16.28.3) in a non-probabilistic way, but later he gave quanta the subjective reading that a quantum of uncertainty, $O(h)$, was unavoidable if one attempted to determine **p** and **q** simultaneously.

These and other issues have excited a massive and often vituperative scientific, mathematical and philosophical discussion that has lasted from that day to this. It is a suitable example of mathematics with which to close the main narrative of this book, which has regularly featured rivalries between approaches, analogies and structure-similarity between theories, and applications of mathematics to physical problems. It also exemplifies the gradual and disparate developments of the ancient and more modern roots of mathematics into the fully fledged theories studied by the time of the Great War – as well as some topics which came to life only later. No doubt other results await their awakening . . .

17

Re-viewing the rainbow

- 17.1 Epilogue and prologue: changes at the *Jahrbuch* during the Great War
- 17.2 Purpose of the chapter
- 17.3 Number and numeral
- 17.4 Number systems and numbering systems
- 17.5 Geometries – in space?
- 17.6 The varieties of algebras
- 17.7 The linear worlds of mechanics and mathematical physics
- 17.8 The calculus and mathematical analysis: style and branch
- 17.9 Logic into mathematics does not go
- 17.10 Axiomatization and the growth of metamathematics
- 17.11 Two late arrivals: probability between mathematics and statistics
- 17.12 The place of axioms
- 17.13 Strategies for effective convolution
- 17.14 Algorithms and approximations
- 17.15 Discrete mathematics and combinatorics
- 17.16 Styles as controls and constraints
- 17.17 Mathematics and religions
- 17.18 The power of the number
- 17.19 Design, decoration and geometry
- 17.20 Fibonacci numbers and the golden number
- 17.21 Recreational mathematics
- 17.22 Mathematical games and game theory
- 17.23 Mathematics and literature
- 17.24 Headquarters for mathematical activity
- 17.25 Mathematics, education and employment
- 17.26 History in mathematics
- 17.27 Mathematics in history

Any science is class-oriented by its nature.
HA-HA-HA!! AND WHAT ABOUT MATHEMATICS?
T.D. Lysenko, with reply by Stalin, 1948
{Rossianov 1993, p. 732}

17.1 *Epilogue and prologue: changes at the* Jahrbuch *during the Great War*

The ever-growing mountain of mathematics was doggedly covered, albeit some years in arrears, by the *Jahrbuch über die Fortschritte der Mathematik,* the abstracting journal founded in the 1860s (§11.5). When the Berlin analyst Leon Lichtenstein (1878–1933) took over as editor from Emil Lampe (1840–1918) around 1914, he retained the size at around 1,400 pages per volume, but he redesigned the classification in a way which reflects well the changing corpus of mathematics.

In Volume 45 (for 1914–15, published 1919–22, beginning with Lichtenstein's obituary of Lampe), molecular physics, capillarity and acoustics were dropped, maybe because of the existence of the companion reviewing journal for physics. But the sections on 'Mechanics' and 'Astronomy, geodesy and geophysics' were retained, and a new one was made for 'Relativity and the theory of gravitation'. Then with Volume 46 (1916–18/1923–4) the other nine sections of the journal were reduced to five. 'Set theory' left 'History and philosophy' to become a new section of its own; 'Arithmetic and algebra' and 'Geometry' each combined most of two old sections, and 'Analysis' took over three. The old section on 'Combinations and probability theory' was split between arithmetic/algebra and analysis, and vector analysis was transferred from algebra to geometry. New headings included 'Galois theory', 'Integral equations', 'Geometry of numbers', 'Analytic number theory', 'The newer theory of real functions' and 'Trigonometric series'. Out went 'Equations', 'Universal algebra' and 'Newer synthetic geometry', among others, although of course the topics were still covered.

The journal ploughed on until closing in 1944 with the 1942 volume {Siegmund-Schultze 1994}; but dissatisfaction with the tardiness of its appearance, and then the exclusion of Jewish authors and reviewers, had already led to two successors. In Germany itself the *Zentralblatt für Mathematik* was founded in 1931 by the Austrian historian of mathematics Otto Neugebauer (1899–1990); then in 1939 in the USA a committee of the American Mathematical Society was set up under the chairmanship of Oswald Veblen, and launched *Mathematical Reviews* the following year, with immigrant Neugebauer as founding editor.

Both journals still function, and publish more promptly than did the *Jahrbuch*. (The Soviets started their own version, *Referativnyi Zhurnal*, in 1953.) They have to cope with a rainbow of mathematics which increases not only in quantity but especially also in ubiquity, as more and more kinds of human activity come under mathematical scrutiny. TABLE 17.1.1 shows their current classification of mathematics. Even this taxonomy is somewhat perfunctory on probability and statistics, which are however covered in detail in *Statistical Theory and Method Abstracts*; and mathematical education, omitted almost entirely, is handled by the *Zentralblatt für Didaktik der Mathematik*. We cannot cope with all that mathematics here; pity its general historian in the 21st century!

TABLE 17.1.1 *Classification of mathematics, 1991.*

00	General
01	History
03	Mathematical logic and foundations
04	Set theory
05	Combinatories
06	Order, lattices, ordered algebraic structures
08	General mathematical systems
11	Number theory
12	Field theory and polynomials
13	Commutative rings and algebras
14	Algebraic geometry
15	Linear and multilinear algebra: matrix theory
16	Associative rings and algebras
17	Nonassociative rings and algebras
18	Category theory, homological algebra
19	K-theory
20	Group theory and generalizations
22	Topological groups, Lie algebras
26	Real functions
28	Measure and integration
30	Functions of a complex variable
31	Potential theory
32	Several complex variables and analytic spaces
33	Special functions
34	Ordinary differential equations
35	Partial differential equations
39	Finite differences and functional equations
40	Sequences, series, summability
41	Approximation and expansion
42	Fourier analysis
43	Abstract harmonic analysis
44	Integral transforms, operational calculus
45	Integral equations
46	Functional analysis
47	Operator theory
49	Calculus of variations and optimal control; optimization
51	Geometry
52	Convex sets and related geometric topics
53	Differential geometry
54	General topology
55	Algebraic topology
57	Manifolds and cell complexes
58	Global analysis, analysis on manifolds
60	Probability theory and stochastic processes
62	Statistics
65	Numerical analysis
68	Computer science
70	Mechanics of particles and systems
73	Mechanics of solids
76	Fluid mechanics
78	Optics, electromagnetic theory
80	Classical thermodynamics, heat transfer
81	Quantum theory
82	Statistical mechanics, structure of matter
83	Relativity and gravitational theory
85	Astronomy and astrophysics
86	Geophysics
90	Economics, operations research, programming, games
92	Biology and behavioural sciences
93	Systems theory, control
94	Information and communication, circuits

17.2 *Purpose of the chapter*

Two tasks are undertaken in this chapter: to note some developments in mathematics which did not fit into earlier chapters; and to reflect on some aspects of the story that were told there, especially concerning similarities and differences across cultures and periods. Most attention naturally falls upon Europe, particularly up to §17.12; but even the later sections cannot cope with the history of mathematics world-wide. What happened, say, among the Cambodians (who had an impressive water technology by the 8th century, incidentally), or the Eskimos, or . . . ? How about the Maltese Islands, from their rich prehistory through several Mediterranean occupations, to the construction of some of the greatest fortifications in Europe?

The name 'ethonomathematics' has been given to studies of mathematics across various cultures, where history interacts with anthropology and sociology {CE 12.1}; they have flourished mainly in the last 30 years, and a general appraisal cannot be attempted here. Indeed, much of the mathematics of this kind is covert (item 6 in §1.6), present at some level but not documented or remembered. Again, this chapter concentrates on mathematics at the level of research (or an equivalent level in some other contexts) and its teaching; little is said about lower levels of activity, since again our overall knowledge there is still very modest.

Clearly, the number of possible comments approaches the actual infinite; I trust that the selection here re-views the rainbow in a reasonably satisfying manner. And one conclusion can readily be drawn: the many branches of mathematics, whether of ancient origins or not, exhibit very different histories, as does any one branch in different countries or cultures. *Mathematics is rich, even dense, with interconnections, but it exhibits no unity.* 'What is mathematics?' is even more misconceived a question than is usual for questions of this form.

17.3 *Number and numeral*

It is appropriate to start with arithmetic, probably the fastest grower among the ancient roots. The distinction between numbers (whether cardinal or ordinal) and numerals needs stressing. If I catch a bus numbered 5, say, I do *not* have an essential experience with the number five; identification of the bus as blue with white dots would suffice, although less conveniently. (Indeed, the bus service in Malta converted only recently from colours to numbers.) Genuine involvements with five come in, say, being the 5th person in the bus queue, or having only 5 pence in my pocket to pay the fare. Bertrand Russell and A.N. Whitehead proved 'the occasionally useful proposition' $1+1=2$, numbered *110.643 in Volume 2 of their *Principia mathematica* (1912); but the difficulties of their logicist grounding of mathematics (§16.6) were not thereby increased, for the 110 and 2 have a different status from the 1 and 2 handled there. Mathematicians concerned with the foundations of arithmetic often used different words for these kinds of integer; for example, Georg Cantor's *Zahl* and *Anzahl* respectively.

However, Cantor was not happy to accommodate zero: he would have had to abstract it from an empty set E, an impossible task. (Russell and Whitehead, and also Frege before them, had no problem – zero was the set containing E.) Zero poses all sorts of difficulties. For example, not understanding the distinction between numbers and numerals, the little boy was rightly puzzled about the correctness of

$$0+0+0=0 \qquad (17.3.1)$$

on the grounds that the left-hand side had three zeros but the right-hand side only one. With due respect to Robert Recorde's justification of his sign '=' (§4.4), equality is neither equal to identity nor identical with it!

At various times sociological reductions of numbers and arithmetic have been offered, but they remain unconvincing. For example, I have just worked out that

$$12{,}987{,}128{,}432 + 34{,}867{,}912{,}237 + 5 = 47{,}855{,}040{,}674; \quad (17.3.2)$$

quite likely this addition has never been performed before, but it would be odd to assert that the act has changed arithmetical theory, or five *as such*.

This calculation was executed longhand, I fear; but some people can do it and indeed far more complicated sums mentally. This gift of 'lightning calculation', still a mystery to mathematicians and psychologists alike, has a fascinating history {S.B. Smith 1983}. Most of these calculators have little idea themselves of what they are doing, and the great majority are not talented in mathematics in general. Furthermore, no other branch of mathematics seems susceptible to such speedy powers of human mastery.

In contrast to fast calculation and the early rise of arithmetic, the progress of number theory was very slow. Euclid presented some properties of prime numbers, for example, and Diophantos went much further (§2.26); but even in European mathematics Pierre Fermat was isolated with his interest in the 17th century (§4.22), and Leonhard Euler for much of the 18th (§6.16). Only with successors such as J.L. Lagrange, A.M. Legendre and C.F. Gauss did it rise, and even then still slowly. It is not a branch of mathematics in the sense in which I have been using the word, but rather an important accretion to arithmetic, then also to combinatorics, algebra and mathematical analysis.

17.4 *Number systems and numbering systems*

{Menninger 1969; CE 1.16}

Many kinds of system have been developed, on various arithmetical bases. The common preference for 10, and its relationship to 60 and 12, were noted in §2.34; but we recall also that traditional systems of money and coinage, and of weights and measures, rarely took 10 as their base.

The binary system, with base 2, interested figures such as Thomas Harriot and G.W. Leibniz in the 17th century, but it has only come into the limelight in recent years because of computing.

Some systems are of relatively recent origin. Here is one, stimulated largely by 19th-century science, that builds upon powers of 10 in multiples of 3 (why 3?):

n=:	−18	−15	−12	−9	−6
name:	atto-	femto-	pico-	nano-	micro-
n=:	−3	3	6	9	12
name	milli-	kilo-	mega-	giga-	tera-.

However, words have never been created from the prefixes for negative indices in order to abbreviate decimal expansions: we always say them as digit strings: 'point three two six one . . .'.

Other systems use numeral representations. The one in T and T′ (§14.4) was pretty baroque! An interesting example, not unique, was the numbering of clauses in the laws of the Austro-Hungarian empire, in inverse proportion to number of digits; clause 6, say, was important, but clause 3.4872 rather special. The Austrian philosopher Ludwig Wittgenstein imitated it in the layout of his 'Logisch-philosophische Abhandlung' (1921; better known in its book printing of the following year as *Tractatus logico-philosophicus*). A wide range of exotic numerations is in use today: car and airplane numbers, for example, and above all computer programmes ('MicroNotch 4.71*a″$_2$', and the like).

A vast but overlooked class of systems concerns streets and property. When towns began to grow in the Middle Ages, their buildings were given *only* a number; for example, the cologne '4711' is named after the number of the factory of its original manufacture. From the late 18th century the further expansion of towns and cities, and the development of postal services, caused this system to be replaced by numberings along the streets. In Paris, houses had been numbered within each *quartier* (that is, quarter) of an *arrondissement* (district): now streets parallel to the River Seine were numbered from east to west following its flow, and streets perpendicular to the river were numbered away from it. Odd and even numbers were often assigned on either side of a street, in preference to consecutive numbers up one side of a street and down the other one, which favoured the postman but nobody else.

An axial system of street and house numbering was used in North America, where cities were planned on virgin territory in rectangular grids of streets. Addresses often read say, '4634 Walnut Avenue', between 46th and 47th Street with 34 as an appropriate demarcator. Now, '4634' is not the number '4,634', with comma, but a digit string: 46–34. Telephone numbers, as they are called, are common examples of strings, and this book has one in its ISBN number on the verso of the title page. A quite different example is given in §17.18.

The development of skyscrapers and high-rise buildings there and elsewhere created numbering problems in the vertical dimension. The difficulty of negotiating large hospitals and hotels bears eloquent testimony to architects' lack of forethought!

Not all arithmetical systems are written or spoken: certain elaborate sign languages have some numerical content. Bede's method of finger-counting was described in §3.11; and others were developed which are still in use, as one could see until recently in those two strikingly similar institutions, stock exchanges and racecourses.

17.5 *Geometries – in space?*

Naturally enough, geometry grew out of consideration of physical space, and it became a dominant branch of mathematics especially among the Greeks, with trigonometry in tow and astronomy as a principal stimulus. However, progress in solid geometry was rather fitful compared with that of line and plane, which must have seemed to possess simpler properties and was certainly easier to use when diagrams had to be drawn or carved on two-dimensional surfaces such as sand, cave walls, skins, papyri, scrolls or (eventually) paper. Indeed, the problem of making clear diagrams for solid geometry or other three-dimensional requirements is perennial; it has always been central for the map- and chart-maker, and played a major role in the development of perspective theory.

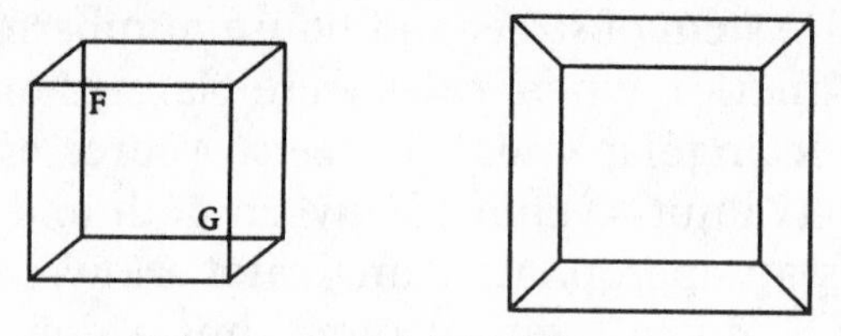

FIG. 17.5.1 *Capturing the cube on paper.*

There are also problems of perception, as FIG. 17.5.1 shows. The normal representation of a cube, on the left, has the disadvantage that points F and G are not real ones; they not only cause the Gestalt effect (does the cube go into or come out of the page?) but also have the drawback that they have to be ignored for mathematical purposes. To overcome the latter difficulty, the 19th-century English mathematician Thomas Kirkman (1806–95) proposed the projective version on the right, in which no false points appear. Interestingly, the Gestalt effect is still there: only the solid going into the page represents a cube, while the one coming out presents a frustum. Or is it the other way round? . . .

From ancient times, geometry developed its study of space in two complementary ways. The practical part dealt with the measurement of length, surface, volume and angle; the speculative side, exemplified among the Greeks by Euclid's *Elements,* avoided calculation and studied properties and relationships of angle, line, curve, region, surface and solid. Then non-Euclidean geometries twisted the tale. During the 1820s János Bolyai and Nikolai

Lobachevsky developed theories for geometries that drew upon alternatives to Euclid's parallel postulate, but they assumed that such geometries could be constructed in the first place (§8.17). More profound was Gauss's contemporary introduction of intrinsic geometry in his formula (8.16.1) for the square of the differential element ds of a surface defined only in terms of the surface itself, and the great extension made 40 years later when Bernhard Riemann proposed his general bilinear form for ds^2 (§13.7). Here nothing was assumed about the enclosing space; in particular, the curvature expressed by ds^2 could be zero or non-zero, so that both Euclidean and non-Euclidean geometries could follow, and the most appropriate geometry for a given physical situation could be chosen. The bearing of non-Euclidean geometries upon axiomatization in general is noted in §17.10.

Geometry connected with all other branches of mathematics. But sometimes it fell under their sways – especially that of algebras, which we consider next.

17.6 *The varieties of algebras*

René Descartes used common algebra with variables as an (important) handmaiden to his *Géométrie* of 1637 (§4.18); the balance became more even when a coordinate axis system was adjoined by Leibniz and others, and even swung the other way when 'analytic geometry' was established by Euler and successors around the mid-18th century (§5.27). Descartes himself showed one way in which algebra was superior to geometry in his use of different letters for known and unknown constants (or variables); for a diagram cannot exhibit such basic distinctions, at least not without clumsy conventions of drawing. This is one of many interesting situations in which common algebra and analytic or coordinate geometry do not match up {Cournot 1847}.

Another example of the swing towards algebra is from algebraic geometry, when in 1750 Euler's contemporary

Gabriel Cramer wondered about the number of points of intersection of curves defined by polynomials (§5.26). A similar change can be seen in the development of trigonometry. Originating in geometry with deep motivations from navigation and astronomy, its functions were understood as lines or lengths; then Franciscus Vieta helped start the move to an algebraic reading (§4.7), and then the functions became functions of variables with Euler's generation (§5.24). After that they were often – but not always – read as dimensionless ratios of sides or quotients of lengths.

The introduction of variables into common algebra in the 1630s constituted a third stage of its convolution out of arithmetic (§3.16), following a gradual introduction of abbreviated words and eventually symbols to denote constants, and also operations such as addition and relationships like inequality. Subtraction was the most problematic operation, for it highlighted the status of negative numbers on their own; the Euclidean preference for taking only 'the lesser from the greater' was to last until the 19th century (§9.14).

By then *algebra* was expanding into *algebras,* with differential operators, functional equations and permutation theory coming through from the 1810s (§9.11–§9.13). This last topic was to help launch general group theory, the fundamental part of abstract algebra which in due course was to aid Felix Klein and others in classifying geometries themselves (§13.10). The other two algebras played a role here, such as motivating F.J. Servois in 1814 to highlight the properties (9.11.3) of commutativity and distributivity. The algebraic world was further enriched during the 1840s by quaternions, probability theory, algebraic logic and determinants.

But matrix theory, the dominant part of linear algebra today, stumbled on fitfully during the 19th century (§9.10, §13.5), unaware of Chinese insights made before the Christian era in (2.31.2)–(2.31.3); even by the Great War it was little publicized (§16.14) and was unknown to Werner Heisenberg in the mid-1920s (§16.28). Textbook-writing in any quantity dates only from the decade following, and general teaching is a tradition merely half a century old.

Another distinguished very-latecomer is topology. It began its rise to theory literally as 'the geometry of place' (§8.18), and was stimulated by the profound but puzzling ideas of Riemann about his surfaces (§13.13), along with curiosities such as the Möbius strip. But then Henri Poincaré led it in the algebraic direction during the 1890s with his homology concept (16.15.2), expressed as a linear combination. During this century the algebraic aspects have become prominent, especially when the Americans took topology as a principal mathematical concern during their remarkable advance from the 1920s onwards (§16.16–§16.17).

17.7 *The linear worlds of mechanics and mathematical physics*

In dynamics and statics, among the most fruitful ancient roots of mathematics, algebras competed with more traditional geometrical approaches. A deep encroachment was made by Lagrange, the all-embracing algebraist, who tried to reformulate the mechanics of his time in his treatise *Méchanique analitique* of 1788 (§6.11) in the form of principles expressible in terms of the calculus and the calculus of variations. 'One will not find Figures in this Work', he wrote in the preface, and he meant it – principles such as (6.6.4) of least action furnished not only rigour but also generality, whereas diagrams could embody only special cases.

As so often, the aspirations were not matched by the fulfilment. One notable limitation was the loss of heuristics. Lagrange could re-prove many results already found and lay them out systematically, but only in relatively few cases, such as the spectacular attempt (6.8.2) to prove that the planetary system was stable, were new results found. Thus the geometrical/spatial style of Newton's laws and of energy conservation or loss retained much allure, especially for astronomers and engineers respectively. These preferences continued in the 19th century when mathematical physics emerged, drawing heavily on analogies from mechanical principles.

A popular means of developing theory in these branches was *desimplification*, as more and more physical factors were put into the mathematical models: gradually and reluctantly abandoning circles and then epicycles as orbits of planets and satellites, treating the Earth as a spheroid rather than a sphere, considering the effects of its rotation on the phenomena in question, allowing for changes of temperature, and so on. The development of the theory of the so-called 'simple pendulum' is a marvellous example from the 19th century: numerous effects of twist, swirl and air resistance were analysed to see if they affected significantly the calculation of its period to the many decimal places then required by geodesy (§4.14, §5.12, §10.6, §14.12).

These strategies were usually being pursued within the framework of *linear* modelling (§10.21): the great power, and often also the generality, of the methods of solution, especially for differential equations, made the necessary simplifications of the phenomena a price worth paying. Fourier analysis, as handled by Joseph Fourier himself and by A.L. Cauchy from the 1810s, greatly encouraged this trend (§8.5–§8.7), which Cauchy advanced further by the use of complex-variable analysis (§8.11). One consequence was the frequent use of associated algebraic forms, especially linear combinations and quadratic forms. Some non-linear work was done; but only in recent years have new methods and, especially, computer power raised the status of non-linear analysis {Israel 1996}. Chatty books on chaos may now be found in airport bookshops; but the deep philosophical questions remain largely unexplored {West 1985}.

17.8 *The calculus and mathematical analysis: style and branch* {Grattan-Guinness 1980}

Cauchy is best remembered today for his founding of mathematical analysis. His improvement in rigour, especially the emphasis on necessary and/or sufficient conditions under which a theorem was held to be true, led to its gradual but world-wide adoption, raising analysis to the status of both a major branch of mathematics and a mathematical style. Karl Weierstrass refined and extended both aspects from the 1860s, and his eminence as a teacher over more than 30 years led to analysis, in both real and complex variables, becoming the dominant branch of pure mathematics on the Continent (§12.2–§12.3). His version was in turn revised in various other hands during the 1900s, with point-set topology at centre stage; the measure approach to integration was introduced, and the axioms of choice highlighted processes of infinite selection hitherto taken for granted (§16.7–§16.9).

Cauchy's mathematical analysis had replaced three ways of handling the calculus: Newton's fluxions and fluents (§5.5), Euler's version of Leibniz's differential and integral methods using differential coefficients (§6.2), and Lagrange's algebraic approach relying upon Taylor's series (§6.5). But a geometrical style was still needed for modelling geometrical, mechanical and physical situations; Euler's approach therefore usually prevailed in applied mathematics, for both traditional areas and newer ones such as mathematical economics. It came to meet up with variational methods in contexts such as the exact differential of a function P:

$$\mathrm{d}P(x, y, z, \ldots) = Q\,\mathrm{d}x + R\,\mathrm{d}y + S\,\mathrm{d}z + \cdots; \qquad (17.8.1)$$

for potential theory would associate the integral of this equation, $P =$ constant, with a surface of equal potential. This subject provided many other services for mechanics and mathematical physics; in particular, thanks to the brilliant insight of George Green in 1828 (§10.18) that the situation on the surface of a solid could be related to the states of affairs inside it. The theory also extended the range of the calculus itself by expanding the roles of line and surface integrals.

Riemann deployed these integrals well, especially in his complex-variable analysis, where they were defined over

Riemann surfaces (§13.12–§13.13), whose weirdness helped significantly in the establishment of topology (§13.17, §16.15). But Weierstrass is reported to have described the approach as a 'geometrical fantasy'; and indeed the contrast between the two men's theories of complex-variable functions is so great that it is not immediately obvious that they were tackling the same subject.

17.9 *Logic into mathematics does not go*

One of the main products of Weierstrass's school was set theory, principally with Georg Cantor in the 1870s; after a quarter-century of gradual and somewhat controversial development, it moved to the centre of many mathematical concerns (§12.15–§12.16, §16.8–§16.14), such as the axioms of choice. It largely replaced the traditional theory of parts and wholes as a means of talking about collections of things. In particular, it was a staple feature of the mathematical logic that Russell and Whitehead developed in the 1900s out of the work of Giuseppe Peano's school.

This was one of many points of difference from the algebraic logic that had been in progress since Augustus De Morgan and George Boole had adapted new algebras to express logical notions in the 1840s. Another difference was quite basic: Boole and his followers applied mathematics *to* logic, whereas Russell sought to obtain mathematics *from* logic. FIG. 17.9.1 sketches the two very separate developments of 'symbolic logic', an umbrella term used at the time; it includes two influences from philosophy (Condillac and Whately) which have not been discussed in this book.

The gap was glaring by 1900, especially with C.S. Peirce and Ernst Schröder on the algebraic side and Peano and Russell on the mathematical, each side not at all appreciating the other. The philosophical irony is great, in that both Peirce and Russell hoped to ground ambitious philosophies of knowledge from their logics; in particular, Russell's logicism profoundly influenced his launch, with the philosopher

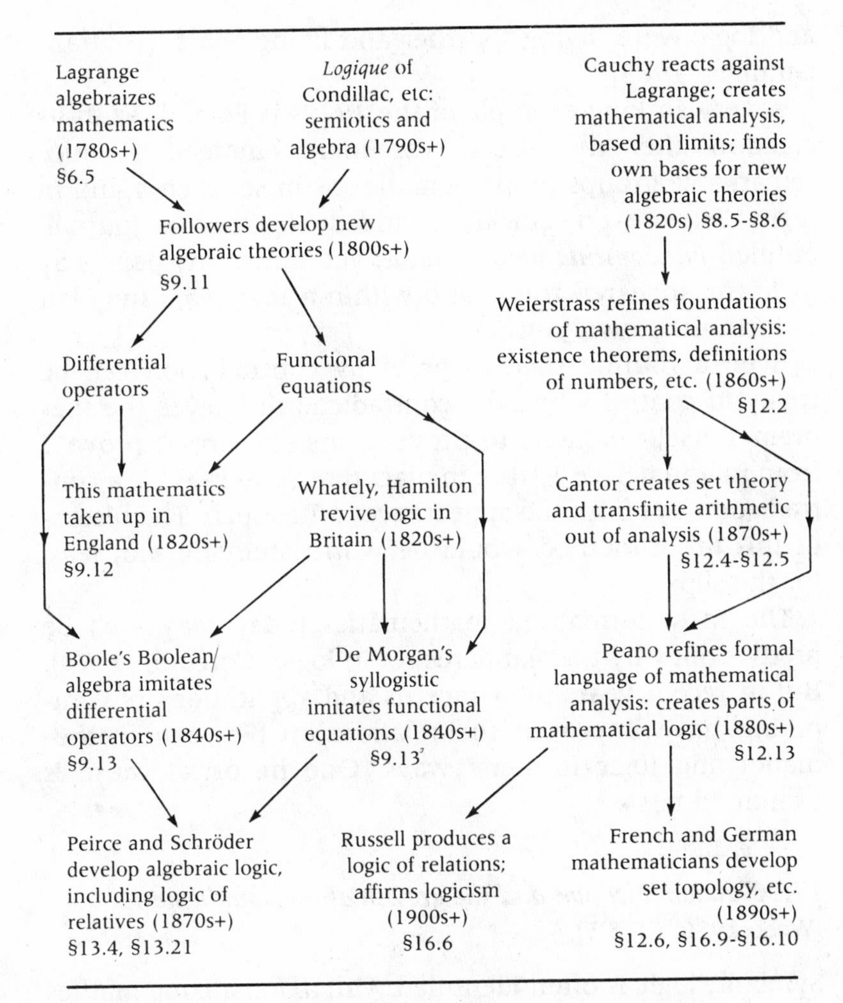

FIG. 17.9.1 *Connections between symbolic logic and mathematics, 1780s–1914.*

sopher G.E. Moore, of Anglo-Saxon analytic philosophy.

Another irony is that both traditions were side-shows to the mathematical community of the time; a typical specialist in, say potential theory or even mathematical analysis took little or no notice of either one. As ever, mathematics and logic were 'living together and living apart' {Grattan-Guinness 1988}.

A very striking example of the 1920s is Poland. With its creation after the Great War there emerged at once remarkable groups of mathematicians in set theory and in logic. They even quickly founded together a journal, entitled *Fundamenta mathematicae*; yet hardly any papers by logicians appeared there, and within a few years they left to form their own journal.

Even a routine such as proof by contradiction can be tricky to control when the contradiction involves the theorem T itself: namely, to prove T, assume not-T; prove T from it; and derive T from the fact that (not-T→T) is a contradiction, *not* from the appearance of T as such. The history of this proof-method would be worth studying, not least for the slips!

The most formalized mathematics today may well be pretty sloppy by the standards set in logic {Corcoran 1973}. But in recent years the concerns and applications of computing have improved the relationship between mathematics and logic in some ways. One historical cause is examined next.

17.10 *Axiomatization and the growth of metamathematics* {Webb 1980; CE 5.5}

Symbolic logic is often identified with axiomatizing mathematical theories; but in fact they are independent activities. The deep concern with the status of Euclid's parallel postulate, and the uninterest in applying syllogistic logic to the *Elements*, shows the distinction very well. Symbolic logic and axiomatization came together only in the 1890s, when Peano imported the latter from geometry. Thereafter the connection was firmly in place, especially after both Whitehead and Russell's logicistic programme and David Hilbert's completion of Euclid's axiom system (§16.4).

With the establishment of this connection, a change in understanding the character of axioms gradually developed. The traditional view was that they ought to be self-evident; for example, some alternatives to the parallel postulate had been put forward on the grounds that they were more 'obvious'. It now became understood that axioms were required *solely* as starting-points, whether obvious or not. Indeed, sometimes they could (and can) be hard to find, and showing that a list is complete for its theory is still harder.

Hilbert's axiomatization, at first with one axiom missing, exemplifies well the struggles involved. His first foray was followed by a long silence; but during the late 1910s he resumed the topic with much greater earnest, and came to outline a convincing programme. Both the propositional and the predicate calculus were axiomatized, and properties of consistency and completeness could be stated, and maybe even proved, in meta-logic. Mathematical theories were also to be axiomatized; in particular, arithmetic including quantification over positive integers and sometimes zero also (hereafter 'AR') was handled Peano-style, and its meta-properties studied. Hilbert seems to have wanted to construct a hierarchy upwards, of metamathematics, metametamathematics, ... with each new level showing a progressively greater simplicity until one was found which contained only unproblematic axioms; the consistency and completeness of (for example) AR would then be *proved* by a flow down the hierarchy from this level to mathematics at the bottom.

However, in 1931 the 25-year-old Austrian logician Kurt Gödel (1906–78) showed that the consistency of AR could be expressed only in a meta-theory more complicated than AR itself. Thus Hilbert's hierarchy was misconceived, and his scheme had to be rethought (initially by Hilbert himself and followers). In addition, in his main theorem Gödel exhibited a proposition P in AR for which neither P nor not-P could be proved, so that the axiomatization of even such a small fragment of mathematics as AR was incompletable. This not only spoiled Hilbert's vision again, but also torpedoed the logicist programme of Russell and

Whitehead; and the relationship between mathematical logic, set theory and arithmetic then required a similar overhaul (carried out initially by W.V. Quine, from the mid 1930s).

Gödel's proof was inspired by Cantor's diagonal argument in set theory (§12.15): P was the proposition in AR corresponding to the meta-theoretic proposition 'this proposition is not provable'. He effected the link by expressing metamathematical properties themselves in arithmetical form, a technique which rapidly increased interest in recursion; one consequence was Alan Turing's conception in 1936 of a general computing machine. In addition, Gödel's analysis required him to observe throughout a rigid distinction between theory and meta-theory in every context; and he brought home to logicians just how *carefully* they had to proceed. For mathematicians, however, his theorem was of marginal interest, since Gödel worked with a far more formal definition of proof than that to which they aspired (or still do); so the separation of logic and mathematics continued largely unchanged.

17.11 *Two late arrivals: probability between mathematics and statistics*

{Krüger and others 1987; CE 10.15}

In 1933, two years after Gödel's theorem appeared, the Soviet mathematician Andrei Kolmogorov (1903–87) furthered the cause of axiomatization by publishing a system for probability theory. This landmark achievement placed the subject at last within the sphere of 'orthodox' mathematics, for he drew upon set theory – another late arrival, but by then impeccably placed in the rainbow. He associated an event with a set, and the probability of its occurrence with a real number between 0 and 1 construed as a set function. He then invoked the axiomatic style, stating clearly the properties that this set function should satisfy: for example, if two events were incompatible, then their corresponding sets were disjoint, and the probability of one or the other occurring was the sum of the two probabilities. This sounds like adding integrals; and indeed, Kolmogorov brought out clearly the way in which probability was working like a measure, amplifying Emile Borel's vision 20 years earlier (§16.21). Thanks to this contribution and several others of the 1930s, this ancient root of mathematics began at long last to run as a major branch along with all the others.

By contrast, much mathematical statistics continued to be practised largely in institutions of industry or government, or in university Departments of Statistics separate from their Department(s) of Pure and Applied Mathematics. Many of its applications to the social and life sciences have blossomed only since the Second World War {CE Part 10 *passim*}. This latest arrival of all is now a gigantic affair, but it is still practised rather outside the mathematical profession – as Table 17.1.1 hints.

17.12 *The place of axioms*

While Euclid's *Elements* shows that the axiomatic formulation of mathematical theories itself has ancient roots, only during this century has it has spread widely. However, the benefits have been mixed. On the one hand, the assumptions are laid bare (or should be, if the axioms are properly formulated) and the general features can be examined more closely. On the other hand, it has encouraged a snobbish attitude: a Fortress Mathematics – a sort of Platonic (§2.15) doctrine of 'let no man unbesotted with axioms enter here' – and oh-so-pure mathematics keeping its nose way above the applications.

In addition, the teaching of theories from axioms, or some close imitation of them such as the basic laws of an algebra, is usually an educational disaster; for whatever the gains in rigour, they are achieved at the expense of a heavy loss of heuristic understanding and intuitive reasoning – the very things that lie at the heart of real educational theory. (What did the poor kids make of Euclid, all down the centuries . . .?) History has much to teach us here, especially that in many cases the axiomatic approach follows the *reverse* of chronology: for example, the Russell–Whitehead logicist programme starts from 1900s logic and set theory, proceeds to 1880s Peano axioms for arithmetic, defines irrational numbers in the manner of the 1870s, and deduces properties of numbers which have been known since antiquity. The teaching of axioms should come *after* conveying the theory in a looser version. It is worth remembering that the word 'prove' comes from the Latin *probare*, meaning 'to test' or 'to strain'.

17.13 *Strategies for effective convolution*

We continue with the theme of testing from a different perspective. A paper by the eminent physicist Eugene Wigner on the mystery of 'the unreasonable effectiveness of mathematics in the natural sciences' {Wigner 1960} has excited much discussion. But surely the mystery is that there is any mystery at all; for this book records that, with great frequency and in all cultures, new mathematics was often created *precisely* to address physical problems. Naturally, the success was not always automatic; at times notional mathematics has been plentiful. But it is hardly amazing that mathematics shows a good record of effectiveness when so much of it was thought up for exactly that purpose.

Better awareness of history would improve the quality of this discussion. It shows that some methods have been elevated to the status of methodology in order to increase effectiveness. An important case is reasoning by analogy, which became a 'method of analogy' with William Thomson in the mid-19th century (§10.19); it was a particularly potent example of reasoning from one branch or area of mathematics to another by structure-similarity. Potential theory, with which Thomson was deeply concerned, often worked by taking a theorem or property found in one application and applying it elsewhere. Greek geometry of curves shows many examples of, say, the quadratrix (FIG.

2.28.2) produced for one reason and then taken over into other ones. A general philosophy of real mathematics awaits development here {Grattan-Guinness 1992b}, closely linked to desimplification (§17.7).

Some strategies rest on intellectual principles. An influential one is the search for *invariants* in a theory, concepts that remain the same while others vary. The algebraic invariant theory (§13.5) is only one case; others include number, volume and mass, and the misnamed relativity theory.

Another principle is symmetries (and antisymmetries), in both geometrical and physical contexts, but also in structural situations such as duality principles in projective geometry (§8.3) and algebraic logic (§13.21). Ever since people began to appreciate pattern, symmetries have not only charmed but even inspired {CE 12.7}.

The final strategy here is generalization. The theory of functions shows a steady widening, from John Bernoulli through Euler and Fourier to Henri Lebesgue (§5.27, §8.7, §16.9); although they all wrote 'any function' or some such phrase, the realms widened from finite expressions and power series, through rational functions and those arising in Fourier analysis, to mappings from one set of values to another. Again, Pierre Fermat's generalization (4.22.2) of Pythagoras's theorem to the nth powers of integers has been a most fruitful conjecture. But there is also the generalization *game*, widely practised: someone has found a lovely result involving A^3, say; so let us generalize it to any A^n for any odd n, where, however, the only case that matters is $n = 3$.

17.14 *Algorithms and approximations*

{Chabert and others 1994; CE 4.13}

The effectiveness of mathematical theories is also to be measured by results, in the arithmetical sense. The history of algorithms is both long and varied. It ranges from ancient methods of arithmetical multiplication and subtraction;

through the abbacists and the algorists, and approximating to roots of polynomial and other equations, to roots of linear systems of equations, and to values of functions and series; to, recently, the general notion of recursion. Much of the history of mathematics 'between mechanics and architecture' {Radelet-de Grave and Benvenuto 1995} is a centuries-long tale of improving approximations and replacing empirical guesses by argued, if oversimple, theories.

The major role of inequalities in mathematics is illustrated clearly in this vast ensemble of results, especially for its role in estimating upper and/or lower bounds on errors. Estimating the efficiency of algorithms in terms of the number of steps needed to attain the desired value or degree of accuracy has itself been studied mathematically. Among branches of mathematics, trigonometry gained much of its importance from the compilation of tables of values of its functions, thereby stimulating the development of logarithms in the 17th century (§4.15).

At times errors can be too small. The numerous properties of π include these near-equalities:

$$\sqrt{2}+\sqrt{3}\approx\pi \quad \text{and} \quad \pi^2\approx 10; \qquad (17.14.1)$$

early workers may have thought that they really were equations, and so were misled. Plato might have been a victim of the first case, which seems to announce the lovely property that the circumference of the circle equals the diagonal of the square on the diameter together with the altitude of the equilateral triangle with side twice the diameter, and the error for π is then only 0.15% in excess. The second case is especially tempting; for if exact, then the circle is squared (§2.28).

17.15 *Discrete mathematics and combinatorics*

{CE 7.13}

One area of application of algorithms is combinatorics. The name comes from properties of permutation and combination, and uses in probability theory and elsewhere; but it is now applied to any topic where totalling the number of discrete situations or procedures is involved; magic squares are an example (FIG. 3.3.1). An attractive example is bell-ringing, where one lists all combinations of 'rounds' in which a given collection of bells can be rung, together with (anti-)symmetries imposed on the order chosen. Many delightful terms, such as 'reverse bob minimus', arose to describe some arrangements.

Sometimes the (best) order of executing the stages of a procedure is sought. A popular case from the Middle Ages, often charmingly illustrated in textbooks, sought the strategy by which a ferryman should carry given numbers of, say, wolves, goats and cabbages across a river in a boat of stated carrying capacity in a minimum number of trips without the goats and the wolves eating the cabbages or each other.

Euler was a father-figure of combinatorics. Among his various studies were the problem of finding a route crossing each of the seven bridges of the city of Königsberg just once (he showed that no such route existed); a similar question about the knight visiting all the squares on the chessboard; and his theorem (8.18.1) on relating the numbers of faces, vertices and edges of a polygon. An important stimulus for theory during the 19th century was the construction of graphs – diagrams representing the salient relationships between 'nodes' as points liked by appropriate networks of lines. FIG. 17.15.1 shows the essentials of the Königsberg

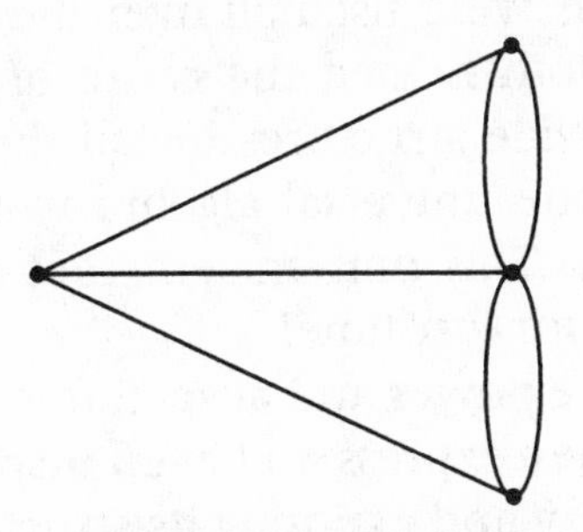

FIG. 17.15.1 *A graph for the Königsberg bridge problem.*

situation; Kirkman proposed his projective representation of the cube in FIG. 17.5.1 as a graph of it in order to avoid the two false nodes F and G in the normal version. Atomic chemistry and its notations provided a rich source of further problems for Arthur Cayley and J.J. Sylvester (§13.17).

17.16 *Styles as controls and constraints*

After all this bonhomie, it is time for a little attrition. As in all other aspects of human activity, one way of doing things may compete with others. The most wide-ranging examples traced in this book are the three traditions in the calculus, followed by a fourth with the introduction of mathematical analysis; and the trio of approaches to mechanics. The competition between geometric, algebraic and analytic styles was particularly strong, for intellectual empire-building was afoot. The style was supposed to ground all the required definitions and means of proof in a theory, and an assumption or axiom in a theory formulated in one style was either restated in another style, or proved as a theorem there – or so its supporters hoped.

Of all styles, algebraic ones are especially marked for their tenacity. With Lagrange, who wished to algebraize all the main theories of his time, their place is very clear and important. But from the early days in the Middle Ages its adherents started to make Euclid into an algebraist (§2.19); trigonometry fell under its sway in the 17th century (§4.7); linear combinations were used all over the place (§13.20); abstract algebras both helped the cause of axiomatization and aspired to provide structures for all theories (§16.19); and maybe even one universal algebra would ground the whole lot (§13.21). This non-ancient root of mathematics certainly made up for lost time!

One source of the power and attraction of algebras is the ease with which they express and even inspire processes of generalizing a theory and exposing its structure. Five stages may be discerned; the proposition $S := (a + b = b + a)$ is used as a case study (with examples):

1. Generalizing within a theory (from $5+2=2+5$ to S in common algebra);

2. Drawing on analogy (and structure-similarity) from other theories, maybe with modifications (Boole's use of S in logic (§9.13) only under his requirement that a and b were disjoint classes);

3. Presenting an abstract algebra of *uninterpreted* objects and operations satisfying laws such as S (group theory in §13.19);

4. Developing the algebra in terms of its structural properties rather than via ancillary notions (Joseph Wedderburn in §16.19); and

5. Setting up a meta-algebraic theory of comparing one mathematical structure with another one (category theory of the 1940s).

17.17 *Mathematics and religions*

One of the reasons that was put forward to deny existence to negative numbers was the claim that God would not admit them. This small detail exemplifies a factor in the development of mathematicians which historians often ignore: the bearing of belief in a religion on the mathematics of its adherents. This factor is present not only in, say, ancient and medieval cultures studying God's universe, or with Arab mathematicians dedicating their works to 'God the creator' (§3.32). It played a role in, for example, guaranteeing the truth and generality of the principle of least action for Euler and others (§6.6). The uniqueness of God over and above the numerous Christian churches was captured by the 1 in the algebra of logic of ecumenist Boole (§9.13). Cauchy the Catholic and Cauchy the mathematician are one and the same, especially in his amazing period of productivity during the epoch 1814–30 of his beloved Bourbons. His colleague A.M. Ampère may well have been drawn to electricity and magnetism in 1820 by seeing God at work pantheistically in the phenomena

(§10.13); this belief has a long history both before and after him in those branches of physics.

Like a social version of Newton's third law of motion, such actions can inspire reactions. Indeed, Newton's celestial mechanics provides an important example, for he was content to allow planetary perturbations to endanger the stability of the system precisely because God would intervene and re-establish order. However, Lagrange's attempt to prove stability from Newton's equations of motion together with the assumption that the planets had to move in the same direction around the Sun (§6.8) involved an elimination of religious factors.

The converse influence, mathematics applied to religions, was less marked; but a striking example is the determination of the *qibla* angle (§3.8), telling Moslems the direction of Mecca at the required times of prayer. In Orthodox Christianity perhaps the most significant mathematical problem was the determination of Easter.

17.18 *The power of the number*

Closely related to religious thought are numerology and gematria. While the mathematics as such is usually very easy, its general cultural significance is far greater than the fob-off as 'folklore mathematics' can convey; and the fact that so much of it was covert should *excite* interest. For example, one should understand that Johannes Kepler was serious when he argued that the planets were spheres because they were defined by the Trinity of centre, radius and surface, and when he associated the Father with the centre of the universe, the Son with its outer sphere and the Holy Ghost with the space in between.

An even more striking case is Newton. Severe criticism of the Biblical account of Christianity led him to Arianism, a sect expounding its faith in a seven-point creed which included a Trinity composed of Christ between God and man rather than Father, Son and Holy Ghost. It is surely no coincidence that both his *Principia* and his *Opticks* have three Books; that the former work builds upon three laws of mechanics which its author may well have known to be insufficient (§5.11); that the latter work, also divided into seven parts, broke tradition further in advocating a spectrum of seven colours in the rainbow and in his rings (§5.13); and that his theory of the motions of the Moon (§5.15) was based on an expression of seven terms {Gouk 1988}.

A socially influential class of cases arose in Freemasonry, which adopted parts of the secret lore of the cathedral builders (§3.22), including their arithmetic. The highest degree of membership is the 33rd, a very important Masonic number probably adopted from the length in years of the life of Jesus in Orthodox Christianity, and as 11 × 3 in the Apocrypha. Several other numbers came from the Masons' presumed origins of civilization, Egypt, especially 6, 18, 42 and 55 (§2.6).

The Masons reached a peak of influence in the late 18th century, especially in the founding of the USA. Traces of their numerology include the Executive Mansion, built in Washington in the 1790s with 33 rooms (it was extended in the 1900s into the White House, with 33 × 4 rooms). The music of that time also shows striking examples {Grattan-Guinness 1996a}. Wolfgang Amadeus Mozart (1756–91), a passionate Mason, built some of the main numbers into several compositions. Even the timing of preparation of symphonies 39–41 (not his numbers) was so governed: 3 principal works written together in the 3 Masonic keys during his 33rd year. The opera *Die Zauberflöte* ('The Magic Flute'), written in 1791 to defend Masonic ideals against political attack, is crammed with numerology (and some gematria). Among many features, 18 controls the part of High Priest Sarastro in amazing detail: numberings of musical items, of notes in his melodies and bars in his arias, 180 bars to sing overall, and so on.

Ludwig van Beethoven (1770–1827), sympathetic to both Christianity and Freemasonry, used 27, 30, 32 and 33 as principal numbers. Some uses were like Mozart's; but in addition, and most unusually for his time, he assigned opus numbers himself, and from the start of his career. In particular, he deliberately chose the Trinitarian digit-string 1-2-3 for the Opus number 123 of his *Missa solemnis* (1823), and built the corresponding number into the work in places: the first section of the 'Credo' has 123 bars, clearly divided into both $4 \times 30 + 3$ and also $33 + 37 + 30 + 30 + 3$. 'Von Herzen – Möge es wieder – zu Herzen gehn!' ('From the heart, may it go again to the heart!'), he wrote at the top of the 'Kyrie' – in a 33-letter motto divided by dashes into 3 parts.

It is well known that numerology (and also gematria) were used extensively in medieval music, by J.S. Bach, and in some composers of this century. But there is great historical ignorance of the use of numerology in the classical and romantic eras, and little effort is made by the experts even to comprehend its existence; for example, Beethoven's motto is almost always mistranscribed by musicologists. This part of the mathematical rainbow is quite out of sight and mind.

17.19 *Design, decoration and geometry*

{CE 12.6–12.8}

Another area of mathematics with a long history, often covert, is that of symbolic geometry. The history of perspective in art, architecture and sculpture (§3.25) is the best-known part, but there are many other mathematical aspects such as proportions, symmetries and tilings. Funerary architecture is an especially rich source {Curl 1993}; in addition, ceremonies relating to death include features such as ancient plinths and daises in the form of the frustum of a pyramid (FIG. 2.7.1), and also the guillotine, introduced during the French Revolution, in which a circle was cut in a rectangle, ready to receive a descending triangle. A Masonic example is the 'Masons' square', the 3 : 4 : 5 triangle from which the ratios 2 : 1 and 3 : 1 can be obtained (FIG. 2.8.1), which was worn by a Mason on his costume as an ornament.

During this century the mathematics of art and design has borne not only upon representational traditions. D'Arcy Wentworth Thompson's classic study *On Growth and Form* (1917) used ideas from projective geometry to illustrate similarities of form in the natural kingdom between skulls and sponges, for example, and the variety of shapes in bubbles and splashes; his analogies fortified non-representational artists in their belief that the world as we see it is often not what it seems.

There are other artistic contexts, some of ancient pedigree, which have been little studied as a source of mathematics. A rich example is dance: the symmetries and other operations involved both in the steps (such as in maypole dancing) and in the gestures and movements of limb and face.

Can such thinking about the arts be elevated into mathematical theory? G.D. Birkhoff made a brave foray in his book *Aesthetic Measure* (1933), where he considered polygons, tilings, ornaments, vases, musical harmony and poetry. The most successful part deals with vases, where he took the vertical section of a vase and defined a number as a function of the numbers of points of inflection, vertical tangents, and ratios 1 : 1 and 2 : 1 in principal horizontal and vertical distances. But even then his results were limited, and aesthetics still largely eludes mathematical dissection.

17.20 *Fibonacci numbers and the golden number*

{Herz-Fischler 1987; CE 12.4}

One area of mathematics with considerable aesthetic implications grew out of the 'division into mean and extreme ratio' (hereafter 'DEMR') (2.13.3) in Greek geometry, where a line with extremities A and B was split into two unequal parts by a point C (FIG. 17.20.1) thus:

$$AB : AC :: AC : CB. \qquad (17.20.1)$$

FIG. 17.20.1 *Division into mean and extreme ratio.*

This ratio turns up all over the pentagon (FIG. 2.13.2), and also on the decagon and on the corresponding regular solids. Several properties were reviewed by Fibonacci in his *Practica geometriae* of the 1220s (§3.23), and later by Piero della Francesca, Raffaello Bombelli, and others.

Today, Fibonacci's name is associated primarily with the sequence of integers that starts with 0 and 1 and proceeds by adding the last two together, to obtain

$$0, 1, 1, 2, 3, 5, 8, 13, 21, 34, 55, 89, 144, \ldots. \quad (17.20.2)$$

He had presented it in his *Liber abbaci* (1202) as the successive population sizes of pairs of rabbits breeding each month from one parent pair. His book enjoyed a great influence (§3.15); yet this result aroused no interest. In addition, he did not know that the ratio of successive numbers tended to that of DEMR. These insights came only in the 17th century with Kepler and a few others; but the study remained very fitful until the 1830s, when some Germans took it up. In particular, Martin Ohm, brother of the physicist (§10.20), introduced the name 'golden number' for DEMR and asserted that it played a role in many natural phenomena.

Some claims made for DEMR and Fibonacci numbers are exaggerated. But one area that repaid study was phyllotaxis: for example, the location of buds along the stem of a plant or twig on both left- and right-handed spirals which complete neighbouring Fibonacci numbers of revolutions, or the organization of the florets on cauliflowers. FIG. 17.20.2 shows a fine example of seed-heads on a sunflower; there are 55 white spirals and 89 black ones.

FIG. 17.20.2 *Spirals of seed-heads on a sunflower.* (photograph by the author)

However, the mathematical interest in Fibonacci numbers remained slight until the 1870s. This is surprising, for they are defined by addition and so multiplicative properties (prime numbers, for example) can be quite difficult (and thereby interesting) to find. An important stimulus was provided in 1878 by the French mathematician Edouard Lucas (1842–91). He helped launch Sylvester's *American Journal of Mathematics* (§16.2) with a long paper dealing both with (17.20.2), which he named the 'Fibonacci sequence', and also with the sequence starting with 1 and 3, which became named after him. Thereafter all such series entered into number theory.

17.21 *Recreational mathematics*

{Rouse Ball and Coxeter 1987; CE 12.3}

The cultural and aesthetic aspects of Fibonacci numbers are strong, and it is no coincidence that Lucas published a large-scale survey of *Recréations mathématiques* (4 vols, 1882–94). I take 'recreational' to refer to any mathematics which primarily sets a poser with a general cultural and maybe educational purpose. All cultures have pursued this kind of mathematics to some extent, and an extraordinary range of ideas has emerged, often deep ones. The connections were strong in probability theory, indeed from its origins (§5.19); other branches include number theory, combinatorics, topology, group theory and logic. Fibonacci numbers show that the line of separation from 'serious' mathematics is not sharp; other examples include magic squares, the Möbius strip (which has been patented!), and codes and ciphers – serious mathematics in the Great War, and even more so during the Second World War. The theories are often written up *only* in surveys such as Lucas's; often they are described in children's comics or manufacturers' leaflets, sometimes in classified documents. Thus much of the material is covert for one reason or another, so that it can be hard to establish origins and lines of influence.

One of the most curious links involves problem 79 of the Egyptian Rhind Papyrus (§2.6). It lists 7 houses, 49 cats, 343 mice, 2,401 measures of spelt-wheat, and 16,807 weight-measures of (perhaps) grain; presumably a question about successive consumption is to be answered. A similar problem occurs in Fibonacci's *Liber abbaci*, and a matching arrangement is presented in a children's nursery rhyme about someone going to St Ives in England and meeting a man with 7 wives, each one with 7 sacks, each one with 7 cats, each one with 7 kittens. However, the similarities are not so striking as they seem, since the answer to the question 'How many were going to St Ives?' is not 2,801, but 1!

17.22 *Mathematical games and game theory*
{CE 12.2}

Games constitute an important and rather special part of recreational mathematics. The mathematical theory includes determining whether a strategy for winning exists, and if so finding an optimal one. Noughts and crosses is very simple, but the three-dimensional version (# # #, imagined as stacked in a cube) can be quite tricky. Similarly, among shape-fitting games tangrams are often not hard; but jigsaw puzzles, which were introduced in the mid-18th century to stimulate interest in maps, are usually too complicated for systematic completion. The same is true of many card games, for example, bridge, upon which Borel co-authored a general study in 1940.

Combinatorics is often useful in analysing board games. In 1710 Leibniz, one of its pioneers, first described Solitaire, where one removes a piece by jumping another one over it and into a neighbouring empty square. Educational needs may also play a role, as in the medieval game 'rithmomachia' to aid teaching arithmetic (§3.14). Chess is too intricate for a complete solution, but many combinatorial problems have been formulated; the knight's moves were mentioned in §17.15, and a problem dating from 1850 seeks to place eight queens on the board such that none of them can take any other.

Chess theory is a good example of the ways in which games presaged concerns of "serious" mathematics. In particular, decidability (whether or not a game can be won, or an arrangement such as the eight queens can be effected) and decision procedures (how many moves are needed to achieve, say, checkmate?) were addressed long before logic and metamathematics began – slowly – to bring these meta-properties to the general attention of mathematicians during the 1920s (§17.10).

But just at that time a mathematical theory of games was developing, in the very general sense of determining strategies for maximizing some planned gain or minimizing a loss. John von Neumann pioneered research in the 1920s, and with the American economist Oscar Morgenstern he produced a major book on the *Theory of Games and Economic Behavior* in 1944. To see whether a best strategy exists at all, they drew diagrams of convex regions just like FIG. 16.22.1; but a beautiful example of mathematical Gestalt is involved (like the cube in FIG. 17.5.1). For von Neumann interpreted such diagrams in terms of the (non-)existence of fixed-point theorems in topology, which had come into prominence in the 1920s (§16.17); but since the rapid rise of linear programming since the Second World War, which he helped to inspire, mathematicians now think of a line moving through a convex region to determine the optimal value of the corresponding function.

17.23 *Mathematics and literature*

{CE 12.9–12.10}

Literary form has concerned mathematicians, even in their work itself. Indian Vedic mathematics was often written in verse, sometimes embodying numerical properties. A few major British mathematicians of the 19th century were enthusiastic poets (Hamilton, Boole, Sylvester, Maxwell), and Einstein's superb tribute to Newton was quoted in §5.10; but the only figure eminent in both fields is (Omar) al-Khayyam. In prose, Sonya Kovalevskaya and Felix Hausdorff wrote plays, and she also produced short novels and the best mathematician's autobiography.

The influence of mathematics on literature has been more fruitful. We noted William Shakespeare's concern with zero and '0' as the Hindu-Arabic numeral system came into his culture (§4.4), and Edwin Abbott's study of *Flatland* inspired by Hermann von Helmholtz's philosophy of geometry (§13.8). Newtonianism excited masses of rhetoric in prose and poetry, although normally and mercifully his mathematics was spared. Various authors have deployed non-Euclidean geometry or $\sqrt{-1}$ as symbols of fictions.

Mathematics and logic have sometimes appeared together in literature. Lewis Carroll (1832–98), practitioner of all three activities, is a remarkable case. His logic books are amusing exercises in syllogistic logic (although an uncompleted Part 2 of *Symbolic Logic* pioneered the use of trees in logic); but the *Alice* stories (1865, 1872) anticipate wonderfully many concerns which logicians and set theorists such as Russell were to face decades later {Jourdain 1918}.

Literature has invoked Euclidean rigour on occasion. A particularly witty example appears in *Literary Lapses* (1910) by the Canadian economist and writer Stephen Leacock (1869–1944). He reprinted there his short Euclidean analysis of 'boarding-house geometry', with definitions such as 'A single room is that which has no parts and no magnitude', and the splendid theorem that 'Any two meals at a boarding-house are together less than two square meals'. Mathematical literature of this quality deserves a good future!

17.24 *Headquarters for mathematical activity*

At various times one culture has been predominant, at least in its level of achievement: the Greeks for a time at Athens and then Alexandria; then the Arabs, especially at Baghdad; Renaissance Italians, largely in Tuscany; the French from the 1780s to the 1830s, mostly in Paris; the Germans (by a short head) from the 1870s to the Great War, especially in the former Prussia. During this century the profession has been pretty international, though with the USA and the USSR as the two major countries from the 1930s. Often the headquarters have moved from east to west; will mathematics complete an orbit during the next century, and settle back in the Pacific rim, where it was practised with great distinction long ago?

Within a culture, groups of mathematicians have often exercised power and patronage. However, only occasionally can one point to a *school* in the strict sense, with a leader (not necessarily beloved) and geographical centre, a specified programme of work (maybe not only in mathematics), settled means of diffusing or even publishing, and a strong sense of bonding among its members. The Pythagoreans (§2.11) and Plato's Academy (§2.15) may have been of this kind, and also the Oxford Calculators at Merton College in the 14th century (§3.13); but clearer evidence comes from later times when more institutions were established (monasteries, and especially colleges and universities) and a profession was evidently in place. During the early 18th century the Newtonians and the Leibnizians formed a pair of schools for both the calculus and mechanics, decidedly not in collaboration (Table 5.3.1). Weierstrass's students (§11.2) and Peano's followers (§12.13) in the late 19th century present a more agreeable pairing, with the aims at Turin explicitly animated by Berlin.

Klein doubtless aspired to lead a rival German school out of Göttingen (Table 11.3.1), and he certainly became a central figure in the group that formed around his mathematical and educational interests, with a massive route of publication to Teubner. This included the *Mathematische Annalen*, a journal of major importance of which he was an editor. Journals increased in number from Klein's time onwards, some as commercial enterprises and others as organs of societies of mathematicians {CE 11.12}.

These examples of schools and groups, and their consequences, exemplify national differences. They are often found, especially across the various parts and countries of Europe, although not involving national*ism* in a notable manner. One major cause is the fact that national institutions and educational policies were formed and developed in markedly different ways. For example, France has always been dominated by Paris, and since the Revolution it had placed its engineering schools above the universities; in England Cambridge was centre, but with a snobbery in the opposite sense, against engineering; and the existence of many states in Germany and Italy made their structures much less centralized, though Berlin and Rome tried at times to be national headquarters. Many of these differences are still quite active.

Opposite to groups and schools, and maybe opposed to them, are the loners; they include some of the greatest mathematicians. The early 19th century shows two remarkable examples in Gauss and Cauchy, whose work could have launched several cliques each; moreover, Cauchy worked in Paris, the mathematical world-capital of his time. However, neither man took any initiatives of this kind.

The outsiders also deserve mention; for example, from that time, Bolyai and Lobachevsky working on fringe mathematics in fringe locations (§8.17). And we can add to the list the schoolteacher Weierstrass until his 40th year – and almost all women at all times and in all cultures {CE 11.11}.

17.25 *Mathematics, education and employment*

Instruction at some level, whether elementary or advanced for its time, has been the single main provider of employment for mathematicians. The texts of Babylonian mathematics may have partly been educational, and many Chinese and Arab ones certainly were, for the training of professionals was expressly recorded (§2.31, §3.6). European mathematics shows numerous cases of the central place of education, with many figures working as teachers. The case of Cauchy (§8.6), though, reminds us that more might be taught than learnt, and that the idea of notional mathematics extends beyond textbooks to teaching.

Cauchy also exemplifies another feature: education motivating research. The students and staff at the Ecole Polytechnique rightly objected to his inappropriate lecture courses on the calculus, but the ideas they contained both constituted a mathematical style and led to mathematical analysis as a major branch of mathematics. This is especially clear with Weierstrass's school, which exhibited further educational motivations of its own. One can see similar effects in other branches and times, for example in mechanics among Cauchy's contemporaries (§10.2, §10.8).

Indeed, Euclid's *Elements* may have been intended as a manual of some kind, but it led to new mathematics. Likewise, the creation of methods of multiplication and division, and the gradual emergence of algebra out of arithmetic (§3.6, §3.14–§3.19), must have received some impetus from the classroom needs of the algorist and the abbacist teacher.

Those figures found another source of employment – in *commerce*, where they worked as consultants for the moneyed. From the 17th century business also provided jobs for probabilists with insurance companies (§5.21), and in the 20th century statisticians were taken on in many industries. One of those was *engineering*, when at last industry woke up to the statistical aspects of production (§16.22); long before, though, the military had needed the services of constructors of machines and of maps and charts, and of experts on embankments and projectiles. Prior to the movement towards professionalization in the early 19th century, *monarchs* of a society often engaged professional mathematicians, from astrologers in ancient and medieval times to court functionaries such as Euler and Lagrange in the 18th century.

Opportunities for employment in all these professions were available by that time, and in many countries education was tailored to fit them. In Britain, the best studied country in this respect, from the mid-17th century many 'philomaths' instructed across the range of mathematics, usually though not always at an elementary level {Wallis and Wallis 1986}; their name related to the low status of 'mathematician' as opposed to 'geometer' described at FIG. 4.8.3. FIG. 17.25.1 shows a schema of their concerns in the 18th century; the distance of a topic from the centre increases roughly as its prominence in their teaching decreases. Thomas Simpson (§5.16) was their most eminent member; during the 1750s he was the man editing the *Ladies' Diary*, the principal among a remarkable range of British mathematical journals for the public at that time {Archibald 1929}. The schema includes accountancy, which

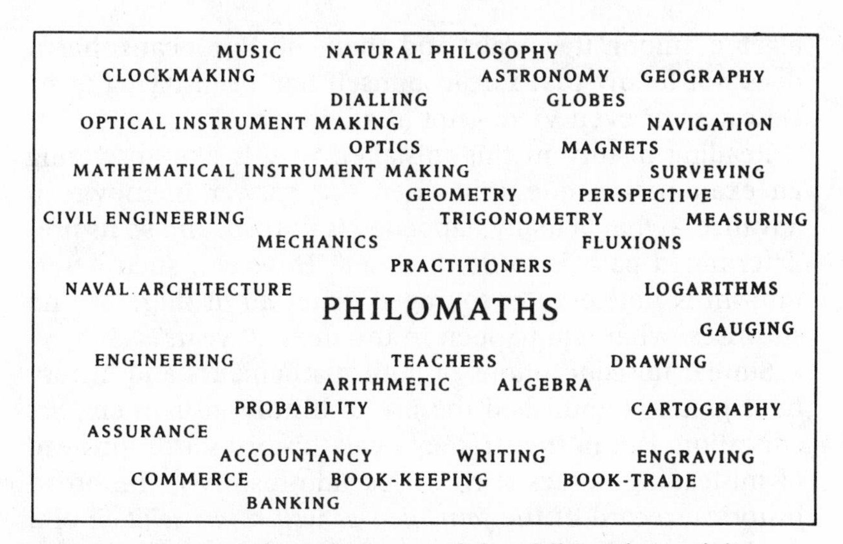

FIG. 17.25.1 *The wide range of the philomaths' activity* (figure kindly provided by Ruth Wallis).

grew as a profession through the centuries in many countries {Brown 1905}.

17.26 *History in mathematics*
{CE 12.13}

The study of philomaths is rather new in the history of mathematics, which is our concern here as an activity. One might expect that mathematicians' interest in history would be deep, for their subject has shown long continuous concerns; after all, all the ancient roots are still with us. However, mathematicians often treat history with contempt (unsullied by any practice or even knowledge of it, of course). One main reason lies in this continuity itself; since our predecessors had similar concerns to ours, let us rewrite their work as if we had done it. Such a re-reading can lead to perfectly respectable mathematics, but of course history itself is quite mutilated; time-warp takes over, with cause and effect changing roles. The case of geometrical algebra (§2.19) is both typical, and also important: without doubt Euclid's *Elements* influenced the development of algebra among the Arabs and then the Europeans, but it does not follow that Euclid himself had been trying to be Descartes or even Vieta – not at all, in fact.

Reading history in this mistaken way is like answering an examination question when the answer is known in advance – what actually happened is known, and so its predetermined path is to be disclosed. However, such determinism is not to be encouraged. After all, if there are no surprises, what will happen in the next 20 years?

Similar misconceptions of both mathematics and history have greatly diminished the use of history in mathematics education. But in the strategy I call 'history-satire' (instead of misleading names such as 'pseudo-history'), the broad historical record of the *formation as well as the solution* of a mathematical problem or theory is respected and followed, though most of the details and complications are left out. Otto Toeplitz (1881–1940) pioneered such a study in the 1920s, for the calculus {Toeplitz 1963}; a wider presentation of this kind appears in {Kramer 1970}, including much applied mathematics. The rich roles played by education in employment and research make especially ironic as well as disappointing the failure of teachers to respond to history.

Despite the indifference, historians of mathematics have developed new methods and changed their concerns and priorities. In recent years the subject's bearing upon education has created a lot of interest, for many modern historians were driven to their research by negative reaction to their own educational experience, as student or teacher, or sometimes both.

A major difference between today's historical method and that of the past concerns the status of information. Many writings of a century ago, pre-eminently in the *Encyklopädie der mathematischen Wissenschaften*, are distinguished above all by the enormous number of facts recorded, authors describing for example masses of minor or neglected texts alongside the well-known literature. By comparison, much modern historical work is sketchy. However, past authors usually provided little or no interpretation: the data were the beginning, middle and end of the enterprise. For example, from the late 19th century, Isaac Todhunter saw himself as writing the histories of his chosen topics; to most historians today, he provided wonderful databases upon which we can build, but not histories as such. In the more questioning world of today, historians are much more sensitive to the contexts and motives behind the creation of those data.

17.27 *Mathematics in history*

The converse application, of mathematics to history, is our final topic. It constitutes mathematical history, a small but interesting dapple in the rainbow {Rashevsky 1968}. One of its main components is called 'cliometrics', where statistical and related techniques (maybe modern ones) are applied to questions in social history {Fogel 1982}. The applications can include the history of mathematics itself: for example, {Eccarius 1989} on the teachers of mathematics in the German state of Thüringen in the 19th century. Another component is the use of combinatorial methods to study seriation (the determination of the chronology of the findings); originally pioneered in archaeology, it has potential for mathematical history in general {Kendall 1975}.

The topic has its own history. Very striking examples are provided by Karl Pearson, whose programme of biometry (§16.21) led him to detailed comparisons of historical questions, which he published in his journal *Biometrika*. For example, he found in favour of the genuineness of the embalmed heads of Lord Darnley (1928) and of Oliver Cromwell (1934), the latter to within a 'moral certainty'.

It is appropriate to end our story with this intriguing detail. For not only do we see mathematics and history in mutual convolution, we also stop as we started with Pearson, whose exchange with Rayleigh over random walks in 1905 in §1.1 set us on our quest for the multi-coloured but distant rainbow of the mathematical sciences.

BIBLIOGRAPHY

The principal bibliographies for the history of mathematics are {May 1973}, which lists biographies and notices of mathematicians as well as topics and institutions; and {Dauben 1985}, a rather more systematic survey of non-biographical sources. {Gottwald and others 1990} is the best concise catalogue of biographies; the *Dictionary of Scientific Biography* {*DSB*} lists fewer mathematicians but the articles are much longer. Among other sources worthy of mention is {Peano 1908}, a compilation of many principal branches of mathematics in a logical format described in §12.13; incidentally but remarkably, it contains bibliographical references to numerous results.

For all recent material the prime source is the journal *Historia mathematica*, which since its inception in 1974 has specialised in providing information on activity in the field. Of especial note is its abstracts department; the items listed in the first 13 volumes were indexed there by name and in the next volume by subject. For the history of science in general, the chief corresponding source is the annual bibliography published by the journal *Isis*; its bibliographies since its founding in 1913 have been reassembled in classified cumulative volumes cited here as {*Isis* 1971– }.

Many original and some historical works have been named in the text and footnotes. Texts written in Russian have been cited only in translation into a European language; the best guide to Russian work is the journal *Istoriko-matematicheski issledovaniya* ('Historical-Mathematical Writings').

In the list below some titles have been shortened. *Extensive* reprint histories of books is not recorded. These abbreviations are used:

AHES	*Archive for History of Exact Sciences*
AS	*Annals of Science*
CE	Grattan-Guinness, I. 1994. (Ed.) *Companion Encyclopedia of the History and Philosophy of the Mathematical Sciences*, 2 vols., London and New York (Routledge).
DSB	1970-1980, 1990. (Ed.) *Dictionary of Scientific Biography*, 16+2 vols., New York (Scribners). [Vols. 1-16 ed. C.C. Gillispie, Vols. 17-18 (largely on 20th-century

	figures) ed. F.L. Holmes. All articles on mathematicians repr. in 4 vols. as *Biographical Dictionary of Mathematicians*, 1991.]
EMW	Klein, F. and others 1898–1935. (Eds.) *Encyklopädie der mathematischen Wissenschaften*, 6 parts in 23 vols., Leipzig (Teubner).
HM	*Historia mathematica*
JDMV	*Jahresbericht der Deutschen Mathematiker-Vereinigung*
UP	University Press

Aclocque, P. 1981. *Oscillations et stabilité selon Foucault*, Paris (CNRS).

Aiton, E.J. 1972. 'Leibniz on motion in a resisting medium', *AHES*, 9, 257–74.

Aiton, E.J. 1981. 'Celestial spheres and circles', *History of Science*, 19, 75–114.

Aiton, E.J. 1984. *Leibniz. A Biography*, Bristol (Hilger).

Aiton, E.J. 1987. 'Peurbach's *Theoricae novae planetarum*: a translation with commentary', *Osiris*, (2) 3, 5–43.

Aiton, E.J. 1988a. 'Polygons and parabolas: some problems concerning the dynamics of planetary orbits', *Centaurus*, 31, 207–21.

Aiton, E.J. 1988b. 'The solution of the inverse-problem of central forces in Newton's *Principia*', *Archives Internationales d'Histoire des Sciences*, 38, 271–6.

Albarn, K. and others 1974. *The Language of Pattern. An Enquiry Inspired by Islamic Decoration*, London (Thames and Hudson).

Alexander, D.S. 1994. *A History of Complex Dynamics*, Braunschweig (Vieweg).

Alexandrov, P.S. 1971. (Ed.) *Die Hilbertschen Probleme*, Leipzig (Geest and Portig). [Text of Hilbert's lecture of 1900 with commentaries on later developments. Russian original 1969.]

Andersen, K. 1987. 'The problem of scaling and choosing parameters in perspective construction, particularly in the one by Alberti', *Analecta Romana Instituti Danici*, 16, 107–28.

Archibald, R.C. 1926. 'A rare pamphlet on Moivre and some of his successors', *Isis*, 8, 671–84.

Archibald, R.C. 1929. 'Notes on some minor English mathematical serials', *The Mathematical Gazette*, 14, 379–400.

Argand, J.R. 1874. *Essai sur une manière de représenter les quantités imaginaires dans les constructions géométriques* (ed. J. Houël), Paris (Gauthier-Villars). [Repr. 1971, Paris (Blanchard). Contains also related papers; 1st edn of book 1806.]

Ariotti, P.E. 1972. 'Aspects of the conception and development of the pendulum in the 17th century', *AHES*, 8, 330–410.

Artin, E. 1925. 'Theorie der Zöpfe', in *Collected papers*, 1965,

Reading, Mass. (Addison Wesley), pp. 416–41.

Atzema, E.J. 1993. 'The structure of systems of lines in 19th century geometrical optics', Utrecht (University doctoral dissertation).

Auroux, S. 1982. *L'illuminismo francese e la tradizione logica di Port-Royal*, Bologna (CLUEB).

Aveni, A.F. 1989. *Empires of Time. Calendars, Clocks and Cultures*, New York (Basic Books).

Bacharach, M. 1883. *Abriss zur Geschichte der Potentialtheorie*, Würzburg (Thein).

Bachmann, F. 1911. 'Über Gauss's zahlentheoretische Arbeiten', in Gauss, *Werke*, Vol. 10, Part 2, No. 1, 74 pp.

Baker, K.M. 1975. *Condorcet: From Natural Philosophy to Social Mathematics*, Chicago (University of Chicago Press).

Barbeau, E.J. 1979. 'Euler subdues a very obstreperous series', *American Mathematical Monthly*, 86, 356–72.

Barbour, J.B. 1989. *Absolute or Relative Motion?*, Vol. 1, *The Discovery of Dynamics*, Cambridge (Cambridge UP).

Barker, A. 1984, 1989. (Ed.) *Greek Musical Writings*, 2 vols., Cambridge (Cambridge UP).

Barnard, F.P. 1917. *The Casting-Counter and the Counting-Board*, Oxford (Oxford UP). [Repr. 1981, Castel Cary (Fox).]

Barrow-Green, J. 1996. *Poincaré and the three body problem*, Providence and London (American and London Mathematical Societies).

Bateman, H. 1911. 'Report on the history and present state of the theory of integral equations', *Report of the British Association for the Advancement of Science*, pp. 345–424.

Bennett, J. A. 1991. 'Geometry and surveying in early-seventeenth century England', *AS*, 48, 345–53. [In a special issue on early surveying.]

Benoit, P. and others 1992. (Eds.) *Histoire des fractions, fractions d'histoire*, Basel (Birkhäuser).

Berggren, L. 1987. 'Mathematical methods in ancient science', in I. Grattan-Guinness (ed.), *History in Mathematics Education*, Paris (Belin), pp. 14–32, 33–49. [Two articles, on spherics and on astronomy.]

Bernkopf, M. 1968. 'A history of infinite matrices', *AHES*, 4, 308–58.

Bernoulli, Jacob 1713. *Ars conjectandi*, Basel (Thurnis). [Various reprs. and translations.]

Bertoloni Meli, D. 1990. 'The relativization of centrifugal force', *Isis*, 81, 23–43.

Bertoloni Meli, D. 1993. *Equivalence and Priority: Newton Versus Leibniz*, Oxford (Clarendon Press).

Biermann, K.-R. 1978. 'Martin Bartels – eine Schlüsselfigur in der Geschichte der nichteuklidischen Geometrie?', *Leopoldina*, (3) 21, 137–57.

Biermann, K.-R. 1988. *Die Mathematik und ihre Dozenten an der Berliner Universität 1810–1933*, Berlin, DDR (Akademie-Verlag).

Bigourdan, G. 1901. *Le Système*

métrique des poids et mesures, Paris (Gauthier-Villars)

Björnbo, A.A. 1902. 'Studien zur Menelaos' Sphärik', *Abhandlungen zur Geschichte der mathematischen Wissenschaften,* 14, 1–154.

Blackmore, J. 1972. *Ernst Mach. His Life, Work, and Influence,* Berkeley (University of California Press).

Bollinger, M. 1972. 'Geschichtliche Entwicklung des Homologiebegriffs', *AHES,* 9, 94–166.

Bolza, O. 1916. 'Gauss und die Variationsrechnung', in Gauss, *Werke,* Vol. 10, Part 2, No. 5, 95 pp. [Date conjectured.]

Bonola, R. 1955. *Non-Euclidean Geometry,* New York (Dover).

Booker, P.J. 1963. *A History of Engineering Drawing,* London (Chatto and Windus). [Repr. 1979, London (Northgate).]

Borga, M., Freguglia, P. and Palladino, D. 1985. *I contributi fondazionali della scuola di Peano,* Milan (Franco Angeli).

Bortolotti, E. 1923. 'La trisezione dell'angolo', *Rendiconti della Accademia della Scienze dell'Istituto di Bologna, n.s.,* 27, 125–38.

Bos, H.J.M. 1974. 'Differentials, higher-order differentials and the derivative in the Leibnizian calculus', *AHES,* 14, 1–90.

Bos, H.J.M. 1981. 'The representation of curves in Descartes's *Géométrie*', *AHES,* 24, 295–338.

Bos, H.J.M. 1984. 'Arguments on motivation . . .: the "construction of equations", 1637–*ca.* 1750', *AHES,* 30, 332–80.

Bos, H.J.M. 1988. 'Tractional motion and the legitimation of transcendental curves', *Centaurus,* 31, 9–62.

Bos, H.J.M. and others 1987. 'Poncelet's closure theorem', *Expositiones mathematicae,* 5, 289–364.

Bottazzini, U. 1986. *The Higher Calculus,* New York (Springer).

Bottazzini, U. and Tazzioli, R. 1995. '*Naturphilosophie* and its role in Riemann's mathematics', *Revue d'Histoire des Mathématiques,* 1, 3–38.

Boyer, C.B. 1956. *History of Analytic Geometry,* New York (Scripta Mathematica).

Brackenridge, J.B. 1988. 'Newton's mature dynamics: revolutionary or reactionary?', *AS,* 45, 451–76.

Brackenridge, J.B. 1995. *The Key to Newton's Dynamics,* Berkeley and London (University of California Press).

Brent, J. 1993. *Charles Sanders Peirce. A Life,* Bloomington (Indiana UP)

Brigaglia, A. and Masotto, G. 1982. *Il Circolo Matematico di Palermo,* Bari (Dedalo).

Brill, A. von and Noether, M. 1894. 'Die Entwicklung der Theorie der algebraischen Functionen', *JDMV,* 3, Part 2, i–xxiii, 109–566.

Brock, W.H. 1967. (Ed.) *The Atomic Debates,* Leicester (Leicester UP).

Brouwer, L.E.J. 1910. 'Zur Analysis situs', *Mathematische Annalen,* 68, 422–34. [Also in *Collected Works,* Vol. 2, 352–70.]

Brown, J. 1905. (Ed.). *A History of Accounting and Accountants,* Edinburgh (Jack).

Brush, S.G. 1976. *The Kind of Motion We Call Heat. A History of the Kinetic Theory of Gases in the 19th Century*, 2 vols., Amsterdam (North-Holland).
Bryce, R.A. 1986. 'Paolo Ruffini and the quintic equation', *Symposia mathematica*, 27, 169–85.
Buchwald, J.Z. 1979. 'The Hall effect and Maxwellian electrodynamics in the 1880's', *Centaurus*, 23, 51–99, 118–62.
Buchwald, J.Z. 1981. 'The abandonment of Maxwell's electrodynamics', *Archives Internationales d'Histoire des Sciences*, 31, 135–80, 373–438.
Burkert, W. 1972. *Lore and Science in Ancient Pythagoreanism*, Cambridge, Mass. (Harvard UP).
Burkhardt, H.F.K.L. 1908. 'Entwicklungen nach oscillirenden Functionen und Integration der Differentialgleichungen der mathematischen Physik', *JDMV*, 10, Part 2, xii + 1804 pp.
Burkhardt, H.F.K.L. 1912. 'Untersuchungen von Cauchy und Poisson über Wasserwellen', *Sitzungsberichte der Akademie der Wissenschaften zu München, Mathematisch-physikalische Klasse*, 97–120.
Bursill-Hall, P. 1993. (Ed.) *R.J. Boscovich. His Life and Scientific Work*, Rome (Enciclopedia Italiana).
Butzer, P.L. and Lohrmann, D. 1993. (Eds.) *Science in Western and Eastern Civilization in Carolingian Times*, Basel (Birkhäuser).
Cajori, F. 1928, 1929. *A History of Mathematical Notations*, 2 vols., La Salle (Open Court).
Cantor, G.F.L.P. 1932. *Gesammelte Abhandlungen mathematischen und philosophischen Inhalts* (ed. E. Zermelo), Berlin (J. Springer). [Repr. 1980; also 1962, Hildesheim (Olms).]
Cantor, G.N. and Hodge, M.J.S. 1981. (Eds.) *Conceptions of Ether*, Cambridge (Cambridge UP).
Cantor, M. 1875. *Die römische Agrimensoren*, Leipzig (Teubner). [Repr. 1982, Schaan (Sandig).]
Cantor, M. 1907. *Vorlesungen über Geschichte der Mathematik*, Vol. 3, 2nd edn, Leipzig (Teubner).
Cantor, M. 1908. (Ed.) *Vorlesungen über Geschichte der Mathematik*, Vol. 4, Leipzig (Teubner).
Carslaw, H.S. 1914. 'The discovery of logarithms by Napier of Merchiston', *Journal of Proceedings of the Royal Society of New South Wales*, 48, 42–72.
Carslaw, H.S. 1921. *Introduction to the Mathematical Theory of the Conduction of Heat in Solids*, 2nd edn, London (MacMillan).
Cauchy, A.L. 1821. *Cours d'analyse*, Paris (De Burė). [Repr. 1992, Bologna (CLUEB), with introduction by U. Bottazzini.]
Cavaillès, J. 1938. *Méthode axiomatique et formalisme*, 3 parts, Paris (Hermann).
Caveing, M. 1985. 'La Tablette babylonienne AO 17264 . . . et le problème des six frères', *HM*, 12, 6–24.
Cayley, A. 1854. 'On the theory of groups', in *Collected Papers*, Vol. 2, pp. 123–32.

Chabert, J.L. and others. 1994. *Histoire d'algorithmes,* Paris (Belin).

Clagett, M. 1964–1984. *Archimedes in the Middle Ages,* 5 vols., Madison (University of Wisconsin Press), then Philadelphia (American Philosophical Society).

Cohen, I.B. 1971a. 'Newton's second law and the concept of force in the *Principia*', in *The* annus mirabilis *of Sir Isaac Newton 1666–1966* (ed. R. Palter), Cambridge, Mass. (MIT Press), pp. 143–85.

Cohen, I.B. 1971b. *Introduction to Newton's Principia,* Cambridge (Cambridge UP).

Cohen, I.B. 1984. 'Florence Nightingale', *Scientific American,* 250, No. 3, 98–107.

Condorcet, Marquis de 1994. *Arithmétique politique. Textes rares ou inédits (1767–1789)* (ed. B. Bru and P. Crépel), Paris (Presses Universitaires de France).

Cooke, R. 1984. *The Mathematics of Sonya Kovalevskaya,* New York (Springer).

Cooke, R.L. 1993. 'Uniqueness of trigonometric series and descriptive set theory', *AHES,* 45, 281–334.

Copenhaver, B. and Schmitt, C.B. 1992. *Renaissance Philosophy,* Oxford (Oxford UP).

Corcoran, J. 1973. 'Gaps between logical theory and mathematical practice', in M. Bunge (ed.), *The Methodological Unity of Science,* Dordrecht (Reidel), pp. 23–50.

Corry, L. 1996. *Modern algebra and the rise of mathematical structures,* Basel (Birkhäuser).

Cournot, A.A. 1847. *De l'origine et des limites da la correspondance entre l'algèbre et la géométrie,* Paris and Algiers (Hachette).

Crépel, P. 1988. 'Condorcet, la théorie des probabilités et les calculs financiers', in R. Rashed (ed.), *Sciences à l'époque de la Révolution Française,* Paris (Blanchard), pp. 267–325.

Cross, J.J. 1985. 'Integral theorems in Cambridge mathematical physics, 1830–55', in {Harman 1985}, pp. 112–48.

Crowe, M.J. 1967. *A History of Vector Analysis,* Notre Dame and London (Notre Dame UP).

Culin, S. 1907. *Games of the North American Indians,* Washington (Government Printing Office: Bureau of American Ethnology, No. 24). [Repr. in 2 vols., 1992, New York (Dover).]

Curl, J.S. 1993. *A Celebration of Death,* 2nd edn, London (Batsford).

Czuber, E. 1899. 'Die Entwicklung der Wahrscheinlichkeitstheorie und ihre Anwendungen', *JDMV,* 7, Part 2, 270 pp.

Czyz, J. 1994. *Paradoxes of Measures and Dimensions Originating in Felix Hausdorff's Ideas,* Singapore (World Scientific).

d'Ocagne, M. 1911. 'Calculs numériques', in *Enyclopédie des sciences mathématiques,* Tome 1, Vol. 4, 196–452 (Article I23).

Dale, A.I. 1988. 'On Bayes' theorem and the inverse Bernoulli theorem', *HM,* 15, 348–60.

Dale, A.I. 1991. *A History of Inverse Probability from Thomas Bayes to Karl Pearson*, New York (Springer).

Daniels, N. 1989. *Thomas Reid's Inquiry: the Geometry of Visibles and the Case for Realism*, 2nd edn, Stanford (Stanford UP).

Daston, L. 1988. *Classical Probability in the Enlightenment*, Princeton (Princeton UP).

Dauben, J.W. 1979. *Georg Cantor*, Cambridge, Mass. (Harvard UP). [Repr. 1990, Princeton (Princeton UP).]

Dauben, J.W. 1981. (Ed.) *Mathematical Perspectives*, New York (Academic Press).

Dauben, J.W. 1985. (Ed.) *The History of Mathematics from Antiquity to the Present. A Selective Bibliography*, New York (Garland).

Davis, A.E.L. 1975. 'Systems of conics in Kepler's work', *Vistas in Astronomy*, 18, 673–85.

Davis, A.E.L. 1992. 'Kepler's resolution of individial planetary motion', *Centaurus*, 35, 97–102 [with succeeding papers, pp. 103–91].

De Morgan, A. 1835. *The Elements of Algebra*, London (Taylor and Walton).

De Morgan, A. 1842. *The Differential and Integral Calculus*, London (Taylor and Walton). [Published in parts from 1836.]

de la Vallée Poussin, C.-J. 1962. 'Gauss et la théorie du potentiel', *Revue des Questions Scientifiques*, 133, 314–30.

Dedekind, J.W.R. 1985. *Vorlesungen über Differential- und Integralrechnung* (ed. M.-A. Knus and W. Scharlau), Braunschweig (Vieweg and DMV).

Dee, J. 1570. 'Mathematicall præface', in *The Elements of Geometrie of the Most Auncient Philosopher Euclide of Megara* (translated by H. Billingsley), London (Daye), 50 pp. [Preface repr., with introduction by A. Debus, 1975, New York (Science History Publications).]

Delaunay, C. 1860, 1867. 'Théorie du mouvement de la Lune', *Mémoires de l'Académie des Sciences*, 28–29 [both volumes: xxviii + 883 pp., xi + 931 pp.].

dell'Aglio, L. 1993. 'Tradizione di ricerca nella meccanica celeste classica: il problema dei tre corpi in Levi-Civita e Sundman', *Physis*, 30, 105–44.

Detlefsen, M. and others 1976. 'Computation with Roman numerals', *AHES*, 15, 141–48.

Dhombres, J.G. 1986. 'Quelques aspects de l'histoire des équations fonctionnelles', *AHES*, 36, 91–181.

Diaz-Bolio, J. 1987. *The Geometry of the Maya and Their Rattlesnake Art*, Merida, Mexico (Mayan Area).

Diaz-Bolio, J. 1988. *Why the Rattle-snake in Mayan Civilisation*, Merida, Mexico (Mayan Area). [Also various other related works.]

Dickstein, S. 1899. 'Zur Geschichte der Prinzipien der Infinitesimalrechnung', *Abhandlungen zur Geschichte der Mathematik*, 9, 65–79.

Dieudonné, J. 1978. (Ed.) *Abrégé d'histoire des mathématiques*

1700–1900, 2 vols., Paris (Hermann).

Dieudonné, J. 1989. *A History of Algebraic and Differential Topology 1900–1960*, Basel (Birkhäuser).

Dinsmoor, W.B. 1923. 'How the Parthenon was planned', *Architecture*, 47, 177–80, 241–4.

Diocles 1976. *On Burning Mirrors* (ed. with commentary by G.J. Toomer), Berlin (Springer).

Dold-Samplonius, Y. 1993. 'Practical Arabic mathematics: measuring the muqarnas by al-Kāshi', *Centaurus*, 35, 193–242.

Dornseiff, F. 1922. *Das Alphabet in Mystik und Magie*, 1st edn, Leipzig (Teubner). [2nd edn 1925.]

Dugac, P. 1973. 'Eléments d'analyse de Karl Weierstrass', *AHES*, 10, 41–176.

Dugas, R. 1957. *History of Mechanics*, New York (Central).

Dutka, J. 1981. 'The incomplete beta function – a historical profile', *AHES*, 24, 11–29.

Dutka, J. 1989. 'On the Saint Petersburg paradox', *AHES*, 39, 13–39.

Dyck, W. 1892, 1893. (Ed.) *Katalog mathematischer und mathematisch-physikalischer Modelle, Apparate und Instrumente*, 2 vols., Munich (Wolf).

Eccarius, W. 1976. 'August Leopold Crelle als Herausgeber wissenschaftlicher Fachzeitschriften', *AS*, 33, 229–61.

Eccarius, W. 1977. 'August Leopold Crelle als Förderer bedeutender Mathematiker', *JDMV*, 79, 137–74.

Eccarius, W. 1989. 'Über die Einfluss der Mathematik ... der thüringschen Realschulen', *Wissenschaftliche Zeitschrift der Pädagogischen Hochschule ... Erfurt/-Mühlhausen*, 25, 147–55. [See also 23 (1987), 136–46.]

Echeverria, J. and others 1992. (Eds.) *The Space of Mathematics*, Berlin and New York (de Gruyter).

Edwards, A.W.F. 1987. *Pascal's Arithmetical Triangle*, New York (Oxford UP).

Edwards, A.W.F. 1989. 'Venn diagrams for many sets', *New Scientist*, (7 January), 51–6.

Edwards, H. 1977. *Fermat's Last Theorem*, New York (Springer).

Eisenhart, C. 1964. 'The meaning of "least" in least squares', *Journal of the Washington Academy of Sciences*, 54, 24–33.

Eisenhart, C. 1983. 'Laws of error', in *Encyclopaedia of Statistical Sciences*, Vol. 4, New York (Wiley), pp. 530–66.

Ellison, W. and Ellison, F. 1978. 'Théorie des nombres', in {Dieudonné 1978}, Vol. 1, pp. 165–334.

Engelsmann, S.G. 1984. *Families of Curves and the Origins of Partial Differentiation*, Amsterdam (North-Holland).

Epple, M. 1995. 'Branch points of algebraic functions and the beginnings of modern knot theory', *HM*, 22, 371–401.

Euclid 1886–1916. *Euclidis opera omnia*, 10 vols. (ed J.L. Heiberg and J. Menge), Leipzig (Teubner).

Euler, L. 1770. *Vollständige Anleitung zur Algebra*, as *Opera omnia*, Series 1, Vol. 1.

Euler, L. 1954–5. *Opera omnia*, Series 2, Vols. 11–12, Zurich (Orell Fussli). [Papers on fluid mechanics.]

Euler, L. 1957. *Opera omnia*, Series 2, Vol. 5, Zurich (Orell Fussli). [Papers on the principle of least action.]

Evans, G.R. 1976. 'The rithmomachia: a mediaeval mathematics teaching aid', *Janus*, 63, 257–73.

Fauvel, J.G. 1987. *The Renaissance of Mathematical Science in Britain*, Milton Keynes (Open University Course MA 290, Block 2).

Feigenbaum, L. 1985. 'Brook Taylor and the method of increments', *AHES*, 34, 1–140.

Ferreirós, J. 1993. 'On the relations between Georg Cantor and Richard Dedekind', *HM*, 20, 343–63.

Feuer, L.S. 1971. 'The social roots of Einstein's theory of relativity', *AS*, 27, 277–98, 313–44.

Field, J.V. 1988. *Kepler's Geometrical Cosmology*, Chicago (University of Chicago Press).

Field, J.V. and Gray, J.J. 1987. *The Geometrical Work of Gerard Desargues*, New York (Springer).

Finsterwalder, S. 1899. 'Die geometrischen Grundlagen der Photogrammetrie', *JDMV*, 6, Part 2, 1–41.

Flegg, G. and others 1984. (Ed. and translators) *Nicolas Chuquet, Renaissance Mathematician*, Dordrecht (Reidel).

Fogel, R.W. 1982. ' "Scientific" history and traditional history', in *Logic, Methodology and Philosophy of Science VI (Hannover 1979)*, Amsterdam (North-Holland), pp. 15–61.

Folkerts, M. 1981. 'Mittelalterliche mathematische Handschriften', in {Dauben 1981}, pp. 53–93.

Forsyth, A.R. 1890–1906. *Theory of Differential Equations*, 6 vols., Cambridge (Cambridge UP).

Fowler, D. 1987. *The Mathematics of Plato's Academy*, Oxford (Clarendon Press).

Fox, R. 1974. 'The rise and fall of Laplacian physics', *Historical Studies in the Physical Sciences*, 4, 81–136.

Franci, R. and Toti Rigatelli, L. 1985. 'Towards a history of algebra from Leonardo of Pisa to Luca Pacioli', *Janus*, 72, 17–82.

Franksen, O.I. and Grattan-Guinness, I. 1989. 'The earliest contribution to location theory? Spacio-economic equilibrium with Lamé and Clapeyron, 1829', *Mathematics and Computing in Simulation*, 31, 195–220.

Fraser, C.G. 1983. 'J.L. Lagrange's early contributions to the principles and methods of mechanics', *AHES*, 28, 197–241.

Fraser, C.G. 1985. 'd'Alembert's principle: the original formulation and application in Jean d'Alembert's Traité de dynamique (1743)', *Centaurus*, 28, 31–61, 145–59.

Fréchet, M. 1906. 'Sur quelques points du calcul fonctionnel', *Rendiconti del Circolo Matematico di Palermo*, 22, 1–74.

Freguglia, P. 1988. *Ars analitica*, Busto Arsizio (Bramante).

Freytag Löringhoff, B. von 1987. *Wilhelm Schickard und seine Rechenmaschine von 1623*, Tübingen (Tübingen UP).

Fu, D. 1991. 'Why did Liu Hui fail to find the volume of the sphere?', *HM*, 18, 212–38.

Fuller, A.T. 1982–1986. 'James Clerk Maxwell's Cambridge manuscripts: extracts relating to control and stability', *International Journal of Control*, 35, 785–805; 36, 547–74; 37, 1197–238; 39, 619–56; 43, 805–18, 1135–68. [See also 30 (1979), 729–44; 43, 1593–612.]

Gabbey, A. 1971. 'Force and inertia in seventeenth-century dynamics', *Studies in the History and Philosophy of Science*, 2, 1–67.

Garciadiego, A.R. 1992. *Bertrand Russell and the Origins of the Set-Theoretic 'Paradoxes'*, Basel (Birkhäuser).

Gårding, L. 1989. 'History of the mathematics of double refraction', *AHES*, 40, 355–84.

Gauss, C.F. 1801. *Disquisitiones arithmeticae*, Leipzig (Fleischer). [Also *Werke*, Vol. 1. Translations: French, 1807, Paris; German, 1889, Berlin; English, 1966, New Haven, Conn.]

Gautier, A. 1817. *Essai historique sur le problème des trois corps*, Paris (Courcier).

Gerdes, P. 1994. 'On mathematics in the history of sub-Saharan Africa', *HM*, 21, 345–76.

Giacardi, L. and Roero, C.S. 1979. *La matematica delle civiltà arcaiche*, Turin (Stampatori).

Giacardi, L. and Roero, C.S. 1987. *L'origine della numerazione romana*, Foligno (Edizioni dell'Arquata).

Gilain, C. 1981. (Ed.) *A.-L. Cauchy. Equations différentielles ordinaires. Cours inédit* (*fragment*) (ed. C. Gilain), Paris (Etudes Vivantes) and New York (Johnson). [Cauchy proofs, prepared probably in 1824.]

Gilain, C. 1991. 'Sur l'histoire du théorème de l'algèbre', *AHES*, 42, 92–136.

Gillies, D. 1987. 'Was Bayes a Bayesian?', *HM*, 14, 325–46.

Gillies, D. 1992. (Ed.) *Revolutions in Mathematics*, Oxford (Clarendon Press).

Gillings, R.J. 1972. *Mathematics in the Time of the Pharaohs*, Cambridge, Mass. (MIT Press). [Repr. 1982, New York (Dover).]

Gispert, H. 1983. 'Sur les fondements de l'analyse en France', *AHES*, 28, 37–106.

Gispert, H. 1991. *La France mathématique. La Société Mathématique de France* (*1870–1914*), Paris (Belin).

Glaisher, J.W.L. 1921. 'On the early history of the signs + and − and on the early German arithmeticians', *Messenger of Mathematics*, 51, 1–148.

Goldstein, B.R. 1984. 'Eratosthenes on the "measurement" of the Earth', *HM*, 11, 411–16.

Goldstine, H. 1980. *A History of the Calculus of Variations From the 16th Through the 19th Century*, New York (Springer).

Gottwald, S. and others. 1990. (Eds.) *Lexikon bedeutender Mathematiker*, Leipzig (Bibliographisches Institut).

Gouk, P. 1988. 'The harmonic roots of Newtonian science', in J. Fauvel and others (eds.), *Let Newton Be!* Oxford (Oxford UP), pp. 99–125.

Grant, E. 1974 (Ed.) *A Source Book in Medieval Science*, Cambridge, Mass. (Harvard UP).

Grant, R. 1852. *History of Physical Astronomy, From the Earliest Ages to the Middle of the Nineteenth Century*, London (Bohn). [Repr. 1966, New York (Johnson).]

Grattan-Guinness, I. 1970a. 'Berkeley's criticism of the calculus as a study in the theory of limits', *Janus*, 56, 215–27 [erratum in 57 (1971), 80].

Grattan-Guinness, I. 1970b. *The Development of the Foundations of Mathematical Analysis from Euler to Riemann*, Cambridge, Mass. (MIT Press).

Grattan-Guinness, I. 1970c. 'An unpublished paper by Georg Cantor', *Acta mathematica*, 124, 65–107.

Grattan-Guinness, I. 1971. 'Materials for the history of mathematics in the Institut Mittag-Leffler', *Isis*, 62, 363–74.

Grattan-Guinness, I. 1972. 'A mathematical union: William Henry and Grace Chisholm Young', *AS*, 29, 105–86.

Grattan-Guinness, I. 1974. 'Achilles is still running', *Transactions of the C.S. Peirce Society*, 10, 8–16.

Grattan-Guinness, I. 1975. 'Wiener on the logics of Russell and Schröder', *AS*, 32, 103–32.

Grattan-Guinness, I. 1977. *Dear Russell – Dear Jourdain. A Commentary on Russell's Logic*, London (Duckworth) and New York (Columbia UP).

Grattan-Guinness, I. 1980. (Ed.) *From the Calculus to Set Theory, 1630–1910: An Introductory History*, London (Duckworth).

Grattan-Guinness, I. 1986. 'The Società Italiana, 1782–1815', *Symposium mathematica*, 27, 147–68.

Grattan-Guinness, I. 1988. 'Living together and living apart: on the interactions between mathematics and logics from the French Revolution to the First World War', *South African Journal of Philosophy*, 7, No. 2, 73–82.

Grattan-Guinness, I. 1990a. *Convolutions in French Mathematics, 1800–1840*, 3 vols., Basel (Birkhäuser) and Berlin (Deutscher Verlag der Wissenschaften).

Grattan-Guinness, I. 1990b. 'The varieties of mechanics by 1800', *HM*, 17, 313–38.

Grattan-Guinness, I. 1990c. 'Does History of Science treat of the history of science? The case of mathematics', *History of Science*, 28, 147–73.

Grattan-Guinness, I. 1992a. 'Charles Babbage as an algorithmic thinker', *Annals of the History of Computing*, 14, No. 3, 34–48.

Grattan-Guinness, I. 1992b. 'Structure-similarity as a cornerstone of the philosophy of mathematics', in {Echeverria and others 1992}, pp. 91–111.

Grattan-Guinness, I. 1993. 'European mathematical education in the 1900s and 1910s', in E. Ausejo and M. Hormigon (eds.), *Messengers of Mathematics: European Mathematical Journals (1800–1946)*, Madrid (Siglo XXI), pp. 117–30. [Other articles on various journals.]

Grattan-Guinness, I. 1994. ' "A new type of question": On the prehistory of linear and non-linear programming, 1770–1940', in E. Knobloch and D. Rowe (eds.), *History of Modern Mathematics*, Vol. 3, New York (Academic Press), pp. 43–89.

Grattan-Guinness, I. 1996a. 'Mozart 18, Beethoven 32: hidden shadows of integers in classical music', in J.W. Dauben and others (eds.), *The History of Mathematics: the State of the Art*, Boston (Academic Press), pp. 29–47.

Grattan-Guinness, I. 1996b. 'Numbers, magnitudes, ratios and proportions in Euclid's *Elements*: how did he handle them?', *HM*, 23, 355–375.

Grattan-Guinness I. and Ravetz, J.R. 1972. *Joseph Fourier 1768–1830*, Cambridge, Mass. (MIT Press).

Gray, J.D. 1985. 'The shaping of the Riesz representation theorem', *AHES*, 31, 127–87.

Gray, J.J. 1984. 'Fuchs and the theory of differential equations', *Bulletin of the American Mathematical Society*, 10, 1–26.

Gray, J.J. 1986. *Linear Differential Equations and Group Theory from Riemann to Poincaré*, Basel (Birkhäuser).

Gray, J.J. 1989. *Ideas of Space*, 2nd edn, Oxford and New York (Clarendon Press).

Gray, J.J. 1993. 'Poincaré, topological dynamics, and the stability of the solar system', in P.M. Harman and A.E. Shapiro (eds.), *The Investigation of Difficult Things*, Cambridge (Cambridge UP), pp. 503–24.

Gray, J.J. 1994. 'On the history of the Riemann mapping theorem', *Rendiconti del Circolo Matemataco di Palermo*, (2) 34, supplement, 47–94.

Greenberg, J.L. 1981. 'Alexis Fontaine's "fluxio-differential method"', *AS*, 38, 251–90. [See also 39 (1982), 1–36.]

Greenberg, J.L. 1995. *The Problem of the Earth's Shape from Newton to Clairaut*, New York (Cambridge UP).

Grossmann, M. and Katz, R. 1972. *Non-Newtonian Calculus*, Pigeon Cove, Mass. (Lee Press).

Guicciardini, N. 1989. *The Development of Newtonian Calculus in Britain 1700–1800*, Cambridge (Cambridge UP).

Haas, A.E. 1909. *Die Entwicklungsgeschichte des Satzes von der Erhaltung der Kraft*, Vienna (Hölder). [Italian translation: 1990, Pavia (La Goliardica).]

Hacking, I. 1975. *The Emergence of Probability*, Cambridge (Cambridge UP).

Hacking, I. 1990. *The Taming of Chance*, Cambridge (Cambridge UP).

Hadamard, J. 1922. 'The early

work of Henri Poincaré', *The Rice Institute Pamphlets*, 9, 111–83. [See also the version of 1923 repr. in Poincaré, *Oeuvres*, Vol. 11, 152–242.]

Hald, A. 1990. *A History of Probability and Statistics and Their Applications Before 1750*, New York (Wiley).

Hamburg, R. R[ider-] 1976. 'The theory of equations in the 18th century: the work of Joseph Lagrange', *AHES*, 16, 17–36.

Hamilton, W.R. 1931, 1940. *Mathematical Papers*, Vols. 1 and 2, Cambridge (Cambridge UP).

Hankins, T. 1980. *Sir William Rowan Hamilton*, Baltimore (Johns Hopkins UP).

Hardy, G.H. 1918. 'Sir George Stokes and the concept of uniform convergence', in *Collected Papers*, Vol. 7, pp. 505–13.

Harman, P.M. 1985. (Ed.) *Wranglers and Physicists*, Manchester (Manchester UP).

Harman, P.M. 1987. 'Mathematics and reality in Maxwell's dynamical physics', in R. Kargon and P. Achinstein (eds.), *Kelvin's Baltimore Lectures and Modern Theoretical Physics*, Cambridge, Mass. (MIT Press), pp. 267–97.

Hasse, H. 1930. *Bericht über neuere Untersuchungen und Probleme aus der Theorie der algebraischen Zahlkörper*, Leipzig (Teubner).

Hawkins, T.W. 1970. *Lebesgue's Theory of Integration*, Madison (University of Wisconsin Press).

Hawkins, T.W. 1971. 'The origins of the theory of group characters', *AHES*, 7, 142–70.

Hawkins, T.W. 1972. 'Hypercomplex numbers, Lie groups, and the creation of group representation theory', *AHES*, 8, 243–87.

Hawkins, T.W. 1975. 'Cauchy and the spectral theory of matrices', *HM*, 2, 1–29.

Hawkins, T.W. 1977. 'Weierstrass and the theory of matrices', *AHES*, 17, 119–63.

Heath, T.L. 1926. *The Thirteen Books of Euclid's Elements*, 2nd edn, 3 vols., Cambridge (Cambridge UP).

Helmholtz, H. von 1847. *Über die Erhaltung von Kraft*, Berlin (Reimer). [Numerous reprs. and translations.]

Hensel, S. and others. 1989. *Mathematik und Technik im 19. Jahrhundert in Deutschland*, Göttingen (Vandenhoeck und Ruprecht).

Hentschel, K. 1991. *Interpretationen und Fehlinterpretationen der speziellen und allgemeinen Relativitätstheorie durch Zeitgenosssen Albert Einsteins*, Basel (Birkhäuser).

Herivel, J. 1975. *The Background to Newton's* Principia, Oxford (Clarendon Press).

Hermann, A. 1971. *The Genesis of Quantum Theory*, Cambridge, Mass. (MIT Press).

Herz-Fischler, R. 1987. *A Mathematical History of Division in Extreme and Mean Ratio*, Waterloo, Canada (Wilfred Laurier UP).

Hess, H.-J. and Nagel, F. 1988. (Eds.) *Des Ausbau des Calculs durch Leibniz und die Brüder Bernoulli*, Wiesbaden (Steiner).

Heyman, J. 1972. *Coulomb's Memoir on Statics. An Essay on the History of Civil Engineering*, Cambridge (Cambridge UP).

Hilbert, D. 1897. 'Die Theorie der algebraischen Zahlkörper', *JDMV*, 4, 175–546. [Also in *Gesammelte Abhandlungen*, Vol. 1, 63–363.]

Hilbert, D. 1902. *Foundations of Geometry*, 1st edn (translated by E.J. Townsend), Chicago (Open Court). [German original 1899; many later edns.]

Hilbert, D. 1905. 'Logische Principien des mathematischen Denkens', manuscript versions in *Nachlass*, Göttingen University Library. [Edition to be prepared.]

Hilts, V.L. 1975. 'A guide to Francis Galton's *English Men of Science*', *Transactions of the American Philosophical Society*, *n.s.*, 75, Part 5, 85 pp.

Hoe, J. 1978. 'The Jade Mirror of the Four Unknowns – some reflections', *New Zealand Mathematical Chronicle*, 7, 125–56.

Hoffmann, B. 1972. *Albert Einstein Creator and Rebel*, London (Hart-Davis).

Hofmann, J.E. 1944. 'Neues über Fermats zahlentheoretische Herausforderungen von 1657', *Abhhandlungen der Preussischen Akademie der Wissenschaften, mathematisch-naturwissenschaftliche Klasse*, No. 9, 52 pp.

Hofmann, J.E. 1959. 'Um Eulers erste Reihestudien', in K. Schröder (ed.), *Sammelband der zu Ehren des 250. Geburtstages Leonhard Eulers*, Berlin (Akademie-Verlag), pp. 139–208.

Hofmann, J.E. 1974. *Leibniz in Paris 1672–1676*, Cambridge (Cambridge UP).

Hogendijk, J.P. 1985. [Review of an edition of Diophantos's *Arithmetica* by J. Sesiano], *HM*, 12, 82–5.

Hopper, V.F. 1938. *Medieval Number Symbolism*, New York (Cooper Square). [Repr. 1969.]

Houzel, C. 1978. 'Fonctions elliptiques et intégrales abéliennes', in {Dieudonné 1978}, Vol. 2, pp. 1–113.

Hoyer, U. 1973. 'Stabilitätsbetrachtungen in der Bohrschen Atomtheorie', *AHES*, 10, 177–206.

Høyrup, J. 1990. 'Algebra and naive geometry. An investigation of some aspects of old Babylonian mathematical thought', *Altorientalische Forschungen*, 17, 27–69, 262–354.

Høyrup, J. 1994. *In Number, Measure and Weight*, Albany (State University of New York Press).

Ingrao, B. and Israel, G. 1990. *The Invisible Hand. Economic Equilibrium in the History of Science*, Cambridge, Mass. (MIT Press).

Isis 1971– . *Isis Cumulative Bibliography* (ed. M. Whitrow, then J. Neu), London (Mansell). [Continuing series.]

Israel, G. 1993. 'The two paths of the mathematicisation of the social and economic sciences', *Physis, n.s.*, 30, 27–78.

Israel, G. 1996. *La mathématisation du réel*, Paris (Seuil).

James, J. 1979, 1981. *The Contractors of Chartres*, 2 vols., London (Croom Helm).

Jaouiche, K. 1986. *La Théorie des parallèles en pays d'Islam*, Paris (Vrin).

Johnson, D.M. 1979, 1981. 'The problem of the invariance of dimension in the growth of modern topology', *AHES*, 20, 97–188; 25, 85–267.

Jourdain, P.E.B. 1918. *The Philosophy of Mr. B*rtr*nd R*ss*ll*, London (Allen and Unwin). [Also in *Selected Essays on the History of Set Theory and Logics (1906–1918)* (ed. I. Grattan-Guinness), 1991, Bologna (CLUEB), pp. 245–342.]

Judaica 1971. 'Alphabet', 'Astronomy', 'Gematria', 'Kabbalah', 'Mathematics', 'Numbers', *passim* in *Encyclopaedia Judaica*, 16 vols., Jerusalem (Keter). [See also biographical articles.]

Kanigel, R. 1991. *The Man Who Knew Infinity. A Life of the Genius Ramanujan*, New York (Scribner's).

Katz, V. 1993. *A History of Mathematics. An Introduction*, New York (HarperCollins).

Kaunzner, W. 1987. 'On the transmission of mathematical knowledge to Europe', *Sudhoffs Archiv*, 71, 129–40.

Kendall, D.G. 1975. 'The recovery of structure from fragmentary information', *Philosophical Transactions of the Royal Society*, 279*A*, 547–82.

Kiernan, B.M. 1971. 'The development of Galois theory from Lagrange to Artin', *AHES*, 8, 40–154.

King, D.A. 1973. 'Ibn Yunus' very useful tables for reckoning time by the Sun', *AHES*, 10, 342–94.

Kipnis, N. 1991. *History of the Principle of Interference of Light*, Basel (Birkhäuser).

Klein, C.F. 1882. *On Riemann's Theory of Algebraic Functions and Their Integrals* (translated by F. Hardcastle), 1893, Cambridge (Macmillan and Bowes).

Klein, C.F. 1925a, 1925b, 1928. *Elementare Mathematik vom höheren Standpunkt aus*, 3rd edn, 3 vols., Berlin (Springer). [Various reprs. English translation of Vols. 1 and 2: *Elementary Mathematics from an Advanced Standpoint*, 2 vols., 1932, New York (Macmillan); repr. 1939, New York (Dover).]

Klein, C.F. 1926, 1927. *Vorlesungen über die Entwicklung der Mathematik im 19. Jahrhundert*, 2 parts, Berlin (Springer). [Repr. n.d., New York (Chelsea).]

Klein, J. 1968. *Greek Mathematical Thought and the Origins of Algebra*, Cambridge, Mass. (MIT Press). [Repr. 1992, New York (Dover).]

Kline, M. 1972. *Mathematical Thought from Ancient to Modern Times*, New York (Oxford UP).

Knobloch, E. 1979. 'Musurgia univeralis', *History of Science*, 17, 257–75.

Knobloch, E. 1992. 'Historical aspects of the foundations of error theory', in {Echeverria and others 1992}, pp. 253–79.

Knorr, W.R. 1975. *The Evolution of the Euclidean Elements*, Dordrecht (Reidel).

Knorr, W.R. 1978a. 'Archimedes and the pre-Euclidean proportion theory', *Archives Internationales d'Histoire des Sciences*, 28, 183–244.

Knorr, W.R. 1978b. 'Archimedes and the *Elements*: proposal for a revised chronological ordering of the Archimedean corpus', *AHES*, 19, 211–90.

Knorr, W.R. 1985. 'The geometer and the archaeoastronomers: on the prehistoric origins of mathematics', *British Journal for the History of Science*, 18, 187–212.

Knorr, W.R. 1986a. 'Archimedes' *Dimension of the Circle*: a view of the genesis of the text', *AHES*, 35, 281–324.

Knorr, W.R. 1986b. *The Ancient Tradition of Geometric Problems*, Boston (Birkhäuser).

Knorr, W.R. 1989. *Textual Studies in Ancient and Medieval Geometry*, Boston (Birkhäuser).

Kollerstrom, N. 1992. 'Thomas Simpson and "Newton's method of approximation": an enduring myth', *British Journal for the History of Science*, 25, 347–54.

Körner, T. 1904. 'Der Begriff des materiellen Punktes in der Mechanik des achtzehnten Jahrhunderts', *Bibliotheca mathematica*, (3)5, 15–62.

Kötter, E. 1898. 'Die Entwickelung der synthetischen Geometrie', *JDMV*, 5, Part 2, 128 pp.

Kötter, F.W.F. 1893. 'Die Entwicklung der Lehre vom Erddruck', *JDMV*, 2, 77–154.

Koyré, A. 1957. *From the Closed World to the Infinite Universe*, Baltimore (Johns Hopkins UP).

Kramer, E. 1970. *The Nature and Growth of Modern Mathematics*, New York (Hawthorn).

Krüger, L. and others 1987. (Eds.) *The Probabilistic Revolution*, 2 vols., Cambridge, Mass. (MIT Press).

Kuhn, T.S. 1970. *The Structure of Scientific Revolutions*, 2nd edn, Chicago (University of Chicago Press).

Lakatos, I. 1976. *Proofs and Refutations*, Cambridge (Cambridge UP).

Lam, L.Y. 1994. 'Jiu Shang Suanshu . . . (nine chapters on the mathematical art)', *AHES*, 48, 1–51.

Lam, L.Y. and Ang, T.S. 1992. *Fleeting Footsteps. Tracing the Conception of Arithmetic and Algebra in Ancient China*, Singapore (World Scientific Publishing).

Lambert, J.H. 1977. *Colloque . . . Jean-Henri Lambert*, Paris (Ophrys).

Larmor, J.J. 1900. *Aether and Matter*, Cambridge (Cambridge UP).

Laughlin, B. 1995. *The Aristotle Adventure*, Flagstaff, Ariz. (Albert Hale).

Laugwitz, D. 1989. 'Definite values of infinite sums', *AHES*, 39, 195–245.

Laugwitz, D. 1995. *Bernhard Riemann 1826–1866*, Basel (Birkhäuser).

Lavin, M.A. 1981. *Piero della Francesca's Baptism of Christ*, New Haven and London (Yale UP).

Lewis, A.C. 1977. 'H. Grassmann's *Ausdehnungslehre* and Schleiermacher's *Dialektik*', *AS*, 34, 103–62.

Li, Y. and Du, S. 1986. *Chinese Mathematics. A Concise History*, Oxford (Clarendon Press).

Lindberg, D.C. 1976. *Theories of Vision from al-Kindi to Kepler*, Chicago (University of Chicago Press).

Lindberg, D.C. 1992. *The Beginnings of Western Science*, Chicago (University of Chicago Press).

Lindgren, M. 1990. *Glory and Failure: the Difference Engines of Johann Müller, Charles Babbage and Georg and Edvard Scheutz*, 2nd edn, Cambridge, Mass. (MIT Press).

Lindt, R. 1904. 'Das Prinzip der virtuellen Geschwindigkeiten', *Abhandlungen zur Geschichte der Mathematik*, 18, 145–95.

Lorey, W. 1916. *Das Studium der Mathematik an den deutschen Universitäten seit Anfang des 19. Jahrhunderts*, Leipzig and Berlin (Teubner).

Lounsbury, F.L. 1978. 'Maya numeration, computation, and calendrical astronomy', in *DSB*, Vol. 15, 759–818.

Lund, J.L.M. 1921. *Ad quadratum. A study of the Geometric Bases of Classic and Medieval Religious Architecture*, Vol. 1, London (Batsford).

Lützen, J. 1979. 'Heaviside's operational calculus and the attempts to rigorise it', *AHES*, 21, 161–200.

Lützen, J. 1990. *Joseph Liouville 1809–1882*, New York (Springer).

Lyapunov, A. 1992. *The General Problem of the Stability of Motion* (translated by A.T. Fuller), London (Taylor & Francis). [Original Russian version 1892.]

McClain, E.G. 1976. *The Myth of Invariance*, York Beach, Maine (Nicolas-Hays).

McClain, E.G. 1978. *The Pythagorean Plato*, York Beach, Maine (Nicolas-Hays).

Mach, E. 1883. *Die Mechanik in ihrer Entwicklung historisch-kritisch dargestellt*, 1st edn, Leipzig (Brockhaus). [Later edns to 9th, 1933. Various translations.]

Mackey, G. 1978. 'Harmonic analysis as the exploitation of symmetry – a historical survey', *Rice University Studies*, 64, 73–228.

MacKinnon, E. 1977. 'Heisenberg's models of matrix mechanics', *Historical Studies in the Physical Sciences*, 8, 137–88.

McMullin, E. 1964. (Ed.) *Galileo Man of Science*, New York (Basic Books).

Mancosu, P. 1992. 'Aristotelean logic and Euclidean mathematics', *Studies in the History and Philosophy of Science*, 23, 242–65.

Manning, K.R. 1975. 'The emergence of the Weierstrassian approach to complex analysis', *AHES*, 14, 297–383.

Marshack, A. 1973. *The Roots of Civilisation*, London (Weidenfeld and Nicolson). [2nd edn 1991.]

Martzloff, J.C. 1987. *Histoire des*

mathématiques chinoises, Paris (Masson).

Maula, E. 1981. 'The end of invention', *AS*, 38, 109–22. [On analysis and synthesis.]

May, K.O. 1968. 'Growth and quality of the mathematical literature', *Isis*, 59, 363–71.

May, K.O. 1973. *Bibliography and Research Manual in the History of Mathematics*, Toronto (University of Toronto Press).

Medvedev, F.A. 1993. *Scenes from the History of Real Functions*, Basel (Birkhäuser).

Mehrtens, H. 1979. *Die Entstehung der Verbandstheorie*, Hildesheim (Gerstenberg).

Menninger, K. 1969. *Number Words and Number Symbols*, Cambridge, Mass. (MIT Press).

Miller, A.I. 1973. 'A study of Henri Poincaré's "Sur la théorie dynamique de l'électron" ', *AHES*, 10, 207–329.

Mirowski, P. 1989. *More Heat Than Light. Economics as Social Physics, Physics as Nature's Economics*, Cambridge (Cambridge UP).

Mirowski, P. 1994. (Ed.) *Natural Images in Economic Thought*, Cambridge (Cambridge UP).

Mittag-Leffler, G. 1923. 'An introduction to the theory of elliptic functions', *Annals of Mathematics*, (2) 24, 271–351. [Swedish original 1876.]

Moore, E.H. 1910. *Introduction to a Form of General Analysis*, New Haven (Yale UP).

Moore, G.H. 1982. *Zermelo's Axiom of Choice*, New York (Springer).

Mouret, E.J.G. 1921. 'Antoine de Chézy. Histoire d'une formule d'hydraulique', *Annales des Ponts et Chaussées*, 1 [for 1921], 165–269.

Mueller, I. 1981. *Philosophy of Mathematics and Deductive Structure in Euclid's Elements*, Cambridge, Mass. (MIT Press).

Nagel, E. 1939. 'The foundations of modern conceptions of formal logic in the development of geometry', *Osiris*, 7, 142–224. [Also in *Teleology Revisited*, 1979, New York (Columbia UP), pp. 195–259, 328–39.]

Nauenberg, M. 1994. 'Hooke, orbital motion, and Newton's *Principia*', *American Journal of Physics*, 62, 321–50.

Neuenschwander, E. 1980. 'Riemann und das "Weierstrassche" Prinzip der analytischen Fortsetzung durch Potenzreihen', *JDMV*, 82, 1–11.

Neugebauer, O. 1957. *The Exact Sciences in Antiquity*, 2nd edn, Providence, RI (Brown UP). [Repr. 1962, New York (Harper).]

Neugebauer, O. 1968. 'On the planetary theory of Copernicus', *Vistas in Astronomy*, 10, 89–103.

Neugebauer, O. 1975. *A History of Ancient Mathematical Astronomy*, 3 vols., New York (Springer).

Newton, I. 1934. *Mathematical Principles of Natural Philosophy* (translated by F. Motte, revised by F. Cajori), Berkeley and Los Angeles (University of California Press). [Of 3rd edn, 1726, of the *Principia*.]

Newton, I. 1967–81. *The Mathematical Papers of Isaac Newton* (ed. D.T. Whiteside), 8 vols., Cambridge (Cambridge UP).

Newton, I. 1979. *Opticks*, 4th edn. (1730), with modern commentary, New York (Dover).

Nicholson, J. 1993. 'The development and understanding of the concept of quotient group', *HM*, 20, 68–88.

North, J. 1994. *The Fontana History of Astronomy and Cosmology*, London (Fontana).

Norton, J. 1985. 'What was Einstein's principle of equivalence?', *Studies in History and Philosophy of Science*, 16, 203–46.

Panteki, M. 1992. 'Relationships between algebra, differential equations and logic in England: 1800–1860', C.N.A.A., London (doctoral dissertation).

Pappos 1986. *Book 7 of the Collection*, 2 parts (ed. with commentary by A. Jones), Berlin (Springer).

Pareto, V. 1911. 'Economie mathématique', in *Enyclopédie des sciences mathématiques*, Tome 1, Vol. 4, pp. 591–640 (article I26) [incomplete].

Parshall, K.H. 1985. 'Joseph H.M. Wedderburn and the structure theory of algebras', *AHES*, 32, 223–349.

Parshall, K.H. 1988. 'The art of geometry from al-Khwarizmi to Viète', *History of Science*, 26, 129–64.

Parshall, K.H. 1989. 'Toward a history of nineteenth-century invariant theory', in {Rowe and McCleary 1989}, Vol. 1, pp. 157–206.

Parshall, K.H. and Rowe, D. 1994. *The Emergence of the American Mathematical Research Community*, Providence, RI (American Mathematical Society).

Peano, G. 1908. *Formulaire de mathématiques*, 5th edn, Turin (Bocca). [Repr. 1960, Rome (Cremonese).]

Pearson, K. 1905. 'Letter on random walk, and comment', *Nature*, 42, 294, 342.

Peckhaus, V. 1990. *Hilbertprogramm und Kritische Philosophie*, Göttingen (Vandenhoeck und Ruprecht).

Pedersen, O. 1993. *Early Physics and Astronomy*, revised edn, Cambridge (Cambridge UP).

Pepper, J.V. 1968. 'Thomas Harriot's calculation of the meridional parts as logarithmic tangents', *AHES*, 4, 359–413.

Petruso, K.M. 1985. 'Additive progression in prehistoric mathematics: a conjecture', *HM*, 12, 101–6.

Phillips, E. 1978. 'Nicolai Nicolaevich Luzin and the Moscow school of the theory of functions', *HM*, 5, 275–305.

Picknett, L. and Prince, C. 1994. *Turin Shroud. In Whose Image?* London (Bloomsbury).

Pier, J.-P. 1994. (Ed.) *Development of Mathematics 1900–1950*, Basel (Birkhäuser).

Pierpont, J. 1896. 'Note on C.S. Peirce's paper on "A quincuncial projection of the sphere" ', *American Journal of Mathematics*, 18, 145–52.

Pingree, D. 1978. 'History of mathematical astronomy in

India', in *DSB*, Vol. 15, 533–633.

Poincaré, J.H. 1953. *Oeuvres*, Vol. 6, Paris (Gauthiers-Villars). [Papers on topology.]

Popper, K.R. 1950. 'Indeterminism in quantum physics and in classical physics', *British Journal for the Philosophy of Science*, 1, 117–33, 173–95.

Price, M.H. 1994. *Mathematics for the Multitude? A History of the Mathematical Association*, Leicester (Mathematical Association).

Pulte, H. 1990. *Das Prinzip der kleinsten Wirkung und die Kraftkonzeptionen der rationalen Mechanik*, Stuttgart (Steiner).

Pycior, H. 1983. 'Augustus De Morgan's algebraic work: the three stages', *Isis*, 74, 211–26.

Rabinovitch, N.L. 1973. *Probability and Statistical Inference in Ancient and Medieval Jewish Literature*, Toronto (University of Toronto Press).

Radelet-de Grave, P. and Benvenuto, E. 1995. (Eds.) *Between Mechanics and Architecture*, Basel (Birkhäuser).

Rashevsky, N. 1968. *Looking at History Through Mathematics*, Cambridge, Mass. (MIT Press).

Rayleigh, Lord 1905. 'The problem of the random walk', *Nature*, 42, 318. [Also in *Scientific Papers*, Vol. 5, 256.]

Redondi, P. 1980. *L'Acceuil des idées de Sadi Carnot*, Paris (Vrin).

Redondi, P. 1987. *Galileo Heretic*, Princeton (Princeton UP). [Italian original 1983.]

Reich, K. 1993. 'The American contribution to the theory of differential invariants, 1900–1916', in J. Earman and others (eds.), *The Attraction of Gravitation*, Basel (Birkhäuser), pp. 225–47.

Reich, K. 1994. *Die Entwicklung des Tensorkalküls*, Basel (Birkhäuser).

Reynolds, R.E. 1979. ' "At sixes and sevens" – and eights and nines', *Speculum*, 54, 669–84.

Rhind Papyrus 1987. *The Rhind Mathematical Papyrus* (ed. G. Robins and C. Shute), London (British Museum).

Richenhagen, G. 1985. *Carl Runge*, Göttingen (Vandenhoeck und Ruprecht).

Riemann, G.F.B. 1851. 'Grundlagen der allgemeine Theorie der Functionen einer veränderlichen complexen Grösse. (Inaugural-dissertation)', in *Gesammelte mathematische Werke*, 2nd edn, 1892, Leipzig (Teubner), pp. 3–48.

Riemann, G.F.B. 1857a. 'Theorie der Abel'schen Functionen', in *Ibid.*, pp. 88–144.

Riemann, G.F.B. 1857b. 'Beiträge zur Theorie der durch die Gauss'sche Reihe', in *Ibid.*, pp. 67–87.

Riemann, G.F.B. 1868a. 'Über die Darstellbarkeit einer Function durch eine trigonometrische Reihe', in *Ibid.*, pp. 227–71.

Riemann, G.F.B. 1868b. 'Über die Hypothesen, welche der Geometrie zu Grunde liegen', in *Ibid.*, pp. 272–87.

Rihill, T.E. and Tucker, J.V. 1995. 'Greek engineering: the case of

Eupalinos' tunnel', in A. Powell (ed.), *The Greek World*, London (Routledge) pp. 403–31.

Roberts, D. and others. 1997: (Eds.) *Studies in the Logic of Charles S. Peirce*, Bloomington (Indiana UP).

Robins, G. and Shute, C.C.D. 1985. 'Mathematical bases of ancient Egyptian architecture and graphic art', *HM*, 12, 107–22.

Rodriguez-Consuegra, F.A. 1991. *The Mathematical Philosophy of Bertrand Russell: Origins and Development*, Basel (Birkhäuser).

Roscher, W.H. 1904. *Die Sieben- und Neunzahl im Kultus und Mythus*, Leipzig (Teubner). [Several other important but forgotten works were published by Teubner, also in journals of the Leipzig Academy.]

Rosenfeld, B.A. 1988. *A History of Non-Euclidean Geometry*, New York (Springer).

Rossianov, K.O. 1993. 'Editing Nature. Joseph Stalin and the "new" Soviet biology', *Isis*, 84, 728–45.

Rothenberg, S. 1910. 'Geschichtliche Darstellung der Entwicklung der Theorie der singulären Lösung totaler Differentialgleichungen', *Abhandlungen zur Geschichte der Mathematik*, 20, 315–404.

Rotman, B. 1987. *Signifying Nothing: the Semiotics of Zero*, London (MacMillan).

Rouse Ball, W.W.R. and Coxeter, H.S.M. 1987. *Mathematical Recreations and Essays*, 13th edn, New York (Dover). [1st edn, by Rouse Ball, 1892.]

Rowe, D.E. 1989. 'The early geometrical works of Sophus Lie and Felix Klein', in {Rowe and McCleary 1989}, Vol. 1, pp. 209–73. [See also article following this, by T.W. Hawkins.]

Rowe, D. and McCleary, J. 1989. *History of Modern Mathematics*, Vols. 1 and 2, Boston (Academic Press).

Rühlmann, M. 1881, 1885. *Vorträge zur Geschichte der theoretischen Maschinenlehre*, 2 parts, Brauschweig (Schwetschke).

Russell, J.L. 1995. 'What was the crime of Galileo?', *AS*, 52, 403–10.

Sakarovitch, J. 1997. *Epures d'architecture. De traités de taille des pierres à la géométrie descriptive*, Basel (Birkhäuser).

Scanlan, M.J. 1988. 'Beltrami's model and the independence of the parallel postulate', *History and Philosophy of Logic*, 9, 13–33.

Scanlan, M.J. 1991. 'Who were the American postulate theorists?', *The Journal of Symbolic Logic*, 56, 981–1002.

Schaffner, K.L. 1972. (Ed.) *Nineteenth-Century Aether Theories*, Oxford (Pergamon). [Includes original texts.]

Schlissel, A. 1977. 'The development of asymptotic solutions of linear ordinary differential equations', *AHES*, 16, 307–78.

Schmandt-Besserat, D. 1992. *Before Writing: From Counting to Cuneiform*, 2 vols., Austin (University of Texas Press).

Schneider, I. 1968. 'Der Mathematiker Abraham de Moivre', *AHES*, 5, 177–317.

Schneider, I. 1981. 'Leibniz on the probable', in {Dauben 1981}, pp. 201–19.

Scholz, E. 1990. (Ed.) *Geschichte der Algebra. Eine Ausführung*, Mannheim (Wissenschaftsverlag).

Schönflies, A.M. 1900. 'Die Entwickelung der Lehre von den Punktmannigfaltigkeiten', *JDMV*, 8, Part 2, 251 pp.

Schubring, G. 1996. (Ed.) *Hermann Günther Grassmann (1809–1877)*, Dordrecht (Kluwer).

Seidenberg, A. 1962a. 'The ritual origin of geometry', *AHES*, 1, 488–527.

Seidenberg, A. 1962b. 'The ritual origin of counting', *AHES*, 2, 1–40.

Seidenberg, A. 1972. 'On the area of the semicircle', *AHES*, 9, 171–211.

Sesiano, J. 1985. 'The appearance of negative solutions in medieval mathematics', *AHES*, 32, 105–50.

Shankarachaya 1965. *Vedic Mathematics*, Delhi (Motilal Babarsidass).

Shapiro, A.E. 1973. 'Kinematic optics: a study of the wave theory of light in the seventeenth century', *AHES*, 11, 134–266.

Sheynin, O.B. 1971. 'Newton and the classical theory of probability', *AHES*, 7, 217–43.

Sheynin, O.B. 1977. 'Early history of probability', *AHES*, 17, 201–59.

Sheynin, O.B. 1984. 'On the history of statistical method in meteorology', *AHES*, 31, 53–95.

Shirley, J.W. 1974. (Ed.) *Thomas Harriot Renaissance Scientist*, Oxford (Clarendon Press).

Shoesmith, E. 1986. 'Huygens' solution to the gambler's ruin problem', *HM*, 13, 157–64.

Siegel, D. 1991. *Innovations in Maxwell's Electromagnetic Theory*, Cambridge (Cambridge UP).

Siegmund-Schultze, R. 1982. 'Die Anfänge der Functionalanalysis', *AHES*, 26, 13–71.

Siegmund-Schultze, R. 1994. *Mathematische Berichterstattung in Hitlerdeutschland. Der Niedergang des "Jahrbuch über die Fortschritte der Mathematik"*, Göttingen (Vandenhoeck und Ruprecht).

Simi, A. 1994. (Ed.) *Trattato dell' alcibra amuchabile*, Siena (University: Centro di Studi della Matematica Medioevale).

Simon, M. 1906. *Über die Entwicklung der Elementar-Geometrie im XIX. Jahrhundert*, Leipzig (Teubner).

Sinaceur, H. 1991. *Corps et modèles*, Paris (Vrin).

Smeur, A.J.E.M. 1970. 'On the value equivalent to π in ancient mathematics', *AHES*, 6, 249–70.

Smith, C. 1976. 'William Thomson and the creation of thermodynamics', *AHES*, 16, 231–88.

Smith, C. and Wise, N. 1989. *Energy and Empire. A Biography of Lord Kelvin*, Cambridge (Cambridge UP).

Smith, D.E. 1908. *Rara arithmetica. A Catalogue of the Arithmetics Written Before the Year 1601*, Boston and London (Ginn). [4th edn 1970, New York (Chelsea), including A. De Morgan, *Arithmetical Books from the Invention of Printing to the Present Time* (1847).]

Smith, S.B. 1983. *The Great Mental Calculators*, New York (Columbia UP).

Stahl, W.H. 1962. *Roman Science*, Madison (University of Wisconsin Press).

Sylla, E. 1984. 'Compounding ratios: Bradwardine, Oresme, and the first edition of Newton's *Principia*', in E. Mendelsohn (ed.), *Transformation and Tradition in the Sciences*, Cambridge (Cambridge UP), 11–45.

Sylvester, J.J. 1850. 'Addition to the articles, "On a new class of theorems" and "On Pascal's theorem"', in *Collected mathematical papers*, Vol. 1, pp. 145–51.

Sylvester, J.J. 1883. 'On the equation to the secular inequalities in the planetary theory', in *Collected Mathematical Papers*, Vol. 4, pp. 410–11.

Tanner, R.C.H. 1980. 'The alien regiment of the minus sign', *AS*, 37, 159–78.

Taton, R. 1947. 'Les mathématiques dans le "*Bulletin de Férrusac*"', *Archives Internationales d'Histoire des Sciences*, 26, 100–125.

Taton, R. 1964. (Ed.) *Enseignement et diffusion des sciences en France au XVIII siècle*, Paris (Hermann). [Repr. 1986.]

Taylor, A.E. 1974. 'The differential: nineteenth and twentieth century developments', *AHES*, 12, 355–83.

Taylor, A.E. 1983. 'A study of Maurice Fréchet I', *AHES*, 27, 233–95.

Temple, G. 1981. *100 Years of Mathematics*, London (Duckworth).

Thiering, B. 1993. *Jesus the Man*, London (Corgi). [First published 1992, New York (Doubleday).]

Thomson, W. 1849. 'An account of Carnot's theory of the motive power of heat', in *Mathematical and Physical Papers*, Vol. 1, pp. 113–55.

Thomson, W. and Tait, P.G. 1879, 1883. *Treatise on Natural Philosophy*, revised edn, 2 parts, Cambridge (Cambridge UP). [Cited as 'T and T".]

Thro, E.B. 1994. 'Leonardo da Vinci's solution to the problem of the pinhole camera', *AHES*, 48, 343–71.

Tietze, H. 1908. 'Über die topologische Invarianten mehrdimensionaler Mannigfaltigkeiten', *Monatshefte der Mathematik und Physik*, 19, 1–118.

Todhunter, I. 1861. *A History of the Progress of the Calculus of Variations During the Nineteenth Century*, Cambridge and London (MacMillan). [Repr. 1961, New York (Chelsea).]

Todhunter, I. 1873. *A History of the Mathematical Theories of Attraction and Figure of the Earth*, 2 vols., London (Macmillan). [Repr. 1949, New York (Dover).]

Toepell, M.-M. 1986. *Über die Entstehung von David Hilberts 'Grundlagen der Geometrie'*, Göttingen (Vandenhoeck und Ruprecht).

Toeplitz, O. 1963. *The Calculus. A Genetic Approach*, Chicago (University of Chicago Press). [German original 1949.]

Tooley, R.V. and Bricker, C. 1976. *Landmarks of Mapmaking*, Oxford (Phaidon). [Repr. 1990, Ware (Wordsworth).]

Torretti, R. 1978. *Philosophy of Geometry from Riemann to Poincaré*, Dordrecht (Reidel).

Toth, I, 1967. 'Das Parallelproblem in Corpus Aristotelicum', *AHES*, 3, 215–422.

Toth, I. 1991. 'Le problème de la mesure . . . Zenon et Platon', in *Mathématiques et philosophie de l'antiquité à l'age classique* (ed. R. Rashed), Paris (CNRS), pp. 21–99.

Tropfke, J. 1980. *Geschichte der Elementarmathematik*, Vol. 1, 4th. edn (ed. K. Vogel and others), Berlin and New York (de Gruyter).

Truesdell, C.A., III 1960. *The Rational Mechanics of Flexible or Elastic Bodies 1638–1788*, Zurich (Orell Füssli) [as Euler, *Opera omnia*, Series 2, Vol. 11, Part 2].

Truesdell, C.A., III 1968. *Essays in the History of Mechanics*, Berlin (Springer).

Truesdell, C.A., III. 1975. 'Early kinetic theory of gases', *AHES*, 15, 1–66.

Tucciarone, J. 1973. 'The development of the theory of summable divergent series from 1880 to 1925', *AHES*, 10, 1–40.

Turner, R.S. 1971. 'The growth of professorial research in Prussia, 1818 to 1848 – causes and context', *Historical Studies in the Physical Sciences*, 3, 137–82.

Tweddle, I. 1988. *James Stirling*, Edinburgh (Scottish Academic Press). [Many manuscripts transcribed.]

Unguru, S. 1974. 'Pappus in the thirteenth century in the Latin West', *AHES*, 13, 307–24.

Unguru, S. 1975. 'On the need to rewrite the history of Greek mathematics', *AHES*, 15, 67–114. [Replies in 15, 199–210 and 16 (1977), 189–200.]

Unwin, W.C. 1911. 'Hydraulics', in *Encyclopaedia Britannica*, 11th edn, Vol. 14, pp. 35–110.

van der Waerden, B.L. 1967. (Ed.) *Sources of Quantum Mechanics*, Amsterdam (North-Holland).

van der Waerden, B.L. 1985. *A History of Algebra*, Berlin (Springer).

van Lunteren, F. 1991. 'Framing hypotheses. Conceptions of gravity in the 18th and 19th centuries', Utrecht (University doctoral dissertation).

Vizgin, V.P. 1994. *Unified Field Theories in the First Third of the 20th Century*, Basel (Birkhäuser).

von Braunmühl, A. 1900, 1903. *Vorlesungen über Geschichte der Trigonometrie*, 2 vols., Leipzig (Teubner).

von Koch, H. 1904. 'Sur une courbe continue sans tangente obtenu par une construction géométrique', *Arkiv for Matematik, Astronomi och Fysich*, 1, 681–704.

Wallis, J. 1685. *A Treatise on Algebra*, London (Davis).

Wallis, P. and Wallis, R. 1986. *Biobibliography of British Mathematics and its Applications, Part II, 1701–1760*, Letchworth (Epsilon Press).

Waterhouse, W. 1972. 'The discovery of the regular solids', *AHES*, 9, 212–21.

Watts, P. 1911. 'Ships' and 'Shipbuilding', in *Encyclopaedia Britannica*, 11th edn, Vol. 24, pp. 860–922, 922–81.

Webb, J.C. 1980. *Mechanism, Mentalism, and Metamathematics*, Dordrecht (Reidel).

Weierstrass, K.T.W. 1895. *Mathematische Werke*, Vol. 2, Berlin (Reimer). [Repr. 1967, Hildesheim (Olms) and New York (Johnson).]

Weierstrass, K.T.W. 1988. *Ausgewählte Kapitel aus der Funktionenlehre* (ed. R. Siegmund-Schultze), Leipzig (Teubner). [Text of lecture course of 1886, with reprints of other important material.]

Weil, A. 1984. *Number theory ... from Hammurapi to Legendre*, Boston (Birkhäuser).

West, B.J. 1985. *An Essay on the Importance of Being Nonlinear*, New York (Springer).

Westfall, R.S. 1971. *Force in Newton's Physics*, New York (Elsevier).

Westfall, R.S. 1980. *Never at Rest*, Cambridge (Cambridge UP).

Whiteside, D.T. 1961. 'Patterns of mathematical thought in the later seventeenth century', *AHES*, 1, 179–388.

Whiteside, D.T. 1966. 'Newton's marvellous year: 1666 and all that', *Notes and Records of the Royal Society*, 21, 32–41.

Whiteside, D.T. 1970a. 'Before the *Principia*: the maturing of Newton's thoughts of dynamical astronomy, 1664–1684', *Journal of the History of Astronomy*, 1, 5–19.

Whiteside, D.T. 1970b. 'The mathematical principles underlying Newton's *Principia mathematica*', *Journal of the History of Astronomy*, 1, 116–38.

Whiteside, D.T. 1974. 'Keplerian planetary eggs, laid and unlaid, 1600–1605', *Journal for the History of Astronomy*, 5, 1–21.

Whittaker, E.T. 1951. *History of the Theories of Aether and Electricity. The Classical Theories*, London (Nelson).

Whittaker, E.T. and Watson, G.N. 1927. *A Course in Modern Analysis*, 4th edn, Cambridge (Cambridge UP).

Wigner, E. 1960. 'The unreasonable effectiveness of mathematics in the natural sciences', *Communications in Pure and Applied Mathematics*, 13, 1–14.

Williams, J. 1995. 'Mathematics and the alloying of coinage, 1202–1700', *AS*, 52, 213–63.

Wilson, C.A. 1968. 'Kepler's derivation of the elliptical path', *Isis*, 59, 5–25.

Wilson, C.A. 1970. 'From

Kepler's laws, so-called, to universal gravitation: empirical factors', *AHES*, 6, 92–170.

Wilson, C.A. 1980. 'Perturbation and solar tables from Lacaille to Delambre', *AHES*, 22, 53–188, 189–304.

Wilson, C.A. 1985. 'The great inequality of Jupiter and Saturn: from Kepler to Laplace', *AHES*, 33, 15–290.

Wilson, C.A. 1987. 'D'Alembert versus Euler on the precession of the equinoxes and the mechanics of rigid bodies', *AHES*, 37, 233–73.

Wisan, W.L. 1974. 'The new science of motion: a study of Galileo's *De motu locali*', *AHES*, 13, 103–306.

Wise, N. 1981. 'The flow analogy in electricity and magnetism', *AHES*, 25, 19–70.

Wolf, C.J.E. 1889, 1891. (Ed.) *Mémoires sur le pendule*, 2 parts, Paris (Gauthier-Villars).

Wussing, H. 1984. *The Genesis of the Abstract Group Concept*, Cambridge, Mass. (MIT Press). [German original 1969, Berlin.]

Yates, F.A. 1972. *The Rosicrucian Enlightenment*, London (Routledge and Kegan Paul). [Repr. 1986, 1993.]

Yavetz, I. 1995. *From Obscurity to Enigma. The Work of Oliver Heaviside, 1872–1899*, Basel (Birkhäuser).

Yeldham, F. 1936. *The Teaching of Arithmetic Through Four Hundred Years*, London (Harrap).

Yoder, J.G. 1988. *Unravelling Time*, Cambridge (Cambridge UP).

Yushkevich, A.P. 1976a. 'The concept of function up to the middle of the 19th century', *AHES*, 16, 37–85.

Yushkevich, A.P. 1976b. *Les mathématiques arabes* (*VIIe – XVe siècles*), Paris (Vrin). [Author named here 'Youschkevitch'.]

Zaslavksy, C. 1979. *Africa Counts*, Boston (Prindle, Schmidt and Weber).

Ziegler, R. 1985. *Die Geschichte der geometrischen Mechanik im 19. Jahrhundert*, Stuttgart (Steiner).

PUBLISHER'S ACKNOWLEDGEMENTS

For permission to reproduce or imitate copyright material in certain Figures, thanks are extended as follows: to Edizioni dell' Arquata (2.29.10), *New Scientist* (part of 13.17.1), Routledge (2.8.1, 3.14.2, 12.6.1, 12.11.1, 14.7.1–2, 16.10.1), Birkhäuser Verlag (3.22.1, 8.7.1, 10.2.1), Springer Verlag (2.21.1, 4.14.3, 5.21.1–2), Teubner (16.17.1), MIT Press (2.6.1), Accademia di Danimarca (3.25.1) and the American Council for Learned Societies (2.21.2, 2.28.2–3, 3.4.1, 4.12.1, 8.4.3, 15.10.1).

Every reasonable effort has been made to contact copyright holders in order to obtain their permission to reproduce those images in which they hold the copyright. However, should any party recognise any figure or quotation in this book that has been reproduced without their assent, the author and publishers, upon receiving written notification from the party concerned, will be happy to print a full acknowledgement of copyright ownership in the next edition of this book and pay a reasonable fee for the permission to reproduce it.

INDEX

Dates of birth and death are given for people with significant roles in this history. Page numbers of the biographical tables are followed by a 't'.

ABOUT THE AUTHOR

IVOR GRATTAN-GUINNESS has both the doctorate (Ph.D.) and higher doctorate (D.Sc.) in the history of science at the University of London. He is currently Professor of the History of Mathematics and Logic at Middlesex University, England. He was editor of the history of science journal *Annals of Science* from 1974 to 1981. In 1979 he founded the journal *History and Philosophy of Logic,* and edited it until 1992. From 1986 to 1988 he was the President of the British Society for the History of Mathematics. He is an effective member of the Académie Internationale d'Histoire des Sciences. His latest individual book is *Convolutions in French Mathematics,* 1800–1840 (three volumes, 1990, Basel and Berlin, Birkhäuser). He edited the substantial *Companion Encyclopedia of the History and Philosophy of the Mathematical Sciences* (two volumes, 1994, London, Routledge). He is the Associate Editor for mathematicians and statisticians for the *New Dictionary of National Biography,* scheduled for publication in 2004. Other projects include a history of mathematical logic and the foundations of mathematics from 1870 to 1930, which is supported by the Leverhulme Foundation.